MACHINE TOOLS HANDBOOK
DESIGN AND OPERATION

P H Joshi

Associate Member
Institution of Engineers

McGraw-Hill
New York   Chicago   San Francisco   Lisbon   London
Madrid   Mexico City   Milan   New Delhi   San Juan
Seoul   Singapore   Sydney   Toronto

The McGraw·Hill Companies

Copyright © 2008 by The McGraw-Hill Companies, Inc. All rights reserved. Printed in the United States of America. Except as permitted under the United States Copyright Act of 1976, no part of this publication may be reproduced or distributed in any form or by any means, or stored in a data base or retrieval system, without the prior written permission of the publisher.

1 2 3 4 5 6 7 8 9 0   DOC/DOC   0 1 3 2 1 0 9 8 7

ISBN 978-0-07-149435-9
MHID 0-07-149435-9

The sponsoring editor for this book was Kenneth P. McCombs and the production supervisor was Richard C. Ruzycka. The art director for the cover was Jeff Weeks.

Printed and bound by RR Donnelley.

This book was previously published by Tata McGraw-Hill Publishing Company Limited, New Delhi, India, copyright © 2007.

McGraw-Hill books are available at special quantity discounts to use as premiums and sales promotions, or for use in corporate training programs. For more information, please write to the Director of Special Sales, McGraw-Hill Professional, Two Penn Plaza, New York, NY 10121-2298. Or contact your local bookstore.

This book is printed on acid-free paper.

Information contained in this work has been obtained by The McGraw-Hill Companies, Inc. ("McGraw-Hill") from sources believed to be reliable. However, neither McGraw-Hill nor its authors guarantee the accuracy or completeness of any information published herein, and neither McGraw-Hill nor its authors shall be responsible for any errors, omissions, or damages arising out of use of this information. This work is published with the understanding that McGraw-Hill and its authors are supplying information but are not attempting to render engineering or other professional services. If such services are required, the assistance of an appropriate professional should be sought.

To
My beloved mother
Late Kesarbai Joshi

Foreword

Few books on machine tools are available, as the subject domain warrants lengthy descriptions and hundreds of intricate illustrations.

Few technocrats venture into writing, many of those who do so, do not have flair or inclination for the lonesome vocation of writing. Consequently most of the technical books are penned by academicians with little field experience and often lack the pragmatic touch. In this present scenario Mr. Joshi's book *Machine Tools Handbook: Design and Operation* should be more than welcome.

Like Mr. Joshi's earlier books on tool design, this treatise too has brevity, clarity, flow and logical composition, which only a few other engineering texts can match. The in-depth coverage renders the book useful for practising professionals and the presentation is simple enough for students' comprehension.

Original contributions such as the graph for designing gears for wear and the thumb rules for designing gear boxes deserve special mention. Solved examples from *Machinery Handbook* enhance the worth of this work as they prove accuracy of the cited formulae.

Mr. Joshi deserves encouragement; I wish him success in his philanthropic mission.

A P Srivastava
Former Director
Profile Gauges and Tools India Pvt. Ltd.
Mumbai
Former Works Manager
Chicago Pneumatics Ltd.
Mumbai

Preface

I have spent only a small fraction of my career in academic field. During my tenure as a lecturer in Fr. Agnel Technical College at Mumbai I taught 'machine tools' to production engineering students. I realized that none of the few contemporary books on machine tools was suitable for students. I had a tough time learning and teaching gear box design as no book, even the reference tomes, had guidelines or solved examples.

As a design consultant I found the existing reference tomes inadequate. None were helpful for an experienced designer while designing a new machine tool. All the books had sufficient information on various drives like belts, chains, gears but none compared them for speed range, capacity, and efficiency of transmission. There was ample information on hydraulic elements and drives but none on design of hydraulic systems: determining mundane details such as the specifications (sizes) of hydraulic valves and the piping, pump and tank.

A machine tool designer should have sufficient knowledge of electrical elements—the working and usage of contactors, time relays etc. Being a mechanical engineer, I learnt it the hard way—by listening, enquiring, attempting, and amending designs in light of discussions with fellow field experts. I have endeavoured to pass on the gist of the elementary circuit design to my readers.

In industry we aim at perfection. Errors can be very costly and blunders disastrous. We can neither rely totally on our subordinates nor redo everything the lesser one does to cross-check the correctness. We resort to broad approximations to see that the subordinate's submissions are more or less correct.

Splitting hair can be very costly in terms of time, which is more valuable than money. So we take 1 kg (force) = 10 Newtons (instead of 9.87) during rapid cross-checks. We round off figures to near values convenient for rapid calculations. Usually this does not make much difference to the ultimate result. In any case the calculated gear module or thread diameter must be adjusted to fit into the prevailing standards for economic manufacture.

I have often used such approximations in the solved examples in this book. Torque (kgm) is 1000/RPM (approx) of power in kW in some examples and 955/RPM of kW in others. I hope readers as well as teaching faculty will understand and condone these small anomalies.

A book addressed to students must aim at clarity sometimes even at cost of brevity. Reader must understand working of machine elements, their potential and parameters. Also he should be able to test his comprehension by way of exercises, and for these reference tables are necessary. Their range and number should be adequate for normal applications. Inclusion of all the prevailing standards would make the book unwieldy and more suitable for consultants than students. So I had to strike a balance between the informative text and industrial standards. I have done away with data for very small (miniature) and very large (heavy duty) applications to save space. In any case practising professionals have enterprise to mine such data.

I hope readers would find my text easy to grasp and remember. Even industrial practitioners, and field consultants should find the treatise convenient as a reference for normal applications, and as a manual for training their subordinates.

I will be gratified if my book facilitates comprehension of machine tool working and simplifies design of machine elements. Nothing will please me more than industry and my fellow consultants using my book for reference and training.

P H JOSHI

Acknowledgements

I am thankful to the following organizations for their valuable help in drafting this treatise:

1. M/s Industrial Press, New York, U.S.A. for granting permission to reproduce Figures 2.18, 5.15, 5.18, 5.19 in this book from Machinery Handbook, 23rd Edition.
2. M/s Chand & Co. Ltd., New Delhi, India for granting permission to reproduce Figures 1.10, 1.48, 1.50, 1.71, 1.72, 8.10 (d, e, f), 8.11 (b, c) from their books: (i) Cutting Tools and (ii) Press Tools.
3. Arnold Publishers, New Delhi, India for granting permission to reproduce Figures 1.2, 1.5, 1.6, 1.9, 1.13, 1.14, 1.15, 1.16 (d, g), 1.17 (a, b), 1.18, 1.19, 1.21, 1.24, 1.42, 1.43, 1.69, 1.74, 1.75, 1.77, 1.78 in this book from Workshop Technology Volumes I, II and III.
4. Mrs. Sunanda Suresh Bohra for typing the manuscript.
5. Mrs. Sangita Narendra Kotkar for her help in illustrating the book.
6. Mr. Satish P. Joshi for data processing of some chapters.

P H Joshi

About the Author

P. H. Joshi is currently head of an engineering consulting firm in Calcutta, India. An engineer, teacher, and trainer with extensive experience in production engineering, he previously headed the design offices of Crompton Greaves, ZF Steering Gear, and Arco Whitney. Mr. Joshi is also the author of *Jigs and Fixtures, Press Tools, Cutting Tools,* and *Machine Drawing*.

Contents

Foreword   *v*
Preface   *vii*
Acknowledgements   *ix*

1. INTRODUCTION   1

    1.1   Classification   *1*
    1.2   Machine Tool Requisites   *1*
    1.3   Economic Considerations for the Type of Machine Tool Chosen   *3*
    1.4   Basic Machine Tools   *7*

2. MACHINE TOOL DRIVES AND MECHANISMS   93

    2.1   Rotary Drives   *93*
    2.2   Mechanical Drives   *102*
    2.3   Tensioning Belts   *116*
    2.4   Timer Belts   *123*
    2.5   Chain Drive   *125*
    2.6   Sprocket Design   *130*
    2.7   Tolerances   *132*
    2.8   Gearing   *134*
    2.9   Correction   *138*
    2.10   Gear Design   *145*
    2.11   Pumps   *259*
    2.12   Hydraulic Valves   *263*
    2.13   Piping   *277*

3. RECTILINEAR OR TRANSLATORY DRIVES   282

    3.1   Converting Rotary Drive to Translatory Motion   *282*
    3.2   Generating Intermittent Periodic Motion   *292*

4. DRIVE TRANSMISSION AND MANIPULATION   303

    4.1   Couplings   *303*
    4.2   Clutches   *310*
    4.3   Motion Reversal Mechanisms   *340*

5. MACHINE TOOL ELEMENTS — 343

  5.1 Spindles *343*
  5.2 Guideways for Machine Tools *459*

6. MACHINE TOOL DYNAMICS — 507

  6.1 Stability of a System *507*

7. MACHINE TOOL OPERATION — 520

  7.1 Requisites of a Good Control System *520*
  7.2 Electrical Automation *540*
  7.3 Fluid Power Automation *554*
  7.4 Numerical Control [NC] *567*

8. TOOL ENGINEERING — 643

  8.1 What Tooling Comprises *643*
  8.2 Requisites of Tool Materials *643*
  8.3 Common Tool Materials *644*
  8.4 Common Workpiece Materials *651*
  8.5 Choice of Material *663*
  8.6 Machining *665*
  8.7 Tooling *702*
  8.8 Jigs and Fixtures *706*

BIBLIOGRAPHY — 720

INDEX — 721

CHAPTER 1

INTRODUCTION

Machine tools are devices for cutting materials (mostly metals), to impart them the required shape. They have an in-built arrangement that facilitates the use of various types of detachable cutting tools that can be changed to suit the task at hand, and removed for replacement or resharpening, after wear. The cut-off material obtained is usually in the form of chips. This is the material peeled off the workpiece by the tool. It can be continuous like a ribbon or in broken chips.

1.1 CLASSIFICATION

Machine tools can be classified according to *size:* light duty (weighing less than 1 tonne), medium duty (1 to 10 tonnes), and heavy duty (above 10 tonnes). Some people prefer classification according to the *method of actuation:* manually operated, semi-automatic, and fully automated. Another method of classification segregates machine tools according to *purpose*: general purpose machine tools, which can be used for a wide variety of operations, on a range of sizes of workpieces, and special purpose machine tools, for specific operations, on a limited range of workpiece sizes and shapes.

Another method classifies machine tools according to the *types of motions* used for removing material.

Rotary cutting machines rotate workpiece (turning lathes, capstans, turrets, autos), or the cutting tool (drill, milling cutter, grinding wheel), or both (cylindrical grinder).

Linear cutting machines remove material by moving the tool (shaping, slotting), or the workpiece (planning, surface grinding), in a straight line.

In both the types of machine tools, linear motion is used for moving (feeding) the workpiece or the tool, to mate (engage) them for material removal.

Feed classification can further subdivide machine tools as: The axial feed machines such as drilling machines that move the workpiece/tool, parallel to the axis of the machine spindle, and the transverse feed machines, such as face milling machines that move the workpiece or tool, in a direction that is at right angle to the m/c spindle axis. Combined feed machines, such as center lathes and boring machines, can use both types of feeds.

1.2 MACHINE TOOL REQUISITES

A good machine tool should have high productivity, and precision commensurate with workpiece specifications, and controlling levers, hand-wheels, push buttons, etc. at a position, convenient for the operator. It should ensure operator and machine safety, its manufacturing and running costs should be reasonable and breakdown frequency and duration, minimal.

1.2.1 Productivity can be maximized by

1.2.1.1 Optimum speeds and feeds usage to suit the workpiece size and material.

General purpose machines call for a wider range of speeds and feeds, to accommodate a variety of workpiece sizes and materials.

Special purpose machines, meant for either a single or few specific workpieces and operations, do not warrant flexibility in speeds and feeds, nor wide variety of tool/workpiece motions provided in general purpose machines.

More about the choice of speeds and feeds in the next chapter on machine tool drives.

1.2.1.2 Multi-tool cutting that can help in overlapping of cutting times.

1.2.1.3 Reduction of Non-productive (Non-cutting) time by

- Jigs and fixtures: to reduce workpiece positioning (locating) and clamping time.
- Minimizing maintenance: by reckoning fatigue stress on parts, while designing running parts, providing easy access for quick replacement of broken/worn parts, and adopting a practice of preventive maintenance (replacing vulnerable parts just before they are likely to break), on weekly holidays.

1.2.2 Precision depends upon

Geometrical manufacturing accuracy: straightness, squareness, concentricity, etc. of guideways, carriages, and spindles.
Adept assembly: minimizing clearances between guideway and carriage (or bearing and spindle), for kinetic (motional) accuracy.
Structural deformation (deflection): which can be minimized by reckoning cutting forces and vibrations, while designing the structure and providing adequate stiffness (strength).

When a machine is subjected to high variation in temperature (as in grinders), thermal stress and ensuing deformation should be minimized by diffusing (conducting away) the heat to a solid structure, and/or by providing for an arrangement for cooling the structure.

Precision aids like fine division scales, dials, verniers, and optical means for accurate readout, promote precise setting and operation.

1.2.3 Safety of operator and machine

1.2.3.1 Operator safety can be ensured by

Guards, which can prevent entanglement of operator's clothes in rotating/moving machine parts; and shield him from hot, sharp, flying chips of the cut-off material

- Adequate clamping of workpiece, to withstand cutting forces, to prevent the workpiece from flying off and injuring the operator
- Handling equipment for heavy workpieces, to avoid accidental fall and injury.
- Electrical safety through earthing, to avoid shock to the operator

1.2.3.2 Machine safety can be accomplished by

- Stops (mechanical obstacles), and limit, and proximity switches, to confine carriage travel within safe limits
- Interlocks through mechanical, hydraulic, and electrical elements, to prevent engagement of conflicting and risky motions simultaneously, or in a wrong order, e.g. ensuring that workpiece is clamped, before cutting commences
- Overload prevention through cheap, easy-to-replace shear pins, electrical fuses, etc. to eliminate the possibility of breakage of costly gears, burning of electrical motor, etc.

Appearance of machine tools is not as important as let's say that of cars. Accurate and troublefree functioning is more important (as in a battle tank).

Simplicity of design is also a desirable feature of machine tools.

1.3 ECONOMIC CONSIDERATIONS FOR THE TYPE OF MACHINE TOOL CHOSEN

It is necessary to meticulously consider the economic aspects of various machine tools available for production, to check for their suitability. Recovery of capital investment usually takes 5 to 10 years. Hence, there is no point in buying a new machine tool, if the production cost or time per workpiece will not decrease substantially.

1.3.1 Production cost (P) comprises

1.3.1.1 Material cost (M)

Depends upon the size and shape of the workpiece and condition of the raw material (casting or rolled stock). This will be same for existing and new machine tools, unless there is a change in the production process.

1.3.1.2 Machining cost (C)

Depends on the time the machine takes for producing the workpiece. The time/piece (T) decreases with the passage of time, as the machine operator learns and masters the production cycle. But it stabilizes at the minimum possible, for the operator, after some time. The stabilized time/piece for an average operator, can be estimated fairly accurately from the optimum speeds and feeds for the workpiece, and the operator motions involved.

1.3.1.3 Labour cost (W)

Or operator wages, is higher for general purpose machines, which call for more skillful operators. Special purpose machines, with a higher degree of automation, can do with an unskilled operator, who only has to load and unload workpiece, and press start/stop buttons.

1.3.1.4 Overheads

Comprise all indirect costs such as money spent on the maintenance department, tool room (for resharpening tools), clerical staff, methods engineers, managers, even sales personnel and advertising. Overheads are usually expressed in overall percentage which ranges from 50% for small workshops, to 500% for large establishments with modern amenities and a better qualified workforce.

1.3.2 Investment

Comprises cost of machine and tooling (jigs and fixtures), expenses incurred on the space for setting up the machine, its accessories, and for temporary storage of incoming and finished workpiece banks, near the machine. If installation of a new machine tool involves building additional premises, the cost of the new building, and facilities for maintaining it should also be reckoned.

1.3.2.1 Machine and tooling costs (Im)

Should include transportation and installation costs such as foundation, additional power cable, etc. (which account for about 5% of machine cost). Tooling (jigs and fixtures) cost usually ranges from 10% of the machine cost for general purpose machine tools, to 20% for fully-automated transfer machines, and is about 15% for semi-automatic special purpose machines.

1.3.2.2 Space comprises

- Machine area
- Accessories and tooling storage area (about 25% of machine area)
- Workpieces storage area: Generally there should be enough space near the machine tool to temporarily store input required, and output obtained during a period of 2 to 7 hours. Depending upon the size and shape of the workpiece and the production time/piece, the storage area can range from 50% to 200% of the machine area.

The space-cost (rent) is usually specified in Rs. /Sq. meter per annum (it was approximately Rs. 600/sq. meter per annum, in the year 2000).

1.3.2.3 New building investment cost

This can be taken as equal to the rental space-cost, specified in the preceding para.

1.3.3 Criteria for buying new machine tool

1.3.3.1 Increase in production

Usually, purchase of a new and improved better machine tool is thought to help in meeting the rise in demand for workpieces. It will however, be foolhardy to buy a new machine tool, producing lesser number of workpieces per annum.

1.3.3.2 Reduction in total production cost per workpiece

Is a sound reason for buying a new machine tool or replacing an existing one with a new one. It is an important criterion for buying a new and better machine tool.

Example 1.1 Compare the cost/shaft for an existing center lathe and a semi-automatic auto, with the following specifications:

Table 1.1: Comparative statement for manufacture of shafts on Center Lathe and Auto

Sl. No.	Specifications	Existing Center Lathe	New Auto
1	Machine cost (Rs.)	10,000	50,000
2	Tooling cost (Rs.)	1,000	7,500
3	Floor space along with tooling	4 m^2	5 m^2
4	Workpiece storage area for 4 hours production	2 m^2	2.7 m^2
5	Production rate	30/shift	40/shift
6	Machining cost (Rs.)	200/shift	300/shift
7	Operator wages	150/shift	80/shift

Space cost = Rs. 700/m^2/annum

No. of shifts/annum = 600 Nos.

Useful machine life = 10 years

Disregard material cost and overheads, which will be same for both the machines.

Solution: Using suffixes 'L' and 'A' respectively, to distinguish costs for Lathe and Auto
Production/annum for Lathe = NL

$$NL = 30 \times 600 \text{ (Shifts) Nos.}$$
$$= 18,000 \text{ Nos.}$$

Production/annum for Auto = NA
$$= 40 \times 600$$
$$24,000 \text{ Nos.}$$

Machining cost/annum for Lathe = ML

$$ML = 200 \times 600 = Rs.\ 120{,}000$$

Machining cost/annum for Auto = MA

$$MA = 300 \times 600 = Rs.\ 180{,}000$$

Labour cost/annum for Lathe = WL

$$WL = 150 \times 600 = Rs.\ 90{,}000$$

Labour cost/annum for Auto = WA

$$WA = 80 \times 600 = Rs.\ 48{,}000$$

∴ Production cost/annum for Lathe = PL

$$PL = ML + WL = 1{,}20{,}000 + 90{,}000$$
$$= Rs.\ 2{,}10{,}000$$

Production cost/annum for Auto = PA

$$PA = MA + WA = 1{,}80{,}000 + 48{,}000$$
$$= Rs.\ 2{,}28{,}000$$

Now, let us find the investment costs for both the machines.

Investment cost of Lathe and Tooling = ImL

$$ImL = 10{,}000 + 1{,}000 = Rs.\ 11{,}000 \text{ for 10 years.}$$

∴ ImL/annum = Rs. 1,100

Investment cost of Auto and Tooling = ImA

$$ImA/annum = \frac{50{,}000 + 7{,}500}{10}$$
$$= Rs.\ 5{,}700$$

Space cost/annum for Lathe = IsL

$$IsL = 6 \times 700 = Rs.\ 4{,}200$$

For computing space-cost for Auto [IsA], we must first find out the space required for higher inputs and outputs of the Auto.

Production rate of Auto = 40/shift nos.

∴ 4 h production = 20 nos.

4 h production on Lathe = 15 nos.

Fifteen shafts require 2 m² for temporary storage near the machine.

∴ Twenty nos. will require $\left(\dfrac{20}{15} \times 2\right)$ m² space

i.e. 2.66 m² = 2.7 m²

∴ $IsA \approx (5 + 2.7) \times 700$

$$= Rs.\ 5{,}390/annum$$

Introduction

∴ Total annual cost on Lathe for 18,000 shafts

$$= PL + ImL + IsL$$
$$= 2,10,000 + 1,100 + 4,200$$
$$= Rs. 2,15,300$$

$$\text{Total cost/shaft on Lathe} = \frac{2,15,300}{18,000}$$
$$= Rs. 11.96$$

Total annual cost on Auto for 24,000 shafts

$$= PA + ImA + IsA$$
$$= 2,28,000 + 5,700 + 5390$$
$$= Rs. 2,39,650$$

∴ $\text{Total cost/shaft on Auto} = \dfrac{2,39,090}{24,000}$

$$= Rs. 9.96$$

Saving/shaft $= 11.96 - 9.96$
$$= Rs. 2.00$$

Saving/annum $= 24,000 \times 2.00$
$$= Rs. 48,000$$

The investment of Rs. 50,000 in the

Auto will be recovered in $\dfrac{50,000}{48,000}$ years

i.e. 1.0417 year = 313 working days or 626 shifts

1.4 BASIC MACHINE TOOLS

Basic machining operations comprise turning, drilling, reaming, boring, facing, milling, broaching, grinding, lapping, and honing.

1.4.1 Turning machine

Turning machine or Lathe is the mother of all machine tools. With some ingenuity most of the machining operations can be carried out on lathe.

1.4.1.1 Center lathe

The center lathe [**Fig. 1.1**] has been in use for over 200 years now. A general purpose machine, it comprises a bed with a guideway. On one end of the bed is mounted the **headstock,** which provides drive to the spindle mounted with a work-holder. The other end is mounted with a **tailstock [Fig. 1.2]** which supports longer workpieces on the center. The tailstock can be adjusted axially on the guideway, and clamped in a position suitable for the workpiece length. The handwheel, used for fine adjustment of the center, can be clamped in set position, and the tailstock moved laterally, for taper turning. The tailstock shaft has a tapered hole [self-holding morse taper], which is used to locate the work-supporting center, or a drill for cutting holes in the workpiece. The handwheel is used, to feed the drill manually into the workpiece. Tailstock center is also used to align the threading tap with the spindle center, during manual tapping.

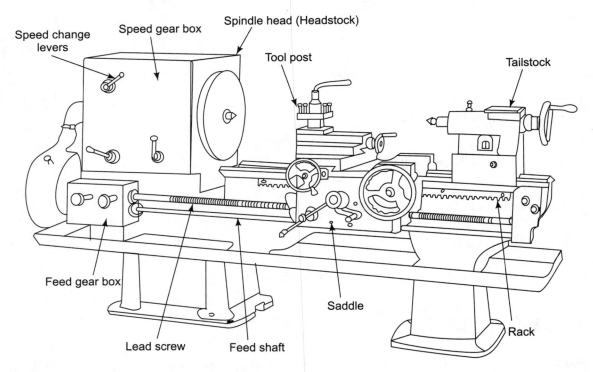

Fig. 1.1: Center lathe

This ensures perfect concentricity of the thread, with the drilled hole. The **headstock** [**Fig. 1.3**] houses the machine spindle in precision bearings. Its speeds should suit the workpiece/tool materials combination, and the workpiece size [diameter]. The change in speed is accomplished by using a back-gear and cone-pulleys combination, which together can give 6–8 speeds. In geared lathes, the gearbox placed between the motor and spindle can give 4–9 speeds. The next section, on machine tool drives, gives comprehensive details of both types of drives.

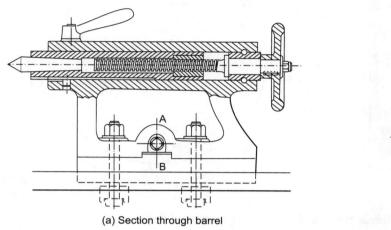
(a) Section through barrel

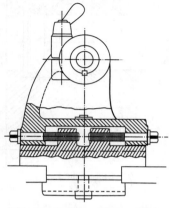
(b) Lower half in section on AB to show method of setting center

Fig. 1.2: Lathe tailstock

The spindle is hollow, to permit passage of the long bar stock, through the **central hole**. The workpiece end of the spindle is furnished with a **tapered bore** to locate the work-supporting center and is also provided with a precisely-machined **locating diameter** [spindle nose], and perfect concentric threads [**Fig. 1.4**]. This facilitates locating and mounting various types of chucks, face plates, etc. on the spindle. Usually, the spindle nose is mounted with a precisely machined **back plate**. It is the back plate, which usually locates chucks or other work-holders on the lathe. The flange diameter of the plate is used to locate the work-holder, while the equispaced holes on the flange of the back plate are used to clamp the work-holder [chuck] to the spindle. This eliminates wear and tear of the spindle nose, due to frequent changes of work-holders.

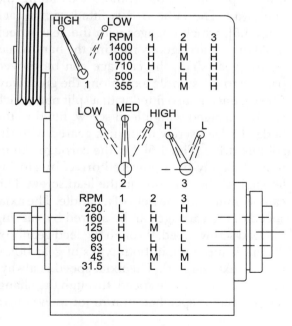

Fig. 1.3: Spindle headstock

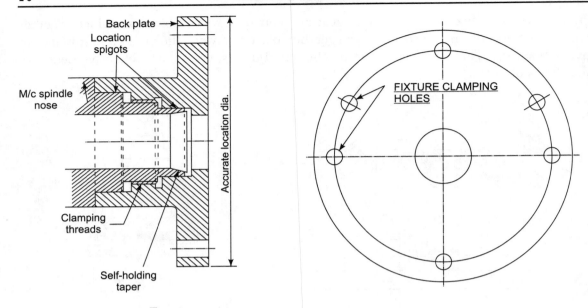

Fig. 1.4: Spindle and back plate for turning fixtures and chucks

Tool carriage: Placed between the headstock and the tailstock, the tool carriage can be moved parallel to the spindle axis along the guideway. The lathe bed is fitted with a rack [**Fig. 1.5**]. which engages with the gear wheel, fitted to the carriage. By rotating the handwheel on the gear shaft, the carriage can be moved parallel to the spindle axis, along the guideway. The carriage is also fitted with a split nut, which can be engaged with a lead screw, fitted to the bed. The lead screw in turn, is geared with the spindle drive [**Fig. 1.6**]. So, the carriage can be moved axially by a powered drive, by closing (engaging) the split nut on the lead screw. This can be done with the help of a handle. The axial speed of the carriage can be altered by varying the lead screw speed, through the feed gear box. (There is more about both types of gear boxes in the next chapter). The axial speed is always related to the spindle speed, through the change gears or feed gear box. Called **feed,** the axial motion rate is specified in mm [or inch equivalent] **per revolution** [of the spindle].

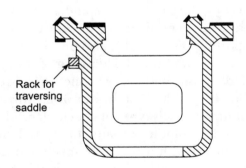

Fig. 1.5: Rack for saddle traversing gear

Introduction

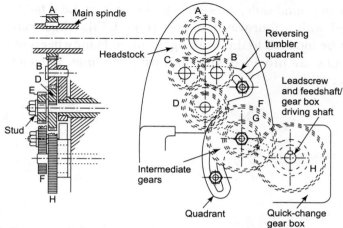

Fig. 1.6: Gearing from spindle to shaft driving leadscrew and feedscrew gear box [shafts for gears A, B & D shown in section]

The carriage has a guideway for cross [lateral] travel of the compound [tool-post] slide **[Fig. 1.7]**. It allows for fine adjustment of the tool-post in longitudinal [axial] as well as lateral [cross] directions. Furthermore, the tool-post can be swivelled to a suitable angle, using the degree markings on the base.

The front portion of the carriage—the apron, has a gear box, and handles, which permit feeding the tool-post in cross [lateral] direction, for facing and parting. There is also a threading dial, which rotates when longitudinal threading feed is used. The markings on the rotating dial allow for a precise closing of the split nut on the lead screw, so that the threading tool is accurately centralized into the rough-cut thread in the preceding threading run. For feed reversal, the spindle rotation is reversed.

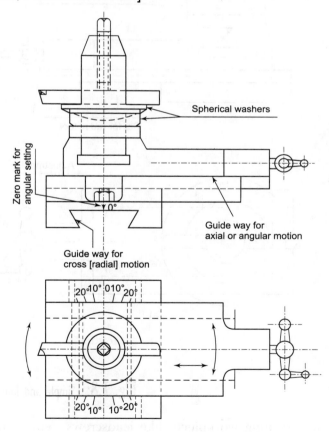

Fig. 1.7: Compound slide

Mounted on the compound slide, the simple tool-post [**Fig. 1.8a**] holds a single tool. Lathes are usually mounted with a four-way tool-post [**Fig. 1.8b**], which can accommodate four tools. Each tool can be indexed to the working position. Handwheels of the carriage, as well as the compound slide, are provided with dials with markings, to facilitate moving the tool within 0.5/1 mm accuracy.

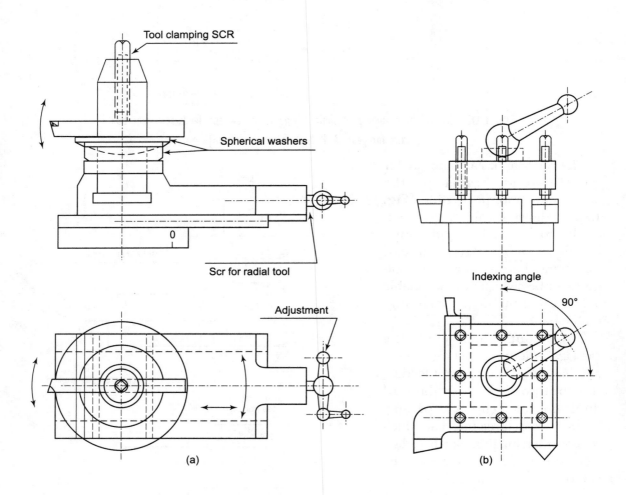

Fig. 1.8: Simple and four-way tool-post

Very long workpieces like leadscrews, require intermediate support **steadies** [**Fig. 1.9**] between the centers. The steadies can either be fixed [**Fig. 1.9a**], or traveling. The traveling steadies are usually fixed to the carriage [**Fig. 1.9b**], and allow for tool access in the supported portion. Steadies are provided with screws to facilitate adjustment of the support, to suit the workpiece diameter.

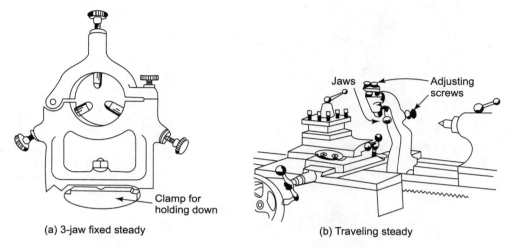

(a) 3-jaw fixed steady

(b) Traveling steady

Fig. 1.9(a): Lathe steadies

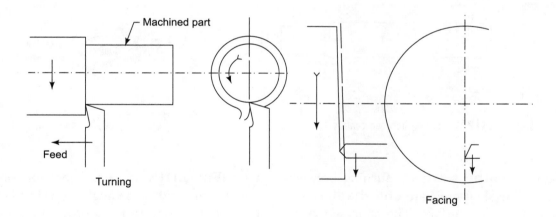

Fig. 1.9(b): Turning and facing

Center lathe operations Figure 1.9(b) and 1.10 show the various operations carried out on center lathes. For turning long tapered shafts, it is necessary to use the arrangement to offset the tailstock center laterally, with the two setting screws [**Fig. 1.2**], as explained earlier. Needless to say, the setting screws can be used to align the centers of headstock and tailstock for parallel turning.

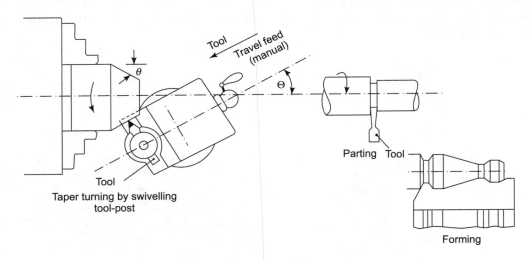

Fig. 1.10: Center Lathe operations

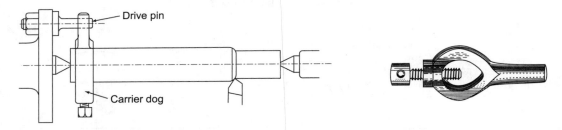

Fig. 1.11(a): Turning between centers   **Fig. 1.11(b):** Carrier dog

Center Lathe setups Turning between centers [**Fig. 1.11**] very convenient for long shafts. The shaft ends are drilled with centers, which provide 60° cone angles, serving as a conical bearing surface. The cone action promotes centralization. In mass production, the centers at both the ends are drilled in one setting, on special centering machines, to ensure perfect alignment of the centers at both the ends.

The drive is provided with a carrier dog, which is clamped to the workpiece. The carrier arm engages, with the drive-pin, fitted on the face plate, which will be described in the following section.

Despite engagement with the centers at both ends, long shafts bend and deflect under cutting tool pressure. This bending can cause convexity on the shaft. It is necessary to use suitable steadies [**Fig. 1.9**] to prevent workpiece deflection.

1.4.1.2 Chuck holding

Instead of using a carrier dog for drive, we can grip one end of the workpiece in a chuck. A self-centering three jaw chuck [**Fig. 1.12a**] can centralize a workpiece within 0.1–0.2 mm. All three jaws can be opened or closed simultaneously, by a single key. For a long shaft, with a hole at one end, the chuck and steady arrangement [**Fig. 1.12b**] can be used. Holding a long shaft in chuck at L.H. end and providing a center support at the other end, is also quite common.

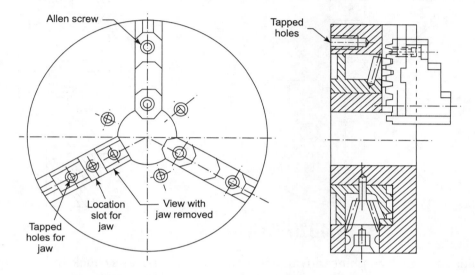

Fig. 1.12(a): Three jaw self-centering scroll chuck

Workpieces which are not very long, can be conveniently held alone in a chuck, without any need for the tailstock center support. Use of **soft jaws** can provide excellent concentricity in second operations. Soft jaws can be machined to suit the workpiece diameter and length, after mounting on a chuck, to ensure perfect concentricity.

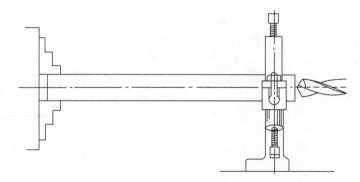

Fig. 1.12(b): Chuck and steady arrangement

Figure 1.12c shows 4 Independent Jaws Chuck. In this, every jaw is operated independently. This is very convenient for holding rectangular workpieces, for facing or boring. Some turners find it convenient to centralize cylindrical workpieces, precisely, within 0.01/0.02 concentricity.

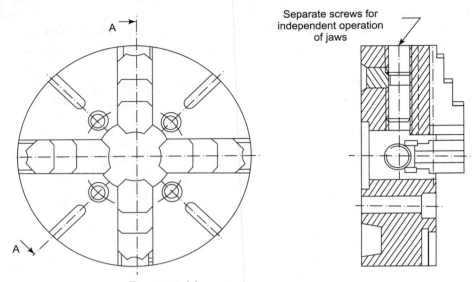

Fig. 1.12(c): Four independent jaws chuck

Combination chuck combines advantages of 3 self centering jaws chuck and 4 independent jaws chuck. Each jaw can be adjusted radially to different positions. This facilitates machining eccentric bores, and odd-shaped workpieces. After adjustment, the jaws are clamped onto the holders. All jaws can be closed/opened simultaneously, by a single key.

Section 8.2.2.0, in the chapter, on tool enginee-ring, gives details of special work holders such as collets, mandrels, and special jaws, for rapid location and clamping of similar cylindrical work-pieces. **Figure 1.13** shows the setup for boring work.

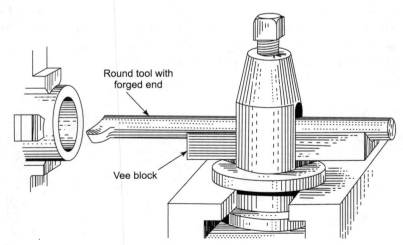

Fig. 1.13: Setup for boring

1.4.1.3 Face plate work [Fig. 1.14]

For workpieces which are not very long or have no cylindrical gripping surface, face plate mounting is convenient. A face plate is provided with many radial slots. These can be used for anchoring clamping studs and heel supports, used for plate clamps. Workpieces with eccentric bore can be machined by centralizing the bore, by suitably moving the workpiece along the face plate, and clamping it in a centralized position.

As the name implies, face plate is very convenient for facing work. In fact, there are special facing lathes which have only a face plate and tool carriage, and no tailstock.

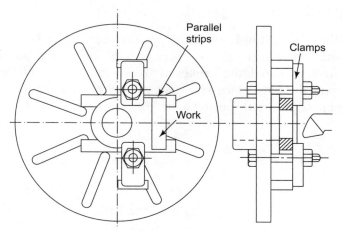

Fig. 1.14: Setup on face plate

Other miscellaneous operations possible on lathe Ingenious turners/engineers can execute many other operations on the lathe, some of which are:

1. **Milling**—The tapered bore of the lathe spindle can be fitted with an end-milling cutter, by using an adaptor sleeve and a make-do draw bolt. Removing the tool-post, the carriage and the compound slide can be used as the work table. A suitable clamping arrangement or fixture is provided to position and clamp the workpiece on the carriage, for facing, slotting, milling keyway, etc. I have converted an old center lathe into a special machine for milling sector of a drafting machine component.

    Horizontal milling cutters can be mounted on a special arbour which can be held in a self-centering chuck at one end, supported by the tailstock center at the other. The carriage can be mounted with a workpiece. The compound slide is used for effecting the manual feed.

2. **Slotting Keyways**—The workpiece is held in chuck and [if necessary] supported by the tailstock center. The keyway tool which is shaped like a parting tool, is mounted on the tool-post. It is mounted with the cutting face facing the headstock. The tool must be centralized with respect to the workpiece, using packings.

    While cutting the keyway in a shaft cross-slide is used to adjust the cut (which should not exceed 0.15 mm), cutting is carried out by moving the saddle longitudinally to execute the stroke. It is like broaching or slotting a keyway, but at a much lesser speed. The shaft keyway must be open at one end to allow longitudinal entry of the keyway tool.

    For keyway in a bore, it is necessary to mount the tool in a boring bar. Inadvertent spindle and workpiece rotation, must be prevented by engaging the back gear of the lathe.

1.4.1.4 Mass production lathes

When a multitude of similar workpieces are to be produced, it is convenient to use lathes which facilitate quick tool change, rapid and precise positioning of tools, and can also repeatedly effect, tool wielding strokes to precise length. The choice of a mass production lathe depends upon the workpiece size [capstan/turret; bar/chucking auto], and the desired productivity [capstan/bar auto; turret/chucking auto].

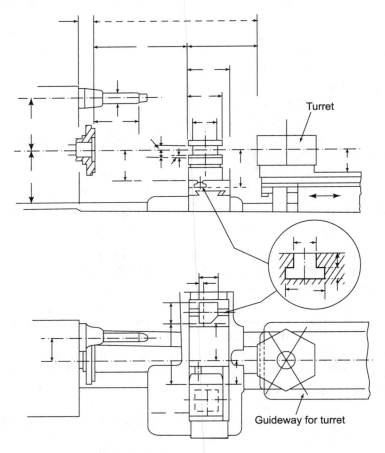

Fig. 1.15(a): Capacity chart for a capstan lathe

Tool turret Mass production lathes have a provision for mounting 4 to 8 tools on an indexible turret [**Fig. 1.15a**], having 4 to 8 [usually 6] tool stations, with a precision locating bore, a pad bolt clamp, and tapped holes, for securing flanged tool-holders. Location fit shanks of the turret tool-holders are clamped by pad bolts, with a cylindrical seat on the clamping area. **Figure 1.15d** shows the shank clamping arrangement. Some stations can hold more than one tool. The combination tool-holder in [**Fig. 1.15b**] can hold 3 turning tools, and a boring bar or a drill/reamer. Long boring bars can be guided by fitting a bush in the spindle bore [**Fig. 1.15c**].

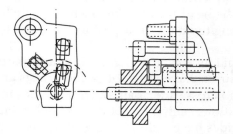

Fig. 1.15(b): Combination tool-holder for 3 turning and 1 boring tool

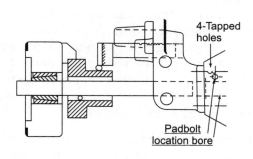

Fig. 1.15(c): Knee tool-holder for 1 turning and 1 boring tool

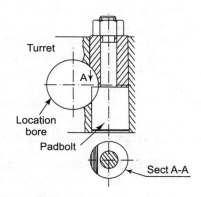

Fig. 1.15(d): Padbolt for clamping shanks

Return stroke of the turret automatically indexes it, to bring the next tool station in a working position. Furthermore, there is an adjustable stopper for every tooling station. In many machines, the stopper also indexes automatically, along with the turret. If not, then the stoppers drum is indexed manually, by the operator.

1.4.1.5 Capstan lathes

Capstan lathes are used for smaller workpieces of up to 100–150 diameter, while turret lathes are used for bigger workpieces. Both have tool-holding turrets. However, in capstan lathes, the turret saddle is mounted on a small, separate guideway, which can be shifted longitudinally, to a position that is suitable for the workpiece length, and is clamped to the bed in set position. The capstan saddle [turret base] is guided along its own guideway [**Fig. 1.15a**]. The cutting stroke is usually executed manually, by the operator.

In **Turret lathes**, the entire turret saddle is guided by the main guides, integrated with the bed. There is no separate guideway for a turret saddle [like in a capstan]. The heavy turret saddle moves in entirety, during the cutting operation. Naturally, the cutting stroke is powered to prevent operator fatigue, which moving the heavy saddle [slide] may entail.

The cross slide between the headstock and the turret usually has two posts: front and rear. The cross slide is moved longitudinally to a position suitable for the workpiece, and then clamped in the set position.

The cross slide is used mainly for facing, grooving, and parting operations by manual/powered cross feed. Some turrets have indexible 4 tools-posts at the front.

Some turret lathes have no cross slide at all. Such machines have facing slides, on turrets. These can be used for operations calling for tool cross feed in lateral direction.

Most of the turret cross slides have a short changeable feed screw to facilitate power feed, and multi-cut threading. The screw can be changed, if necessary, to suit the pitch of the thread to be cut.

The geared headstock usually provides 6–8 reversible spindle speeds. The number can be doubled by using a two-speed electric motor, for drive.

There is no provision for 'between the centers' turning setup. So capstan and turret lathes use chuck, collet, or a turning fixture to locate and secure the workpiece. These can be operated manually, or pneumatically. In pneumatic clamping, the gripping force can be increased during roughing by using higher air pressure. It can be reduced to much lesser value [1/2, 1/3], to reduce clamping distortion of the workpiece, during the finishing cut.

Special supports and tool-holders used on capstans and turrets

1. **Ball bearing center** permits higher bearing pressure on the center for the center can revolve along with the workpiece. This eliminates friction between the center, and the workpiece center hole, altogether.
2. **Ball bearing bush** with the bore, suitable for the workpiece diameter, can be fitted in the turret hole to support the machined diameter of the workpiece. The bush acts as a steady, and reduces workpiece deflection, due to pressure from the tool, mounted on the cross sldide.

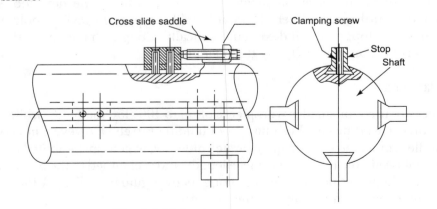

Fig. 1.16(a): Axial stops for cross slide

Introduction

3. **Roller supports [Fig. 1.16b]** also support the machined workpiece diameter, by two/three rollers, mounted on the ball bearings.
4. **Drill/reamer holder** are adaptors with outside diameter close location fit in the turret hole. They are provided with female morse tapers, for mounting drills and reamers.
5. **Boring bar holders [Fig. 1.16c]** are split, bush type holders, with a clamping bolt to tighten the split bush onto the boring bar.

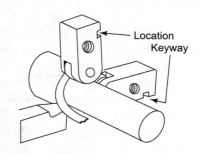

Fig. 1.16(b): Roller steadies

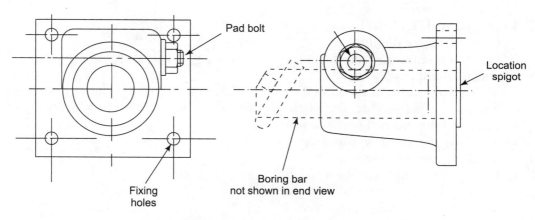

Fig. 1.16(c): Boring bar holder for turrets

6. **Floating reamer holder [Fig. 1.16d]** allows the reamer sleeve to align perfectly with the drilled hole in the workpiece. The clearance between the reamer sleeve and its housing, and the pivot mounting of the reamer sleeve onto the housing, provide the necessary float.
7. **Box turning tools [Fig. 1.16e]** mounted on the turret, are convenient for turning longer workpieces. These also

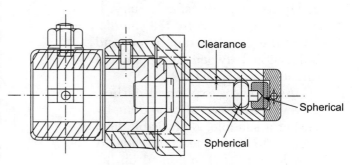

Fig. 1.16(d): Floating reamer holder

have a provision for mounting cutting tools. The machined diameter is supported by rollers, or a solid vee. This arrangement provides for precision and excellent surface finish.

Some box tools are provided with two tandem cutting tools, one for roughing, and the other for finishing. These have two sets of support rollers, for roughed and finished diameters. Naturally, the workpiece spindle should run at either the lower rough-ing speed, or the higher finishing speed.

8. **Die head** [Fig. 1.16f] can be used conveniently, for external threading of short and medium-length workpieces. Four radial chasers cut the thread. After threading, the chaser inserts open up to a diameter, bigger than the threaded part. This facilitates rapid withdrawal of the die head and the turret. There is a lever to adjust the chasers' positions for roughing and finishing cuts.

9. **Collapsible taps** can be used for cutting bigger sizes of internal threads. These taps close to size, lesser than the thread minor diameter, after threading. This permits rapid withdrawal of the collapsible tap. For smaller taps, it is necessary to reverse the spindle, for pushing out the tap, by nut and bolt action, between the tap and the threaded workpiece.

As already explained earlier, screw cutting can also be done by using a suitable, replaceable lead screw, to drive the cross slide. A single-point tool, mounted on the front tool-post, or in a boring bar can be used for external and (bigger) internal threads.

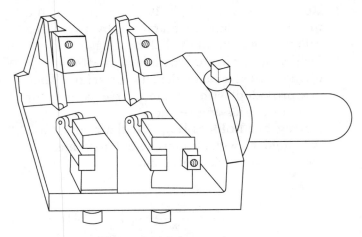

Fig. 1.16(e): Box turning tool

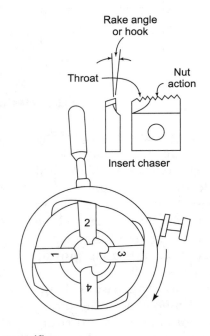

Fig. 1.16(f): Radial die—Head and chaser inserts

Introduction

10. **Taper turning attachment** [Fig. 1.16g] can be used for turning/boring tapered workpieces. The turret face is mounted with a horizontal slide, on which the tool-holder is bolted. The tool holder has a bracket for a guide roller. This roller is guided by the slot in the roller guide plate, mounted at a suitable place on the bed of the machine. The angle of the slot can be set to suit the required taper angle.

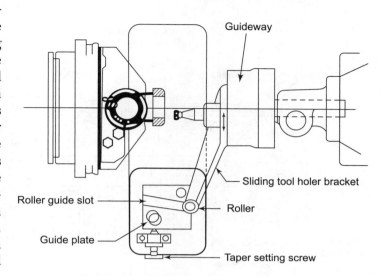

Fig. 1.16(g): Taper turning bore with a slide on turret face

As the turret moves longitudinally towards the headstock, along the angular slot, the guide roller moves the tool holder and the horizontal slide radially, at the rate necessary for the required taper.

Capstan and turret lathes are convenient and economical, for batches running up to 5 or 6 shifts (about 50 working hours). For workpieces calling for longer machine runs, automatic lathes are more economical.

1.4.1.6 Automatic lathes

Automatic lathes or Autos are of two types: bar autos and chucking autos. As the name implies, bar autos are used for workpieces produced from bar-stock. Bar autos can be single-spindle or multi-spindle.

Single-spindle bar auto Figure 1.17 uses a combination of geared and chain drives. The drive comprises simple change gears, instead of a sophisticated gear box, for the time spent on changing gears, is a very small fraction of the long production run. The spindle is driven by two chains, which rotate the two sides of the double-ended clutch, necessary for spindle reversal, while the L.H. chain and clutch drive the spindle at high, clockwise speeds, used for turning, the R.H. clutch and chain drive the spindle in slow-range, anti-clockwise speeds, used for slow speed reaming, threading, etc. Naturally, for clockwise direction, the front cross slide tool should be mounted upside down.

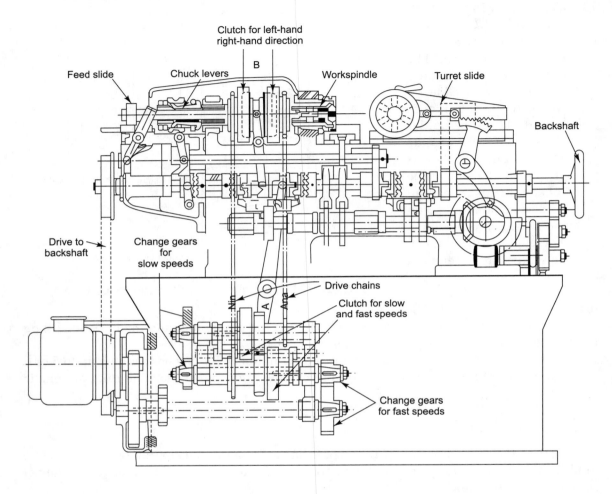

Fig. 1.17(a): Single-spindle automatic—Elevation of drive

There is a belt-driven backshaft, used for mounting the clutches for bar feed, bar chucking, changing speed, and indexing the turret. The clutches are operated by long levers, whose ends are moved by trip cams on the drums, rotating on the front shaft. The handwheel on the backshaft comes handy during machine and tool setting.

Introduction

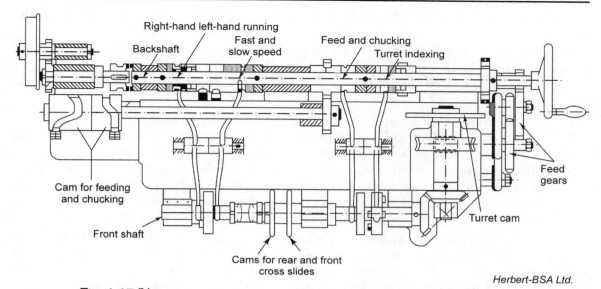

Fig. 1.17(b): Single-spindle automatic—sectional plan showing backshaft and front shaft

The cross slide and the turret are actuated by cams, designed for a particular workpiece. The cutting strokes of the turret and cross slide depend upon the rise and fall of their respective cams. A single turret cam takes care of the cutting strokes of all the turret tools. Front and rear cross slides call for two additional cams [**Fig. 1.18**].

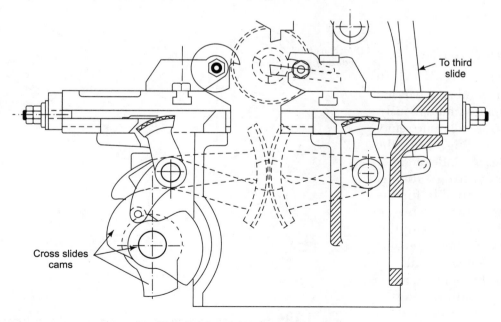

Fig. 1.18: Cams for auto cross slides

In some autos there is a third slide. Used for parting, this slide is mounted close to or on the headstock.

There is a swing stop for feeding the bar. It can be set to suit the workpiece length. The stop is swung clear of the tool paths, before machining commences.

A number of additional accessories are available. The following is a list of the widely used accessories:

1. Thread cutting attachment, mounted on front slide and geared to the spindle
2. Screw slotting [or flats milling] attachment with a motorized arbour, for slotting/flat milling cutters, and a transfer arm for conveying and positioning the parted screw
3. Drilling attachment, with a separate motor to drive the smaller drill in a direction, opposite to the spindle rotation, to obtain higher [relative] drilling speed
4. Chip conveyor belt, to remove the scrap from the working area
5. Magazine feed, for automatic loading of workpieces for second operation

With automated feeding, chucking, slides and turret actuation through cams, the operator only has to feed a new bar, when the previous one has been consumed. It is therfore common for an unskilled/semi-skilled operator, to attend to two automatic lathes. He can even inspect the produced workpieces periodically, by simple gauges. The tool-setter presets various tools, so that a worn out tool can be replaced quickly. Only a slight tool adjustment might be necessary. Ease in replacing a worn tool rapidly, by a preset tool, allows for the usage of 25–50% higher cutting speeds, to enhance productivity.

Multi-spindle bar Autos [Fig. 1.19] In these, a number of bars are machined simultaneously. In the 6-spindle auto setup, shown in **[Fig. 1.19]**, the spindles are indexed after every cutting stroke, which is suitable for the longest turned/bored length. The smaller lengths of other diameters are machined, by suitably shifting the tool-holders for those smaller lengths axially, by using only a part of the set strike.

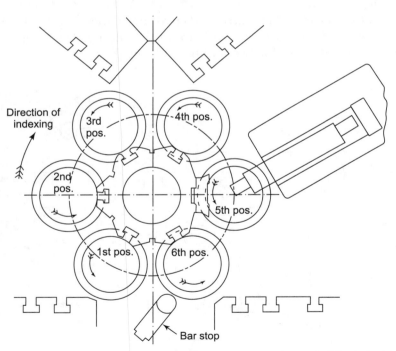

Burton Griffiths & Co. Ltd

Fig. 1.19: Spindle and slide arrangement of a six-spindle automatic

Introduction

The central tool-holding turret has axial tee slots, for locating and clamping the tool-holders, for the operations executed by longitudinal cutting strokes.

Tools calling for radial [cross] motionss are mounted on radial slides around the spindles, e.g. parting slide at the fifth position. The spindles are numbered. Generally, there is one more radial slide, usually in position two for facing, forming, or grooving.

Tables with tee slots around the spindles, can also be used for mounting additional tool-holders. The bar stop is mounted on the arm, shown near the sixth spindle position.

All the tools simultaneously cut the bars in the nearby position. All the bars rotate simultaneously, while the tools remain stationary. Every cutting stroke and indexing, produces one finished workpiece.

Chucking autos   used a chuck instead of a bar collet. They are more suitable for machining bigger odd-shaped workpieces.

The cutting strokes are actuated by standard curved plate cams, mounted on the cams drum. The cams inside, match the drum diameter. Cams with more rise, can be used for cutting strokes, lesser than its rise, by using only part of the cam rise.

1.4.1.7 Special purpose turning machines

Manufacture of cams for autos is a time-consuming process. The precision necessarily makes cams quite costly. Even special tools, such as form tools and workpiece-tailored tool-holders, don't come cheap. Furthermore, when the workpiece becomes obsolete, all these cams, special tools, and holders depreciate instantly, to the cost of scrap metal. The investment necessary for manufacture of a workpiece on a copying lathe, is much lesser than the investment necessary for manufacture of a workpiece on a capstan/turret/auto. Moreover, the time required for the manufacture of a master for a copying lathe, is only a small fraction of the time required, for manufacture of the numerous cams and special tool-holders. There is also far more formal and dimensional consistency in the workpieces, produced on copying lathes.

Copying lathes   For manufacture of very long workpieces, which must be turned between the centers, the copying lathe is the most productive and economical machine. These machines use hydraulic copying devices. **Subhead 7.3.3.4** and **Figure 7.40**, in the chapter on Machine Tool Operation, describes a hydro-copying device.

In hydro-copying, both longitudinal [axial], as well as lateral [cross] feeds, are effected by two separate hydraulic cylinders. The stylus fitted to the ram of the cross-feed cylinder is pressed against a master by hydraulic pressure. The motion of the stylus and the ram are converted to radial tool motions by the copying system. While copying a parallel diameter, the cross feed cylinder ram and the stylus remain almost stationary, during facing cuts, the ram actuating the longitudinal cylinder, remains stationary. A cam-operated valve facilitates rapid approach to the cutting position.

The master-holder is indexible to facilitate roughing. The master is fitted with special keys, whose height corresponds with the roughing cut profile. It is possible to take three roughing and one finishing cut, with four indexing positions. **Figure 1.20** shows a master for three roughing and one finishing cut. The workpiece can be machined in one roughing and one finishing cut, by using double indexing. For workpieces with maximum material removal less than three mm, the workpiece can be machined to finish size, in a single cut. No indexing of the master is necessary, nor the keys for roughing profile. The master is an exact replica of the required workpiece.

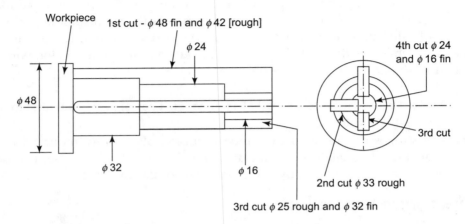

Fig. 1.20: Master for copying lathe

1.4.2 Boring machines

Precise parts like anti-friction bearings, require accurately-machined bores, with close tolerance [sixth grade], for satisfactory assembly. Maintaining close tolerance on bores is more difficult than accomplishing the same accuracy on outside diameters. That is the reason the shaft tolerances in recommended fits, are usually one grade more accurate, than the tolerances on the mating bores, e.g. [H7/g6, H7/p6]. In mass production, thousands of components need to be accurately bored. Consequently, there are many special-purpose machines which bore only similar workpieces, like connecting rods, day in and day out. General type of boring machines with provisions for a variety of related operations, like reaming and facing, are also used widely.

1.4.2.1 Horizontal boring machine

Horizontal boring machine **[Fig. 1.23]** is particularly convenient for machining long precision bores, in a heavy workpiece. The machine comprises a workpiece table with tee slots for clamping bolts. The table can be moved in cross [lateral] direction manually, and by powered drive. The long lead screw in the bed can move the table longitudinally, at a set feed rate [mm/inch per min]. The cross feed comes handy for facing workpiece, by mounting a face mill on the boring spindle. It can also be used for aligning the to-be-machined bore laterally, with the machine spindle. Needless to say, the power feed is used to reach the approximate position rapidly, while fine adjustment is carried out manually, by a handle.

The boring spindle can be adjusted vertically by a handle, actuating the lead screw/nut mechanism. It can also be moved by a powered drive. The spindle is provided with a facing slide, which can feed the mounted tool radially, to face flanges square with the bore.

There is a column at the other end of the horizontal workpiece slide, which houses the boring bar support bracket. Provided with anti-friction bearings, the support can be adjusted vertically by a handle, through the lead screw/nut combination.

The spindle comprises the inner core-spindle, similar to a drilling machine spindle, with a morse taper, and slots for tang and cotter [see the following section on drilling machines]. The outer flanged sleeve-spindle, carries the facing slide. The nine-speed gearbox is driven by toothed belts. The spindle can be fed longitudinally [axially], to machine medium-length bores. For long bores, the workpiece is moved axially to effect the boring stroke. The supporting boring bar at both the ends, minimizes taper, due to boring bar deflection, for deflection of an overhanging, cantilever-type boring bar, is 16 times the deflection of a bar supported at both ends. Tools for long bores are mounted on boring bars **[Fig. 1.24]**. So reamers, spot facing cutters, etc. are shell-shaped. **Figure 1.24** shows some of the tools used on horizontal boring machines.

1.4.2.2 Relieving lathes

For manufacture of a form-relieved milling cutter, it is necessary to withdraw the form tool periodically, by controlled distance [the relief], at a controlled rate. The relief is the radial distance the cutting tool moves between the faces of two adjacent teeth of the cutter being relieved **[Fig. 1.21]**.

Figure 1.21 shows the drive for form-relieving, on the special-purpose lathe.

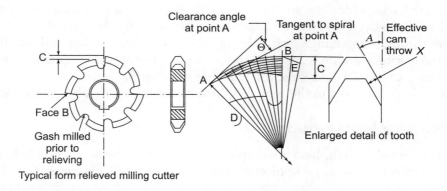

Fig. 1.21(a): Schematic layout of form relieving system

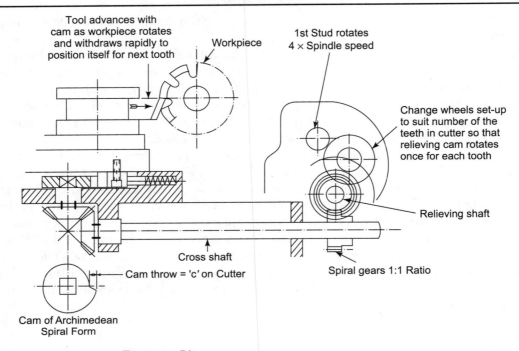

Fig. 1.21(b): Schematic layout of form relieving system

The relieving shaft at the back, is geared to the spindle through the headstock. A cross shaft transmits the drive from the relieving shaft to the cam shaft, through two gear trains. The cam shaft is fitted with a cam of required relief throw [rise]. It is necessary to use different change gears, to suit the number of teeth on the required cutter. The relieving form tool is mounted on the relieving [cross] slide. The spring-return slide is pushed by the cam to cut off the relief. At the end of the relieving stroke, the spring pushes the slide backwards, to the retracted position. Set through the change gears, the number of cutting strokes and withdrawals per rotation of the workpiece spindle, depends upon the number of teeth required on the cutter. It is also necessary to use a lead screw, for relieving gear cutting hobs.

The cutter blank is turned to the required form on a center lathe, before relieving.

1.4.2.3 Vertical spindle turning machines

Large diameter, small length, heavy workpieces like the fly wheel, can be loaded/unloaded more easily on the horizontal table of vertical spindle turrets [Boring mill, **Fig. 1.22**].

These have a large circular work table with radial tee slots. These can be used for clamping workpieces with clamps, grip down jaws, etc. The table rotates the workpiece at the optimum speed. The tool turret is mounted on a cross rail slide and like the radial drill, is radially adjustable, to facilitate taking turning and boring cuts. The turret can be moved axially [vertically], to execute the longitudinal cutting stroke.

There is a side slide for operations like facing, grooving, etc., which need radial feed like the cross-slide in the horizontal, spindle turning machines. All slides can be moved manually for setting, and by powered drive, during manufacture.

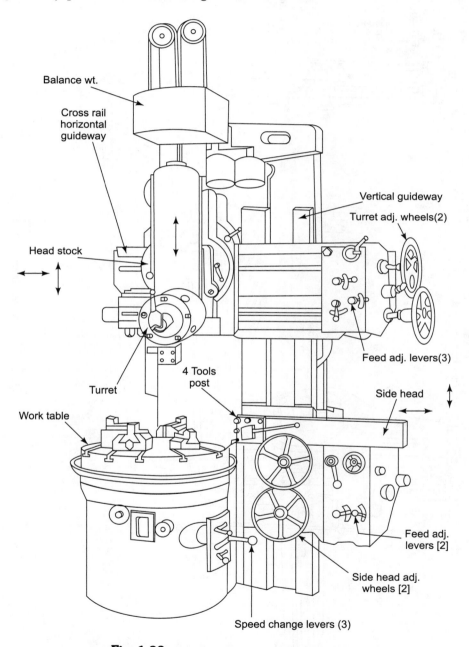

Fig. 1.22: Vertical turret turning/boring machine

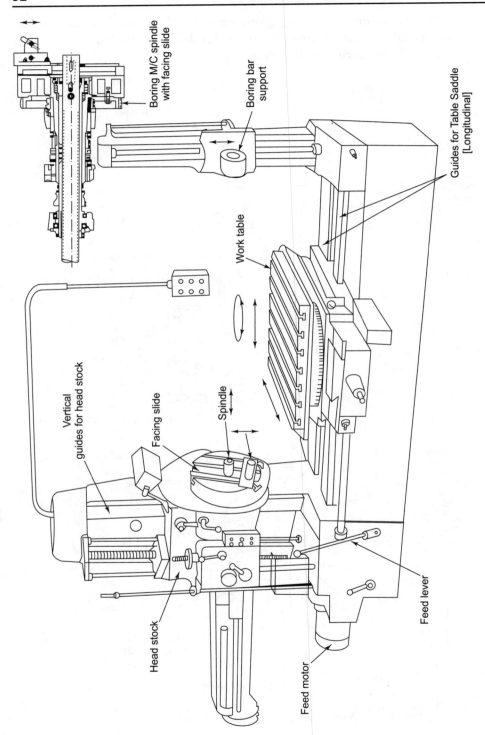

Fig. 1.23: Horizontal boring machine

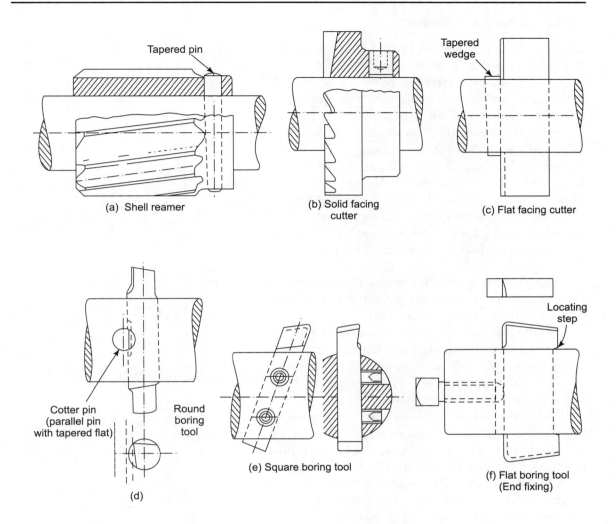

Fig. 1.24: Tools for boring bars

1.4.3 Drilling machines

Almost every machine part has holes, for fixing it with screws. Some flat parts like the cover, require no other operation, except drilling these fixing holes. Drilling machine is a must for almost every engineering workshop.

Although designed mainly for making holes with twist drills, drilling machines are also used for finishing accurate holes, by reaming and facing fixing bolt seats by spot facing. Tap-holders or provision for spindle reversal also facilitates threading holes, by tapping.

1.4.3.1 Column drilling machines

Column drilling machines [**Fig. 1.25a**] comprise a base with a tee-slotted table, mounted with a vertical guide column, for mounting auxiliary work table with radial tee slots, for clamping the workpiece. There is a central hole that provides a passage [overshoot] for the drill. The auxiliary table can be adjusted vertically on the column, to suit the workpiece height. The column carries the spindle head above the auxiliary table. The spindle head is also vertically adjustable and clampable on the column. The geared spindle heads provide 4 to 9 speed, in two [high/low] speed ranges. The belt driven machines on the other hand may only have 3 to 4 speeds.

The spindle can be fed downwards by a levered handwheel. Some machines have 4–8 power feeds at various rates [mm/inch per rev].

In some machines, the cutting stroke is effected by moving the worktable by a handle, through a screw/nut combin-ation. Such machines are called **upright drill press.** They have a rigid rectangular column and a feed gear box.

When a workpiece calls for a number of operations, such as drilling, reaming, spot facing, counter-boring, it is advisable to use a Gang drilling machine, with a number of spindle heads mounted on a common table. Each spindle head carries a different tool, running at its recommended speed. The workpiece mounted in a jig/fixture is moved from one spindle to another, for subsequent operations. This saves time that may be required for changing the tools on a single spindle, and changing the spindle speed.

1.4.3.2 Turret drilling machine [Fig. 1.25b]

Turret drilling machine requires less space than the gang drilling machine. Instead of a single spindle, the head has a 6-spindle turret. Each turret spindle can hold one drilling tool. The 6 spindles

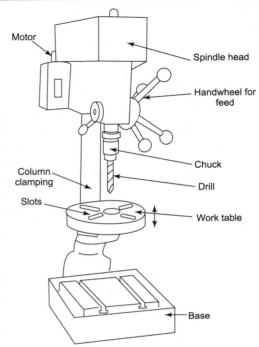

Fig. 1.25(a): Column drilling machine

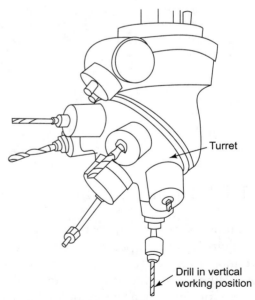

Fig. 1.25(b): Turret-type drilling machine

rotate at different speeds. Some rotate at a speed, slow enough for threading. Each tool can be brought to its working position by indexing the turret.

An ingenious device called the **quick change chuck [Fig. 1.26]** allows for changing a drilling tool, even while the machine is running. The tools are mounted in special holders, which can be coupled/uncoupled rapidly, with the quick change chuck. You only have to raise the outermost sleeve of the chuck to decouple the tool-holder, for it to be removed. In a sleeve-up position, a tool-holder for the next operation can be inserted into the chuck. Releasing the outer sleeve causes its gravitation to fall to a 'down' position, which couples the inserted tool-holder with the quick change chuck, though two balls.

1.4.3.3 Radial drilling machines [Fig. 1.27]

It is not practical to slide heavy workpieces onto a drilling machine table, to align the hole position with the fixed drilling spindle. Under such circumstances, it is necessary to use a radial drilling machine, with a radially adjustable spindle head. In these machines, the spindle head is mounted on a horizontal slide on the radial arm, which can be pivoted around the rigid column guiding it. The arm, along with the spindle head can be raised or lowered on the column by power drive, and if necessary, done manually, for fine adjustment. The radial motion of the spindle head too can be powered or manual, as necessary.

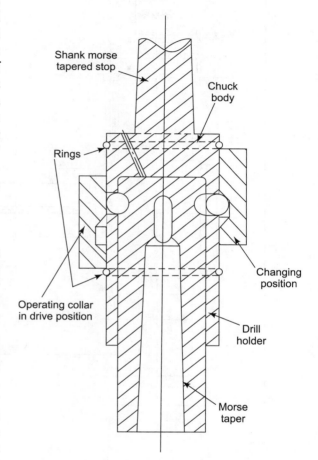

Fig. 1.26: Quick change chuck

The sophisticated speed and feed gear boxes prove eight speeds and feeds in two high/low range. Drilling feed can be powered. There is a handwheel for fine manual feed too.

A radial drilling machine, can drill holes anywhere in the area covered by the circle, with the maximum radius of the spindle center from the column center. So it is a common practice to provide additional tables in the circular area. This allows an operator to drill the workpiece at one table, while another workpiece is being clamped/unclamped on the other table. Sometimes, large workpieces are placed on the ground, and leveled by suitable packings before drilling.

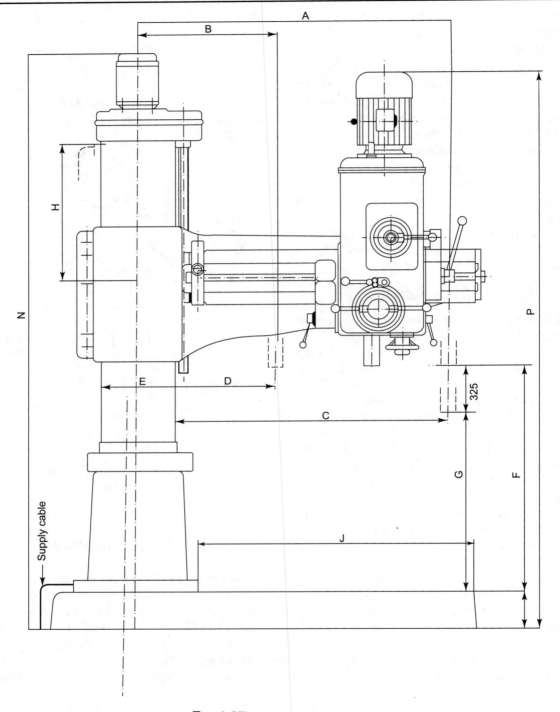

Fig. 1.27: Radial drilling machine

For drilling very large workpieces like a railway engine, it is necessary to use powered hand drills [**Fig. 1.28**], which run at only one speed. But they can be carried to the workpiece, and held manually, at any angle.

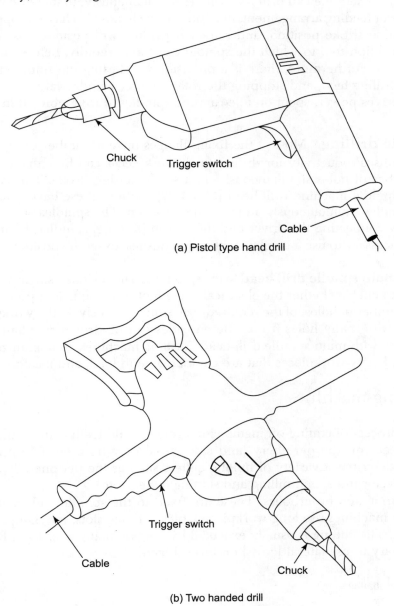

(a) Pistol type hand drill

(b) Two handed drill

Fig. 1.28: Powered hand drills

1.4.3.4 Special purpose drilling machines

In special purpose drilling machines, the drilling spindle head can be mounted in any suitable position, at any angle. One can drill many holes simultaneously, to increase productivity. In addition to power feeding arrangement, the drill spindle heads also have stoppers for forward, end of the cutting stroke position, and retracted [withdrawn] position. Usually there is an intermediate position too, to which the spindle can travel rapidly, before commencing the slow cutting feed. Furthermore, there is a provision for starting the spindle rotation before commencing drilling feed, and stopping the rotation after withdrawal to the clear retracted position. This saves power, spent on idle-running spindles, during non-cutting time.

Multi-spindle drilling Most of the fixing flanges have a number of equispaced holes, for securing bolts. Productivity for drilling such workpieces can be enhanced manifold by drilling a number of holes simultaneously, in a single cutting stroke. This can be done by using a multi-spindle machine/drill-head **[Fig. 1.29]**. Both of these have gear boxes, which drive 2–12 spindles simultaneously, in the same direction. The spindles are made adjustable for position, by connecting the driver and the driven [drilling] spindles, by universal joints. Usually, it is necessary to use a jig plate with guide bushes, to set the positions of the spindles correctly.

We can fit a **multi-spindle drill head** to the spindle of a heavy-duty, single-spindle machine. It is advisable to check whether the electrical drive motor has sufficient power [KW] to drill the required number of holes, of the required size, simultaneously, in the workpiece material.

Workpieces with many holes from different sides can be drilled simultaneously, from a number of sides, with multi-spindle drill heads/drilling machines, to maximize productivity, particularly for heavy workpieces that are cumbersome to load, and locate.

1.4.4 Milling machines

Milling is the process of cutting-off metal with a rotary multi-teeth cutter [mill]. A variety of cutters, with teeth on the periphery [and side faces] may be used. **Chapter 8 on Tool Engineering**, describes a variety of milling cutters such as the peripherally-cutting sawing mill, side and face cutters, endmilling and slotting cutters, etc.

Cutters are mounted on arbours. These are fixed to the spindle and rotated at suitable speeds. Milling machines use long workpiece tables with tee slots, for clamping workpiece/milling fixtures. The cutting is usually executed by longitudinal power feed. But sometimes, it is also necessary to use lateral [cross] or vertical feed.

1.4.4.1 Horizontal milling

Figure 1.31 shows a horizontal milling machine along with the arbour. The spindle is provided with a Quick Release ISO taper and two drive keys on face. The arbour is clamped to the spindle by a draw bar bolt.

Introduction

The table, gear box, and power feed mechanisms below are collectively called **knee**. It can be moved vertically on the guideways on the upright column, housing the spindle geared drive. This type of machine is often called Column and Knee Milling Machine. There are stops [cams] to set the longitudinal stroke to suit the workpiece. The stops disengage the power feed automatically, by moving the feed engaging lever.

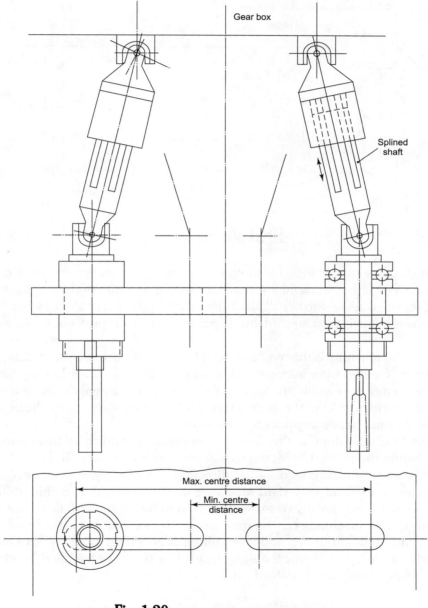

Fig. 1.29: Two spindle drilling machine

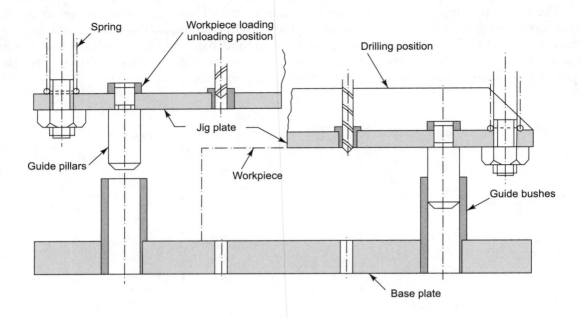

Fig. 1.30: Multispindle jig

After engaging the power feed by moving the feed lever in the required direction, the table can be moved faster, in rapid approach, to within 3–5 mm of the cut-starting position, by pressing the feed lever further against the return spring. As soon as the feed lever is released, the return spring pushes it from the rapid approach position to slow cutting feed position.

At the top of the machine is the overarm, which carries the arbour support, fitted with anti-friction bearings. Guided by a horizontal dovetailed guideway, the arbor support should be as close to the cutter as possible, reckoning the milling fixture width. In gangmilling, slab-milling, or other wide-cut operations generating severe vibrations, it might be necessary to provide an additional **brace support** for the arbor [**Fig. 1.31**].

Usually, the feeding is done in the direction, opposite to cutting. In this method called up milling, the cutting forces tend to lift the workpiece/milling fixture, off the table.

Used rarely, **down milling** feed, in the direction of cutting, generates severe vibrations. However, it can be used on very rigid machines to mill cylindrical or thin, difficult-to-hold workpieces, for deep-cut sawing, or for wide cuts, and for better surface finish.

Horizontal milling machines can be mounted with swivellable vertical milling heads, to enable use of end milling, slotting, and face milling cutters [**Fig. 1.32**]. We can machine slots, keyways, and faces using the vertical milling head. The swivel of the head, facilitates angular slotting and facing, using the cross feed for the cutting stroke.

Introduction

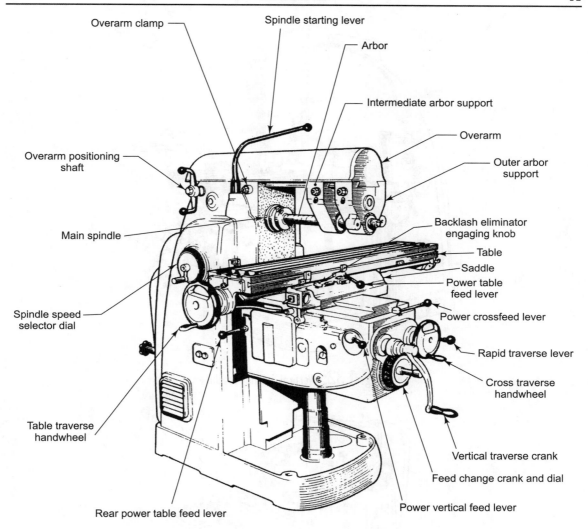

Fig. 1.31: Horizontal milling machine

1.4.4.2 Vertical milling machines [Fig. 1.33]

Vertical milling machines are designed for use of drills, reamers, endmills, slot drills, and face mills. As such, the spindle is often bored to a self-holding morse taper, slotted for tang, and sometimes even for cotter. It is necessary to use a draw bolt for holding the cutters, which must have a tapped hole for the draw bolt, for the positive rake angle on the cutter teeth, tends to push end milling cutters downwards, out of the tapered spindle bore. The extraction and fall, can break the costly cutter.

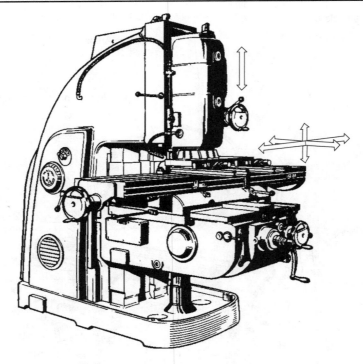

Fig. 1.32: Vertical milling machine

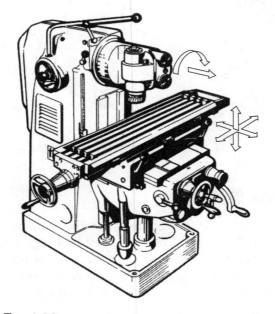

Fig. 1.33: Vertical milling machine with swivel head

Introduction

Cutter fall can be prevented by using a collet-holder, secured to the spindle with the draw bold. The cheaper, straight shank end mills and slot drills can be used if the collet-holder is available. The vertical milling spindle can be fed [moved] axially and manually.

Vertically milling machines are convenient for drilling, end and face milling, and boring. They can also be used for less accurate [within ± 0.1 center distance] jig-boring work, by using slip and dial gauges.

1.4.4.3 Universal milling machine

Basically a horizontal milling machine, the universal milling machine has a swivellable table. This facilitates milling spiral gears and gullets in cutters, with a helix angle. For helical milling, it is necessary to swing the table to the required helix angle.

Universal milling machines are usually equipped with a dividing head to facilitate accurate indexing of the workpiece. Dividing head is also used for rotating the workpiece at a precise angular rate, with respect to the longitudinal feed rate. This is done by gearing the longitudinal feed screw with the dividing head spindle, in correct ratio, through suitable change gears.

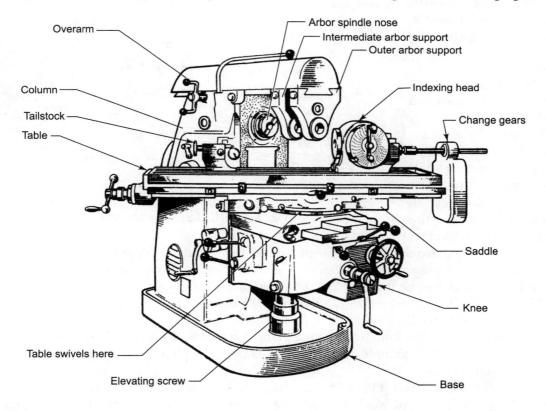

Fig. 1.34: Universal milling machine

Dividing head The dividing head is a gear box with worm-wheel reduction gearing [**Fig. 1.35a**]. It is used for precise angular movement [indexing] of a workpiece for milling squares, hexagons, splines, gears, chain ratchets, etc. The worm is fitted with an index-ing plate, while the wheel shaft is coupled with the workpiece, held between centers [**Fig. 1.35b**]. There is also a tailstock, for supporting the long workpiece at the other end.

The indexing plate has a number of equispaced indexing holes, at different pitch diameters. The slot in the indexing lever, permits radial adjustment of the indexing plunger, to coincide with the required pitch diameter

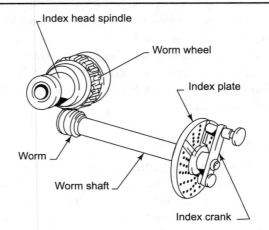

Fig. 1.35(a): Simple indexing head mechanism

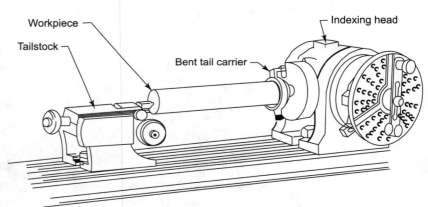

Fig. 1.35(b): Center to center setup

of the holes: the pitch diameter with the number of holes, suitable for indexing the workpiece. For example, 25 equispaced holes can provide 5 or 25 equal divisions of a 360° circle.

28 holes P.C.D. can be used for 2, 4, or 7 equal divisions.

30 holes P.C.D. can be used for 2, 3, 5, 6, or 10 equal divisions.

40 holes P.C.D. can be used for 2, 4, 5, 8, 10, 20, or 40 equal divisions.

For executing the division, the indexing plunger is moved by a suitable number of spaces between the holes. For example, for milling a hexagon, the plunger must be moved 5 spaces on the 30 holes P.C.D.

Universal dividing head [**Fig. 1.35c**] can be suitably swivelled in the vertical plane to mill the bevel gears. The head workpiece spindle can be swivelled up to 90° for milling cams. Needless to say, the tailstock center axis is also tilted to match the dividing head tilt.

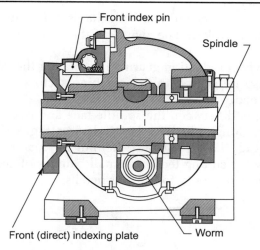

Fig. 1.35(c): Universal indexing head

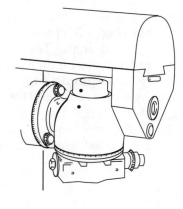

Fig. 1.36: Universal spiral attachment

Other accessories Universal spiral milling attachment enables milling spirals. The housing, swivellable in the vertical plane is mounted on the milling machine spindle. The swivel of the spindle in the horizontal plane facilitates spiral milling on machines, without a horizontal table swivel. The small-length, sturdy cutter spindle can be swivelled to any compound angle [combination of angles with horizontal and vertical planes].

Motorized overarm attachment [**Fig. 1.37a**] has a captive reversible drive motor, gear box, and swivelling head. We can use vertical milling cutters. The independent motor also permits the simultaneous usage of the horizontal milling drive, if feasible. This attachment is fitted in place of the overarm [arbor support].

Fig. 1.37(a): Motorized Overarm

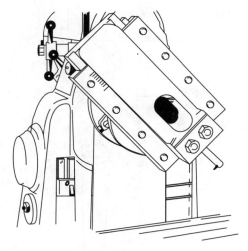

Fig. 1.37(b): Slotting attachment

Slotting attachment [Fig. 1.37b] comprises the wheel and crank mechanism. It converts the rotary drive to reciprocate cutting and return strokes. The swivel of the head enables machining keyways, slots, etc., at any angle.

Circular table [Fig. 1.37c] with worm and wheel reduction gearing for rotating the table slowly to the required angle. The table can also be geared to the machine sectors of the arched slots. This can also be done by manual feed.

Rack milling attachment [Fig. 1.37d] Mounted between the spindle face and overarm support, this device uses cross feed for cutting, and longitudinal feed for indexing [Fig. 1.37e]. Used along with universal spiral attachment [Fig. 1.37f], it is also helpful for machining broach teeth. Milling machines are used widely for cutting teeth on small batches of gears. Refer to the section titled 'Gears manufacture' in Chapter 2, for further details.

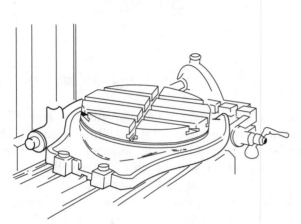

Fig. 1.37(c): Circular table attachment

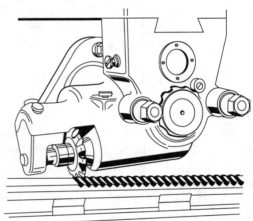

Fig. 1.37(d): Rack milling attachment

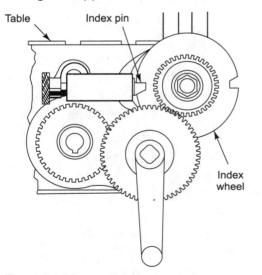

Fig. 1.37(e): Rack indexing attachment

Fig. 1.37(f): Broach milling with universal spiral attachment

Introduction

1.4.5 Jig boring machine

Used for precision positional drilling/boring/grinding, jig boring machines are like rigid, high-precision, sophisticated universal milling machines.

1.4.5.1 MAHO MAH800P Jig boring machine

It has a diagonally-ribbed, torsionally rigid column, with spindle adjustment and reversal gear for headstock, at its top. The feed motor is built into the column.

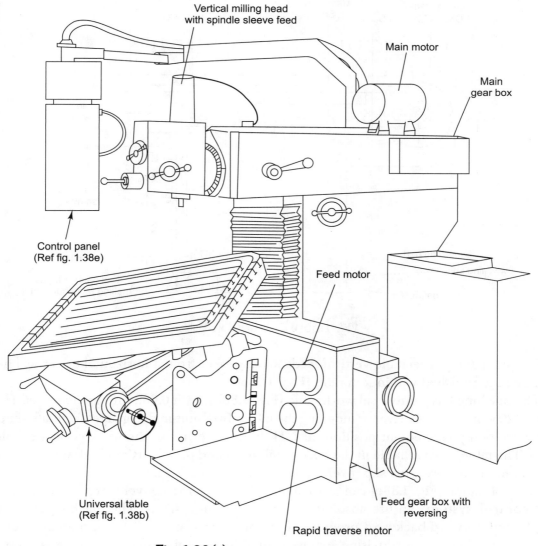

Fig. 1.38(a): MAHO 800 Jig boring machine

The compound table can be clamped hydraulically. The electrical interlock prevents inadvertent feed when the slide is in the clamped position. There is a monitor for the pressure of the lubrication oil, which automatically stops the machine, in case of inadequate lubrication.

The stepless [infinitely variable] feed drive is powered by a D.C. motor and coupled/decoupled by an electo-magnetic clutch. The spindle head has a pre-selectable, 18-stage precision transmission powered by a brake motor, with instant stoppage. The concentricity of the roller bearings mounted spindle, is contained within 0.002 mm. The swingable vertical milling head can be swung accurately and repeatedly, to the same angle.

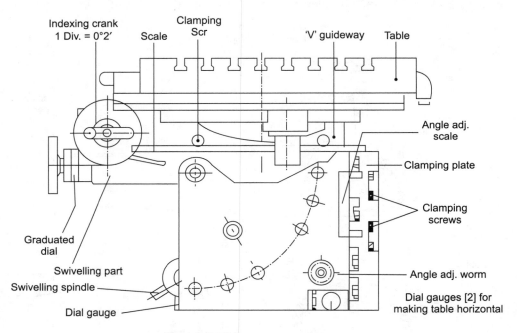

Fig. 1.38(b): Universal worktable

There are 6 pre-selectable vertical feeds. Changeover from a horizontal to vertical head can be accomplished within a minute [**Fig. 1.38c**].

The machine has a universal work table [**Fig. 1.38b**], with 0.002 mm graduations. There is a 1:120 worm/wheel drive for indexing the table. Two built-in dial gauges facilitate shifting the zero to any convenient position: at L.H. bottom corner of the workpiece: workpiece zero. The table is swivellable in the horizontal plane and tiltable. Rigid hydraulic clamping at a large radius, prevents slip, even under heavy milling loads.

Control station [**Fig. 1.38e**] can be rotated and/or tilted to a convenient position. **Straight cut control system** can use absolute/chain dimension systems. The machine also has a single-step, forward-backward program. The computer can store 128 positional values in its memory. There is a provision for using or skipping sub-programs. There is a photo-electric linear measurement system, with digital display. The glass scale has grating, with lines and gaps of 0.02 mm width, and an additional reference mark. The measurement is carried out by photo-electric sensing of the glass scale, using a mating grid and photo-electric cells. The motion of the scale, generates periodic signals. These are converted to 0.005 mm pulses. The reference mark generates an additional peak during the passage over it, to distinguish it from the periodic scale signals. In the event of a power failure, the reference [datum] point facilitates quick re-location of the starting point.

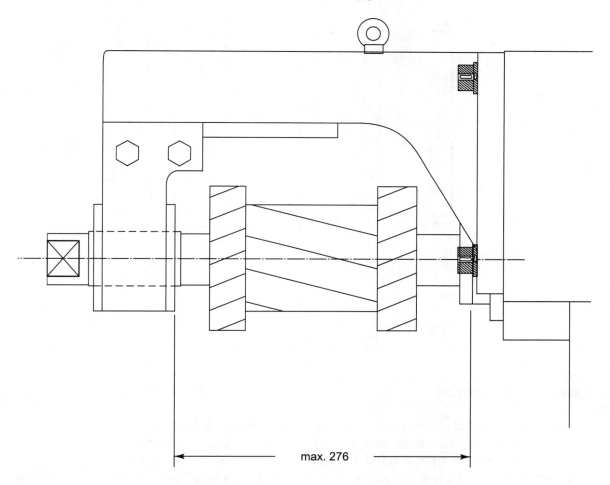

Fig. 1.38(c): Quick change overarm for horizontal milling

The relevant workpiece dimensions can be directly keyed. Keeping the magnetic tape printer on, automatically records the keyed programme on miniature cassettes, used widely in compact dictating machines.

Operation Manufacturers advise manual operation of the machine, for producing the first workpiece. Accordingly, the magnetic tape printer should be kept 'on', to record all the manually executed motions. Manual operation ensures perfect safety. It prevents inadvertent damage due to the workpiece or tool through programming errors. Manual operation also facilitates program corrections, during the course of the operation.

In addition to the slotting head [**Fig. 1.38d**], available on the universal milling machine, the Maho jig boring machine can be equipped with the following exclusive attachments:

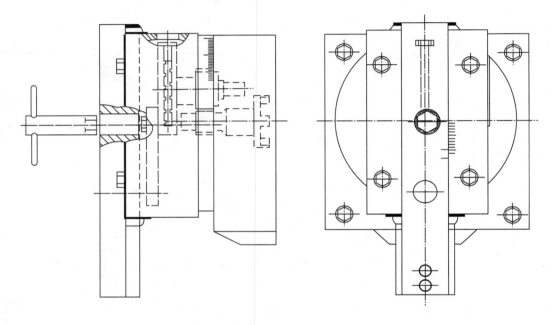

Fig. 1.38(d): Slotting head

Exclusive accessories

1. Punch milling attachment with compound slide about the rotary table
2. Corner milling attachment with provision for swivelling the conical cutter
3. Centering microscope with a dial gauge and slide for eccentricity removal
4. Facing and boring head, which facilitates turning short lengths of workpiece, in addition to facing and boring.

1.4.5.2 Face milling machines

A major portion of machining work comprises producing a flat face or facing. Many mass production facing machines have been developed for the purpose.

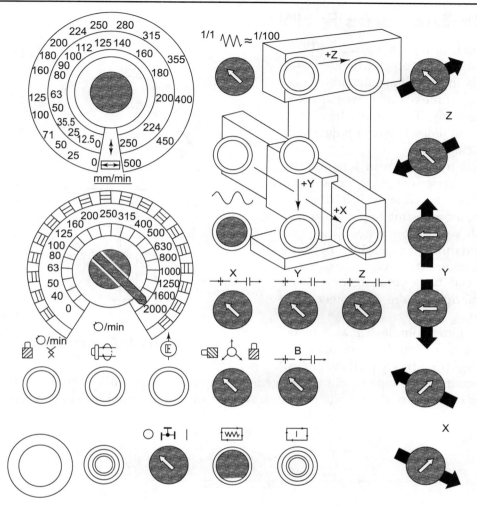

Fig. 1.38(e): Control panel of MAHO jig boring machine

1.4.5.3 Rotary milling machine [Fig. 1.40a]

These machines use a rotary table for mounting workpieces. The table is rotated to generate suitable linear feed at the center of the facing cutter [face mill]. The feed rate is slow enough to permit replacement of milled workpieces with unmachined ones, even while the table is rotating. A long periphery of the table can be used for the replacement.

Some large machines have two milling heads one for a roughing cut, and the other for the finishing cut. There is of course, a workpiece loading/unloading zone as well.

1.4.5.4 Straddle facing machines [Fig. 1.40b]

These machines have two milling heads, facing each other. Mounted with face mills, they facilitate machining two parallel faces of workpiece, simultaneously. The milling heads are mounted on slides, to enable adjustment of the facing cut.

The central table with longitudinal tee slots, carries clamped workpieces or fixtures. The table has trip dogs on one side, for adjusting the table stroke. The table reciprocation [feed] rate can be altered through a gear box. In drum type machines, the reciprocating table is replaced by a rotating drum, on whose periphery the workpieces are clamped.

The milling spindle R.P.M. can be varied, if at all (rarely) necessary, by changing the drive pulley/gears. Reciprocating table type machines are generally used for large workpieces like engine crank cases, or machine bodies; while the drum type machines are used for smaller workpieces.

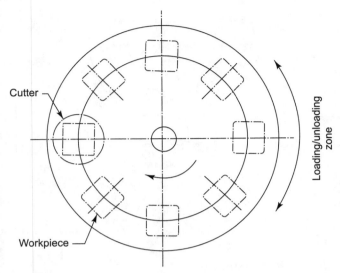

Fig. 1.40(a): Rotary milling

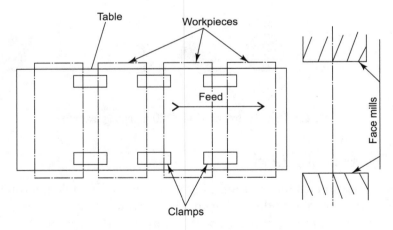

Fig. 1.40(b): Straddle milling machine

Introduction

1.4.5.5 Plano-milling machines

These are used for machining flat surfaces of large workpieces like machine beds/bodies.

The work table is similar to milling machine tables. But it can be moved only in a longitudinal direction [no cross travel]. It has tee slots and stroke-setting trip dogs. The work table slides on solid rigid guideways. There are two strong columns on both sides of the guideways. These columns have vertical guideways for side-heads and a cross-rail. The side-heads have axial adjustment for setting the cut of the face mills, mounted on the milling spindles. The side-heads can be moved vertically, by power feed. They also have a provision for fine manual adjustment.

The cross rail is mounted with two milling heads. They can be moved horizontally, either manually or by powered feed, on the guideways on the cross-rail. The entire cross-rail can be moved vertically, either manually or by powered drive, for setting the cutters. The milling heads have an arrangement for manual axial adjustment, for setting the cut.

Even the return stroke of table can be used for face milling. The machine can be operated on a pendulum cycle. The machined workpieces at one table end can be replaced by unmachined ones, while the workpieces at the other end of the table are being machined.

1.4.6 Rectilinear cutting machines

Shaping, planing, slotting, and broaching machines, use linear cutting strokes for material removal. The feed is usually in the cross direction square with the cutting strokes. These machines are useful for facing, slotting, and rectilinear profile machining work.

1.4.6.1 Shaping machines [Fig. 1.42a]

These use a reciprocating ram for effecting the cutting stroke. The return, non-cutting stroke is faster than the cutting stroke. The ram can be longitudinally adjusted to a position, suitable for the workpiece length. It can be clamped in the set position by a handle. Guided in dovetail slide, the ram has a vertical guideway, for the tool post at the front. A handle and a screw/nut combination facilitate adjustment of the cut. The tool-post is swivellable in the vertical plane [**Fig. 1.42b**] to facilitate shaping plate edges and vertical faces, by manual feed through the cut adjusting handle. The tool-post is free to swivel upwards during the return stroke, to prevent the edge of the cutting tool from rubbing against the machined workpiece. The length of the cutting stroke can be altered by varying the eccentricity of the rams drive crank pin. Similarly, the cross feed rate can be adjusted by adjusting the eccentricity of the feed drive crank-pin. The rotation of the pin reciprocates the rams. This moves the pawl at the other end of the rod, to rotate the mating ratchet through a certain angle. The angle depends upon the number of ratchet teeth the pawl traverses. So the cross feed can be adjusted only in steps of 1, 2, 3 ratchet teeth. The ratchet rotates the keyed cross-feed screw, to move the workpiece table laterally, by a fixed distance, corresponding with the number of teeth the ratchet rotates through.

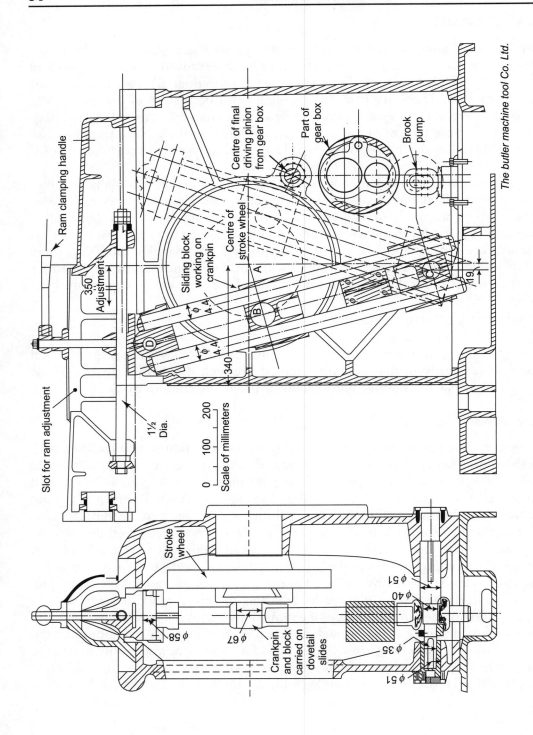

Fig. 1.42(a): Shaping machine drive—Adjustment of stroke

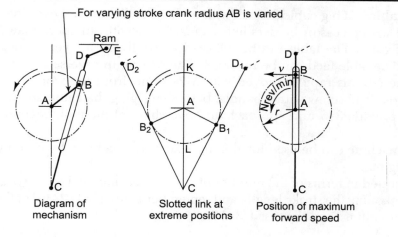

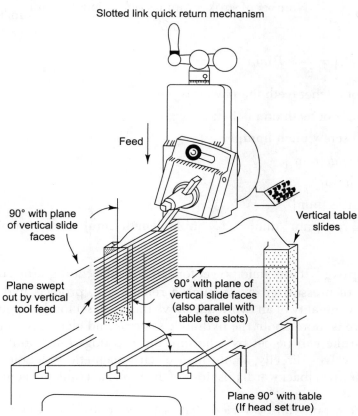

Fig. 1.42(b): Tool-post for shaping machine

The **work table** is a big cubical block with tee slots at the top and two sides. These can be used to mount workpieces or holders like standard vice with std/special jaws, circular table or a shaping fixture. The table can be adjusted vertically by a lever, through a screw/nut combination. The table feed can be reversed. The reverse transverse can also be used for a facing cut. The table is rapidly moved within 2–3 mm from the workpiece manually, by rotating the cross feed screw with a handle, before engaging the automatic feed. The cutting stroke should provide 3–5 mm approach and overshoot of the tool, beyond the workpiece end.

A shaping machine can be used for slotting a keyway in a bore, by mounting the slotting tool in a boring bar.

Speed is adjusted in terms of the number of strokes per minute. Linear speed [m/min] can be found by reckoning the stroke length, and remembering that only 2/3rd of the reciprocation time is used for cutting, and the rest 1/3rd for the return.

$$\text{Shaping speed } [V_s] = \frac{\text{Number of strokes / min} \times \text{Stroke length [mm]}}{667} \text{ m/min}$$

$$\text{Chip thickness } [t] = \frac{N}{T} \times P \text{ mm}$$

N = Number of ratchet teeth the pawl traverses;

T = Total number of teeth on ratchet;

P = Cross-feed screw pitch [mm]

Chip Area [A] = dt [mm^2]

d = Cut depth [mm]

t = Chip thickness [mm]

Material removal rate [mm^3/min] = A × Stroke length [mm] × no. of strokes/min

Hydraulic shaping It uses fluid power for actuating the cutting stroke through a hydraulic cylinder. Bypass of pressurized oil through a relief valve, protects hydraulically-operated elements, in the eventuality of an overload, above the level, set for the opening pressure of the relief valve. There is more about the hydraulic systems and elements, in the next chapter.

The cutting stroke can be set by adjusting the trip dogs longitudinally. These operate direction-control valves directly, or through electro-magnetic [solenoid] coils, to limit the forward and withdrawn [backward] positions of the ram. No crank pin adjustment is necessary here.

The speed of the ram can be varied steplessly by controlling flow to the ram-actuating cylinder, through a flow control valve. The ram return speed is 2–3 times faster than the cutting speed.

The cross-feed rate can be varied using a hydraulic piston, whose stroke can be varied by an adjusting screw. The stroke-setting determines the cross-feed rate [mm/stroke]. Even the

Introduction

vertical feed can be actuated by hydraulic power. Furthermore, the tool-post is tilted by hydraulic power, to clear the tool off the workpiece, during the rapid return stroke.

Although much costlier than mechanical machines, hydraulic shapers are suitable for heavy work, due to their built-in safety measures and infinitely variable speed and feed drives.

Gear shaper uses a circular, gear-like cutter for shaping gear teeth. Refer to the section on 'gears manufacture' in Chapter 2.

1.4.6.2 Planing machines [Fig. 1.43]

They also use linear longitudinal cutting strokes and lateral cross-feed. But in planing machines, the cutting stroke is obtained by reciprocating work table during the cutting while the cutting tool remains stationary. Tee slots on the table facilitate clamping of the workpiece. There are adjustable trip dogs on one side, for adjustment of forward and backward table positions, i.e. the cutting stroke. The table is driven by worm and wheel reduction gearing, through the rack and pinion mechanism.

Planers use a number of tool slides. They can be fed simultaneously to the machine top, and edges or vertical faces. For this, there are two columns on both sides of the work table. The columns have guideways for saddles, which can be fed vertically through a screw/nut combination, driven by a vertical splined shaft, powered by a separate feed motor. The same shaft also drives the lead screw for horizontal [cross] feed of the two tool-saddles, on the horizontal cross-rail. The cross-rail can be moved vertically, in rapid approach, to an approximate position before the manual fine adjustment of a cut. The feed rate can be altered through a gear box which gives 15 feeds, ranging from 0.31–2.0/stroke. Feeds are engaged/disengaged by sliding a driven pinion to mate with/separate from the driver. A safety slip clutch stops the machine in the eventuality of an overload.

There is a provision for controlled approach called **inching**, by an electric push button.

The solid support available to the reciprocating work table, reduces deflections and vibrations. A number of faces can be machined simultaneously. Planing machines are ideal for heavy facing, and rectilinear profiling work. The stroke of a planer is rarely less than one meter.

Some planers use hydraulic power for table reciprocation, and horizontal and vertical feeds. The hydraulic system is similar to the arrangement in the hydraulic shaping machine.

1.4.6.3 Slotting machines [Fig. 1.44a]

Comparable to vertical shapers, these machines have a vertically reciprocating ram, with a tool-post with tee slots. The machine is particularly suitable for machining keyways in cylindrical bores, of workpieces that are difficult to secure on shaping machines. Depending upon the machine size, the cutting stroke capacity ranges from 200 to 1000 mm. The ram can be shifted vertically and clamped in a position suitable for the workpiece. The variable radius crank provides stroke adjustment through an adjustable screw. The gear box provides only 4 speeds. There is a clutch to start/stop ram reciprocation, and a built-in brake for its instant stoppage. There is also a compound slide for moving the workpiece longitudinally and laterally, to position it precisely with respect to the slotting tool, i.e. the slotting ram.

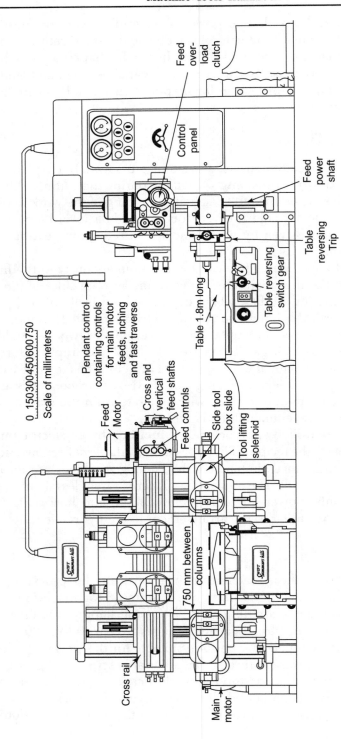

Fig. 1.43(a): Planing machine

Introduction

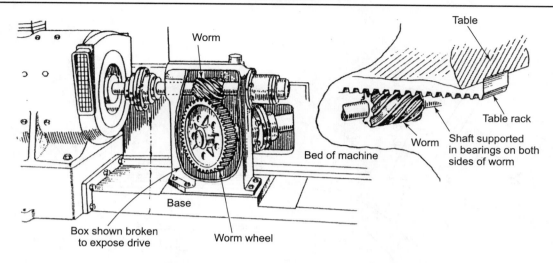

Fig. 1.43(b): Table drive for planing machine

Slotting machines are usually equipped with a circular table and a provision for effecting an accurate angular motion [indexing], for machining equispaced keyways, splines, etc. The circular table can be automatically rotated by power drive to transverse a cut. This facilitates machining a cylinder on the workpieces that are almost impossible to machine on turning machines. For keyway work, the circular table must be clamped, to prevent its rotary motion during slotting.

The circular table has angular divisions for less precise work such as spanner squares/hexagons on lead screws. For more accurate work, an indexing plate and pin arrangement, similar to the dividing head indexing plate, is necessary.

It should be remembered that the cutting direction of slotting tools is parallel to the square shank. Therefore the rake angle is given on the end face [**Fig. 1.44b**]. During keyway cutting a bore the tool enters a confined bore. Hence no part of tool entering the workpiece should protrude beyond the cutting edges. Furthermore, the part entering the bore must be longer than the keyway length. Also there must be additional shank length for clamping the tool.

1.4.6.4 Broaching machines

Broach [**Fig. 1.45**] is a multi-tooth, rectilinear cutter. It is pulled or pushed through or onto the workpiece. Each tooth removes a small amount [0.02–0.2] of workpiece material. The tool [broach] is quite long and its length is divided among roughing, semi-finishing and finishing teeth [**Fig. 1.45**]. The material removed is maximum in the case of roughing teeth, and minimum in the case of finishing teeth.

Broaches for internal cutting [keyway/splines on bores] are also provided with a guide, suitable for unbroached workpiece size at the entry, and another to suit the broached workpiece size at the tail end. Pull type machines also require a grooved locator to suit the pull head [**Fig. 1.46**] of the machine.

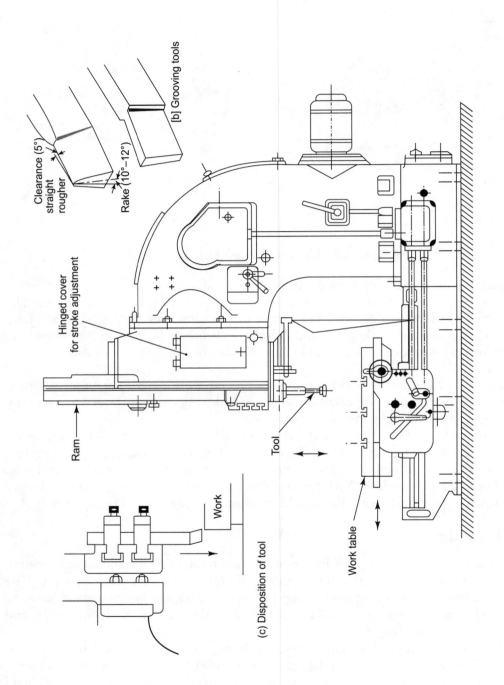

Fig. 1.44: Slotter

Introduction

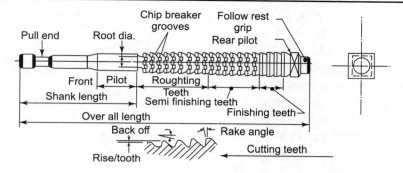

Fig. 1.45: Broach for a square from round hole

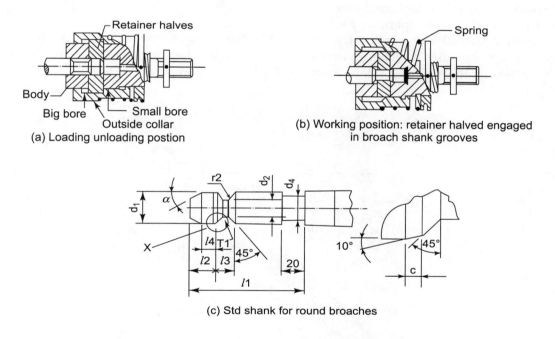

Fig. 1.46: Pull head and shank for round/sq./hex. internal broaches

The inordinate length of the broach calls for a long broaching stroke. Broaching machines therefore have long slides, running up to 6 meters. Many machines have two rams: one for roughing, and the other for finishing. This reduces the broach and stroke length.

The work table has a bore for location of the fixture or jig, for in addition to locating the workpiece, the fixture also guides the broach [**Fig. 1.47**].

Most of the modern broaching machines are operated by hydraulic power although some old screw/nut actuated mechanical machines are still in use. A hydraulic cylinder is used to pull/push the ram, fixed with broach puller. Pull type machines are better for long, slender, internal broaches. Push type machines can be used only for short broaches, for slender broaches can buckle under pressure.

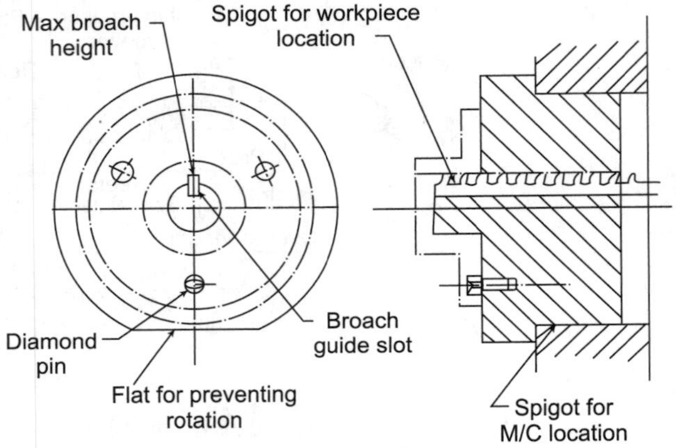

Fig. 1.47: Keyway broaching fixture

Horizontal broaching machines

They can be hydraulically, mechanically, or electro-mechanically operated. Electro-mechanical machines use a rack and pinion combination, along with worm and wheel reduction gearing, for actuating the cutting stroke. Some internal broaching machines are provided with indexing turret [**Fig. 1.48**], to speed up the broach changeover.

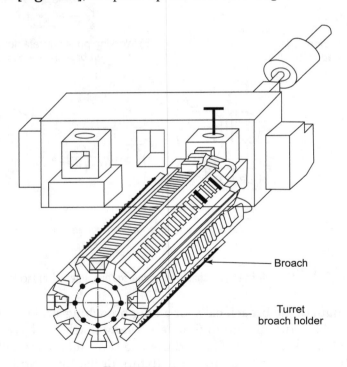

Fig. 1.48: Turret broaching

External surface broaching machines can produce a variety of surface profiles [**Fig. 1.49**].

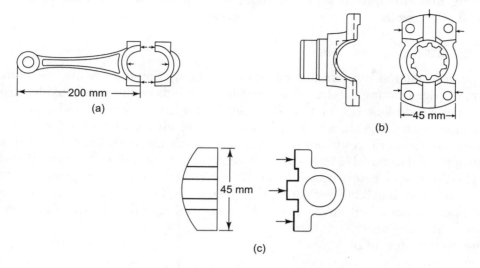

Fig. 1.49: Examples of surface broaching

Vertical internal broaching machines They can be of a pull [up/down] or push down type. The surface broaching machines usually have a long, rigid column with guideways for the saddle carrying workpiece. The broach is mounted on the column.

Mass production face broaching machines They have rotary feed tables [**Fig 1.50a**], which permit loading/unloading of workpieces even when the machine is running. Even in a linear/continuous broaching machine [**Fig 1.50b**], workpiece loading/unloading can be done, while the machine is running.

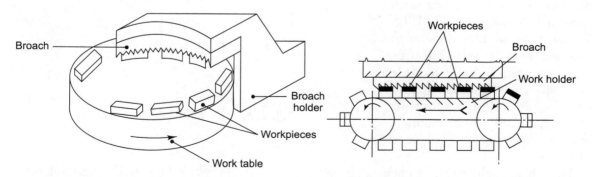

Fig. 1.50(a): Rotary continuous broaching machine

Fig. 1.50(b): Linear continuous broaching

Chapter 8 on Tool Engineering gives more details on broach and broaching fixture design.

Gear cutting machines, tools and processes are described in the section titled 'Gears Manufacture' in Chapter 2.

1.4.6.5 Grinding machines

Grinding removes material by abrasion rather than cutting. Some academicians consider grinding as multi-edged cutting. It is true that hard silicon carbide or aluminium oxide particles in a grinding wheel, act like a cutting tool. But they do not have a solid, rigid support like a tool-post or a cutter body. The abrasive particles of grinding wheels are embedded in silicate, shellac, rubber, or synthetic resin bonds, which are much more elastic than solid steel. Consequently, material removal is achieved through high speed abrasion. Although the wheel for internal [10–20 m/sec], external cylindrical [28–33 m/sec], and surface grinding [20–25 m/sec], functions at less than 33 m/sec speed, the speeds of rubber, shellac and bakelite bond cutting wheels range from 45–61 m/sec. It should be remembered that the speeds mentioned are per second, and are nearly 60 times the speeds for high speed steel cutting tools.

Balanced wheels High speed makes grinding wheels vulnerable to vibrations. Consequently, grinding wheels must be balanced carefully by chipping off some of the mounting bush at its center. Sometimes, it is also necessary to dig out a little portion of the grinding wheel. Some grinding machines are provided with special mounting collets [**Fig. 1.51b**], with circular dovetailed groves and small balance weights, which have a male dovetail. These balance weights can be adjusted circumferentially, to aid wheel balance, and for clamping in a balanced position.

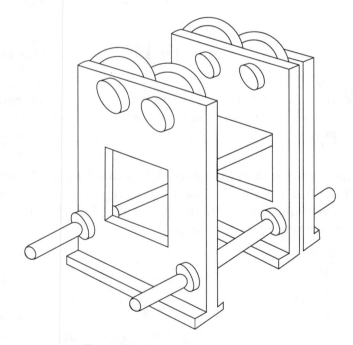

Fig. 1.51(a): Wheel balancing stand

The wheel balance can be checked, by placing the wheel-mounted balancing mandrel, on 4 knife-edged rollers [**Fig. 1.51a**], two at each end. If there is an imbalance, the wheel will swing

Introduction 65

to bring the heavier part of the wheel to the lowest position. A better method of checking the balance is to mount the grinding wheel on the machine spindle, and check the face run-out.

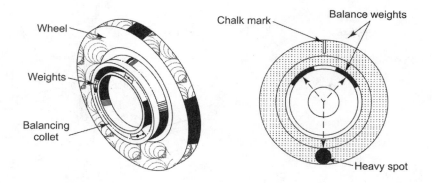

Fig. 1.51(b): Balancing wheel with collet balance weights

Mounting [Fig 1.52] Grinding wheel flanges are relieved in the central portion. They hold the wheel at the flange's outermost annular area. The driving side flange is keyed to the drive shaft, but the wheel mounting bush has no keyway. The wheel is held by frictional force between the flanges and the wheel. The shaft is provided with threads and a nut for clamping the wheel. The direction of the thread [RH/LH] should be such that the rotational direction of the wheel will tend to tighten the nut onto the wheel. Naturally, a pedestal grinder, with two wheels at both ends of a common shaft, will have threads of the opposite hand at the opposite ends.

Wheel truing and dressing [Fig. 1.53] After mounting, the grinding wheel is trued by a light cut on the diameter, and a short length of the face.

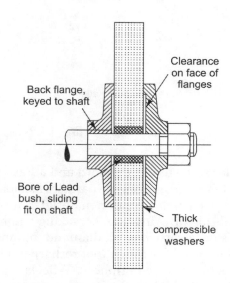

Fig. 1.52: Clamping wheel near outer periphery by relieving clamping flange faces

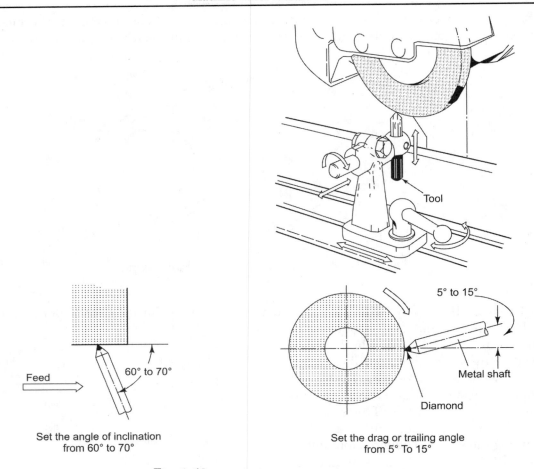

Fig. 1.53: Truing (dressing) grinding wheel

After some usage, the hard and abrasive particles in the wheel become blunt. The flexible bond between the abrasive particles facilitates the loosening and escape of the blunted abrasives. Blunted particles hinder material removal. Therefore, it is necessary to remove them, periodically, by dressing, i.e. taking a light cut on the diameter and faces of the wheel.

It is necessary to use a diamond, mounted in a holder, to dress a grinding wheel. Pedestal grinders, used for rough tool-resharpening, can be dressed by a metal dresser with sprocket-like teeth on their periphery. Wheels for rough work can be dressed by the metal dresser, whose teeth dig into the wheel periphery to generate roughness.

Cut limitations Grinding wheels cannot cut more than 0.04 mm. If one experiments with bigger cuts, say 0.5 mm, by advancing the wheel by 0.5 mm through the machine dial, the person would find that the wheel is removing only 0.02–0.04 material. In fact, the flexible bond, the matrix of the wheel yields, to allow radial compression of the wheel.

Introduction

Spindles used to grind small internal bores, have to be slender to permit manipulation in small bore sizes. Cantilever-type mounting aggravates the situation. So the cut for internal grinding can be as less as 0.005 mm, and at most 0.01 mm.

Consequently, in comparison with cutting, grinding is a very slow process. But the slow material removal aids precision. Moreover, there is no other alternative for finishing hardened workpieces.

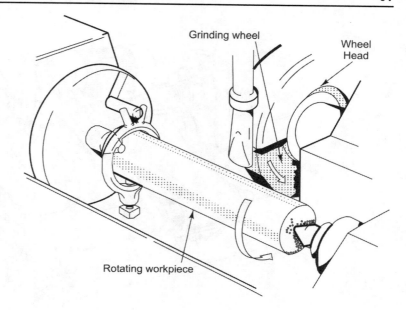

Fig. 1.54(a): Plain cylindrical grinding

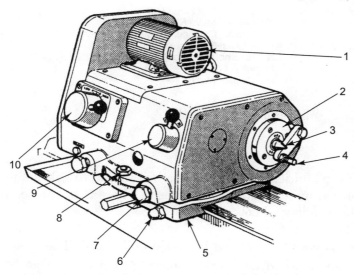

1. Headstock Drive Motor
2. Drive Plate
3. Centre
4. Driving Dog
5. Table Mounting Base
6. Base Clamping Bolt
7. Base Adjusting Screws
8. Swivel Clamping Nut
9. Clutch Control
10. Headstock Speed Selector

Fig. 1.54(b): Headstock with drive plate in position

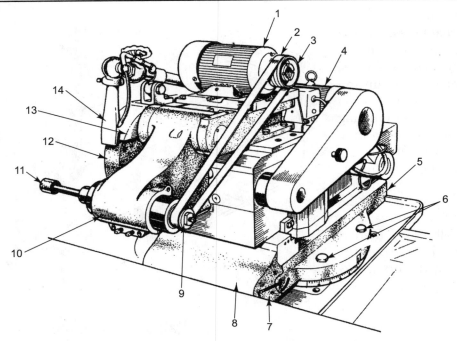

1. Internal Grinding Spindle Motor
2. Drive Belt
3. Motor Pulley
4. External Grinding Spindle Motor
5. Universal Turret
6. Locking Screws
7. Wheel Slide
8. Wheel Slide Water Guard
9. Wheel Sleeve Right
10. Internal Grinding Spindle Unit
11. Internal Grinding Wheel
12. External Grinding Wheel
13. Wheel Guard
14. Coolant Nozzle

Fig. 1.54(c): Wheel head of universal grinder

1.4.6.6 Cylindrical grinding machines

These can be simple/universal. Simple cylindrical grinders have a bed similar to lathes. But the tool-post and cross slide are replaced by a [grinding] wheelhead. There is of course, a lateral slide for adjusting the cut of the grinding wheel.

Workpiece speed The workpiece is held and rotated by the spindle head. The speed depends upon the arc of contact between the wheel and the workpiece. As the arc is longer in bigger workpieces, the linear speed for bigger workpieces should be higher. Finishing is done at slower feed [contrary to metal cutting]. Thin workpieces must also be ground at a higher speed to contain the heating. Glazing of a wheel is a clear indication that the wheel is harder than necessary. The situation can be remedied to some extent by increasing the workpiece speed. Rapid wear of the wheel on the other hand, indicates that the wheel is softer than necessary. This can be compensated for to some extent, by reducing the workpiece speed.

Introduction

Generally, the workpiece should be rotated to generate **25 m/min** linear [surface] speed. The speed should be increased if the wheel glazes or the workpiece gets overheated. Rapid wear of the wheel calls for speed reduction.

The machine setup in a cylindrical grinding machine is similar to that in turning machines.

1. **Center to center [Fig. 1.55a]** supporting the workpiece with a drive dog [carrier] for long jobs, which call for good concentricity. Accessories like steadies are used for intermediate supports.

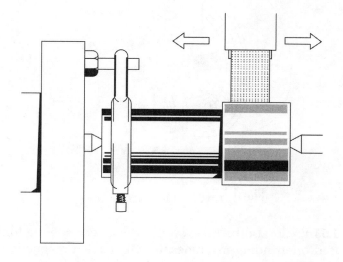

Fig. 1.55(a): Cylindrical grinding between centers

For **Taper grinding [Fig. 1.55b]**, the table is swiveled to a suitable angle by trial and error, using the angle markings at the tailstock end of the table.

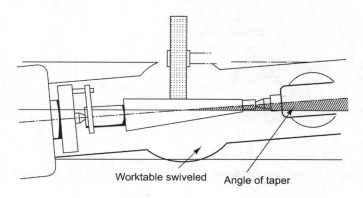

Fig. 1.55(b): Grinding a taper

2. **Chuck [Fig. 1.55c]** which is fitted to the workpiece spindle head, can be used conveniently for short-length jobs. The workpiece is usually trued up within 0.01 mm, by using paper packing between the workpiece and chuck jaws.

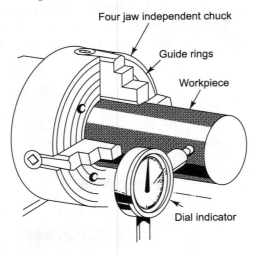

Fig. 1.55(c): Truing work in chuck

3. **Collets [Fig. 1.55d]** can also be used on the workpiece spindle for bar work.
4. **Face plates**, used on grinding machines, usually have a magnetic workpiece holding; naturally, only for ferrous magnetic materials.
5. **Swivelable workpiece spindle head [Fig. 1.55e]** in universal grinders, facilitates angular grinding.

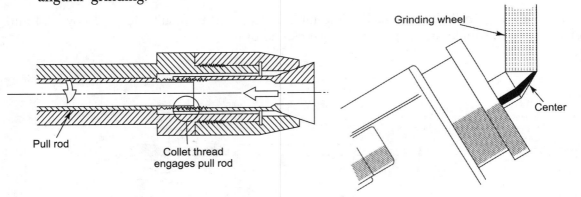

Fig. 1.55(d): Fitting collet to spindle   **Fig. 1.55(e):** Headstock tilt for angle grinding

Introduction

6. **Swivelable wheel head** [Fig. 155f] in universal grinders, comes handy while grinding angular surfaces.

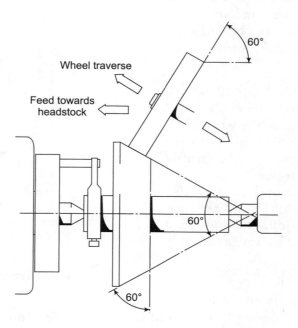

Fig. 1.55(f): Grinding an angle with wheel head swiveled

The table has one central 'L' slot, which is used for locating and clamping the axially adjustable tailstock steadies, etc. The table is swivelable on the saddle, calibrated with angle markings.

The saddle is reciprocated on longitudinal guideways to bring the entire length of the workpiece in contact with the grinding wheel. Like milling machines there are dogs to adjust the extreme left and extreme right positions of the table. Not only these dogs limit the longitudinal stroke, but also reverse the direction of the longitudinal stroke automatically, by striking the central direction lever and changing its position. This lever can be operated manually to change the table traverse direction during setting.

Longitudinal feed The reciprocation speed or the **longitudinal feed** however is much higher than a milling feed. The workpiece moves 1/3 to 2/3 grinding wheel width, for one revolution of the workpiece.

Cross feed At the end of every longitudinal stroke, the wheel moves onto the workpiece by 0.005 to 0.025 mm cross feed, to apply a cut. It is necessary to have 8 to 10 idle passes, even after the workpiece has been ground to finish size. These idle passes are called 'sparking off', for they are continued to be used till there is no sparking between the workpiece and the grinding wheel.

Plunge [grinding] feed   [**Fig. 1.56**] is used when the length to be ground is lesser than the width of the grinding wheel. It is also used for form grinding, in which the wheel form corresponds to the profile required on the workpiece.

Plunge grinding [cross] feed is also specified in mm/rev of the workpiece. Depending upon the width of the wheel, it can be as low as 0.0025 / rev, and rarely more than 0.01/rev.

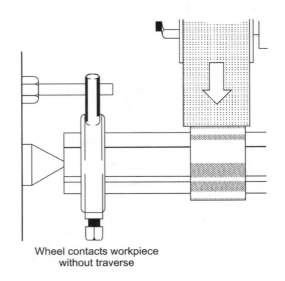

Wheel contacts workpiece without traverse

Fig. 1.56: Plan view plunge grinding

Coolant flow   must be ample to carry away the heat, generated during the process of abrasive material removal. It should be related to the grinding wheel diameter.

$$\text{Coolant flow [Lits/min]} = \frac{\text{Wheel Dia [mm]}}{15}$$

The coolant is usually a cocktail of soda, machine oil, cutting oil, and water in a 1:1:0.4:20 proportion (by weight), respectively.

Internal grinding   [**Fig. 1.57**] Many cylindrical grinders are provided with an internal grinding attachment, which can be swung into/off the working position.

Internal grinding wheels are much smaller than external grinding wheels. In fact, the wheel diameter must be smaller than the workpiece bore, to be ground by at least 2–3 mm, to allow entry and radial motion of the wheel, to apply a grinding cut. The small, wheel diameter calls for a high rotary speed. The range of 10,000 to 20,000 RPM is quite common for grinding diameters ranging between 12 and 25 mm.

The spindle [**Fig 1.57**] is mounted in anti-friction, angular contact bearings, with a spring for axial clamping. It is provided with a female taper for locating the adaptor and a tapped hole, for securing it. The adaptor diameter depends upon the wheel diameter, which in turn depends upon the bore to be ground.

Introduction

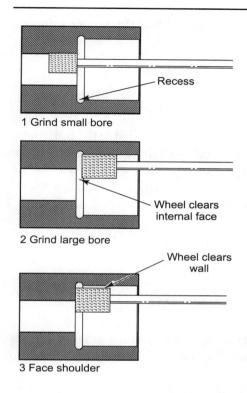

Fig. 1.57: Internal grinding

Facing wheel Machines designed exclusively for internal grinding are usually provided with an auxiliary facing head. Mounted with a cup-shaped grinding wheel, it can be used for facing workpiece surfaces square, with the ground bore.

Slender shape and cantilever action deflects the adaptors mounted with grinding wheel. Consequently a longer '**spark off**' period [more idle passes] is necessary at the end of the internal grinding operation.

1.4.6.7 Special purpose grinding machines

Designed for a range of similar workpieces such as bearing races, these have separate wheel heads for external, internal, and face grinding operations. The number of grinding passes and rates of longitudinal and cross feeds, are controlled automatically, after the machine has been set. Wheels dressing is also automatic.

The operator only has to load and unload the workpiece in such types of sophisticated automated grinders. In some machines, even loading/unloading is automated through magazine feed and a transfer arm.

1.4.6.8 Centerless grinding machines [Fig. 1.58]

These machines are very convenient for grinding long slender workpieces, without supporting centers or steadies. Since there are no centers, therefore these machines have been named centerless.

Centerless grinders have 2 grinding wheels, rotating in same direction. The bigger one effects most of the material removal. The smaller wheel is called the regulating wheel. These wheels rotate the workpiece in the direction opposite to the rotation of the grinding wheels. The bigger grinding wheel should be suitable for the workpiece material. The last part of this section gives recommendations for grinding wheel abrasive, grit, bond, etc. for centerless grinding of various workpiece materials.

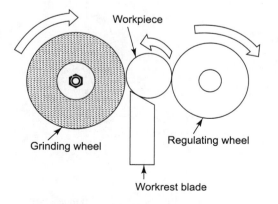

Fig. 1.58: Centerless grinding principle

The regulating wheel, driven at a much slower [surface] speed [than the bigger grinding wheel], usually has a rubber bond.

There are 3 methods of operation:
1. Through feed
2. In-feed
3. End feed

1. **Through feed [Fig. 1.59a]**: In this, the regulating wheel axis is swivelled in the horizontal plane by a few degrees. This exerts an axial force on the workpiece. It pushes the workpiece from the front [operator side] to the rear. The emerging, ground workpieces can be collected via a chute, in a receptacle. The axial feed rate depends upon the surface speed of the regulating wheel, and the angle [α] between the two wheels.

   Feed speed $[V_f] = \pi\, d\, N \sin \alpha$ [mm/min]

   d = Regulating wheel dia [mm]

   N = Regulating wheel RPM

   α = Angle between regulating and grinding wheel axes

2. **In-feed [Fig 1.59b]:** This method is convenient for facing the shoulders of collared workpieces. There is a small angle between the grinding and regulating wheels. It pulls the workpiece towards the rear end of the machine. But the workpiece collar limits the motion [feed].

   A stop is also provided to limit the axial travel of the workpiece, between the wheels. The stop limits material removal from the shoulder face.

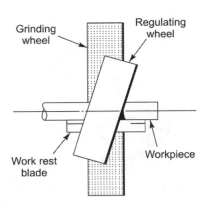

Fig. 1.59(a): Through feed

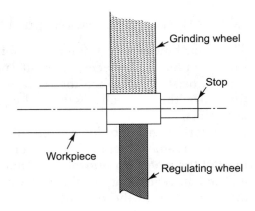

Fig. 1.59(b): In-feed

3. **End feed**: In this method, axes of grinding and regulating wheels are kept parallel to each other. For grinding, the workpiece is placed on the work rest, and located axially by stop [**Fig. 1.59c**]. For grinding, the regulating wheel is moved towards the grinding wheel. Touching the workpiece, the wheel rotates the workpiece against the grinding wheel, to effect material removal. There is a settable stop for travel of the regulating wheel. It ensures repeated accuracy during mass production.

It is needless to say that the workpiece length to be ground must be lesser than the width of the grinding wheel.

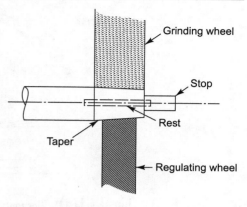

Fig. 1.59(c): End feed

1.4.7 Surface grinders

About one-third of the grinding work in tool rooms entails grinding surfaces of important plates parallel and at least two edges square to the parallel surfaces and to each other [**Fig 1.60**]. The edges serve as X–Z and Y–Z planes, of datum faces for marking and/ or machining of important components of press tools, jigs/fixtures, and special purpose machines.

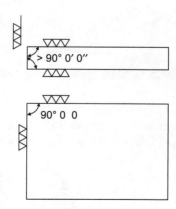

Fig. 1.60: Grinding two edges and two faces of important plates for marking or jig boring

1.4.7.1 Surface grinder types

1. Horizontal disk wheel and reciprocating work table [**Fig. 1.61a**]

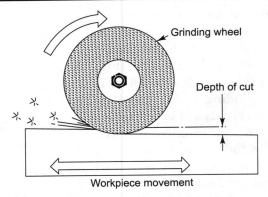

Fig. 1.61(a): Horizontal wheel surface grinding with reciprocating table

2. Horizontal disk wheel and rotating work table [**Fig 1.61b**]

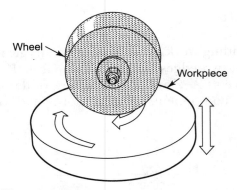

Fig. 1.61(b): Horizontal wheel, surface grinding with rotary work table

3. Horizontal ring wheel and reciprocating work table.
4. Vertical cup/segmental wheel and reciprocating work table.
5. Vertical cup/segmental wheel and rotating work table [**Fig. 1.62**]

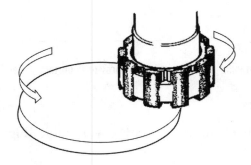

Fig. 1.62: Vertical wheel rotating table surface grinder

Introduction

Reciprocating work tables are moved laterally [cross feed], by an appropriate distance, at the end of every longitudinal stroke. This distance is called **cross feed/stroke**. It ranges from 0.6 mm to $1/10^{th}$ of the wheel width, for workpieces with RC 48–64 hardness. It can be up to $1/4^{th}$ of the wheel width for workpieces with hardness up to RC 45, and up to $1/3^{rd}$ wheel width for non-ferrous metals. But it is necessary to reduce the cross feed to 0.6 to 1.25 mm/stroke, while grinding hollow, ring-like workpieces, with insufficient resting area, to provide good magnetic clamping. Such workpieces would fly off the magnetic table if high cross feed is used.

Magnetic work table [**Fig. 1.63**] Surface grinders use a magnetic table for securing workpieces. Electromagnetic tables [**Fig. 1.63**] have coils which create opposite, north-south poles in adjacent coils. These are separated by non-magnetic separators which force the magnetic flux to pass through the connecting workpiece. This clamps the workpiece to the table by magnetic force. Rings made of non-magnetic materials like brass, are used to separate the north and south poles.

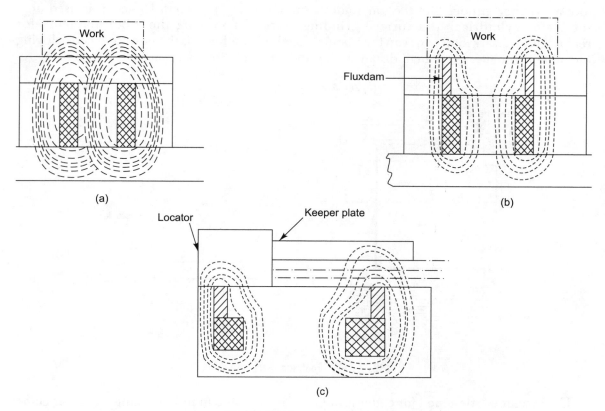

Fig. 1.63: Magnetic clamping

A handle is provided for switching on/off the coils and reversing the direction of the current. After finishing grinding, the handle can be moved to the extreme opposite side to reverse the current and polarities. This demagnetizes the workpieces, well. For unclamping the workpiece, the handle is moved to central position, which switches off the current supply to the magnet coils.

There are also permanent magnet tables. In these, the magnetic flux to the workpiece is blocked by non-magnetic material shields for unclamping. The non-magnetic shields are moved aside, to clear the flux path to and through the workpiece, to effect magnetic clamping.

The table reciprocating speed usually ranges from 18–30 m/min.

For grinding non-magnetic materials it is necessary to make some sort of fixture for locating [positioning] and securing [clamping] the workpiece. It might be necessary to remove the magnetic table and use the tee-slotted table below, for mounting the workpiece or fixture.

Tool and cutter grinders

This is a multi-purpose grinding machine [**Fig. 1.64**] in which the wheel head can also be moved vertically, in addition to the lateral [cross feed] direction. One can also perform cylindrical [external and internal] and surface grinding on a tool and cutter grinder. But the longitudinal motion is not powered. Hence, it is used only rarely for cylindrical and surface grinding work, for moving the table manually for reciprocation, is irksome, nor can the manual feed be as uniform as the powered feed. Moving the table manually, by a small distance, to grind small cutting teeth faces, is not so tiring.

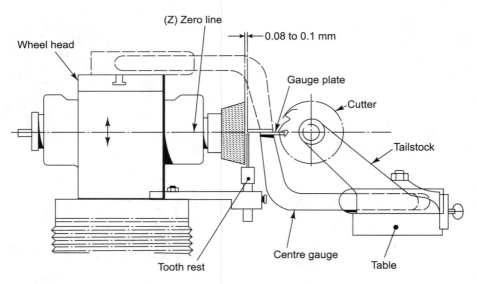

Fig. 1.64: Tool and cutter grinder

The machine is designed for cutter grinding. The first step in resharpening a circular cutter is usually grinding its diameter. Hence, the cylindrical grinding facility.

The workhead spindle can be swivelled in a horizontal plane. It can also be tilted in a vertical plane. A central tee slot on the work table facilitates location and clamping of tailstock and cutter grinding accessories.

Introduction

The clearance angle is obtained by keeping the center of the cutter spindle below the wheel center [**Fig. 1.65**]. The vertical offset [h] depends upon the grinding wheel radius [R] and the required clearance angle.

$$H = R \sin C.$$

If a cup wheel is used [**Fig. 1.66**], the offset depends upon the cutter radius [r]

$$h = r \sin C.$$

A separate section on tool geometry, in **Chapter 8** on Tool Engineering, gives recommended land and primary and secondary clearance angles for machining soft [S]/normal [N]/hard [H] workpiece materials with high-speed steel cutters.

Tooth rests While grinding clearance the cutter tooth is supported on a rest to prevent rotation of cutter due to grinding force. A variety of tooth rests [**Fig. 1.65**] are used for grinding various surfaces of cutters. The tooth rest acts as a make-do indexing plunger, using the cutting [rake] face as a location.

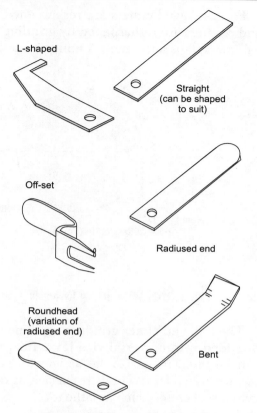

Fig. 1.65(a): Tooth rest blades

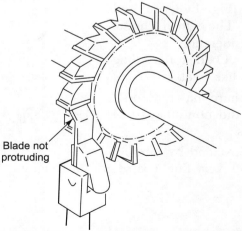

Fig. 1.65(b): Positioning tooth rest blade clear of grinding wheel while grinding radial a clearance on cutter

Form relieved cutters are resharpened by grinding the cutting faces only. Similarly, taps and reamers are resharpened by grinding the cutting faces [flutes] only. Cylindrical grinding will make taps and reamers undersize and useless.

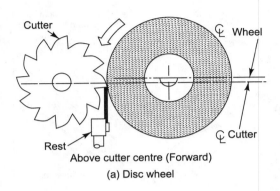 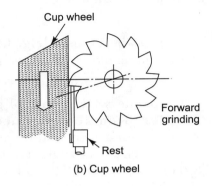

Fig. 1.66: Wheel setting for grinding radial clearance on horizontal milling cutters

The tool and cutter grinding machine is also provided with universal vice [**Fig. 1.67**] which can be tilted in horizontal as well as vertical planes, to facilitate accurate resharpening of non-cylindrical cutters such as lathe tools, broaches, and threading chasers.

1.4.7.2 Grinding wheel choice

A variety of wheel shapes [**Fig. 1.68**] are used for various applications. The wheel choice involves the following considerations:

1. Abrasive – Aluminium Oxide for grinding high tensile materials like steels
   - Silicon carbide for hard materials, non-ferrous materials and non-metals like stone and ceramics.
2. Grain [Grit] size – Specified by number 10 [most coarse] – 600 [very fine] coarse grit for roughing. Fine grit for finishing.

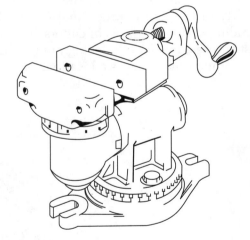

Fig. 1.67: Universal vice

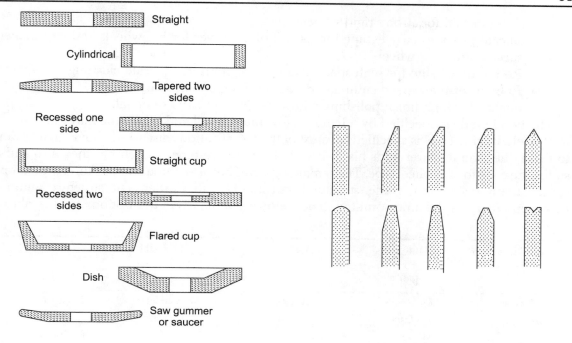

Fig. 1.68(a): Wheel shapes

Fig. 1.68(b): Wheel faces

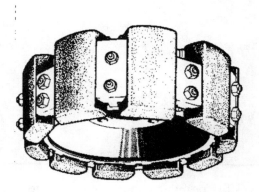

Fig. 1.68(c): Segmental wheel used for surface grinding

3. Structure–Spacing between abrasive grains; specified by 1 [close, dense] to 12 [wide, open] numbers. Structure number being optional, is omitted by some manufactures.

4. Bond—Vitrified for strong rigid wheel.
   - silicate provides easy escape [break off] of abrasives for big wheels and cup-shaped surface grinding wheels.
   - Resinoid/Bakelite for high-speed cutting, rough grinding, scale cleaning.
   - Rubber for high speed cutting and fine finishing of ball bearing races.
   - Shellac for high finish polishing with very little material removal.

The bond grade is specified by a letter 'A' for softest bond and 'Z' for the hardest.

Wheel marking code is specified by most of the specifications discussed above. In addition to these, the manufacturer adds his own code for abrasive type. Sometimes they also add a suffix: the code for the wheel type. The remaining terms between these manufacturer's additions specify the abrasive, grit, grade, structure, and bond, in that order. The structure number being optional is sometimes omitted. The following table gives the options for the above choices.

ABRASIVE	GRAIN SIZE [No]	GRADE	STRUCTURE [OPTIONAL] [No]	BOND
A- Aluminium Oxide	8 to Coarse 24	E - Soft F G	1 - Most Dense	V - Vitrified
				B - Resinoid
C - Silicon Carbide	30 to Medium 60	H I J	8 - Medium	R - Rubber
				E - Shellac
		K		
	80 to Fine 180	M - Medium N O	15 - Most Open [Least dense]	S - Silicate
		P		BF - Resinoid Reinforced
	220 to Very Fine 600	Q R T		Mg - Magnesium
		U - Hard		

The following table gives the popular choices of wheels for various applications.

1. **Cylindrical grinding** – A60KV - Steels up to HB250
   - A60JV – Hard steels up to RC 56 – 65
   - A60LV – C. I. up to RC45
   - C46JV – Magnesium alloys HB 40 – 90
   - C46KV – Aluminium alloys HB 30 – 150
   - C60JV – Titanium alloy HB300 – 380
   - C60KV – Copper alloys RB10 – 100

Introduction

2. **Internal grinding**
   - A60JV – Steel up to HB275
     – C.I. up to RC45
   - Aluminium alloys HB30 - 150
   - Titanium alloys HB300 - 386
     A69KV – Steel up to RC65
       – Magnesium alloys HB - 40 - 90
     A60MV – Steel up to RC48
     A80JV – Tool steel RC56 - 65
     A80KV – Tool steel HB 100 - 150
     C60KV – Copper alloys RB 10 - 100

3. **Surface grinding**
   [Horizontal spindle
   Reciprocating table]
   - A46V – Steel RC48 - 65
     C.I. up to RC45

   - A46JV – Steel up to RC48
            – Stainless steel HB 135 - 275
   - C46KV – Aluminium, Copper, Magnesium, and
              Titanium alloys HB 100 - 150

4. **Surface grinding**
   [Vertical spindle
   Rotary table]
   - A80IV   – Steel HB100 - 350
              C.I. up to RC45
   - A100G H – Tool steel RC 48 - 62
   - C60I   – C.I. up to RC45
              Aluminium alloys HB30 - 150

5. **Centerless grinding**
   - A60JV – Magnesium alloys HB - 40 - 90
              Titanium alloys HB300 - 380
              C.I. up to RC45
   - A60LV – Copper alloys RB10 - 100
   - A60MV – Steel up to RC66
   - A80LV – Tool steel RC56 - 65
              Aluminium alloys HB 135 - 275
   - C60KV – Stainless steel HB 135 - 275
              Magnesium alloys HB40 - 90

1.4.8 Electrical machining

1.4.8.1 Electrolytic grinding

Grinding difficult materials is a tough task. Expensive diamond-impregnated wheels are necessary. Even with these, there is such intense heat generation that it can damage some workpieces. For such work, electrolytic grinding offers a dear, but feasible alternative.

Grinding wheel is impregnated with diamonds in a metal bond. The working portion [diameter and small length of face] of wheel is coated with semi-conductive oxide layer [**Fig. 1.69**]. It promotes electrolyte action.

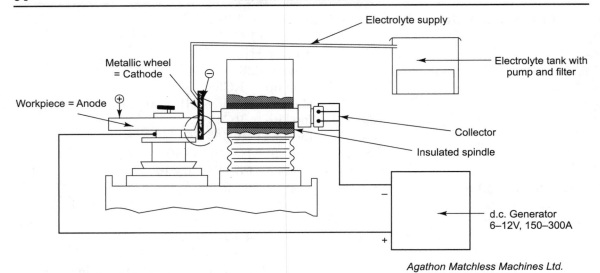

Fig. 1.69: Electrolytic grinding

Naturally, the wheel, spindle, its bearings, etc. must be insulated from the machine body by using slip rings for current supply. Current density of 100 amperes/cm^2 of grinding contact area at 6–12 variable D.C. voltage gives satisfactory results. **Figure 1.69** shows a typical electrolyte grinding machine. Similar to a tool and cutter grinder, it has 150–300 amperes current range. The electricity conducting electrolyte is much more than a coolant. Electrolytic action removes 75% of workpiece material. There is no risk of damage to the workpiece due to overheating. For electrolytic action is cool. Wheel wear is negligible, less than 0.01 mm for 30 working hours.

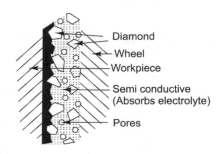

Fig. 1.70: Wheel face

1.4.8.2 Spark erosion machines [electrical discharge machining]

A gap between two electrodes energized with electric current causes sparking which erodes both the electrodes. Control of the gap and the passing current facilitates control of the rate of spark erosion. One of the electrodes is substituted with the workpiece to be eroded. The other electrode serves as an eroding, Electrical Discharge Machining [EDM], tool.

The metal is removed by the thermal effect of the current. The workpiece and the electrode, the discharge interface, is submerged in di-electric liquid like paraffin or transformer oil **[Fig. 1.71]**.

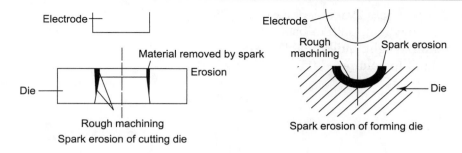

Fig. 1.71: Spark erosion

The eroded surface is a female replica of the sparking surface of the electrode. Hardened dies can be shaped, as required by shaping the electrode suitably. There is no risk of thermal distortion, as in quenching, during heat treatment. Furthermore, erosion produces 0.025–1.125 thick layer Six to Seven points [on Rockwell 'C' scale] harder than the hardness before spark erosion. The life of a spark-eroded surface is much more than the life of nitrided steel workpiece.

Finish and dimensional accuracy Spark erosion can produce sharp corners and fine details, eliminating handwork, and reducing polishing time. Appearance and dimensional accuracy is of better quality than conventional machining methods. With proper care, a finish of 0.5 micron can be obtained. Cavities can be produced within ± 0.05. In press tools, die clearances can be easily controled within 0.02 mm to 0.08 mm, by using the correct gap between the electrode and the die. Even finer clearances can be achieved by using undersized electrodes. For clearances above 0.08 mm, it is necessary to use oversized electrodes.

Electrodes Wear of an electrode is an inherent hazard in spark erosion machining. As a result, the end of the electrode, near the workpiece, diminishes in cross-section. It will produce an undersized cavity if used again. The worn electrodes are therefore, used for roughing. Very little material is removed by the finishing electrode [**Fig. 1.72**].

The wear depends upon the electrode material. Brass, a popular electrode material, has a poor wear factor. Copper and copper-tungsten have better wear factors. Tips of these materials are widely used for spark erosion of press tools. For diecasting dies, large electrodes are made of zinc or tin alloys. They are sprayed with a layer of copper. Carbon electrodes have a good wear factor. But they taper down 0.12 mm after eroding a 50 mm deep cavity.

Spark erosion machines are shaped like vertical spindle drilling machines. The Wickman erodomatic machine has following capacities:
1. Maximum quill stroke   145 mm
2. Worktable area   445 × 375

3. Area of top of base available for work   760 × 555
4. Maximum daylight, quill to base   505
5. Maximum daylight, quill to table   315
6. Throat, quill to column   305

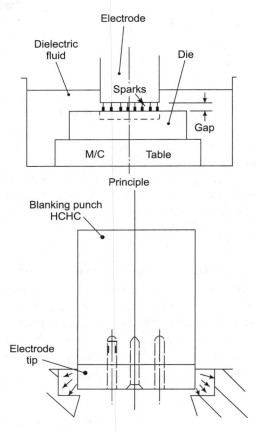

Fig. 1.72: Tipped electrode

1.4.8.3 Electrode design

1. Length: This depends on the depth of the penetration required. Normally electrode length = three times thickness of the workpiece + 35 (for mounting).

Spark gap allowance The electrodes are made undersize to ensure proper spark gap between the electrode and the workpiece. The spark gap depends upon the penetration range selected. Seven ranges are available on the machine. The coarsest, denoted by a single triangle, is used for roughing at a high penetration rate. The finest, denoted by seven triangles, is used for the best surface finish obtained by eroding at the minimum penetration rate.

Introduction

The following table gives the spark gaps required for eroding a round hole in a hardened steel plate with a brass electrode.

Range no	Diametrical spark gap
1	0.60
2	0.50
3	0.30
4	0.15
5	0.10
6	0.05
7	0.025

1.4.8.4 Electrical discharge wire-cutting machine

Electrical discharge wire-cutting machines are like electrical discharge vertical band saws. A 0.1 to 0.25 diameter wire is guided along a programmed track **[Fig. 1.73]**. Electrical discharge between wire and workpiece, in the gap filled with dielectric fluid, effects erosion of workpiece material along the programmed path, to produce the desired shape of die cavity/punch shape.

Wire as a cutting tool The brass or superalloy wire is subjected to wear; so it is run continuously. The wire is unwound from one reel and runs around a system of pulleys, which stretches the wire. The wire can be used only once and the worn wire is discarded after use. The speed of the wire depends upon the thickness of the workpiece to be cut. The wire is guided between two nozzles. The top nozzle can be adjusted vertically to suit the thickness of the workpiece.

The workpiece is mounted on a coordinate table which can be programmed by numerical control to move along the required path to produce the desired profile. The path can be monitored by a pantograph mechanism which can be fitted with a pencil or a pen. The pencil draws the path followed by table (wire) to desired scale. The profile traced by the monitor can be checked by scale for correctness.

Cutting draft angle It is possible to cut the draft angle (fall through taper) up to 30° in the workpiece. This facilitates finishing the die cut-out on wire cutting machine. It is also possible to wire cut parallel punches from a solid plate, thicker than the punch length, by keeping the draft angle zero.

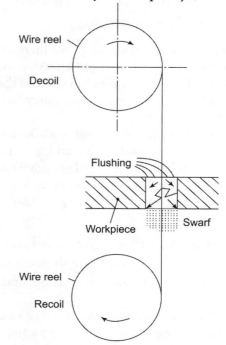

Fig. 1.73: Electrical discharge wire cutting

Water as dielectric fluid In addition to serving as a dielectric fluid, water also serves as a coolant. During erosion, water is contaminated with swarf of tiny eroded particles of the workpiece. So these wire cutting machines are provided with water treatment equipment, with coarse and fine filters to remove particles bigger than 0.005 mm.

Wire cutting machines of smaller sizes have 'C' type cantilever machine frames while bigger machines are provided with portal frames supported at both the ends.

Operation As in a vertical band saw, it is necessary to provide a hole for passage of wire before work can be commenced. **Figure 1.73** shows a hole for wire passage. The wire first moves linearly to the nearest point of required cut-out periphery, before traversing along the programmed path.

The electronic control system of wire cutting machines displays various parameters on the numerous dials and screens, and automatically stops the machine if something goes wrong. The displays point the error, and a hooter is sounded if the error is not corrected within a certain time limit.

Wire cutting machines function like an electrical discharge band saw. Precision numerical control for table movement assures high accuracy. Once a program is recorded on tape it can be used again and again. Sophisticated male and female profile can be produced precisely, repeatedly eliminating human error altogether.

1.4.8.5 Ultrasonic machining

This method uses vibratory action of abrasive particles to remove even non-magnetic workpiece material. The worked out abrasive particles as well as the workpiece swarf particles must be cleaned quickly, continuously from the working area. This is done by adding water to abrasive particles, converting it into slurry, which will flow through the working area at appropriate rate.

Slurry comprises water and Boron nitride/carbide, Silicon carbide, or Aluminium oxide abrasive particles. Boron Carbide gives satisfactory material removal rate. Boron Nitride though costlier is better. The ideal abrasive particle size [grit] for maximum material removal depends upon the amplitude of vibrations. For example for 0.033 amplitude 100 grit will give maximum material removal rate. The grit should preferably be within 90–110 [100 ± 10].

Vibrations frequency and amplitude An electronic oscillator-amplifier drives piezomagnetic transducer to generate vibrations ranging from 15–30 kHz of 0.03–0.1 mm amplitude. Ultrasonic means beyond human hearing range of 12–15 kHz.

The vibrations system comprises a transducer which can be of piezoelectric crystal/ceramic, piezomagnetic or electro-dynamic. **Figure 1.74** shows a piezomagnetic transducer. It comprises two laminated yokes with polarized windings. The vibrating member between the yokes is wound with the energizing coil.

Introduction

Toolholder and tool   The vibrating member is fitted with a tool-holder in the threaded hole in the plate brazed/soldered at the bottom of the transducer. Called 'Velocity - Focussing Transformer' the tool-holder size [dia] and shape [length] should be such that it resonates to amplify the transducer vibrations.

Hard non-magnetic material like brass, aluminium bronze, and K-Monel are used widely for the tool-holder cum Velocity Focussing Transformer.

The **tool**, made of toughened [RC25 - 30] steel, is screwed or brazed to the lower end of the tool-holder. The shape should be similar to the hole/depression to be machined. But it should be of smaller cross-section reckoning the oversize the process produces [**Fig 1.75**].

The tool gets tapered during use. Consequently, the hole machined is also tapered unless the tool is shaped suitably to compensate for tapering.

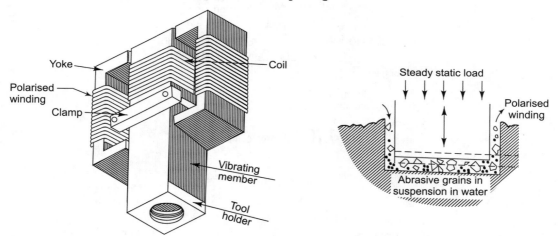

Fig. 1.74: Transducer   **Fig. 1.75:** Tool for ultrasonic machining

Material removal rates for soft ceramic is 20/mm/min. It is lesser for harder, tougher materials. Precision of ± 0.0005 mm and 0.1–0.125 μ surface finish are normal.

1.4.8.6 Ultrasonic drilling machines

Ultrasonic drilling machines [**Fig. 1.76**] resemble vertical milling machines. The vibrator is mounted on a vertical slide to facilitate height adjustment. The compound table can be moved longitudinally and laterally by handwheels to position workpiece accurately with respect to the tool fitted to the transducer.

Generally, increases in vibrations amplitude and frequency, increase the penetration [material removal] rate.

1.4.8.7 Electro-chemical machining

In this process, the tool acts like a [negative] cathode while the workpiece acts like the [positive] anode. Electrical current is passed through the gap between the two. The gap is flooded by electrolyte [brine] which in addition to conducting the current carries away particles removed from the positive [anode] workpiece [**Fig. 1.77**].

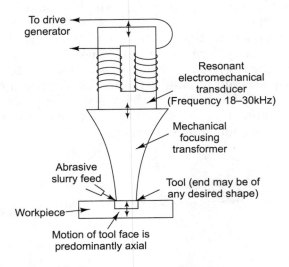

Fig. 1.76: Ultrasonic machining

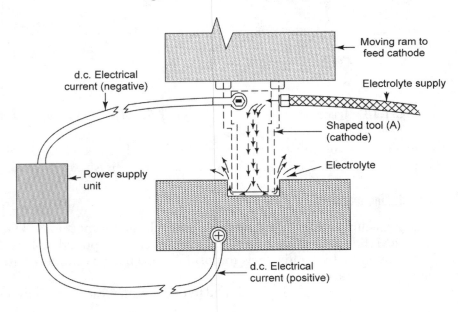

Fig. 1.77: Electrochemical machining

The material removal rate depends upon the current density and the electro-chemical equivalent of the workpiece material. One thousand amperes current at 8–18 D.C. volts supply can remove 1.6 cm^3 of steel per minute. Vertical feed rates range for 0.25–25mm/min.

Introduction

This method is used for machining hardened steel, nimonic alloys, and carbides. Finish is usually of 0.15 μm grade.

Electro-chemical machines They have a vertical ram, guided by roller bearings. It is actuated by a lead screw/nut mechanism, protected from brine by polyurethane seals. Work table is made of stainless steel base, mounted with granite segments. The ram is fitted with stainless steel plate, with locating keyways and tapped holes for securing tool.

There is a tank and circulation system for the electrolyte [brine].

Capacity for 10,000 amperes current at 8–18 volts D.C. supply is quite common. There is a provision for automatic cycle with settable stops for rapid approach, forward and retracted [withdrawn] ram positions.

As there is no contact between the tool [cathode] and workpiece, there is no tool wear at all.

1.4.8.8 Chemical machining

Uses acids or alkalies to remove material by dissolution-etching. Curved, oddly-profiled workpieces can be machined [etched] on selected surfaces by dissolving the workpiece material at the face, into the etchant acid/alkali.

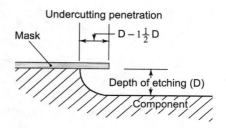

Fig. 1.78: Undercutting in chemical milling

The surfaces not to be etched must be masked by vinyl, neoprene, etc. by spray coating, dipping or flow coating. After drying the coat, the mask on the portion to be machined chemically, is peeled off using a template and a knife. It should be remembered that there is some undercutting at the mask edge [**Fig. 1.78**].

For chemical etching workpiece is suspended in the solvent chemical tank with positive clearance from tank bottom and solvent top level.

Solvents

workpiece material	Solvent	Etching rate mm/min	After etching
Aluminium and its alloys	10% Caustic soda [NaOH] 1% Aluminium	0.03	Wash with water. Then soak in 10% Nitric acid, and rinse in water
Magnesium and its alloys	8% Nitric acid	0.035	Wash with water
Titanium and its alloys	25% Hydrofluoric acid 0.8% Titanium		Cooling Jacket Necessary

This method is used for tolerances of ±1 mm on length and ± 0.1 mm on thickness [depth]. The finish ranges from 1– 4 μm.

Safety hazards—Nitric, chromic, and hydrofluoric acids produce toxic, suffocating fumes. splash of acid can be perilous for skin and eyes. Hydrogen produced in soda, and sulphuric acid etching is highly explosive. Exhaust fans, protective gloves, clothing, goggles etc. and extreme caution is necessary during chemical machining operations.

At the dawn of the third millennium…

A visit to India Machine Tool Exhibition [IMTEX 2004] was very enlightening. The all pervading, computerized controls have contributed immensely to machine tools operation and automation. Even impacting machines like mechanical presses now sport electronic displays recording die heights, stroke, ram position etc. Precise controls have facilitated fine manipulation of tools thru accurate strokes. In-process measurement of workpiece dimensions has eliminated post mortem quality control almost altogether. Mass production machines can be manipulated precisely repeatedly to achieve not altogether impossible 'Zero Rejection'.

What is more, most of the constituents necessary for a precision Computerized Numerical Control [CNC] Machines are commercially available. You can buy ready precision ball/roller screws, even slides and guideways; electronic scales, readers, displays for stroke measurements. Accurately indexible rotary tables and precision guideway, what not? With skill and ingenuity a competent tool room can build [assemble] its own special purpose CNC machine.

Fluid mechanics has advanced by leaps and bounds. Now you can buy a rod-less cylinder with stroke control thru axially adjustable proximity switches anchored to the cylinder itself. There are air cylinders, which need neither piston nor a rod. They can take misalignment up to twenty percent of their self adjusting non-linear bore. There are special cylinders, which can give you controlled rotary motion of rod in addition to axial stroke. They can be very handy for actuating simultaneously hook clamps used widely in turning fixtures [refer author's *Jigs & Fixtures* Indian/International edition].

Turning, milling-cum-drilling machining centers are common sights in modern machine shops. Programable robots have enhanced safety in hazardous operations like welding in mass production.

Computers have become programable eyes [readers], feelers [transducers] and brain [corrective action] of modern machine tools.

The machine tools muscles [screws, transmission elements, carriages, tool, and workholders] have, however, not changed much. Of course, ground ball/roller screws, gears, antifriction guideways have enhanced precision; and fine count displays and dials have reduced drastically the human skill earlier necessary for precise measurements. In-process measurement and automatic tool [slide] adjustment thru feed-back has almost eliminated workpiece variations due to tool wear.

I have endeavored to update this treatise right up to February 2004. Most of other contemporary books discourse the theme at most up to 1996.

CHAPTER 2

MACHINE TOOL DRIVES AND MECHANISMS

Machine tools generally draw power from a single source, such as an electric motor or engine. Even when an engine is used, it is useful to convert thermal power into electrical energy by using a generator. Mostly, a single-speed electric motor is used. Although the speed (R.P.M.) of a motor decreases marginally under load, it is considered constant. The maximum torque available, depends upon the motor power-rating (kW/H.P.), and rotary speed [R.P.M.]. Costly two-speed motors can be used conveniently in some applications.

The spindle speed must be increased or decreased to suit the size and material of the workpiece. The following table gives the recommended linear speeds for commonly used materials for workpieces, and cutting tools.

Table 2.1: Linear cutting speeds 'V' (m/min) for commonly used workpiece materials and cutting tools

Sr. No.	Workpiece material	Cutting tool			
		High speed steel		Carbides	
		roughing	finishing	roughing	Finishing
1	Mild Steel	30 [98]	40 [131]	80 [262]	150 [492]
2	Cast Iron	25 [82]	40	70 [230]	80 [262]
3	Aluminium	70 [230]	100 [328]	80	150
4	Copper alloys	60 [197]	100	60 [197]	120 [394]

Bracketed values give equivalents in feet. 1m = 3.2808ft.

2.1 ROTARY DRIVES

For rotary cutting, it is necessary to compute revolutions per minute (R.P.M.) of the machine spindle, using the following equation:

$$V = \frac{\pi DN}{100}$$ (Eqn. 2.1)

V = Linear (cutting) speed (m/min);
D = Workpiece diameter (mm);
N = Revolutions per minute (R.P.M.) of machine spindle.

Example 2.1: A machine spindle is to be used for roughing and finishing a mild steel workpiece of 60 ⌀. Find the R.P.M.s for roughing and finishing.

Solution: For roughing: (from Table 2.1 and Eqn. 2.1)

$$V = 30 = \frac{\pi DN}{1000} = \frac{\pi \times 60 \times N}{1000}$$

$$N = \frac{30 \times 1000}{\pi \times 60} = 159.1 \text{ [R.P.M.]}$$

For finishing:

$$N = \frac{40 \times 1000}{\pi \times 60} = 212.2 \text{ [R.P.M.]}$$

2.1.1 Cutting speed variation

A few rotary the speeds (N) suffice for special purpose machines that are used for a few operations, on workpieces of the same size and material. But general purpose machines, used for a range of workpiece sizes and materials, call for a much wider variation in rotary speeds (N).

The variation can be of two kinds: stepped or stepless. Stepped variation provides a limited number of fixed rotary speeds. For the case, in Example 2.1, stepped variation with 2 fixed rotary speeds will suffice.

Stepless variation can provide infinite number of speeds, within the maximum and minimum values. For the case in Example 2.1, the speed can have any value between 159.1 to 212.2 R.P.M., in stepless variation. Stepless variation is necessary when there is a wide variation in the hardness of the workpiece material. Infinitely variable, stepless drives are generally unsuitable for low-speed, high-torque applications.

2.1.1.1 Stepped variation: ray diagrams

Graphical representations called ray diagrams (or speed charts) can be conveniently used for designing rotary stepped drives. They show the input source as a single point which is connected to the points for rotary output speeds (N), by lines similar to the rays emanating from a light source. **Figure 2.1** shows the ray diagram for Example 2.1, where the input source is an electric motor, running at 750 R.P.M.

General purpose machines have to use different rotary speeds for different workpieces. As the R.P.M. of the spindle varies inversely with the workpiece (cutting) diameter, it is convenient to organize the intermediate rotary speeds in a geometric progression. In this arrangement, the next, higher rotary speed is a constant multiple of the preceding speed. The multiple is called 'Step (progression) ratio' (ϕ). If the minimum rotary speed is 'N' R.P.M., the higher speeds will be as below:

$N, N\phi, N\phi^2, N\phi^3, N\phi^4 \ldots N\phi(2-1)$ for
No. of speeds $= z$

Also Step (progression) ratio $(\phi) = \sqrt[(z-1)]{\dfrac{N_{max}}{N_{min}}}$ **(Eqn. 2.2)**

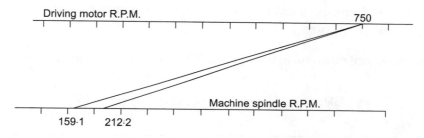

Fig. 2.1: Ray diagram for Example 2.1

Geometric progression gives smaller increments at lower rotary speeds, and bigger steps at higher rotary speeds. This is very convenient and economical, due to the following reasons:

1. The cutting diameter decreases with each cut which progressively decreases the actual linear cutting speed (V).

    The decrease in the linear cutting speed, as well as the resulting decrease in surface area machined per minute, in a single step, depends upon the step (progression ratio 'ϕ')

    $$\dfrac{dv}{V} = \dfrac{dA}{A} = \dfrac{\phi - 1}{\phi} \quad \textbf{(Eqn. 2.3)}$$

    dv : Decrease in speed in a single step;

    dA = Corresponding decrease in area;

    A : Area machined/ min; V = Max. cutting speed (m/min).

2. As smaller bars are available in smaller increments, the difference between the rough and finish size is less. Therefore, the decrease in linear cutting speed (V) is lesser for smaller diameters and higher speeds (N). Consequently, the bigger steps for higher rotary speeds (N), do not have much effect on the optimum linear cutting speed, or productivity.

3. The material removal rate depends upon the power (kW/HP) available. As the maximum power of an electric motor is constant, it is necessary to use smaller cuts for bigger cutting diameters, machined at lower rotary speeds (N). Machining times are hence, longer for bigger diameters. Thus, lower increments at lower rotary speeds, facilitate the usage of near-optimum cutting speeds for bigger diameters. This makes the geometric progression more productive.

Choice of step (progression) ratio 'ϕ'

Two-speed electric motors are available commercially. Their usage can be facilitated by keeping one of the intermediate rotary speeds at double the level of the minimum rotary speed (N), i.e. $\phi = \sqrt[E]{2}$.

It is also convenient to make the rotary speeds (N) equal to the preferred numbers. These numbers are derived from the number 'ten', as per the equation below:

$$\phi = \sqrt[E]{10} = 10^{1/E}$$

The step (progression) ratios (ϕ) in the following table, combine both the requisites explained above.

Table 2.2: Preferred step (progression) ratios ϕ

Step Ratio ϕ	Relation with No. 2	Relation with No. 10	Max Speed Decrease $\dfrac{\phi - 1}{\phi} \times 100$
1.222	$2^{1/6}$	$10^{1/20}$	10%
1.26	$2^{1/3}$	$10^{1/10}$	20%
1.414	$2^{1/2}$	$10^{3/20}$	30%
1.585	$2^{1/1.5}$	$10^{1/5}$	40%

Note: The lower $\left(1.06 = 2^{\frac{1}{12}} = 10^{\frac{1}{40}}\right)$ and higher $\left\{1.78 = 2^{\frac{1}{1.2}} = 10^{\frac{1}{4}} \text{ and } 2 = 2^1 = 10^{\frac{3}{10}}\right\}$ values of f are rarely used in machine tools.

Range ratio (R)

The relation between maximum and minimum rotary speeds (N max/N min), depends upon the maximum and minimum values of linear cutting speeds (V max/V min), and workpiece/cutting diameters (D max/D min).

$$R = \text{Range ratio} = (N\max/N\min) = (V\max/V\min) \times (D\max/D\min) \quad \textbf{(Eqn. 2.4)}$$

For a given range ratio 'R' of rotary speeds, the number of intermediate rotary speeds depends upon the speed (progression) ratio ϕ. The number of rotary speeds 'z' can be found from the graph in **[Fig. 2.2]**. Needless to say, 'z' should be an integer. Furthermore, the number of rotary speeds 'z' must strictly be multiples of 2 or/and 3, for, in gearboxes used for stepped changes of rotary speed, the gear shifting lever should have only 2 or 3 positions—left or right; or left, center, or right (or up or down; or up, horizontal, or down).

If 'N' is the lowest rotary speed, the higher rotary speeds will be: $N\phi$, $N\phi^2$, $N\phi^3$, $\cdots N\phi^{(z-1)}$. Rotary speeds can be conveniently selected from Table 2.3, for standard spindle speeds.

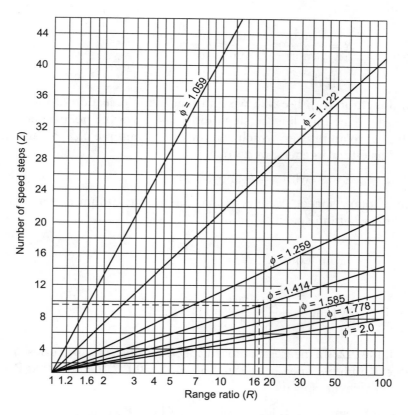

Fig. 2.2: Finding speed (progression) ratio graphically, from the number of speeds (Z), and range ratio (R)

Example 2.2: Find range ratio 'R' and rotary spindle speeds for roughing and finishing mild steel and cast iron workpiece faces, with carbide facing cutters, ranging from 25 to 200 in diameter. Take $\phi = 1.414$.

Note: In milling and drilling machines, cutting diameter 'D' is equal to the cutter diameter.

Solution: From Table 2.1, linear cutting speeds for roughing and finishing mild steel with carbide tool are: 80 and 150 m/min. For cast iron, the corresponding speeds are 70 and 80 m/min.

∴ Maximum linear cutting speed $V\max = 150$

and Minimum linear cutting speed $V\min = 70$

$$\text{Maximum cutter dia} = D\max = 200$$

$$\text{Minimum cutter dia} = D\min = 25$$

From Eqn. 2.3:

$$R = \frac{V\max}{V\min} \times \frac{D\max}{D\min} = \frac{150}{70} \times \frac{200}{25}$$

$$= 17.14$$

Referring to the graph in **[Fig. 2.2]**, we find that the intersection of the vertical line from range ratio 17, and the inclined line for step ratio 1.414, gives the no. of speeds 'Z' = 9.5. This must be rounded off to an integer, which can only be a multiple of 2, or/and 3, i.e. $Z = Z^x 3^y$ with 'X' and 'Y' as either integers, or zero.

∴ $Z = 9$ [with $x = 0$ and $y = 2$].

The minimum rotary speed 'N/min', is that for the maximum cutter diameter '$D\max$', and minimum linear cutting speed 'V' min.

From Eqn. 2.1, $N = \dfrac{1000\, V}{\pi D}$

∴ $N\min = \dfrac{1000 \times 70}{\pi \times 200} = 111.41$

Referring to Table 2.2, the nearest standard speed for the '$\phi = 1.41$ series', is 125. The table also gives the next 6 higher speeds as 180, 250, 355, 500, 710, and 1,000 R.P.M. The table however, gives only seven speeds.

The 8th speed = $1,000 \times 1.414 = 1,414$ R.P.M.

and the 9th speed = $1,000 \times 1.414^2 = 1,999.4$ R.P.M.

Alternatively, the speeds can be calculated. These will be 125, 125×1.414, 125×1.414^2, 125×1.414^3, 125×1.414^4, 125×1.414^5, 125×1.414^6, 125×1.414^7.

Needless to say, the calculated speeds will be equal to the speeds found by referring to the table.

Table 2.3: Standard spindle speeds for various step ratios (ϕ)

$\phi \rightarrow$	1.26	1.414	1.585	1.78
	10		10	10
		11.2		
	12.5			
	16	16	16	
	20			18
		22.4		
	25		25	
	31.5	31.5		31.5
	40		40	
		45		
	50			
				56
	63	63	63	
	80			
		90		
	100		100	100
	125	125		
	160		160	
		180		180
	200			
	250	250	250	
	315			315
		355		
	400		400	
	500	500		
				560
	630		630	
		710		
	800			
	1,000	1,000	1,000	1,000

Note: The actual spindle speed [R.P.M.] can be 6% higher or 2% lower, as per IS 2218 [1962], i.e. variation in nominal speed of 1,00 R.P.M. should be within +6 [106] and –2 [98].

Example 2.3: Find 6 speeds for a lathe, to be used for rough turning and finishing mild steel workpieces, ranging from 22 to 65, in diameter. Use standard speeds and standard step ratios (ϕ). Roughing speed = 25 m/min; Finishing speed = 30 m/min.

Solution: $$R = \frac{V \max}{V \min} \times \frac{D \max}{D \min} = \frac{30}{25} \times \frac{65}{22}$$

$$= 3.545$$

$$N\min = \frac{1000V\min}{\pi D\max} = \frac{1000 \times 25}{\pi \times 65}$$

$$= 122.4 \text{ R.P.M.}$$

$$N\max = \frac{1000V\min}{\pi D\max} = \frac{1000 \times 30}{\pi \times 22}$$

$$= 434 \text{ R.P.M.}$$

For a geometric progression:

$$N\max = N\min \times \phi^{(z-1)}$$

$$\phi = \sqrt[z-1]{R} = 3.545^{\left(\frac{1}{6-1}\right)}$$

$$= 1.288$$

The nearest higher standard step ratio (ϕ) = 1.414, will give a slightly higher rotary speed. From Table 2.3, one can observe that the nearest standard minimum rotary speed ($N\min$) is 125 R.P.M. The higher rotary speeds will be: 125×1.414, 125×1.414^2, 125×1.414^3, 125×1.414^4, 125×1.414^5.

Therefore, for $\phi = 1.414$, the speeds are: 125, 180, 250, 355, 500, and 710 R.P.M.

If we choose the lesser but nearer step ratio 1.26, the rotary speeds will be: 125, 160, 200, 250, 315, and 400 R.P.M.

Comparing rotary speeds of both the speed ratios it can be noted that:

1. Smaller speed ratio (ϕ) gives lower rotary speeds, suitable for bigger workpieces. Lesser increments in rotary speeds make smaller speed ratios convenient for optimizing the linear cutting speed (V) in special purpose machines, with lesser workpiece size variation, and automatic machines, in which long runs of similar workpieces are common.

2. Bigger speed ratios give higher rotary speeds, suitable for smaller workpiece/cutting diameters.

Table 2.4: Recommended speed ratios (ϕ) for machine tools

Speed ratio ϕ	Machine size and load	Machine Tool type
1.06	Very heavy duty	
1.122	Heavy duty	Mass production with change gear train
1.26	Heavy duty	Mass production with change gears, General purpose
1.414	Medium duty	General purpose
1.585	Small size Light duty	
1.78	Very small size Light duty	

2.1.2 Power source: Electrical motors

The spindle is usually powered by a 3-phase, squirrel cage, induction motor, using alternating current. Motors are specified by rotary speed (R.P.M.) and power rating (kW/HP), and class (continuously operated, frequently started, with reverse current electrical braking, or having suitable slip, to cope with high inertia load). Section, 6.3.3.0, covering electrical automation, lists further considerations such as speed requisites (single/multi-speed), power needs (constant/variable), torque requirements (constant/variable) and duty demands (on load/no load starting, inertia load, load cycle, acceleration time, starting current limitations, starting and braking methods, and needs for reversal), site constraints (maximum temperature, humidity, alkalinity/acidity of atmosphere, other hazards), type of protection and cooling, and of course, the insulation strength (class) and electric supply compatibility (single/three phase, voltage and frequency of supply along with maximum variation).

2.1.3 Motor Speed (R.P.M.)

The theoretical speed depends upon the number of poles $\left(N = \dfrac{6000}{\text{No. of poles}}\right)$. The nominal speed (the declared speed) is the minimum speed under load: the input speed used for computations. It depends upon the power rating and the number of poles. The actual speed (under load) can be 2.8–4.5% higher than the nominal speed, but never less than the declared speed.

Table 2.5: Electrical motor speeds ('N' R.P.M.)

No. of Poles	2		4		6		8	
Theoretical speed R.P.M.	3,000		1,500		1,000		750	
Power rating range KW	0.37	5.5	0.37	5.5	0.37	3.7	0.37	2.2
	3.7	37	3.7	37	2.2	37	1.5	37
Actual speed Min.	2,785	2,870	1,370	1,410	910	950	680	700
Max.	2,870	2,955	1,430	1,483	940	980	705	720

In two-speed motors, the lower speed is usually half of the higher speed. The speed is decreased by doubling the number of poles through suitable switches.

Std. power ratings (kW): 0.06, 0.09, 0.12, 0.18, 0.25, 0.37, 0.55, 0.75, 1.1, 1.5, 2.2, 3.7, 5.5, 7.5, 11, 15, 18.5, 22, 37, 45, 55, 75, 90, 100.

Finding motor kW rating: Power consumed by a machine tool depends upon the rate of material removal (cc/sec).

Specific power 'K' (measured in kW), for removing 1 cc material per second from a workpiece, depends upon the hardness of the workpiece material. Table 2.6 gives the specific power for commonly used materials.

Table 2.6: Specific power (K) for commonly used workpiece materials (kW/CC/sec)

Material	Hardness BHN	Specific Power KW
M.S.	137	1.88 [30.81]
	170	2.08 [34.08]
C.I.	150	1.04 [17.04]
	241	2.51 [41.13]
Aluminum	100	1.64 [26.87]
Copper alloys	60	1.64
	180	2.54 [41.62]

Bracketed values give kW/Inch^3/sec. 1 Inch^3 = 16.38707 cm^3

Usage of worn-out cutting tools can increase the power requirement by 50% in drilling and milling, and as much as 100% in rough turning.

The efficiency (deficiency) of transmission of drives increases the power requirement further.

2.2 MECHANICAL DRIVES

These can be broadly classified as:
1. **Positive drives:** In these, the driver and the driven elements, mesh with each other (gears), or with the power transmitting elements (chains).
2. **Frictional drives:** Belts and clutches rely on friction for power transmission. There is always a possibility of slip, under high speed (above 30 m/sec in belts), or due to overload. Timer belts with a toothed profile on the inside, are used for light load. They provide almost slip-free transmission.

Positive drives are suitable for low-speed (below 6 m/sec), high-torque applications, while frictional drives are more convenient and economical for high-speed (above 15 m/sec), low-torque applications.

Table 2.7 outlines the important features and efficiencies of both the types of drives.

Table 2.7: Comparison of machine tool drives

Sr. No.	Drive Type	Speed Range m/sec Max. Ideal	Transmission Ratio $i = \dfrac{N_{in}}{N_{out}}$	Efficiency %	Remarks
1	Belts	5–30	17–22.5		For 180° Arc of Contact
	Flat		$25 - \dfrac{1}{4}$	97%	Flat Belts

Cont.

Sr. No.	Drive Type	Speed Range m/sec Max. Ideal		Transmission Ratio $i = \frac{N_{in}}{N_{out}}$	Efficiency %	Remarks
	Vee			$20 - \frac{1}{3}$	92%	'V' Belts
					85%	Variable speed drives [Refer Table 2.8 for other angles]
2	Roller Chains	Up to 20	Up to 12	$7 - \frac{1}{3}$	80	Sprocket Engagement should be above 120°
3	Toothed Gears	Up to 20	Up to 6	$\frac{1}{10} - 14$	70	for open gears
	Straight Spur			$4 - \frac{1}{2}$		For
	Helical Spur			$4 - \frac{1}{2.5}$		Compact gear boxes
4	Hydraulic system	Up to 20	Suction 0.8 Pipes 5–6	— — —	60 75 85	Gear pumps Vane pumps Piston pumps
5	Electric motor	750 to 3000 R.P.M.	1500 R.P.M.	—	90 95	Min Max

The resultant efficiency of a composite drive, will be a multiple of the efficiency of all the constituent drives.

$$\eta = \eta_1 \times \eta_2 \times \eta_3 \qquad \text{(Eqn. 2.5)}$$

$$\text{Power rating of motor} = \frac{\text{Cutting power (kW)}}{\eta} \qquad \text{(Eqn. 2.6)}$$

Example 2.4: A turning lathe is required to machine M.S./C.I. workpieces of 100 diameter, with 4 mm roughing cut, at 25 m/min speed, at 0.5/revolution feeds. Belt is used for drive, from motor to the gearbox input shaft at overall efficiency of 90%. The gearbox reduces the spindle speed, further to required value.

Solution: $$N = \frac{1000\,V}{\pi D} = \frac{1000 \times 25}{\pi \times 100} = 79.6$$

The nearest standard spindle speed is 80 R.P.M. (Table 2.3).
In turning axial linear speed of the cutting tool = Nf; where N = R.P.M. and f = feed rate/rev.
As specific power is stated in cc/sec, all related values will be converted to cms and seconds.

∴   Linear Feed Rate = Nf = 80 × 0.5/10 per min.

$$= 4 \text{ cms/min}$$

$$\text{Linear feed/sec} = \frac{4}{60} = 0.067 \text{ cms/sec}$$

The 4 mm cut removes material in the annular area between 100 and 92 (100 − 2 × 4) diameters.

$$\text{Material removal/sec (cc)} = 0.7854 \times (10^2 - 9.2^2) \times 0.067$$

$$= 0.808 \text{ cc}$$

Among M.S. and C.I., hardness can be higher for C.I. Specific power for C.I. (241BHN) is 2.51 kW/cc. sec.

∴   Cutting power = 2.51 × 0.808

$$= 2.03 \text{ kW}$$

Increase in power requirement due to worn-out tool can be 100% in rough turning.

∴   Max. cutting power = 2.03 × 2 = 4.06 kW

overall $\eta = \eta_1$ (belt) × η_2 (gear) × η_3 (motor)

$$= 0.9 \times 0.7 \times 0.9 = 0.567$$

∴   Rating of motor = $\frac{4.06}{0.567}$ = 7.16 kW

$$= 7.5 \text{ [nearest higher standard rating]}$$

Motor speed should preferably be at the mean value of the speeds (R.P.M.)

Example 2.5: A machine tool spindle has six speeds: 355, 500, 710, 1,000, 1,400, and 2,000. Select suitable speed for a 7.5 kW motor. Draw the ray diagram for the arrangement, assuming belt drives.

Solution: Preferable electric motor speed = Nm

$$N_m = \frac{N_{min} + N_{max}}{2} = \frac{355 + 2000}{2} = 1177.5$$

From Table 2.5, it can be seen that the nearest minimum motor speed is 950 R.P.M., for six pole motors above 3.7 kW.

It will be convenient to use a 2 stage transmission. The two intermediate speeds can be equal to the middle terms of the lower three speeds (355, 500, 710), i.e. 500 R.P.M., and the upper three speeds (1,000, 1,400, 2,000), i.e. 1,400 R.P.M. Then the intermediate shaft should have two speeds: 500 and 1,400 R.P.M.

For drawing the ray diagram [Fig. 2.3a], draw three parallel vertical lines with suitable spacing (say 50 mm). The left line will represent the input motor shaft, the middle—the intermediate shaft, and the right one—the output at the machine spindle.

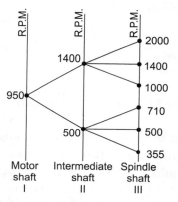

Fig. 2.3(a): Ray diagram for Example 2.5

On the right, on the spindle line, mark six equidistant points with suitable spacing: say 20 mm. These six points represent six spindle speeds. Write the speeds on the right of the line: the lowest speed (355) at the lowest point, and each higher speed near the next higher point, with 2,000 at the uppermost point.

Draw from the points representing speeds 500 and 1,400 R.P.M., horizontal lines towards the left, to meet the middle line for the intermediate shaft. Mark intercept points with speeds 500 and 1,400 R.P.M.

Join the point for 500 R.P.M. on the intermediate shaft, with the points representing lower speeds 355 and 710 R.P.M. Similarly, join the point for 1,400 R.P.M. on the intermediate shaft, with points for 1,000 and 2,000 R.P.M. on the spindle shaft. We now have the ray diagram for transmission between the spindle and the intermediate shaft.

Now, mark a point on the left input line, representing motor R.P.M. 950. It will be almost in line with the point representing 1,000 R.P.M. on the right (spindle) shaft line. Write the motor speed near the point. Join the motor speed point with the intermediate speeds points 500 and 1,400, on the middle line to finish the ray diagram.

Let us suppose that the 950 R.P.M. motor is not immediately available. We can use the commercially popular 1,410 R.P.M. motor, or 700 R.P.M. motor. Still, the intermediate shaft speed can remain the same. **Figure 2.3b** shows the ray diagram for a 1,410 R.P.M. motor, while [**Fig. 2.3c**] show the speed chart for a 700 R.P.M. motor. Ray diagrams or speed charts are blue-prints, plans for speed changes to be effected.

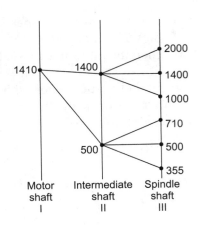

Fig. 2.3(b): Speed chart for 1500 R.P.M. motor   **Fig. 2.3(c):** Ray diagram for 750 R.P.M. motor

The most important parameter in a speed chart is the transmission ratio (i). As each stage of speed change might involve a maximum of three transmission ratios, we will distinguish them by the suffixes 'h' (higher), 'm' (middle), and 'l' (lower). Similarly, shafts involved in the changes will be distinguished by the stage suffixes for input I, II, III, etc. for the following stages, i.e.:

$$\text{Transmission ratio (i)} = \frac{\text{Input R.P.M.}}{\text{Output R.P.M.}}$$

$$= \frac{N_1}{N_2} \left(\text{or } \frac{N_2}{N_3}, \frac{N_4}{N_5} \text{ etc} \right)$$

Thus, for the transmission ratios between motor shaft I and intermediate shaft II, in [**Fig. 2.3a**] are:

$$^i\text{I II h} = \frac{950}{1400} = 0.678 \text{ and } ^i\text{I II l} = \frac{950}{500} = 1.9$$

Ratios for [**Fig. 2.3b**] will be:

$$^i\text{I II h } \frac{1410}{1400} = \frac{141}{140} = 1.007$$

$$\text{I II l} = \frac{1410}{500} = 2.82$$

You can calculate transmission ratios for [**Fig. 2.3c**] as an exercise.

2.2.1 Changing speed

2.2.1.1 Stepped change

- **Slow Changeover:** General purpose machines are used for a wide variety of workpiece sizes and materials. Higher cost, entailed in quick change of speed is not warranted. Belt-driven cone pulleys or change gears are quite adequate for speed change. Cone pulleys are combined, composite pulleys with 2 to 4 steps. **Figure 2.4(a)** shows a cone pulley with 4 steps. Any of the available four speeds can be obtained, by shifting the belt to the required step in a minute or two. Belt drives are more convenient for higher speeds. The transmission capacity decreases with a decrease in the angle of the arc of contact (Table 2.8). At speeds exceeding 10 m/sec, there is a slip between the belt and the pulley (Table 2.9). It affects the torque capacity (Table 2.10). V belts with 40° included angle and thicker cross section must have bigger diameter pulleys, for proper contact.

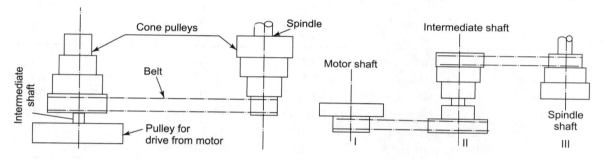

Fig. 2.4(a): Cone pulley belt drive

Fig. 2.4(b): Cone pulley belt drive for **[Fig, 2.3c]**

Table 2.8: Coefficients for various arc or contact angles in belt transmission

Angle ($a°$)	Coefficient (A)	Angle (α)	Coefficient (A)
180°	1	190°	1.05
160°	0.95	200	1.10
140°	0.89	210	1.15
120°	0.83	220	1.20
100°	0.74	$\alpha = \sin^{-1}\dfrac{(D-d)}{(2C)}$	
80°	0.64		

D: Big pulley ϕ (mm); $d =$ Small pulley ϕ (mm); $C =$ Center distance (mm)

Table 2.9: Velocity coefficients for belts

Velocity (m/sec)	Coefficient (V)	Velocity (m/sec)	Coefficient (V)
1 [3.28]	1.04	20 [65.62]	0.88
5 [16.4]	1.03	25 [82]	0.79
10 [32.81]	1.00	30 [98.42]	0.68
15 [49.21]	0.95		

Bracketed values give feet equivalents. 1m = 3.28084ft.

Table 2.10: Standard (40°) 'V' belts with minimum pulley ϕ and recommended torque

Designation	Belt Width × Height ($b \times h$)	Min. pulley (ϕ)	Recommended torque (kgm)	Tensile force (kg)
0	10 × 6 [.394″ × .236″]	63 [2.48″]	3 [22.1 Lbsft]	95 [209.4 Lbs]
A	13 × 8 [.512 × .314]	90 [3.543]	6 [44.2 Lbsft]	133 [293.2 Lbs]
B	17 × 11 [.669 × .433]	125 [4.921]	15 [110.6 Lbsft]	240 [529.1 Lbs]
C	22 × 14 [.866 × .551]	200 [7.874]	60 [442 Lbsft]	600 [1322.8 Lbs]
D	32 × 19 [1.26 × .748]	355 [13.976]	240 [1770 Lbsft]	1352 [2980.6 Lbs]
E	38 × 23 [1.496 × .905]	500 [19.685]	600 [4420 Lbsft]	2400 [5291 Lbs]

Bracketed values give Inch/Lbs/Lbsft equivalents

Table 2.10 gives the minimum pulley diameters, and **[Fig. 2.5]** gives the groove size, margins, and angles, for various standard sections. Bigger pulley diameter depends upon the transmission ratios (i).

For V belts, the ideal center distance can be found from the following equation:

$$\text{Center distance} = C = \frac{1.5 D}{\sqrt[3]{i}} \qquad \textbf{(Eqn. 2.7)}$$

$$C\min = 0.55\,(d + D) + h \qquad \textbf{(Eqn. 2.8)}$$

$$C\max = 2\,(d + D) \qquad \textbf{(Eqn. 2.9)}$$

The center distance (C) must be adjusted, to permit usage of standard belt lengths available. Belts are specified by inside length (Li). The pitch line length (Lp), used for finding the center distance between pulleys, passes through the neutral axis of the belt.

$$Lp = Li + P \qquad \textbf{(Eqn. 2.10)}$$

Table 2.11: Difference 'P' between inside and pitch lengths of 'V' belts

Section	Size b × h	Difference P
0	10 × 6 [.394" × .236"]	28 [1.102"]
A	13 × 8 [.512 × .314]	35 [1.378]
B	17 × 11 [.669 × .433]	43 [1.732]
C	22 × 14 [.866 × .551]	56 [2.205]
D	32 × 19 [1.26 × .748]	79 [3.11]
E	38 × 24 [1.496 × .905]	92 [3.622]

1mm = 0.03937"

Table 2.12(a) gives the standard inside lengths for 'V' belts.

Table 2.12(a): Standard inside lengths for V belts

Inside Length L_i	Belt Designation						Pitch Length Variation
	0	A	B	C	D	E	
395 to 2500 In steps of 25 [15.551" to 98.42" steps of 0.984"]	*	—	—	—	—	—	
610 [24.016"]	*	—	—	—	—	—	+11.4 [.449"]
660 [25.984]	*	—	—	—	—	—	−6.4 [.252]
711 [27.992]	*	—	—	—	—	—	
787 [30.984]	*	—	—	—	—	—	
813 [32.008]	*	—	—	—	—	—	+12.5 [.492]
889 [35]	*	*	—	—	—	—	−7.5 [.295]
914 [35.984]	*	*	—	—	—	—	
965 [37.992]	*	*	—	—	—	—	
991 [39.016]	*	—	—	—	—	—	+14.0 [.551]
1016 [40"]	*	*	—	—	—	—	−8.9 [.35]
1067 [42]	*	*	—	—	—	—	
1092 [42.992]	*	*	—	—	—	—	
1168 [45.984]	*	*	—	—	—	—	
1219 [47.992]	*	*	*	—	—	—	
1295 [50.984]	*	*	*	—	—	—	+16.0 [0.63]
1372 [54.016]	—	*	—	—	—	—	−9.0 [0.354]
1397 [55]	*	*	*	—	—	—	
1422 [55.984]	*	*	—	—	—	—	
1473 [57.992]	*	—	—	—	—	—	
1524 [60]	*	*	*	—	—	—	

Inside Length Li	Belt Designation						Pitch Length Variation
	O	A	B	C	D	E	
1600 [62.992"]	*	—		—	—	—	+17.8 [.701"]
1626 [64.016]	*	*		*	—	—	−12.5 [.492"]
1651 [65]	*	*		—	—	—	
1727 [67.992]	*	*		*	—	—	
1778 [70]	*	*		*	—	—	
1905 [75]	*	*		*	—	—	
1981 [77.992]	*	*		—	—	—	+30 [1.181]
2032 [80]	*	*		*	—	—	−16 [.63]
2057 [80.984]	*	*		*	—	—	
2159 [85]	*	*		*	—	—	
2286 [90]	*	*		—	*	—	
2438 [95.984]	*	*		*	—	—	
2464 [97]	—	*		—	—	—	
2540 [100]	—	*		—	—	—	+34 [1.339]
2667 [105]	*	*		*	—	—	−18 [.709]
2845 [112]	*	*		*	—	—	
3048 [120]	*	*		*	—	—	
3150 [124]	—	—		*	—	—	
3251 [127.992]	*	*		*	*	—	+38 [1.496]
3404 [134]	—	—		*	—	—	−21 [.827]
3658 [144.015]	*	*		*	*	—	
4013 [157.992]	—	*		*	*	—	
4115 [162.007]	—	*		*	*	—	+43 [1.693]
4394 [172.992]	—	*		*	*	—	−24 [.945]
4572 [180]	—	*		*	*	—	
4953 [195]	—	*		*	*	—	
5334 [210]	—	*		*	*	*	+49 [1.929]
6045 [237.992]	—	—		*	*	*	−28 [1.102]
6807 [267.992]	—	—		*	*	*	+56 [2.205]
7569 [297.992]	—	—		*	*	*	−32 [1.26]
8331 [327.992]	—	—		*	*	*	+65 [2.559]
9093 [357.991]	—	—		*	*	*	−37 [1.457]
9855 [387.991]	—	—		—	*	*	
10617 [417.991]	—	—		—	*	*	+76 [2.992]
12141 [477.991]	—	—		—	*	*	−43 [1.693]
13665 [537.991]	—	—		—	*	*	+89 [3.504]
15189 [597.991]	—	—		—	*	*	−50 [1.968]
16713 [657.991]	—	—		—	*	*	+105 [4.134]
							−59 [2.323]

Bracketed values give inch equivalents.

Note: 'V' belts are graded in length. The grade unit is 2.5 in inside (or pitch) lengths 610–1168, 5 in lengths 1219–1778, 7.5 in lengths 1905–1168, 10 in lengths 2464–4013, 12.5 in lengths 4115–6807, 15 in lengths 7569–9093, and 17.5 in lengths 9855–16713. Suffix 50 means the actual length is equal to the specified standard length. Lesser value indicates that the actual length is less than the standard. Higher value of the suffix indicates that the actual length is more than the specified standard length.

B 1016–49 means that the B section (17 × 11) belt has length, one unit (50–49) less than the standard length. The unit is 2.5 for lengths 610–1168. It means the belt is shorter by 2.5, i.e.

Actual length = 1016 − 2.5 = 1013.5

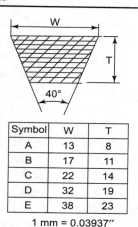

Symbol	W	T
A	13	8
B	17	11
C	22	14
D	32	19
E	38	23

1 mm = 0.03937"

Fig. 2.5(a): Standard 'V' belts cross-section

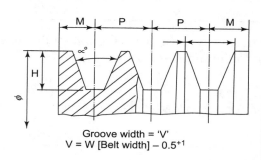

Groove width = 'V'
V = W [Belt width] − 0.5^{+1}

Belt	ϕ	$\alpha°$	H	M	P
A	73 – 81 – 32° 82 – 125 – 34° > 125 – 38°		12	+ 2 − 1 10	± 0.3 15
B	126 – 134 – 32° 135 – 209 – 34° > 209 – 38°		15	+ 2 − 1 12.5	± 0.4 19
C	195 – 220 – 34° 221 – 320 – 36° > 320 – 38°		20	+ 2 − 1 17	± 0.5 25.5
D	328 393 } 34° 394 530 } 36° > 530 38°		28	+ 3 − 1 24	± 0.6 37
E	478 – 610 – 36° > 610 – 38		33	+ 4 − 1 29	± 0.7 44.5

1 mm = 0.03937"

Fig. 2.5(b): Pulley grooves for 'V' belts

Similarly, C 3404–52 means the C section (22 × 14) belt is 2 units (52−50) longer than the standard length 3,404.

Actual length = 3,402 + 2 × 10 (Unit for lengths 2,464–4,013) = 3,422.

Even if matched lengths of belts are used, there can be variations up to ± $\frac{1}{2}$ unit, i.e. ±1.25 in belt lengths 610–1168. This variation causes unequal tensions in various belts in a multi-belt drive. Shorter belts will be stretched more, to a higher tension level than longer belts.

This variation can be compensated, in power computations, in various ways. Most of the contemporary text and reference books use the length factor 'Fc' to take care of length variation. Although the approach is logical, some of the results in the tables for the length factor are ridiculous. The factor exceeds the 1.0 value for belt lengths more than 1,778, and is as high as 1.14 for belt lengths more than 3,251.

When more than one belt is used, the belts longer than the shortest, are stretched less. They will transmit lesser power than the shortest belts. None of the additional belts can transmit as much power as the most stretched, and shortest belt. The power transmitted by most of the individual belts in a multi-belt drive will be lesser than the capacity of a single belt. How can the length factor exceed one?

Machine design by Berezovsky, Chernilevsky, and Petroy gives a simpler and more logical method for length variation compensation in a multi-belt drive. This method uses the coefficient 'M' to take care of unequal tensions due to variation in belt length.

Table 2.12(b): Coefficient 'M' for multi-belt drives

No. of Belts	2–3	4–6	More than 6
M	0.95	0.9	0.85

For very long center distances, it is necessary to increase the pulley diameters, to satisfy equations 2.8 and 2.9 for maximum and minimum center distances.

Table 2.13: Recommended small pulley ø (d) for a given Center distance (C) and Transmission ratio $\frac{(D)}{(d)}$

$\frac{(D)}{(d)}$	1	2	3	4	5	6–9
$\frac{C}{d}$	1.5	2.4	3	3.8	4.5	5.1–7.6

The center distance (C) for given belt length can be found from the following equation:

$$C = 0.25 \left[L_p - 1.57(d + D) + \sqrt{(L_p - 1.57(d + D))^2 - 2(D - d)^2} \right] \quad \textbf{(Eqn. 2.11)}$$

As V belts wedge in pulley grooves, they transmit more power than flat belts. Lesser tension necessarily decreases the load on the shaft and bearings. 'V' belts provide a more compact drive than flat belts. However, higher deformation of V belts makes them less efficient (92%) than flat belts (97%). Also, the shafts used for 'V' belt transmission should be parallel.

Flat belts are more convenient for drives between non-parallel shafts [**Fig. 2.6g, h, i**]. The driven shaft can be rotated in a direction, opposite to the rotation of the driving shaft, by crossing flat belt or using 2 idlers [**Fig. 2.6d, e, f**]. The thinner, rectangular section of flat belts also enables usage of smaller pullies. The width of flat belts varies from 20 to 500 and the thickness from 3 to 13.5. The pullies are crowned, made bigger at the center of the width, by making them convex [**Fig. 2.7a**] or conical, to facilitate centralizing the flat belts on ungrooved pullies. The pullies are made about 10% wider than the belt width. As there is a tension-prone arrangement, the pullies have wider tolerances, ranging from ± 0.5 for 40ϕ, to ± 10 for 2000 ϕ. Even the width of the pullies have wide tolerances from ±1 for 20 width, to ± 3 for 315 width.

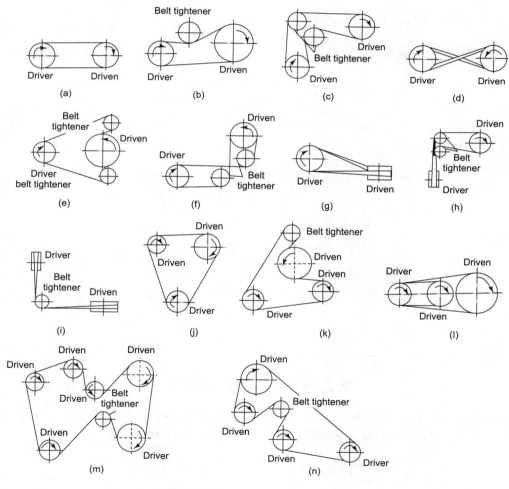

Fig. 2.6: Belt drive configurations

Standard pulley diameters: 40, 45, 50, 56, 63, 71, 80, 90, 100, 112, 125, 140, 160, 180, 200, 224, 250, 280, 315, 355, 400, 450, 500, 560, 630, 710, 800, 900, 1,000, 1,120, 1,250, 1,400, 1,600, 1,800, 2,000. [1mm = 0.03937″]

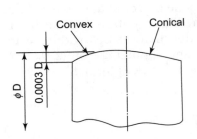

Fig. 2.7(a): Crowning flat belt pulleys

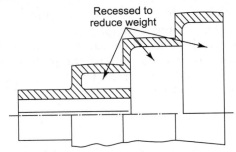

Fig. 2.7(b): Cone pulley for lathe

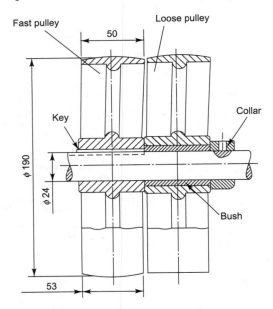

Fig. 2.7(c): Fast and loose pulley drive

Std. pulley widths: 20, 25, 32, 40, 50, 63, 71, 80, 90, 100, 112, 125, 140, 160, 180, 200, 224, 250, 280, 315, 355, 400, 450, 500, 560, 630. [1mm = 0.03937″]

Table 2.14(a): Crown for small pullies [Fig. 2.7a]

ϕ From to	40 [1.57″] / 112 [4.41]	125 [4.92″] / 140 [5.51″]	160 [6.3″] / 180 [7.09″]	200 [7.87″] / 224 [8.82″]	250 [9.84″] / 280 [11.02″]	315 [12.4″] / 355 [13.98″]
Crown 'h'	0.3 [.012″]	0.4 [.016″]	0.5 [.02″]	0.6 [.024″]	0.8 [.032″]	1.0 [.039″]

For bigger pulley diameters, the width of the pulley is also taken into account.

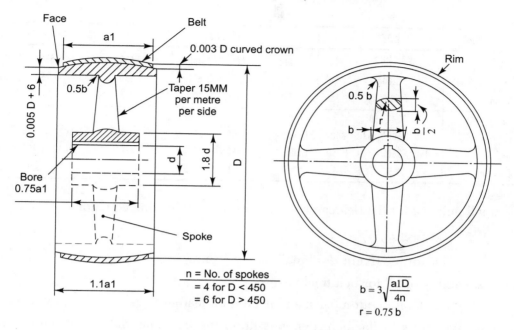

Fig. 2.7(d): Cast iron pulley

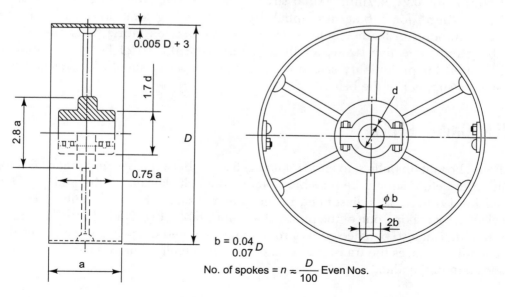

Fig. 2.7(e): Mild steel pulley

Table 2.14(b): Crown for bigger Pullies

Pulley Width	125		140–180			180–100			
Pulley φ	400 / 1,000		400 / 450	500 / 1250	1,400 / 1,600	400 / 450	500 / 560	630 / 1,250	1,400 / 1,600
Crown h	1.0		1.2	1.5	2.0	1.2	1.5	2.0	2.5

Pulley width	224–250					above 280					
Pulley φ	400 / 450	500 / 560	630 / 710	800 / 1,250	1,400 / 1,600	400 / 450	500 / 560	630 / 710	800 / 900	1,000 / 1,250	1,400 / 1,600
Crown h	1.2	1.5	2.0	2.5	3.0	1.2	1.5	2.0	2.5	3.0	3.5

1mm = 0.03937″

The small pulley φ for flat belt can be found from the following equation.

$$d = 53 - 65\sqrt[3]{T}$$ (Eqn. 2.12)

T = Torque (Nm); d = Small pulley φ (mm)

Check suitability of 'd' for belt thickness (t)

$d \geq 30t$ for cotton belts; and $d \geq 35t$ for leather belts (Eqn. 2.13)

The effective stress in flat belt, depends upon the belt material and the initial stress. Cotton belt effective stress ranges from 0.153 kg/mm² for 0.16 kg/mm² initial stress, to 0.183 kg/mm² for 0.24 kg/mm² initial stress. The effective stress of leather belts ranges from 0.18 kg/mm² (for 0.16 kg/mm² intial stress) to 0.247 kg/mm² (for 0.24 kg/mm² initial stress). But the joint in leather belts cause jerks. Rubberized fabric belts are suitable for 0.2–0.26 kg/mm² effective stress, for initial stress from 0.16–0.24 kg/mm². These values should be used for preliminary selection. The stress capacity should be checked with the manufacturer/supplier of the belts.

2.3 TENSIONING BELTS

Belts should be taut on pullies, to ensure suitable friction between belt and pulley. But for easy installation of belt, it should be possible to make the belt slacken on the pulleys. Thick 'V' belts call for 1% to + 3% adjustment, in the center distance between pullies. This can be accomplished by making one of the pulley shafts adjustable [**Fig. 2.8 a, b, c**]. The slackening and tensioning can also be effected by providing a weighted or spring loaded idler [**Fig. 2.8 d, f**]. The idler reduces the transmission capacity, marginally. The reduction depends upon the disposition of the idler.

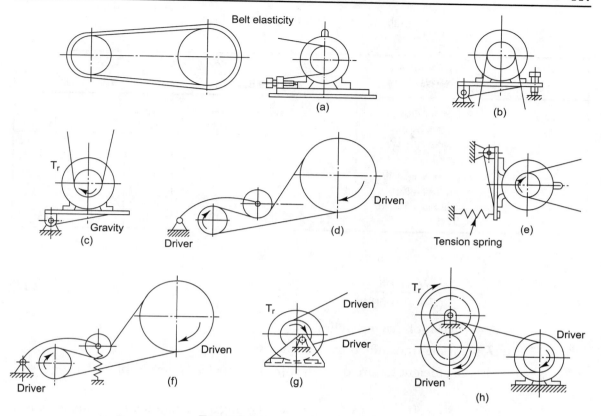

Fig. 2.8: Belt tensioning arrangements

Table 2.15: Correction for idler position

Idler position	Slack side	Tight side
Inside belt loop	0	0.1
Outside belt loop	0.1	0.2

The idler correction factor is usually added to the service factor, which depends upon the daily running time (hrs), the type of driving unit, and the type of duty.

Table 2.16: Service factor for belts

Duty	Machine	Driving unit					
		Normal torque			High torque		
		Up to 10 hrs.	10–16 hrs.	Over 16 hrs.	Up to 10 hrs.	10–16 hrs.	Above 16 hrs
Light	Light and Medium M/C tools, pumps, compressors up to 7.5 kW	1	1.1	1.2	1.1	1.2	1.3
Medium	Medium to Heavy M/C Tools Positive Displacement Pumps, Punches, Shears.	1.1	1.2	1.3	1.2	1.3	1.4

Pn = Power transmitted by 'n' belts (kW)

$$= \eta \times \frac{NTAVMn}{(S+I)1000} \qquad \text{(Eqn. 2.14)}$$

N = R.P.M. of driving pulley

T = Torque transmitted by one belt (kgm)

A = Coefficient for arc of contact [α] for smaller pulley [Table 2.8]

$$\alpha = \left[180 - 60\,\frac{D-d}{C}\right]$$ where D and d are big and small pulley diameters and C the center distance. 'A' from Table 2.8

V = Velocity coefficient from Table 2.9

M = Multi-belt drive factor from Table 2.12(b)

n = No. of belts

S = Service factor from Table 2.16

I = Idler correction I from Table 2.15

η = Efficiency of transmission = 92% for V belts

97% for flat belts

Example 2.6: Design belt drives between shafts I and II for a 2.2 kW transmission system, shown in **[Fig. 2.3c]** that would run for about 8 hrs. every day.

Solution: From the figure the motor speed = 700 R.P.M.

The intermediate speeds are 1,400 and 500 R.P.M., and the spindle speeds are 2,000, 1,400, 1,000, 710, 500, and 355 R.P.M.

Assuming a single 'V' belt transmission between shafts I and II and arc of contact = 180°; $\eta = 100\% = 1.0$, $N = 700$ (at motor); $A = 1$, assume $V = 1$, for preliminary torque computations. From Table 2.16, for a daily 8 hrs. run, service factor $S = 1$, for normal torque light-duty drive (without idler).

From Eqn. 2.14:

$$Pn = 2.2 = \frac{700 \times T \times 1 \times 1 \times 1}{1 \times 1{,}000}$$

$$\therefore T = \frac{2.2 \times 1000}{700} = 3.143 \text{ kg M}$$

$$= 3{,}143 \text{ kg mm}$$

From Table 2.10 'A' size belt with 6 kgm torque capacity, would be suitable.

Note: Calculating pulley diameters first, for higher transmission ratio saves time.

$$\text{Transmission ratio} = i_{\text{I Iih}} = \frac{1{,}400}{700} = 2$$

As the intermediate shaft has a higher speed (1,400), it will carry the smaller pulley, i.e. $\phi 90$ [Table 2.10]. The motor shaft will have a bigger pulley.

$$\therefore d_1 = 90;\ D_1 = \frac{1{,}400}{700} \times 90 = 180$$

While using stepped pullies the sum of the small (d) and big (D) diameters should be kept equal, to facilitate the usage of the same belt length.

$$d_1 + D_1 = d_2 + D_2 = d_3 + D_3 \qquad \textbf{(Eqn. 2.15)}$$

Transmission ratio for 500 R.P.M., on intermediate shaft II = $i_{\text{I II 1}}$

$$i_{\text{I II 1}} = \frac{500}{700} = \frac{5}{7}$$

$$\therefore d_2 + D_2 = d_2 + \frac{7}{5} d_2 = d_1 + D_1 = 90 + 180 = 270$$

$$\therefore d_2 = \frac{270}{2.4} = 112.5 \text{ and } D_2 = 112.5 \times 1.4 = 157.5$$

From Eqn. 2.7:

$$C = \frac{1.5 D}{\sqrt[3]{i}} = \frac{1.5 \times 157.5}{\sqrt[3]{\dfrac{5}{7}}} = 264.3$$

From Eqn. 2.11:

$$C = 264.3 = 0.25 \{L_p - 1.57(112.5 + 157.5) + \sqrt{\{L_p - 1.57(112.5 + 157.5)\}^2 - 2(157.5 - 112.5)^2}$$

$$= 0.25 \{L_p - 423.9 + \sqrt{L_p^2 - 847.8 L_p + 179691.21 - 4050}\}$$

$$= 0.25 \{L_p - 423.9 + \sqrt{L_p^2 - 847.8 L_p + 175641.21}\}$$

$$1057.2 - L_p + 423.9 = \sqrt{L_p^2 - 847.8 L_p + 175641.21}$$

$$1481.1 - L_p = \sqrt{L_p^2 - 847.8 L_p + 175641.21}$$

Squaring both sides,

$$2193657.2 - 2962.2 L_p + L_p^2 = L_p^2 - 847.8 L_p + 175641.21$$

$$2193657.2 - 175641.21 = (-847.8 + 2962.2) L_p$$

$$L_p = \frac{1478942.4}{2114.4} = 954.4$$

From Eqn. 2.10

$$\text{Inside length} = L_i = L_p - P$$

From Table 2.11 for A belt $P = 35$

∴
$$L_i = 954.4 - 35 = 919.4$$

Reference to Table 2.12 shows that A914 belt can be used for the drive.

$$\text{Actual } C = 0.25\{(914 + 35) - 423.9 + \sqrt{[949 - 423.9]^2 - 4050}\}$$

$$= 261.5$$

Now, let us take other factors into account. The problem of arc of contact will be acute for the highest transmission ratio, i.e. 2.

$$\alpha = 180 - 60 \frac{D - d}{C}$$

$$= 180 - 60 \frac{180 - 90}{262.8} = 159.45°, \text{ for smaller pulley}$$

From Table 2.8, $A = 0.95$ for 160° Arc

Velocity at the smaller pulley $= \dfrac{\pi dn}{60{,}000}$

$$= \dfrac{\pi \times 90 \times 1400}{60{,}000} = 6.6 \text{ m/sec.}$$

From Table 2.9, for 5 m/sec. $V = 1.03$
$V \approx 1.02$ for 6.6 m/sec.
From Eqn. 2.14

$\therefore$
$$P_n = \eta \times \dfrac{NTAVMn}{(S+I)\,1000}$$

$$= \dfrac{0.92 \times 700 \times 3.143 \times 0.95 \times 1.02 \times 1}{(1+0)\,1{,}000}$$

$$= 1.96 \text{ kW}$$

The power capacity can be increased by placing an idler within the belt loop, and pushing the slack side. The idler increases the arc of contact, and the power capacity. However, an idler outside the belt loop or/and on the taut side, decreases power capacity by increasing the service factor (Tables 2.15, 2.16).

Let us use flat belts for the three-stepped pulley, for the next stage [**Fig. 2.4b**].

The para following Eqn. 2.13 states that rubberised fabric belts can be used for 0.2–0.26 kg/mm^2 effective tension.

From Eqn. 2.12:

$$d = 53 - 65\sqrt[3]{T} \quad (T \text{ in Nm})$$

$$= 53 - 65\sqrt[3]{31.43} = 167.2 - 205.1$$

Nearest std. pulley diameters are 160 and 200.
For ϕ 160 pulley on the intermediate shaft,

$$\text{Tensile force on belt} = \dfrac{T}{d/2} = \dfrac{3143}{160/2} = 39.28 \text{ kg}$$

$$\text{Belt cross sect area} = \dfrac{39.28}{0.2} = 196.4 \text{ mm}^2$$

From Eqn. 2.13, $d \geq 30\,t$

$\therefore$
$$t \approx \dfrac{d}{30} = \dfrac{160}{30} = 5.33 \approx 5$$

$$\text{Belt width} = \dfrac{196.4}{5} = 39.28 \approx 40$$

$$i_{\text{II III h}} = \frac{2{,}000}{1{,}400} = 1.428 \left(\text{or } \frac{710}{500} = 1.42\right)$$

$$i_{\text{II III l}} = \frac{1{,}000}{1{,}400} = 0.7143 \left(\text{or } \frac{355}{500} = 0.71\right)$$

For $i_{\text{II III h}}$,

$$D_1 = 1.42 \times 160 = 227.2$$
$$(d_1 + D_1) = 160 + 227.2 = 387.2)$$
$$d_2 + D_2 = 2d_2 = 387.2$$
$$d_2 = D_2 = 193.6$$

for $i_{\text{II III l}}$,
$$D_3 + d_3 = D_3 + 0.71\, D_3 = 387.2$$

$\therefore$
$$D_3 = \frac{387.2}{1.71} = 226;\ d_3 = 0.71 \times 226 = 160.76$$

From Eqn. 2.7:

$$C = \frac{1.5D}{\sqrt[3]{i}} = \frac{1.5 \times 227.2}{\sqrt[3]{1.42}} = 303.2$$

$$\approx 303$$

$$\text{Arc of contact} = 180 - 60\,\frac{227.2 - 160}{303} = 166.7$$

From Table 2.8,
$$A \approx 0.967$$

Linear speed at ϕ 160 pulley, at 1,400 R.P.M.

$$= \frac{\pi \times 160 \times 1400}{60 \times 1000} = 11.72 \text{ m/sec.}$$

From Table 2.9, for 11.7 m/sec, $V = 0.98$

No. of belts = 1; $M = 1$, $n = 1.0$, $s = 1$ [from Table 2.16]

$$T = \frac{2200}{1400} = 1.57 \text{ kgm}$$

$$P_n = \eta \times \frac{NT\,AV\,Mn}{(S+I)1000}$$

$$= 0.97 \times \frac{1{,}400 \times 1.571 \times 0.967 \times 0.98 \times 1 \times 1}{(1+0)\,1{,}000}$$

$$= 2.02$$

It should be noted that flat belts transmit more power than 'V' belts, due to a 5% higher efficiency.

From Table 2.14, crown values for φ 160, 193.6, and 227.2 pullies are 0.5, 0.6, and 0.6, respectively. Std. pulley width = 50.

Example 2.7: Convert a single V belt drive (A914), transmitting 1.96 kW at 700 R.P.M., in Example 2.6, into a multi-belt drive of '0' size 'V' belt.

Solution: From Table 2.10, maximum recommended torque for '0' belt = T = 3 kgm, whereas the torque to be transmitted = 3.143 kgm.

Using two belts of '0' size (from Table 2.12), for 2 belts, $M = 0.95$ and $n = 2$. Other factors (A, V, S, I) and R.P.M. 700 (N) remain unchanged, i.e. $\eta = 0.92$; $N = 700$, $T = 3.0$ (Table 2.10); $A = 0.95$; $V = 1.02$; $S = 1$; $T = 0$

$$\therefore \quad P_n = \eta \times \frac{NTAVMn}{(S+I)1000}$$

$$= \frac{0.92 \times 700 \times 3.0 \times 0.95 \times 1.02 \times 0.95 \times 2}{(1+0) \times 1,000}$$

$$= 3.557 \text{ kW}$$

This value gives the capacity of two 'O' size belts. Acutal power to be transmitted is only 2.2 kW.

2.4 TIMER BELTS

These have moulded cogs on the inside [**Fig. 2.9**], which act as flexible internal teeth, by engaging with the axial grooves (teeth) in mating pullies. Although the drive is almost positive, there is the possibility of some lag due to belt flexibility. It's like a flexible gear/chain drive.

Timer belts can transmit power up to 1,000 kW at linear speeds from 5–50 m/sec. The transmission ratio can be as high as 12, and the efficiency is higher (98%) than in flat belts. Made from oil resistant rubber, timer belts can operate even in pool of oil.

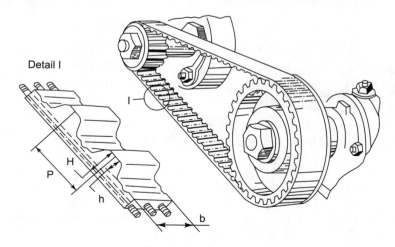

Fig. 2.9: Timer belt

Timer belts are specified by module (m). The following table gives the important specifications of the standard timer belts.

Table 2.17: Standard timer belts

	Specifications	Module m(mm)		
		3	5	7
1	Torque 'T' (kgm)	0.24 [1.77 Lbsft]	0.49 [3.61 Lbsft]	1.9 [14.01 Lbsft]
2	Permissible tensile (tangential) force/mm of width (kg)	1.0 [2.2 Lbs]	3.5 [7.72 Lbs]	4.5 [9.92 Lbs]
3	Min no. of teeth (cogs) on pulley (n)	14–20	18–24	22–36
4	Belt width 'b' (mm)	12.5–25 [.492"/.984"]	25–50 [.984"/1.97"]	50–80 [1.97"–3.15"]
5	Overall thickness 'H' (mm)	4.0 [.157"]	6.0 [.236"]	8.0 [.315"]
6	Cog pitch 'p' (mm)	9.42 [.371"]	15.71 [.618"]	21.99 [.866"]
7	Cog depth 'h' (mm)	1.8 [.071"]	3.0 [.188"]	4.2 [.165"]
8	Belt length range in no. of cogs (nb)	36–160	45–140	45–125

One of the pullies is provided with a 1.5–4 high flange, to prevent axial slip of the belt.

$$\text{Timer belt pulley P.C.D.} = d_p = mn \quad \textbf{(Eqn. 2.16)}$$

m = module (mm); n = No. of teeth (grooves)

Transmission ratio should not exceed 6 $\left(\text{or be less than } \dfrac{1}{6}\right)$.

Driving pullies can run without trouble for 100 to 4,000 R.P.M., even up to 5,000 R.P.M., for 5 module, and 8,000 R.P.M. for 3 module. Like gears, the grooved pullies are specified by the number of teeth (grooves) and module.

The pitch ϕ is measured at mid-thickness (above cogs). If the pulley outside is d_0,

$$\text{P.C.D.} = d_p = d_o + (H - h) \quad \textbf{(Eqn. 2.17)}$$

$$\text{Pitch distance of cogs} = p = \pi m \quad \textbf{(Eqn. 2.18)}$$

The approximate center distance can be found from the pitch ϕ (D_p) of the big pulley.

The belt length can be found using Eqn. 2.11. The length must be rounded off to the nearest integer multiple of the cog pitch p.

Example 2.8: Design timer belts transmission for 5.5 kW at 1,400 R.P.M. driver speed, and 700 driven R.P.M. The system should run continuously for 24 hours.

Solution:

$$\text{Torque} = \frac{5.5 \times 100}{1,400} = 0.393 \text{ kgm}$$

From Table 2.17, 5 module belt with a 0.49 kgm torque capacity, appears suitable. The minimum number of teeth = 18–24. Taking the higher value,

$$\text{Driver pulley } \phi = d_p = mn = 5 \times 24 = 120$$

$$\text{Driven pulley } \phi = D_p = 120 \times \frac{1,400}{700} = 240$$

From Eqn. 2.7:

$$C = \frac{1.5D}{\sqrt[3]{i}} = \frac{1.5 \times 240}{\sqrt[3]{0.5}} = \frac{360}{0.794} = 453.6$$

From Eqn. 2.11:

$$453.6 = 0.25\, L_p - 1.57\, (d + D) + \sqrt{[L_p - 1.57(D + d)]^2 - 2(D - d)^2}$$

$$= L_p - 565.2 + \sqrt{[L_p^2 - 1,130.4\, L_p + 3,19,451 - 2 \times 120^2]}$$

$$= L_p - 565.2 + \sqrt{[L_p^2 - 1,130.4\, L_p + 2,90,651]}$$

$$2379.6 - L_p = \sqrt{[L_p^2 - 1,130.4\, L_p + 2,90,651]}$$

Squaring both sides,

$$56,62,496 - 4,759.2\, L_p + L_p^2 = L_p^2 - 1130.4\, L_p + 2,90,651$$

$$L_p = \frac{56,62,496 - 2,90,651}{4,759.2 - 1,130.4} = 1,480.33$$

$$\text{No. of cogs} = \frac{L_p}{p} = \frac{1,480.33}{15.71} = 94.22 \approx 94 \text{ (integer)}$$

$$\therefore \quad \text{Belt pitch length} = 94 \times 15.71$$

$$= 1476.74$$

The belt length is usually specified by the number of pitches (cogs), i.e. 94 pitches of 5 module, in this case.

2.5 CHAIN DRIVE

Instead of belts, we can use a more compact and positive roller chain and sprockets if the linear speed is less than 12 m/sec (at the most 20 m/sec) and transmission ratio is less than 7 (Table 2.7). The flexibility of the chain makes the drive shock absorbent. The number of teeth on the smaller sprocket should not be less than 17 [preferably 21]. The number of teeth on the other sprocket can be determined from Eqn. 2.19.

A. Roller chains [Fig. 2.10c]

Chain No. ISO/DIN	Rolon	Pitch mm	Roller dia max D_r mm	Width between inner Plates min W mm	Pin Body dia max D_p mm	Plate Depth max mm	Transverse pitch P_t mm	over joint max A_1, A_2, A_3 mm	Bearing Area cm²	Weight per meter kgf	Breaking load min kgf
04–1	R628	6.0 [.236"]	4.00[.157"]	2.80	2.85[.11"]	5.45[.214"]	—	8.00[.315"]	0.07[.011in²]	0.12[.264Lbs]	300 [661 Lbs]
05B–1	R830	8.0 [.315"]	5.00[.197"]	3.10	2.31[.091"]	7.05[.277"]	—	11.10[.437]	0.11[.017]	0.18[.397]	460 [1014 Lbs]
05B–2	DR830	8.0	5.00	3.10	2.31	7.05	5.64[.222"]	15.80[.622]	0.22[.034]	0.33[.727]	800 [1764 Lbs]
05B–3	TR830	8.0	5.00	3.10	2.31	7.05	5.64	21.50[.846]	0.33[.051]	0.52[1.15]	1140[2513 Lbs]
	R940	9.525[.375"]	6.35[.25"]	4.00	3.28[.129"]	8.15[.321"]	—	14.40[.567]	0.22[.034]	0.37[.816]	900 [1984 Lbs]
06B–1	R957	9.525	6.35	5.90	3.28	8.15	—	16.40[.646]	0.28[.043]	0.41[.904]	910 [2006 Lbs]
06B–2	DR957	9.525	6.35	5.90	3.28	8.15	10.24[.403"]	26.60[1.043]	0.56[.087]	0.77[1.7]	1730[3814 Lbs]
06B–3	TR957	9.525	6.35	5.90	3.28	8.15	10.24	36.70[1.445]	0.84[.13]	1.09[2.4]	2540[5600 Lbs]
082–1	R1224	12.7[.5"]	7.75[.305"]	2.40	3.68[.145"]	9.90[.39"]	—	8.2 [.323]	0.17[.026]	0.28[.617]	1000[2205 Lbs]
081–1	R1230	12.7	7.75	3.48	3.68	9.90	—	11.00[.433]	0.21[.032]	0.30[.66]	830 [1830 Lbs]
	R1230H	12.7	7.75	3.48	3.68	9.90	—	12.10[.476]	0.22[.034]	0.35[.75]	1000[2205 Lbs]
	R1248	12.7	7.75	4.90	3.68	9.90	—	12.80[.504]	0.26[.04]	0.34[.75]	820 [1808 Lbs]
084–1	R1248H	12.7	7.75	4.90	4.09[.161"]	11.10[.437"]	—	16.30[.642]	0.36[.056]	0.58[1.28]	1600[3527 Lbs]
083–1	R1249	12.7	7.75	4.90	4.09	10.25[.403"]	—	14.40	0.32[.05]	0.49[1.08]	1200[2645 Lbs]
	R1253	12.7	7.75	5.25	3.68	10.35[.407"]	—	13.80[.543]	0.29[.045]	0.38[.838]	1400[3086 Lbs]
	R1264	12.7	7.77	6.40	3.97[.156"]	11.70[.461"]	—	15.90[.626]	0.39[.06]	0.47[1.04]	1500[3307 Lbs]
	R1252	12.7	8.51[.335"]	5.30	4.45[.175"]	11.70	—	18.10[.713]	0.39[.06]	0.58[1.28]	1820[4012 Lbs]
08A–1	R40	12.7	7.95[.313"]	8.00	3.97	11.70	—	20.90[.823]	0.44[.068]	0.69[1.52]	1410[3108 Lbs]
08A–2	DR40	12.7	7.95	8.00	3.97	11.70	14.38[.566"]	35.30[1.389]	0.88[.136]	1.20[2.64]	2820[6217 Lbs]
08A–3	TR40	12.7	7.95	8.00	3.97	11.70	14.38	49.70[1.957]	1.32[.205]	1.80[3.97]	4230[9325 Lbs]
08B–1	R1278	12.7	8.51	8.00	4.45	11.70	—	20.50[.807]	0.50[.077]	0.70[.011]	1820[4012 Lbs]

Bracketed values give equivalents in inches, inch², and kg.

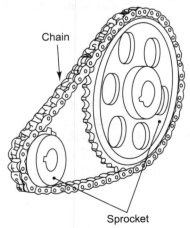

Fig. 2.10(a): Chain transmission

$$i = \frac{N_h}{N_l} = \frac{\text{No. of teeth of bigger sprocket}}{\text{No. of teeth of smaller sprocket}} = \frac{t_n}{t_l} \qquad \text{(Eqn. 2.19)}$$

t_h = No. of teeth of bigger sprocket t_l = No. of teeth on the smaller sprocket.
Center distance between the sprockets = C

$C \min = 20 \text{ p or } 1.5 \text{ D}$; $C \max = 80 \text{ P}$

For high-pulsations load, $C = 20 - 30 \text{ P}$

P = Chain roller pitch; D = Bigger sprocket ϕ.

The length of the chain should preferably be equal to an even number of pitches. The smaller the chain pitch, and more the number of sprocket teeth; the smoother will be the chain drive. Therefore, it is usually more convenient to use a duplex or even triplex chain, instead of a single big-pitch chain. **Figure 2.10b** gives the dimensions and breaking loads of standard ISO chains.

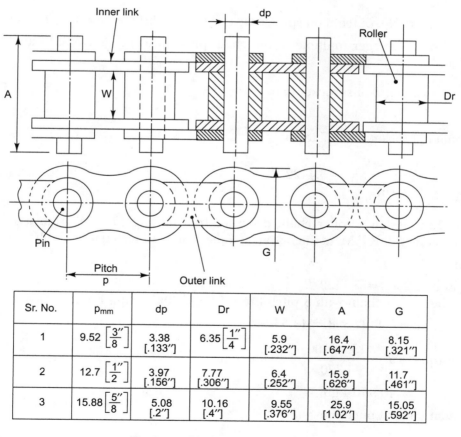

Sr. No.	P_{mm}	dp	Dr	W	A	G
1	9.52 $\left[\frac{3''}{8}\right]$	3.38 [.133"]	6.35 $\left[\frac{1''}{4}\right]$	5.9 [.232"]	16.4 [.647"]	8.15 [.321"]
2	12.7 $\left[\frac{1''}{2}\right]$	3.97 [.156"]	7.77 [.306"]	6.4 [.252"]	15.9 [.626"]	11.7 [.461"]
3	15.88 $\left[\frac{5''}{8}\right]$	5.08 [.2"]	10.16 [.4"]	9.55 [.376"]	25.9 [1.02"]	15.05 [.592"]

Fig. 2.10(b): Standard roller chains

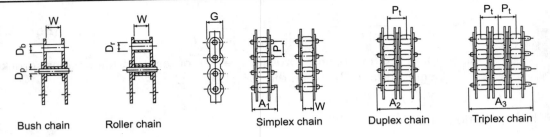

Fig. 2.10(c): Dimensions and breaking loads of ISO chains [Dimension in Fig. 2.10a]

$$\text{Chain length} = L = 2\frac{C}{P} + \frac{t_1 + t_2}{2} + \left(\frac{t_2 - t_1}{2\pi}\right)^2 \frac{P}{C} \qquad \textbf{(Eqn. 2.20)}$$

[No. of pitches]

t_1, t_2 = No. of teeth on sprockets; C = Center distance P = Chain pitch.

L should be an integer multiple of the chain pitch (preferably even), to avoid a special (cranked) link.

$$\text{Pitch } \phi \text{ of sprocket } (d \text{ or } D) = \frac{P}{\sin\frac{180}{t}} \qquad \textbf{(Eqn. 2.21)}$$

For smaller sprocket with 17 teeth,

$$d = \frac{P}{\sin\frac{180}{17}} = 5.4422 \, P \qquad \textbf{(Eqn. 2.22)}$$

Example 2.9: Design a roller chain drive for transmitting 150 kW from a 1,370 R.P.M. motor, to obtain 400 R.P.M. at the output spindle that is subjected to a highly pulsating load.

Solution:

From Eqn. 2.22, $d = 5.4422 \, P$

For smallest 05 B chain with 8 pitch (from [Fig. 2.10a], using Eqn. 2.22),

$$d = 5.4422 \times 8 = 43.53$$

$$\text{Torque} = T = \frac{150 \times 100}{1,370} = 10.95 \text{ kgm}$$

$$= 10,950 \text{ kgm}$$

Tangential force at pitch line = F

$$F = \frac{10950}{43.53/2} = 503.05 \text{ kg, i.e. much more than the breaking load 460 kg, of a single (simplex) chain.}$$

Although a duplex chain [**Fig. 2.10**] with an 800 kg breaking load will suffice, it will be safer to use a triplex chain with a 1,140 kg breaking load. Then, Factor of safety will be

$$= \frac{1140}{503} = 2.266$$

Alternatively, we can use a 06 B chain with a 9.525 pitch.

Then $d = 5.4422 \ P = 5.4422 \times 9.525 = 51.84$

$$F = \frac{10950}{51.84/2} = 422.45, \text{ i.e. much lesser than the breaking load 910 of a single (simplex) chain.}$$

For the driven sprocket teeth $= \dfrac{t_{out}}{t_{in}} = i$

$$i = \frac{N_h}{N_l} = \frac{1370}{400} = 3.425$$

$t_{out} = t_{in} \times i = 17 \times 3.425 = 58.225 \approx 58 \text{ (integer)}$

Min. Center distance $= 20P$ or $1.5D$

$20P = 20 \times 9.525 = 190.5$

From Eqn. 2.21:

$$D = \frac{P}{\sin\dfrac{180}{t}} = \frac{9.525}{\sin\dfrac{180}{58}} = 175.93$$

$\therefore \quad 1.5 \ D = 1.5 \times 175.93 = 263.9.$

Selecting the higher value of 263.9 from Eqn. 2.20.

Chain length $(L) = 2\dfrac{C}{P} + \dfrac{t_1 + t_2}{2} + \left(\dfrac{t_2 - t_1}{2\pi}\right)^2 \dfrac{P}{C}$

$$= 2\frac{263.9}{9.525} + \frac{17 + 58}{2} + \left(\frac{58 - 17}{2\pi}\right)^2 \frac{9.525}{263.9}$$

$= 55.41 + 37.5 + 1.537$

$= 94.446 \approx 95$

This can be evened to 96 pitches, to avoid a cranked link. The center distance 'C' will increase marginally.

$$96 = 2\frac{C}{9.525} + 37.5 + \frac{405.57}{C}$$

Solving the equation,

$$C = 285.36, \text{ i.e. } \frac{285.36}{9.525} = 29.95 \text{ pitches.}$$

As this is less than 30 pitches, it will be suitable for a highly pulsating load. The overall length of the chain will be $96 \times 9.525 = 914.4$ mm.

If the application calls for more compactness, it would be better to use a triplex 05 B chain. Then,

$$D = \frac{8}{\sin\frac{180}{58}} = 147.77$$

$$C = 1.5 \times 147.77 = 221.65$$

The reader can find chain length 'L' for the triplex chain and adjusted center distance, as an exercise.

2.6 SPROCKET DESIGN

Figure 2.11 shows a sprocket profile and other important proportions. The number of sprocket teeth should be limited as below

10–100 R.P.M. – 9 minimum
$17 < t < 125$ (to prevent excessive wear due to chain ride)
preferred – 21 minimum

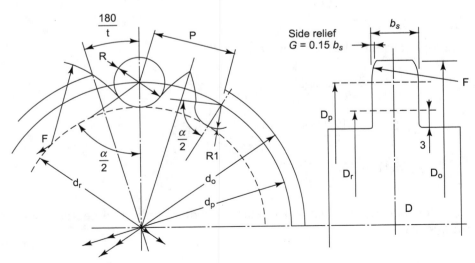

Fig. 2.11a: Sprocket tooth profile

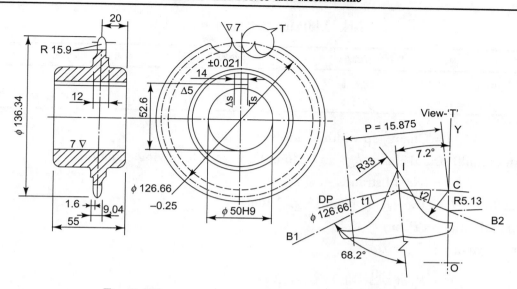

Fig. 2.11b: Twenty-five teeth sprocket for 15.875 pitch chain

$$\text{Root } \phi = d_r = d_p - R$$
$$R = \text{Roller } \phi$$
$$\text{Tip (outside) } \phi = d_o = d_r + 1.25\ P \text{ maximum}$$
$$= d_r + \left(1 - \frac{1.6}{t}\right) P \text{ minimum}$$

Roller seat radius $(R_1) = 0.505\ R$ ... minimum
$$= 0.505 R + 0.069 \sqrt[3]{R} \quad \text{... maximum}$$

Roller seat Angle $[\alpha] = 140° - \dfrac{90}{t}$ for minimum tooth gap

$$= 120 - \frac{90}{t} \text{ for maximum tooth gap}$$

Tooth flank radium $(F) = 0.12\ R\ (t+2)$ for minimum gap
$$= 0.008\ R\ (t^2 + 180) \text{ for maximum gap}$$

The width of the sprocket tooth (b_s) depends upon the chain pitch (P), and the number of strands of the chain: simplex, duplex, or triplex.

Table 2.18: Sprocket tooth width (b_s)

P	Simplex	Duplex (single)	Triplex
Up to and incl 12.7	0.93 b	0.91 b	0.88 b
More than 12.7	0.95 b	0.93 b	0.93 b

b = Chain gap width (from [**Fig. 2.10a**])

Width of multi-strand chains b_{sm} can be found using the following equation:

$$b_{sm} = (\text{No. of strands} - 1)(P_t + b_s)$$

P_t = Transverse pitch [**Fig. 2.10c**]

Tooth end radius = P min

Hub ϕ (max.) = d_n (or D_h)

$$= P \cot \frac{180}{t} - 1.05\, G - 1 - 2\, r_f$$

G = maximum link width [**Fig. 2.10 a/b**]

r_f = Fillet radius between hub ϕ and tooth end face

Measurement over rollers (MR), having diameter equal to the chain roller [ϕR] depends upon the number of teeth (t). For an even number of teeth:

$$MR_e = d + R$$

For odd number of teeth:

$$MR_O = d \cos \frac{90}{t} + R$$

2.7 TOLERANCES

Pitch, root, and tip (outside) diameters can be up to –0.25 mm for root diameters (d_r) of up to and including 127 mm. For 128 to 250 root diameters, the tolerances can be up to –0.3. Above 251 root diameter, the diameters should have h 11 tolerance.

Tooth width (b_s) can have an even wider h 14 tolerance.

Concentricity between the mounting bore and root diameter (d_r/D_r) should not exceed (0.0009 d_r + 0.08) or 0.76 mm, whichever is smaller. The same formula can be used for checking axial (face) run-out, which should not exceed 1.14 mm.

The breaking load, stated in [**Fig 2.10a**], should be divided by service factor outlined in the table below, to account for shock and daily running time, i.e. fatigue.

Table 2.19: Service factors for chains

Daily running time (hours)	Load		
	Uniform	Moderate shock	Severe shock
8	1	1.2	1.4
16	1.1	1.3	1.55
24	1.2	1.4	1.7

Table 2.20: Sprocket materials

Speed m/sec	No. of teeth t	Material
≤3	≤25	M.S.
above 600 R.P.M.	≤30	Steel hardenable to RC 40–45
	>30	Cast iron

Generally, sprockets with pilot bore less than 0.7 of hub ϕ, are available commercially. These can be bored and slotted (for keyway), to suit the mounting shaft.

2.7.1 Compensation for chain elongation

The slack resulting from sprocket wear or/and chain elongation can be countered by making the center distance between sprockets adjustable up to 1.5 times the chain pitch (P). Alternatively, an idler roller can be provided on the slack side to remove the slack. When the sprocket axes are horizontal, the slack should be on the lower side. At least 3 teeth of the idler should engage with the chain.

2.7.2 Lubrication

Feeding lubricant to the driven sprocket, between the chain link plates, enhances chain and sprocket life. Light drop lubrication of up to 14 drops per minute is adequate for chain speeds up to 4 m/sec. The drop rate should be increased to 20 drops/min for 4–7 m/sec chain speeds. Speeds above 7 m/sec call for forced feed lubrication.

Changing sprockets to change spindle speed takes a longer time than shifting a belt on a stepped (cone) pulley. But if only a few speed changes are required, we can mount sets of 2 or 3 sprockets on the driver and driven shafts, and shift the chain from one set of sprockets to the other, just like the cone pulley. The sum of the number of teeth of all the driving and driven sprockets should be equal, to facilitate usage of the same chain.

2.8 GEARING

Some mass production machines like autos, produce similar workpieces for long runs, lasting many shifts. A little extra setting time, spent on changing the pullies, sprockets, or gear does not make much difference in the overall economy. Even in machines with very long, cutting times or slower operations, such as thread-cutting, the machine setting time is only a small fraction of the total running time. Under such circumstances, slow, manual replacement of change gears is quite satisfactory. **Figure 2.12a** shows an arrangement using change gears, for effecting a change in the feed rate in lathes. The straight slot in the quadrant, facilitates center distance adjustment for the intermediate gear mating with the gear on the feed shaft. The curved slot simplifies center distance adjustment for the intermediate gear, engaging with the gear on the main spindle. The curved slot is also used for clamping the swivel quadrant in set position, by a screw and a hexagonal nut.

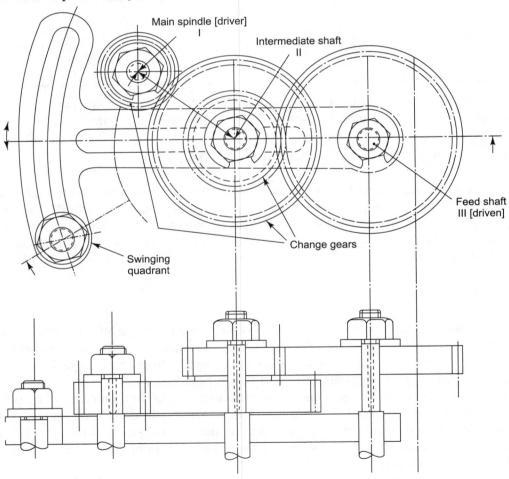

Fig. 2.12(a): Change gears and swinging quadrant for changing speed of feed shaft

The teeth of the gears engage and intermesh with the teeth of the mating gears. Spur gears have teeth on the cylindrical portion. When the teeth are parallel to the axis of rotation, the gears are called straight spur gears or simply spur gears. Making the teeth twisted with the gear axis, helical [**Fig. 2.12b**], increases the load capacity and promotes a smooth, and gradual engagement. Straight and helical spur gears are used for transmission between parallel shafts.

The teeth usually have involute flanks. The tooth profile is often modified to improve engagement. Usually, the addendum of a gear with lesser number of teeth, i.e. the pinion, is increased, while the addendum of the bigger mating gear is decreased by an equal amount.

A plane surface with gear teeth is called a rack [**Fig. 2.12e**]. Considered a gear wheel with an infinite number of teeth, rack teeth have straight angular sides (flanks), instead of the involute curve. The included angle between the angular flanks of the rack tooth is double the *pressure angle* $[\alpha]$. Usually, the twenty degrees pressure angle and forty degrees included angle, are more widely used than the $14\frac{1}{2}°$ pressure angle, mostly used for worm gears with an acme thread form.

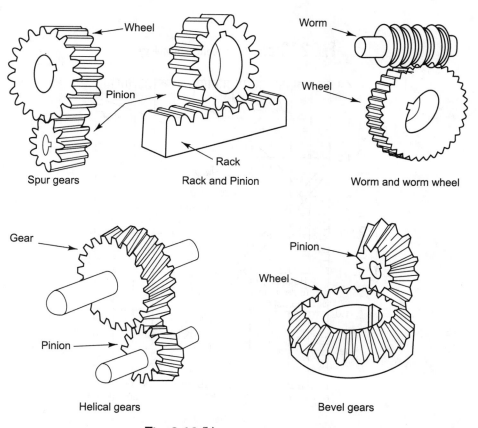

Fig. 2.12(b): Types of gear drives

The size of the tooth is specified by module (m). The other important parameter in spur gears is the pitch diameter [**Fig. 2.12f**]; the diameter where the meshing gear pitch ϕ touches tangentially, and where the tooth thickness is equal to the gap between adjacent teeth. In racks, the tooth thickness and the gap are equal at the pitch line.

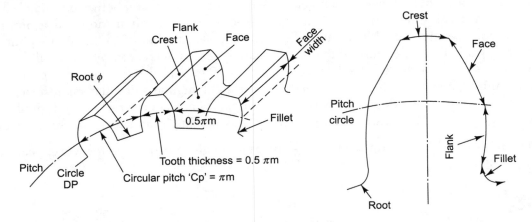

Fig. 2.12(c): Gear tooth terminology

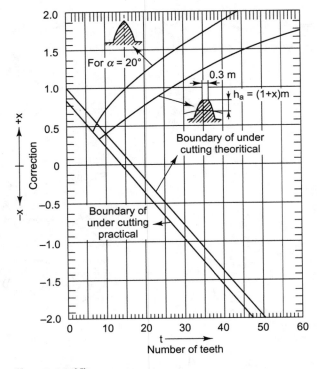

Fig. 2.12(d): Tooth thinning due to addendum correction

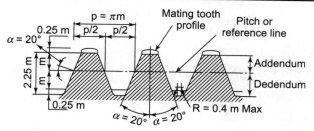

Fig. 2.12(e): Standard 20° pressure angle rack and preferred modules

I	0.6,	0.8,	1,	1.25,	1.5,	2,	2.50,	3,	4,	5,	6,	8,	10,	12,	16,	20
II	0.7,	1.75,	2.25,	2.75,	3.5,	4.5,	7,	9,	14,	18						

The pitch diameter (d_p or D_p) of any spur gear is an integer multiple of the gear module.
d_p (or D_p) $= t_m$ **(Eqn. 2.23)**

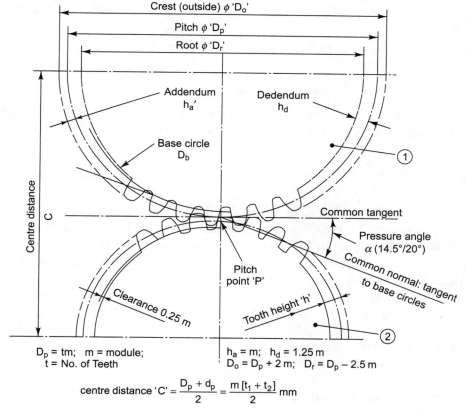

Fig. 2.12(f): Important dimensions of mating gears

d_p = pinion pitch ϕ (mm); D_p = gear pitch ϕ (mm)
t = No. of teeth; m = module (mm)

For standard, unmodified tooth profile:

Tooth height above pitch ϕ = Addendum = $h_a = m$ **(Eqn. 2.24)**

Tooth height (depth) below pitch ϕ = Dedendum = $h_d = 1.25\ m$ **(Eqn. 2.25)**

Tooth height = $h_a + h_d = h = 2.25\ m$ **(Eqn. 2.26)**

Tip clearance = $h_d - h_a = 0.25\ m$

Circular pitch = $P = \pi m$ **(Eqn. 2.27)**

Circular tooth thickness (or gap) = $0.5\ \pi m$ **(Eqn. 2.28)**

∴ Outside or tip ϕ of gear = $d_o = m(t + 2)$ **(Eqn. 2.29)**

Root ϕ of gear = $d_r = m(t - 2.5)$ **(Eqn. 2.30)**

Another important parameter is the base circle [d_b, D_b], which is the base diameter for the involute curve, used for gear flanks profile. The common tangent to the base circles of two intermeshing gears, intersects the common tangent to their pitch circles at the pressure angle (α) [**Fig. 2.12f**]

$$\text{Base circle } \phi = d_b \text{ (or } D_b) = d_p \cos \alpha \text{ (or } D_p \cos \alpha)$$

$$\alpha = \text{Pressure angle (usually 20°)}$$

$$\text{Center distance} = C = m(t_1 + t_2)\,2 \quad \textbf{(Eqn. 2.31)}$$

Variation in the center distance changes the pressure angle. The pressure angle (α) will increase if the center distance is more than the value given in Eqn. 2.31.

2.9 CORRECTION

If the number of teeth is less than 17 (for $\alpha = 20°$), the addendum [h_a] and the outside ϕ (d_o) are increased for the smaller gear, to avoid interference. This reduces the tip thickness [**Fig. 2.12d**], which should not be less than $0.3\ m$. The dedendum of the pinion is reduced by an equal amount. Thus, the outside ϕ of the pinion is increased and enlarged, without increasing the tooth depth (h). The increase as well as the decrease, depends upon the correction coefficient 'x'. It depends upon the number of teeth (t) and the transmission ratio (i). **Figure 2.13** gives correction coefficients for speed reduction ($i = 1 - 14$), while [**Fig. 2.14**] gives correction coefficients for increasing speed $\left(i = 1 - \dfrac{1}{9}\right)$.

$$\text{Correction} = mx \quad \textbf{(Eqn. 2.32)}$$

Addendum of the (smaller) pinion is increased by 'mx', while the addendum of the (bigger) gear is reduced by the same amount.

For pinion, $h_a = m(1 + x)$ (Eqn. 2.33)
For gear, $h_a = m(1 - x)$

The dedendum is adjusted by an equal amount, i.e.
For pinion, $h_f = m(1.25 - x)$
For gear, $h_f = m(1.25 + x)$

The outside and root diameters will change accordingly, i.e. for pinion:
$$d_o = d_p + 2m(1 + x) \quad \text{and}$$
$$d_r = d_p - 2m(1.25 - x)$$

For the gear:
$$D_o = D_p + 2m(1 - x) \quad \text{and}$$
$$D_r = D_p - 2m(1.25 + x)$$

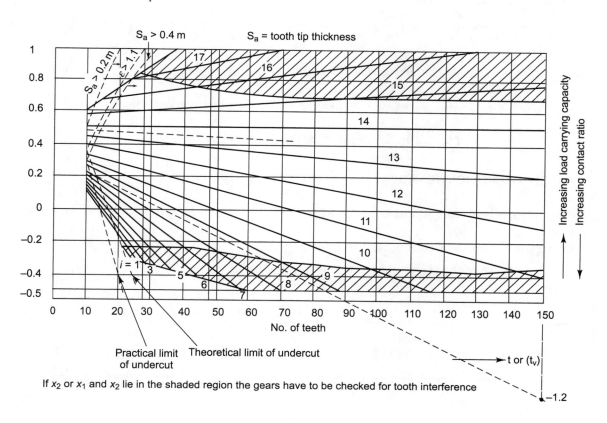

Fig. 2.13: Distribution of correction factor for reduction drives

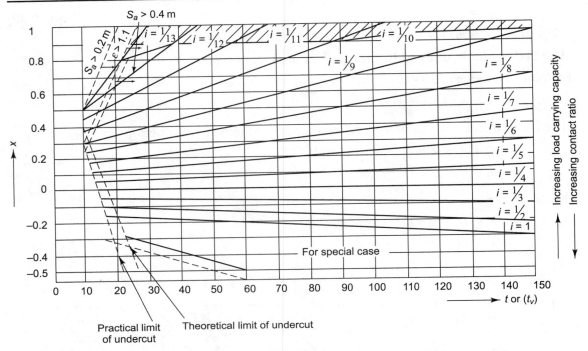

Fig. 2.14: Distribution of correction factor for step-up drives

2.9.1 Tooth thickness and bases tangent length

Tooth thickness is usually found by measuring the base tangent length W_k over 'K' (number of) teeth.

For standard, unmodified (uncorrected) teeth:

$$W_k = m \cos \alpha \, (K - 0.5) \, \pi + t \, \text{inv} \, \alpha)$$ **(Eqn. 2.34)**

α	No. of teeth (t)					
20°	12–18	19–27	28–36	37–45	46–54	55–63
14½	12–18	19–37	38–50	41–62	63–75	76–87
K	2	3	4	5	6	7

$$\text{inv. } \alpha = \tan \alpha - \alpha$$ **(Eqn. 2.35)**

$$= 0.014904 \text{ for } 20°, \text{ and } 0.00554 \text{ for } 14\tfrac{1}{2}°$$

For corrected teeth, correction coefficient 'X' must be reckoned.

$$W_{kc} = W_k + 2\, m\, X \sin \alpha \qquad \text{(Eqn. 2.36)}$$

Chordal tooth thickness Measuring the chordal tooth thickness of a single tooth, at a specific depth [**Fig. 2.15**], is easier than measuring the base tangent length. The special vernier gear tooth callipers required, are commercially available.

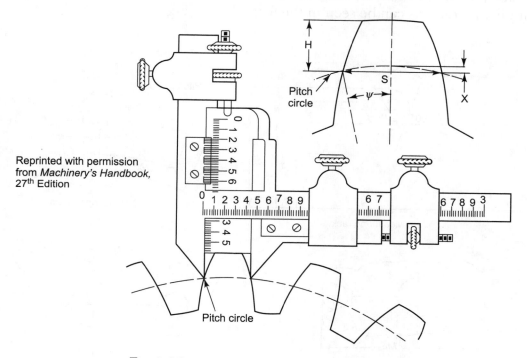

Reprinted with permission from *Machinery's Handbook*, 27th Edition

Fig. 2.15: Measuring chordal pitch by special vernier

$$\text{Chordal thickness} = \overline{S} = m_n\, t \sin \psi \qquad \text{(Eqn. 2.37)}$$

$$\psi = \frac{\pi}{2t} + \frac{2 \times \tan \alpha_n}{t_v} \qquad \text{(Eqn. 2.38)}$$

Depth at which at the tooth thickness is to be measured = H

Chordal addendum = H

$$H = m_n \left(1 + x + \frac{t_v}{2}\, (1 - \cos \psi)\right) \qquad \text{(Eqn. 2.39)}$$

m_n = normal module (measured at right angle to the tooth flanks in helical gears).

= m for straight spur gears

x = correction coefficient; t = No. of teeth; t_v = Virtual no. of teeth = $\dfrac{t}{\cos^3 \beta}$

β = Helix angle [mean]

Measurement over rollers (R) Measurement over rollers placed in tooth gaps [**Fig. 2.16**], is simple. The measurements over the gears with even and odd teeth are naturally different, as can be seen in the formulae in [**Fig. 2.16**].

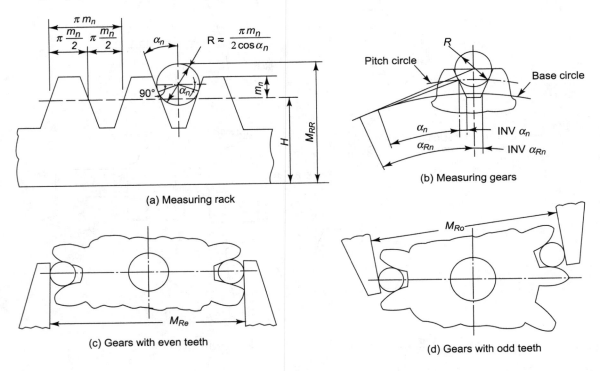

Fig. 2.16: Measuring gear teeth form over rollers

The recommended roller sizes can be found from the following equation:

$$R = \dfrac{\pi m_n}{2 \cos \alpha_n} = 1.67\, m_n \text{ for } \alpha_n = 20°$$

Figure 2.16a shows the arrangement for measuring rack tooth over rollers. The length of the rack (L) is used to find the equivalent pitch diameter (d_e)

$$d_e = \dfrac{L}{\pi}$$

Referring to [**Fig. 2.16a**],

$$M_{RR} = H + 0.5 \left\{ R\left(1 + \frac{1}{\sin \alpha_n}\right) - \frac{\pi m_n \cot \alpha_n}{2} \right\}$$

Tolerance on $M_{RR} = \delta M_{RR} = \dfrac{\delta_S^-}{2 \tan \alpha_n}$

δ_S^- can be found from [**Fig. 2.17**] for equivalent diameter (d_e) of the rack, explained above.

Figure 2.16b shows disposition of roller on spur and helical gears. It is necessary to calculate the normal pressure angle at the center of the roller (α_{Rn}), for finding the measurement over rollers.

$$\text{Inv } \alpha_{Rn} = \text{Inv } \alpha_n + \frac{R}{m_n t \cos \alpha_n} - \frac{\pi}{2t} \pm \frac{2x \tan \alpha_n}{t} \qquad \begin{array}{l} + \text{ for pinion} \\ - \text{ for gear} \end{array}$$

Absolute value of x should be used. If $[x]$ is positive radially outwards.

For helical gears it is necessary to find the transverse pressure angle (α_{Rt}) at the center of the roller.

$$\text{Inv } \alpha_{Rt} = \text{Inv } \alpha_t + \frac{R}{m_n t \cos \alpha_n} - \frac{\pi}{2t} \pm \frac{2x \tan \alpha_n}{t} \qquad \begin{array}{l} + \text{ for pinion} \\ - \text{ for gear} \end{array}$$

Figure 2.16c shows an arrangement of rollers, for gears measuring over an even number of teeth. Note that for straight spur gear, normal module (m_n), normal pressure angle (α_n), and normal roller pressure angle (α_{Rn}) are used. While dealing with helical gears, these are substituted by tansverse module (m_t), transverse pressure angle (α_t), and transverse roller pressure angle (α_{Rt}), respectively.

The tolerance on the measurements over rollers can be found using the following equations, and the tolerances on base tangent length (W_k) observed from [**Fig. 2.17**].

<u>Tolerances on measurements over rollers (δM_r)</u>

Straight spur gears	Helical gears
Even No. of teeth	
$\delta M_{Re} = \dfrac{W_k}{\sin \alpha_{Rn}}$	$\delta M_{Re} = \dfrac{W_k}{\sin \alpha_{Rt} \cos \beta b}$
Odd no. of teeth	
$\delta M_{RO} = \dfrac{W_k \cos\left(\dfrac{90}{t}\right)^\circ}{\sin \alpha_{Rn}}$	$\delta M_{Ro} = \dfrac{W_k \cos\left(\dfrac{90°}{t}\right)}{\sin \alpha_{Rt} \cos \beta b}$

βb: Helix angle at base circle; δM_R negative for external gear and positive for internal gear
W_k: Tolerance on base tangent length [**Fig. 2.17**]

Of the twelve standardized classes of precision, 5 to 10 classes are more widely used. The first criterion for the choice of 'accuracy class' is the **running speed**.

Table 2.20: Speed ranges for gear classes

Accuracy class	5	6	8	10
Peripheral speed m/s	12–20	6–12	3–6	Up to 3

The manufacturing tolerance depends on the module, and pitch circle diameter. [**Fig. 2.17**] gives the tolerances for tooth-to-tooth spacing error, chordal thickness, and base tangent length

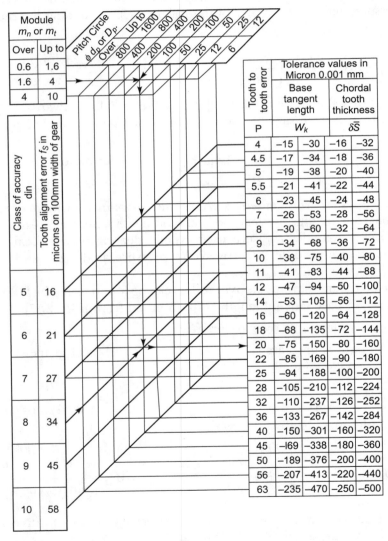

Fig. 2.17: Manufacturing tolerances for straight and helical spur gears

for 6–1600 pitch diameters of 0.6 to 10 modules. **Table 2.21** outlines the permissible errors for various modules and pitch diameter ranges. **Table 2.22** lists the tolerances for the center distances, for various accuracy classes of spur gears. **Table 2.23a** specifies the popular materials for gears.

2.10 GEAR DESIGN

Module is selected to suit the bending load. For straight spur gears:

$$m = 12.6 \sqrt[3]{\frac{T}{Y f_b \psi_m t_1}}$$ (Eqn. 2.40)

T = Torque (kg cm)

Y = Form factor (from Table 2.26)

f_b = Permissible bending stress (kg/cm^2) (Table 2.23)

ψ_m = Width factor = $\dfrac{b}{m}$ (Table 2.25)

t_1 = No. of teeth of pinion (smaller gear)

For Helical Gears:

$$m = 11.5 \cos \beta \sqrt[3]{\frac{T}{Y_e f_b \psi_m t_1}}$$ (Eqn. 2.41)

Y_e = Form factor based on equivalent No. of teeth t_v

$t_v = \dfrac{t}{\cos^3 \beta}$; β = Helix angle, generally 8°–45°.

The module can also be found from the graph in **[Fig. 2.18a]**. The intersection of the (dotted) horizontal line for 1,313 kg cms torque, intersects the curve for, f_b = 1,400 kg/cm^2, at a point which gives module 3.6 ≈ 3.5 (std).

$$\text{For straight spur gears, } f_b = \frac{10(i \pm 1)T}{CmbY}$$ (Eqn. 2.42a)

$$\text{For Helical spur gears, } f_b = \frac{7(i \pm 1)T}{Cbm_n Y_v}$$ (Eqn. 2.42b)

Note in both these equations $i \geq 1$. For in gears pair,

$$i = \frac{\text{No. of teeth of (bigger) gear}}{\text{No. of teeth of (smaller) pinion}} \text{ ; the sign + for external gears}$$

$$- \text{ for internal gears}$$

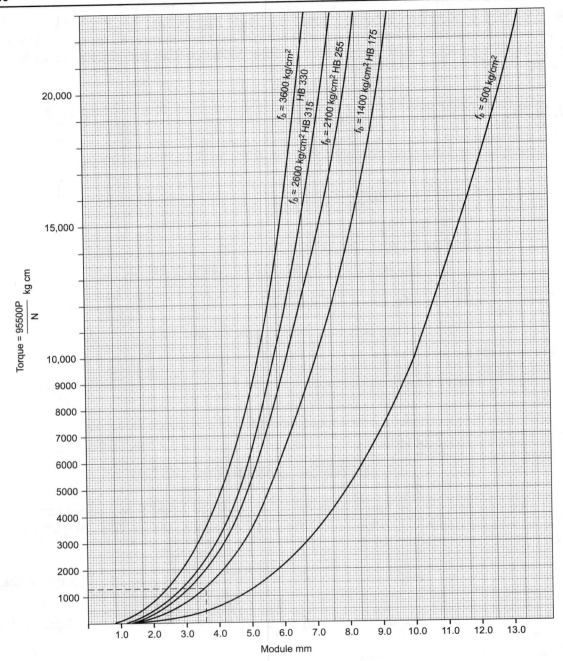

Fig. 2.18(a): Finding module from torque bending stress [f_b]

Note: The torques & modules are for 20 teeth pinion with $b = 6$ m i.e., $\psi_m = 6$; $P =$ Power in kW

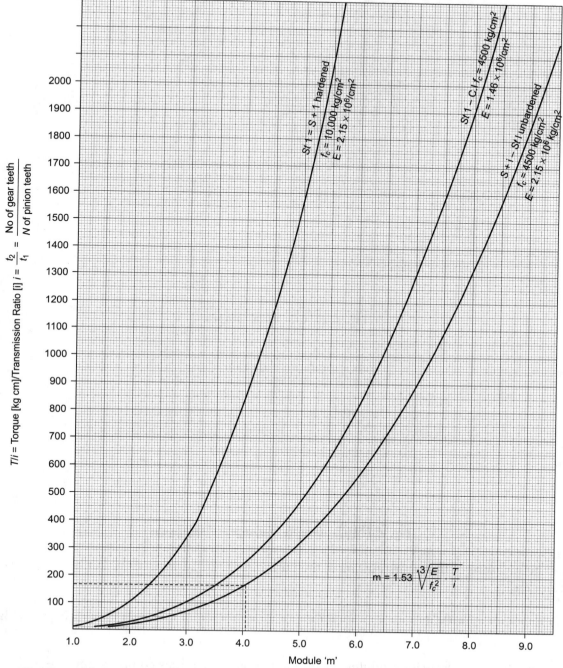

Fig. 2.18(b): Module of gear for long life [> 10^7 cycle] from torque [T] and transmission ratio [i] fro gear width [b] = 0.15 C [centre distance]

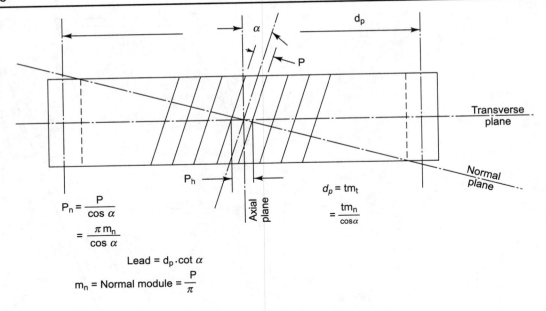

Fig. 2.18(c): Helical spur gears

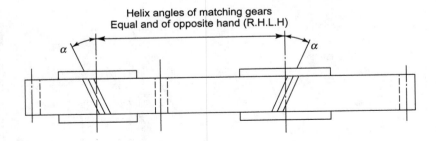

Fig. 2.18(d): Opposite helix angles of mating gears

The measure of the wear resistance, the compressive or surface stress can be found from Eqn. 2.43(a).

$$f_c = 0.74 \left\{ \frac{i \pm 1}{C} \right\} \sqrt{\frac{i \pm 1}{ib} ET} \qquad \textbf{(Eqn. 2.43a)}$$

Note that E = Elasticity modulus for pinion and gear materials combination = $\dfrac{2 E_1 E_2}{E_1 + E_2}$

The value of the surface stress should be less than the permissible (Table 2.23) limits. For helical spur gears,

$$f_{ch} = 0.7\left\{\frac{i\pm 1}{C}\right\}\sqrt{\frac{i\pm 1}{ib}ET} \quad \text{The sign + for external gears} \qquad \textbf{(Eqn. 2.43b)}$$

Example 2.10: Design a pair of gears for speed-up (or step-up) transmission ratio $(i) = 8$, power to be transmitted = 1.1 kW. Driver R.P.M. range 80–500 Revs/min

Solution: Power = 1.1 kW; At the slowest level of 80 R.P.M., the torque will be the highest.

$$\therefore \quad T = \frac{1.1 \times 955.0}{80} = 13.13 \text{ kgm}$$

$$= 1313 \text{ kgcm}$$

From Table 2.25 for unground gears, $\psi_m = \dfrac{b}{m} = 6$

From Table 2.23 for C40, [continued on page 153]

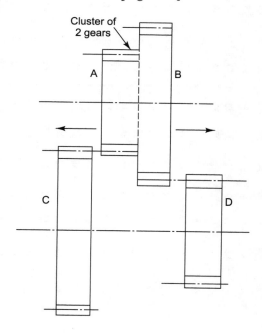

Fig. 2.19: Cluster of 2 gears for getting 2 speeds

Table 2.21: Permissible pitch (Pt), Base pitch (Pe), Profile (Pf), Cumulative pitch (P_T), Radial runout (r), and total composite double flank (T) errors, for 5–10 accuracy classes. Unit = 0.001 mm

Class of accuracy DIN	Kind of error	Module 'm' over 0.6 up to 1.6 mm						Module 'm' over 1.6 up to 4 mm						Module 'm' over 4 up to 10 mm					
	PITCH CIRCLE DIA 'd' (mm) Over	12	25	50	100	200	400	12	25	50	100	200	400	25	50	100	200	400	800
	Upto	25	50	100	200	400	800	25	50	100	200	400	800	50	100	200	400	800	1600
5	Pt, Pe, Pf	4	4.5	5	5.5	7	8	4.5	5	5.5	6	7	9	6	7	8	9	10	12
5	P_T	14	16	18	20	25	28	16	18	20	22	25	32	22	25	25	32	36	45
5	r, T	12	14	16	18	20	22	14	16	18	20	22	25	18	20	22	25	28	32
6	Pt, Pe, Pf	6	6	7	8	9	11	6	7	8	9	10	12	9	10	11	12	14	18
6	P_T	20	22	25	28	36	40	22	25	28	32	36	45	28	32	36	40	50	63
6	r, T	18	20	22	25	28	32	20	22	25	28	32	36	25	28	32	36	40	45
7	Pt, Pe, Pf	8	9	10	11	12	16	9	10	11	12	14	18	12	14	16	18	20	25
7	P_T	28	32	36	40	50	63	32	36	40	45	50	63	40	45	50	63	71	90
7	r, T	25	28	32	36	40	45	28	32	36	40	45	50	36	40	45	50	56	63
8	Pt, Pe, Pf	11	12	14	16	18	22	12	14	16	18	20	25	18	20	22	25	28	36
8	P_T	40	45	50	56	71	80	45	50	56	63	71	90	63	63	71	80	100	125
8	r, T	36	40	45	50	56	63	40	45	50	56	63	71	56	56	63	71	80	90
9	Pt, Pe, Pf	16	18	20	22	25	32	18	20	22	25	28	36	25	28	32	36	40	50
9	P_T	56	63	71	80	100	125	63	71	80	90	100	126	80	90	100	125	140	180
9	r, T	50	56	63	71	80	90	56	63	71	80	90	100	71	80	90	100	110	125
10	Pt, Pe, Pf	25	28	32	36	40	50	28	32	36	40	45	56	40	45	50	56	63	80
10	P_T	90	100	110	125	160	180	100	110	125	140	160	200	140	140	160	180	220	280
10	r, T	71	80	90	100	110	125	80	90	100	110	125	140	100	110	125	140	160	180

Table 2.22: Tolerances for center distance [f_a]

Class of center distance accuracy DIN	center distance						
	over	6.3	16	40	100	250	630
	upto	16	40	100	250	630	1600
	Fit				J		
5	Center distance tolerance values ($\pm f_a$) in μ [microns]	9	11	14	18	22	28
6		12	16	20	25	32	40
7		18	22	28	36	45	56
8		25	32	40	50	63	80
9		36	45	56	71	90	110
10		50	63	80	100	125	160

Table 2.23: Gears materials

Material	I. S. Designation	Hardness HB core	UTS kg/cm² core	Design stress kg/cm²		
				Reversible bending f_b gear life No. of cycles		Compressive stress f_c
				< 20,000	> 10^7	
Cast Iron	Grade-20	190	2,000	900	460	5,000
	Grade-25	210	2,500	1,300	550	6,000
Direct hardening steel	C40	175	6,300	3,000	1,400	4,500
		200	6,700	3,400	1,600	
		230	7,700	3,800	1,800	
	C45	190	6,500	3,200	1,500	5,000
		200	6,700	3,400	1,600	
		230	7700	3,800	1,800	
Case hardening steel	30Ni4Cr1	445	15,500	3,200	1,500	10,000
	40Ni2CrIMo28	255	8,700	2,900	2,100	
		315	10,700	3,200	2,600	
		335	11,700	3,500	2,800	11,000
		370	12,700	3,800	3,100	
		445	15,500	4,700	3,800	
	C14	140	5,000	2,500	1,500	
	13Ni3Cr80	255	8,500	3,700	2,800	
	15Ni2CrIMo15	330	11,000	5,000	3,600	9,500
	17Mn1Cr95	240	8,000	3,500	2,600	

Table 2.24: Gear width [b] in terms of module [m] and center distance [c]

	Applications, accuracy, mounting	$\psi_m = \frac{b}{m}$	$\psi_c = \frac{b}{c}$
1	Unground gears, mounting shaft supported at one end only [canti-lever]	6–8	Upto 0.1
2	Low accuracy [9th grade], low speed [3–6m/s], mounting shaft supported at both ends	10–15	0.12
3	General purpose [7th grade], medium speed [6–10 m/s], mounting shaft supported at both ends	15–20	0.15
4	High precision [6th grade], High speed [10–30 m/s], hardness above 350 BHN, mounting shaft supported at both ends	16–24	0.3
5	Very rigid mounting, high precision [6th grade], hardness above 350 BHN, low speed [<3 m/s]	Up to 45	up to 0.8

Table 2.26: Form factor, Ψ for $\alpha = 20°$

(Ψ includes cos α term)
External pinion, wheel and internal pinion
Addendum modification coefficient, X

t or t_v	−0.6	−0.4	−0.2	0	+0.2	+0.4	+0.6	+1.0	Annulus
12	—	—	0.239	0.308	0.378	—	—	—	—
14	—	—	0.266	0.330	0.392	0.458	—	—	—
16	—	—	0.302	0.355	0.408	0.461	—	—	—
18	—	—	0.330	0.377	0.424	0.470	—	—	—
20	—	—	0.348	0.389	0.431	0.471	0.503	—	—
22	—	—	0.367	0.402	0.437	0.473	0.504	—	—
24	—	0.355	0.384	0.414	0.445	0.475	0.504	—	—
26	—	0.373	0.400	0.427	0.455	0.481	0.509	—	—
28	0.358	0.383	0.408	0.434	0.458	0.484	0.509	—	0.942
30	0.369	0.392	0.416	0.440	0.464	0.486	0.511	—	0.912
35	0.390	0.411	0.431	0.452	0.473	0.494	0.514	0.556	0.863
40	0.406	0.426	0.445	0.465	0.485	0.503	0.523	0.562	0.825
45	0.415	0.434	0.452	0.471	0.490	0.509	0.528	0.565	0.795
50	0.423	0.441	0.459	0.477	0.495	0.513	0.531	0.568	0.769
60	0.440	0.456	0.474	0.490	0.507	0.523	0.540	0.574	0.731
80	0.457	0.471	0.485	0.499	0.512	0.526	0.541	0.569	0.688
100	0.464	0.481	0.490	0.505	0.517	0.530	0.542	0.566	0.660
150	0.492	0.499	0.508	0.515	0.521	0.531	0.543	0.556	0.620
300	0.517	0.519	0.521	0.521	0.523	0.533	0.544	0.526	0.585
RACK				0.550					

$f_c = 45$ kg/mm² $= 4{,}500$ kg/cm²

If we take a pinion having 20 teeth, the gear will have $20 \times 8 = 160$ teeth. From Table 2.26, form factor for a standard 20 teeth pinion $Y_{20} = 0.389$, and for 160 teeth $Y_{160} \approx 0.516$

From Eqn. 2.40:

$$m = 12.6 \sqrt[3]{\frac{T}{Y f_b \psi_m t_1}}$$

$$= 12.6 \sqrt[3]{\frac{1313}{0.389 \times 1400 \times 6 \times 20}} = 3.4 \approx 3.5$$

The module can also be found from the graph in **[Fig. 2.18a]**. The (dotted) horizontal line for 1,300 kg/cm torque intersects the curve for 1,400 kg/cm², bending stress at the point, which gives 3.6 mm as the module. It can also be noted that the intersection of the torque line with the 500 kg/cm² stress curve gives 5 mm module.

For a gears having 20 and 160 teeth, standard center distance $= C$.

$$C = \frac{m(t_1 + t_2)}{20} = \frac{3.5(20 + 160)}{20} = 31.5 \text{ cms}$$

$$b = \psi_m \times m = 6 \times 3.5 = 21 \text{ mm} = 2.1 \text{ cms}$$

For external straight spur gears, the sign for bracketed term is +

$$f_b = \frac{10(i+1)T}{C m b Y} = \frac{10(8+1)1313}{31.5 \times 3.5 \times 2.1 \times 0.389}$$

$$= \frac{1{,}18{,}170}{90.06}$$

$$= 1{,}312$$

Let us check up the gear for compressive stress.

For the elastic modulus for the combination of C.I. and steel gear and pinion

$$E = \frac{2 E_1 E_2}{E_1 + E_2} = \frac{2 \times 2.15 \times 10^6 \times 1.1 \times 10^6}{2.15 \times 10^6 + 1.1 \times 10^6} = 1.46 \times 10^6$$

From Eqn. 2.43a:

$$f_c = 0.74 \left\{ \frac{i \pm 1}{C} \right\} \sqrt{\frac{i \pm 1}{ib} ET}$$

$$= 0.74 \left(\frac{8+1}{31.5} \right) \sqrt{\frac{8+1}{8 \times 2.1} \times 1.46 \times 10^6 \times 1313}$$

$$= 6775.5 \text{ i.e.} > 4500 \text{ for C40 (Table 2.23)}$$

We can find the center distance (C) suitable for obtaining 4500/cm² compressive stress from the above [2.43a] equation. Using the present gear width center distance ratio, $\psi_c = \dfrac{2.1}{31.5} = 0.0666$

$$f_c = 4500 = 0.74 \left(\frac{i+1}{C}\right) \sqrt{\frac{i+1}{ib} ET}$$

$$= \frac{0.74 \times 9}{C} \sqrt{\frac{9}{8 \times 0.0666 C} \times 1.46 \times 10^6 \times 1313}$$

∴ $C = 41.4$ cms

$$= \frac{m}{20} (20 + 160)$$

∴ $m = 5.3 \simeq 5.5$ the next higher std. module. The module can also be found from the graph in [Fig. 2.18b] (dotted line). It is 4.05, but for $b = 0.15 C$.

Example 2.11: Design a pair of gears for transmitting 11 kW at 1:2 speedup, at driver R.P.M.s ranging from 200–630.

Use steel for both the gears. Desired Life = 10^7 cycles.

Solution: Power = 11 kW = 11,000 Watts/sec.

Min R.P.M. = 200 for driver

$$\text{Max torque } (T) = \frac{11 \times 95{,}500}{200}$$

$$= 5252.5 \text{ kgcm}$$

We will design the gears for wear-load first. From Table 2.23a, let us choose C 40 (175 BHN) for wheel and C45 (230 BHN) for the pinion.

For 175 BHN steel and 10^7 cycles life permissible,

$f_b = 1400$ kg/cm², and $f_c = 4500$ kg/cm².

For 230 BHN steel with same life, $f_b = 1800$ and $f_c = 5000$. The pinion hardness should be at least 50 BHN higher than the wheel hardness, to balance the wear of the two gears. Higher R.P.M. wears out the pinion more than the lower R.P.M. gear.

From Eqn. 2.43a for external gears:

$$f_c = 0.74 \left(\frac{i+1}{C}\right) \sqrt{\frac{i+1}{ib} ET}$$

For steel to steel combination $E = 2.15 \times 10^6$/cm²

Width (b) = $0.15 C$ for medium speeds [Table 2.24]

$$f_c = 0.74\left(\frac{2+1}{C}\right)\sqrt{\frac{2+1}{2\times 0.15C}\times 2.15\times 10^6 \times 5{,}252.5}$$

$$= \frac{746028}{C^{1.5}}$$

Taking $f_c = 4{,}500$, $C^{1.5} = \dfrac{746028}{4{,}500}$

$\therefore \quad C = 30.18/\text{cm} = \dfrac{m(t_1 + t_2)}{20}$

Taking $t_2 = 40$ and $t_1 = = 20$ for $i = 2$

$$\frac{m(20+40)}{20} = 30.18$$

$\therefore \quad m = \dfrac{30.18 \times 20}{60} \approx 10.06 \approx 10$, i.e. very high

$b = 0.15\,C = 0.15 \times 30.18 = 4.53$ cms ≈ 45 mm

The gears to be used must be ground. Therefore, the unhardened, unground, and unlapped gears of the 9th accuracy class cannot be used.

The gear sizes can be reduced by using considerably hardened gears. Referring to Table 2.23, it can be observed that case hardening steels can allow the usage of a much higher (9,500 kg/cm^2) value for compressive stress (f_c). For $f_c = 9{,}500$.

$$C = \sqrt[1.5]{\frac{746028}{9{,}500}} = 18.34$$

Module $= \dfrac{18.34 \times 20}{60} = 6.11 \approx 6.5$

We can use 13 Ni 3 cr $\underline{80}$ with 255 BHN for the gear, and 15 Ni 2 cr 1 Mo $\underline{15}$ with 330 BHN for the pinion, to provide the required difference in hardness [Min. 50 BHN] between the pinion and the gear.

2.10.1 Contact ratio (E)

For minimizing noise and impact, the meshing gears must always be in contact. The following pair of teeth should engage before the preceding pair separates. Contact ratio (E) is a measure of such contact. It can be measured along the profile (E_α) or along the face (E_β). The latter, being of little significance, is disregarded.

$$E_\alpha = \frac{\text{Angle of contact of one tooth}}{\text{Angular pitch, i.e. } \frac{2\pi}{t}}$$

For straight spur gears with t_1 and t_2 teeth,

$$E_\alpha = 1.88 - 3.2 \left(\frac{1}{t_1} \pm \frac{1}{t_2} \right)$$

Note: The minus sign is used for the internal gear. Contact ratio (E_α) must be equal to or more than 1.2.

Example 2.12: Find the profile contact ratio for pairs of straight spur gears with a standard profile [with no correction] and no. of teeth:

(a) 20 and 160
(b) 20 and 40.

Solution:

(a) $E_\alpha = 1.88 - 3.2 \left(\frac{1}{20} + \frac{1}{40} \right) = 1.64$

(b) $E_\alpha = 1.88 - 3.2 \left(\frac{1}{20} + \frac{1}{160} \right) = 1.67$

For helical gears, with helix angle (β):

$$E_{\alpha h} = \frac{E_\alpha}{\cos \beta} = \frac{1.88 - 3.2 \left(\frac{1}{t_1} \pm \frac{1}{t_2} \right)}{\cos \beta}$$

Note: The minus sign is used for internal gears.

As $\cos \beta$ will always be less than one, it is obvious that an increase in the helix angle will increase the contact ratio up to 45°.

2.10.2 Helical gears

Slanting teeth with a rotation axis, facilitates gradual engagement. The module in the plane normal to the tooth flanks is called the normal module (m_n). The inclination of the tooth with the axis changes the module in the plane square to the axis. This transverse module (m_t) determines the pitch diameter (d_p), the transverse pressure angle (α_t), and lead (L). The angle with the axis (β), the helix angle, determines the transverse and overall dimensions.

$$\text{Transverse module } m_t = m_n \left(\frac{1}{\cos \beta} \right) \quad \text{(Eqn. 2.44)}$$

$$\text{Pitch } \phi \, (d_p) = tm_t = tm_n \left(\frac{1}{\cos \beta} \right) \quad \text{(Eqn. 2.45)}$$

$$\text{Base circle } \phi \, (d_b) = d_p \cos \alpha_t$$

$$\tan \alpha_t = \tan \alpha_n \times \left(\frac{1}{\cos \beta}\right) \quad \text{(Eqn. 2.46)}$$

α_t = Transverse pressure angle

$$\text{Virtual no. of teeth} = t_v = \frac{t}{\cos^3 \beta} \quad \text{(Eqn. 2.47)}$$

Center distance also changes with the pitch diameters.

$$C = \frac{d_p + D_p}{20} = \frac{(t_1 + t_2)m_n}{20 \cos \beta} \text{ cms} \quad \text{(Eqn. 2.48)}$$

$$\text{Lead} = L = \pi d_p \cot \beta \quad \text{(Eqn. 2.49)}$$

Note: C is in cms while d_p, D_p, and L are in mms.

It should be remembered that the helix angle does not affect addendum or dedendum, which correspond with the normal module (m_n), the module of the cutting tool: milling, shaping, or hobbing cutters.

Example 2.13: Find the important dimensions for a 40 teeth, 2 module gear, with a 10° helix angle. The width (b) can be taken as 7 times the module. Normal pressure angle (α_n) = 20°.

Solution: From Eqn. 2.44:

$$m_t = m_n \frac{1}{\cos \beta} = 2 \cdot \frac{1}{\cos 10°} = 2.0308$$

$$d_p = t\, m_n \cdot \frac{1}{\cos \beta} = 40 \times 2 \times \frac{1}{\cos 10°}$$

$$= 81.23$$

Addendum = m_n = 2

Dedendum = 1.25 m_n = 2.5

Outside $\phi = d_0 = 81.23 + 2 \times 2 = 85.23$

Root $\phi = d_r = 81.23 - 2 \times 2.5 = 76.23$

Transverse pressure angle = α_t

$$\alpha_t = \tan^{-1}\left(\tan \alpha_n \frac{1}{\cos \beta}\right) = \tan^{-1}\left(\tan 20 \cdot \frac{1}{\cos 10°}\right)$$

$$= 20.283° = 20°17'$$

$$\text{Virtual no. of teeth} = t_v = \frac{40}{\cos^3 10°} = 41.88$$

$$\approx 42$$

$$\text{Lead} = L = \pi \, d_p \cot \beta$$

$$= \pi \times 81.23 \cdot \cot 10°$$

$$= 1447.84 \text{ mm}$$

2.10.3 Designing gears for given center distance

When space constraints call for a change in the center distance, addendum modifications can be used to accomplish the same.

The following parameters define such changes:
1. Working (required) center distance C' cms
2. Average no. of teeth, 't_m' $= \dfrac{t_1 + t_2}{2}$
3. Center distance modification coefficient $= K_c$

$$K_C = \frac{(C' - C)10}{m} \qquad \text{(Eqn. 2.50)}$$

$$C = \text{Std. center distance} = \frac{m(t_1 + t_2)}{20} \text{ cms}$$

4. Working pressure angle $= \alpha'$

$$\cos \alpha' = \frac{C \cos \alpha}{C'} \qquad \text{(Eqn. 2.51)}$$

5. Sum of modification coefficients $(x_1 + x_2)$ for pinion (x_1) and gear (x_2). The value of the sum can be found from Table 2.27.
6. Addendum reduction coefficient $= K_{ar}$

$$K_{ar} = x_1 + x_2 - K_c \qquad \text{(Eqn. 2.52)}$$

7. Addendum $= h_a = (1 + x - K_{ar}) \, m,$ \qquad (Eqn. 2.53)

where $x = \dfrac{x_1 + x_2}{2}$

\qquad\qquad\qquad\qquad\qquad\qquad\qquad\qquad\qquad\qquad\qquad (Eqn. 2.54)

8. Dedendum $= h_d = (1.25 - x) m$
9. Outside $\phi = d_o = (t + 2 + 2x - 2K_{ar}) m$
10. Root $\phi = d_r = (t - 2(1.25 - x)) \, m_n$
11. Contact ratio (E) is the angle of action of one of the gears, divided by its angular pitch. A decrease in the center distance, decreases the pressure angle $[\alpha]$ and contact ratio.

$$E = \frac{1}{2\pi}\left(t_1(\tan\theta_1 - \tan\alpha') + t_2(\tan\theta_2 - \tan\alpha')\right) \quad \text{(Eqn. 2.55)}$$

$\theta_1 = \cos^{-1}\left\{\dfrac{d_b}{d_o}\right\}$; $\theta_2 = \cos^{-1}\left\{\dfrac{D_b}{D_o}\right\}$ D_b = Base circle dia of bigger gear
D_o = Outside dia of bigger gear

Contact ratio (E) for helical spur gears is calculated in the transverse plane, with transverse module (m_t) and mean virtual number of teeth (t_{mv}).

$$t_{mv} = \frac{t_{v1} + t_{v2}}{2}$$

$$\cos\alpha' = \frac{C}{C'}\cos\alpha$$

As $C = \dfrac{d_p + D_p}{20}$,

$$\frac{x_1 + x_2}{t_m} = \frac{\text{inv}\alpha' - \text{inv}\alpha}{\tan\alpha}, \quad \frac{K_c}{t_m} = \frac{\cos\alpha}{\cos\alpha'} - 1$$

Table 2.27: Modification coefficients for various working pressure angles $[\alpha']$

α'	$\dfrac{K_c}{t_m}$	$\dfrac{x_1 + x_2}{t_m}$	α'	$\dfrac{K_c}{t_m}$	$\dfrac{x_1 + x_2}{t_m}$
16° 00	−0.02244	−0.02036	40′	−0.00210	−0.00208
10′	−0.02162	−0.0197	50′	−0.00105	−0.00105
20′	−0.02079	−0.01902	20° 00′	0.00000	0.00000
30′	−0.01995	−0.01833	10′	0.00106	0.00107
40′	−0.0191	−0.01762	20′	0.00214	0.00216
50′	−0.01824	−0.01689	30′	0.00323	0.00326
17° 00′	−0.01737	−0.01615	40′	0.00432	0.00439
10′	−0.01649	−0.0154	50′	0.00543	0.00554
20′	−0.01560	−0.01463	21° 00′	0.00655	0.00670
30′	−0.01471	−0.01384	10′	0.00768	0.00789
40′	−0.01380	−0.01304	20′	0.00882	0.00910
50′	−0.01288	−0.01222	30′	0.00997	0.01033
18° 00′	−0.01195	−0.01138	40′	0.01113	0.01158
10′	−0.01101	−0.01053	50′	0.01230	0.01285
20′	−0.01006	−0.00966	22° 00′	0.01349	0.01415
30′	−0.00910	−0.00878	10′	0.01469	0.01546
40′	−0.00813	−0.00787	20′	0.01589	0.01680
50′	−0.00715	−0.00695	30′	0.01711	0.01816
19° 00′	−0.00616	−0.00601	40′	0.01835	0.01954
10′	−0.00516	−0.00506	50′	0.01959	0.02094
20′	−0.00415	−0.00408	23° 00′	0.02084	0.02238
30′	−0.00313	−0.00309	10′	0.02211	0.02383

α'	$\dfrac{K_c}{t_m}$	$\dfrac{x_1+x_2}{t_m}$	α'	$\dfrac{K_c}{t_m}$	$\dfrac{x_1+x_2}{t_m}$
20'	0.02339	0.0253	40'	0.05154	0.06015
30'	0.02468	0.0268	50'	0.05309	0.06218
40'	0.02598	0.02833	27° 00'	0.05464	0.06424
50'	0.0273	0.02987	10'	0.05621	0.06633
24° 00'	0.02862	0.03145	20'	0.05780	0.06845
10'	0.02996	0.03304	30'	0.05939	0.07060
20'	0.03131	0.03466	40'	0.06100	0.07278
30'	0.03267	0.03631	50'	0.06263	0.07499
40'	0.03405	0.03798	28° 00	0.06427	0.07724
50'	0.03544	0.03968	10	0.06592	0.07951
25° 00'	0.03684	0.04141	20'	0.06759	0.08182
10'	0.03825	0.04316	30'	0.06927	0.08816
20'	0.03967	0.04494	40'	0.07097	0.08654
30'	0.04111	0.04674	50'	0.07268	0.08894
40'	0.04256	0.04857	29° 00	0.07440	0.09138
50'	0.04403	0.05043	10'	0.07614	0.09385
26° 00'	0.04550	0.05232	20'	0.07790	0.09636
10'	0.04700	0.05423	30'	0.07966	0.09889
20'	0.04850	0.05618	40'	0.08145	0.10147
30'	0.05001	0.05815	50'	0.08325	0.10408
			30°00'	0.08506	0.10673

$$\cos \alpha'_t = \frac{d_b + D_b}{20 C'} \qquad \text{(Eqn. 2.56a)}$$

Note: C' is in cms, while d_b and D_b are in mm

$$\frac{K_C}{t_m} = \frac{1}{\cos \beta}\left\{\frac{\cos \alpha_t}{\cos \alpha'_t} - 1\right\} \qquad \text{(Eqn. 2.56b)}$$

Example 2.14a: Design gears for 6.8 cms center distance and 1:3 step-up transmission. $\alpha = 20°$.

Solution: For 1:3 ratio $t_1 = 3t_2$

$$\frac{(t_1+t_2)m}{20} = \frac{(3t_2+t_2)m}{20} = \frac{4t_2 m}{20}$$

If we take $\quad t_2 = 20;\quad \dfrac{4 \times 20 \times m}{20} = 6.8$

$\therefore \qquad m = 1.7$

If we take $\quad m = 1.75$,

Machine Tool Drives and Mechanisms

$$C = \frac{4 \times 20 \times 1.75}{20} = 7 \text{ cms}$$

For $\quad C' = 6.8$

From Eqn. 2.50
$$K_C = \frac{(6.8 - 7)10}{1.75} = -1.1428$$

$$t_m = \frac{60 + 20}{2} = 40$$

$$\alpha' = \cos^{-1}\left\{\frac{C}{C'}\cos\alpha\right\} = \cos^{-1}\left\{\frac{7}{6.8}\cos 20°\right\}$$
$$= 14.6858 \approx 14° \, 41.15'$$

$$x_1 + x_2 = t_m\left\{\frac{\text{inv}\,\alpha' - \text{inv}\,\alpha}{\tan\alpha}\right\} = 40\left\{\frac{\text{inv}\,14°41' - \text{inv}\,20°}{\tan 20°}\right\}$$

$$= 40\left(\frac{0.0057647 - 0.014904}{0.36397}\right) = -1.0044$$

$$x = \frac{-1.0044}{2} = -0.5022$$

Referring to **[Fig. 2.14]**, mark point for $t_m = 40$ and $x = -0.5$.

Draw a dotted line in the mean position (Refer to **[Fig. 2.14]**). The intersection points of this line with the vertical lines for 20 and 60 teeth, give pinion correction coefficient (x_1) and gear correction coefficient (x_2), as -0.33 and -0.57, respectively.

$$K_{ar} = x_1 + x_2 - K_c = -1.0044 - (-1.1428)$$
$$= 0.1384$$
$$h_{ap} = m(1 + x_1 - K_{ar}) = 1.75\,(1 + (-0.33) - 0.1384)$$
$$= 0.93103$$
$$d_p = mt_2 = 1.75 \times 20 = 35$$
$$d_o = 35 + (2 \times 0.9303) = 36.86$$
$$d_b = d_p \cos\alpha' = 35 \cos 14.686° = 33.856$$
$$D_p = mt_1 = 1.75 \times 60 = 105$$
$$D_b = 105 \cos 14.686° = 101.57$$
$$h_{ag} = m(1 + x_2 - K_{ar}) = 1.75(1 - 0.57 - 0.1384)$$
$$= 0.9303$$
$$D_o = 105 + (2 \times 0.5103) = 106.021$$

$$\theta_1 = \cos^{-1}\left(\frac{d_b}{d_o}\right) = \cos^{-1}\left(\frac{33.856}{36.86}\right) = 23.29°$$

$$\theta_2 = \cos^{-1}\left(\frac{D_b}{D_o}\right) = \cos^{-1}\left(\frac{101.57}{106.021}\right) = 16.661°$$

$$E = \frac{1}{2\pi}(t_1(\tan\theta_1 - \tan\alpha') + t_2(\tan\theta_2 - \tan\alpha'))$$

$$= \frac{1}{2\pi}[20(\tan 23.29° - \tan 14.686) + 60(\tan 16.661 - \tan 14.686))]$$

$$= \frac{1}{2\pi}(20(4302 - 0.2621) + 60(0.0373)$$

$E = 0.89 \le 1.2$, and therefore, very low.

The contact ratio can be increased by increasing the number of teeth. This will entail reducing the module. If we change the module to 1.25, then,

$$C = 6.8 = \frac{4t_2 m}{20} = 6.8 = \frac{4 \times t_2 \times 1.25}{20}$$

$\therefore$ $t_1 = 27.2 \approx 27$ an integer

$t_2 = 3 \times 27 = 81$

$$C = \frac{m(t_1 + t_2)}{20} = \frac{1.25(81 + 27)}{20} = 6.75 \text{ cms}$$

$$\alpha' = \cos^{-1}\left(\frac{C}{C'}\cos\alpha\right) = \cos^{-1}\left(\frac{6.75}{6.8}\cos 20°\right) = 21.127° = 21°7.63'$$

$$t_m = \frac{81 + 27}{2} = 54$$

Referring to Table 2.27, for $21.127° = 21°7.63'$, $\dfrac{x_1 + x_2}{t_m} = 0.0076$

$$x_1 + x_2 = 54 \times (+0.0076) = 0.4104$$

$$x = \frac{x_1 + x_2}{2} = 0.2052$$

Refer to [Fig. 2.14]. Mark point for 54 teeth and +0.2 correction. The correction line should be between $i = \dfrac{1}{7}$ and $\dfrac{1}{6}$. The intersection points of the vertical lines, for 27 teeth and 81 teeth with the line, give the pinion and gear correction as: $x_1 = +0.16$ and $x_2 = +0.23$, respectively.

From Table 2.27 for $21°7.63'$; $\dfrac{K_c}{t_m} = +0.007412$

$$K_c = 54 \ (+ 0.00741) = 0.4002$$
$$K_{ar} = x_1 + x_2 - K_c = (+0.4104) - (0.4002)$$
$$= 0.010152$$
$$h_{ap} = 1.25 \ (1 + (+0.16) - 0.01015) = 1.437$$
$$d_o = (1.25 \times 27) + (2 \times 1.437) = 36.624$$
$$d_b = d_p \cos \alpha' = 33.75 \cos 21.127° = 31.48$$
$$\theta_1 = \cos^{-1}\left(\frac{d_b}{d_o}\right) = \cos^{-1}\left(\frac{31.48}{36.624}\right) = 30.121°$$
$$D_o = (1.25 \times 81) + 2(1 + 0.23 - 0.01015) \ 1.25$$
$$= 104.30$$
$$D_b = D_p \cos 21.127° = 94.444$$
$$\theta_2 = \cos^{-1}\left(\frac{D_b}{D_o}\right) = \cos^{-1}\left(\frac{94.444}{104.3}\right) = 25.108$$
$$E = \frac{1}{2\pi} \ \{20(\tan 30.121 - \tan 21.127) + 60 \ (\tan 25.108 - \tan 21.127)\}$$
$$= \frac{1}{2\pi} \ (3.875 + 4.931) = 1.401.$$

The contact ratio is more than the recommended minimum of 1.2.

Example 2.14b: Find out the necessary tooth modifications if 20 teeth and 40 teeth, 1.75 module reduction gears have to operate at a center distance of 54, instead of the standard center distance of 52.5. Teeth pressure angle $(\alpha) = 20°$.

Solution: Center distance modification coefficient (K_c) should be found first.

From Eqn. 2.50:
$$K_C = \left(\frac{C' - C}{m}\right) = \frac{54 - 52.5}{1.75}$$
$$= 0.857$$

From Eqn. 2.51:

Pressure angle $\alpha' = \cos^{-1}\left(\frac{C}{C'} \cos \alpha\right) = \cos^{-1}\left(\frac{52.5}{54} \cos 20°\right)$
$$\alpha' = 24°$$

From Table 2.27 for $\alpha' = 24°$,

$$\frac{x_1 + x_2}{t_m} = 0.03145; \text{ and } \frac{K_c}{t_m} = 0.02862$$

$$t_m = \frac{20 + 40}{2} = 30$$

$$\frac{x_1 + x_2}{30} = 0.03145; \text{ Thus, } x_1 + x_2 = 30 \times 0.03145 = 0.9435$$

$$d_b = d_p \cos 24° = 20 \times 1.75 \times 0.913$$
$$= 31.974$$

$$D_b = D_p \cos 24° = 40 \times 1.75 \times 0.913$$
$$= 63.91$$

$$\therefore \quad x_1 + x_2 = 0.9435$$

$$x = \frac{0.9435}{2} = 0.4717$$

$$\frac{K_C}{30} = 0.02862$$

$$K_C = 0.8586.$$

This approximately matches the value 0.857, found earlier.

Refer to **[Fig. 2.13]**. Following the vertical line for t_m, i.e. 30 teeth, we mark the intersection point for $x = 0.47$ and draw a dotted line, as shown in the mean position. The intersection of the dotted line with the line for 20 teeth gives us $x_1 = 0.47$ for the pinion. Similarly, for a 40 teeth gear, $x_2 = 0.45$.

$$\therefore \quad K_{ar} = x_1 + x_2 - K_C = 0.9435 - 0.858$$
$$= 0.0855$$

Pinion addendum $[h_a] = (1 + x_1 - K_{ar}) \, m = (1 + 0.47 - 0.0855)1.75$
$$= 2.423$$

$$d_o = 20 \times 1.75 + 2 \times 2.423$$
$$= 39.846$$

$$\theta_1 = \cos^{-1}\left(\frac{d_b}{d_o}\right) = \cos^{-1}\left(\frac{31.974}{39.846}\right) = 36.636°$$

Gear addendum $= (1 + x_2 - K_{ar}) \, m = [1 + 0.45 - 0.0855] \, 1.75 = 2.38$

$$D_o = 40 \times 1.75 + 2 \times 2.38$$
$$= 74.776$$

$$\theta_2 = \cos^{-1}\left(\frac{D_b}{D_o}\right) = \cos^{-1}\left(\frac{63.91}{74.776}\right)$$
$$= 31.275°$$

$$E = \frac{1}{2\pi} [t_1 \tan \theta_1 - \tan \alpha'] + t_2 (\tan \theta_2 - \tan \alpha')$$

$$= \frac{1}{2\pi} (20 (\tan 36.636° - \tan 24°) + 40 (\tan 31.275° - 0.4452)$$

$$= \frac{1}{2\pi} (5.968 + 6.487)$$

$$= 1.98, \text{ i.e. more than 1.2, which is necessary for a smooth, impact-free transmission.}$$

It can be concluded from the preceding two examples that,
1. When the center distance is reduced below the standard ($C' < C$), the addendum of the gears is decreased, while the dedendum is increased. Reduction in the center distance, reduces the pressure angle, and the contact ratio.
2. When the center distance is increased above standard ($C' > C$), the addendum of the gears is increased, while the dedendum is decreased. Increase in center distance also increases the pressure angle, and the contact ratio.

Increase in addendum thins the tooth and decreases the tip thickness. The tip thickness should not be less than 0.3 m. Referring to [**Fig. 2.12d**], it can be noted that the point for 0.47 correction and 20 teeth, lies well below the line for 0.3 m tip thickness.

2.10.4 Gear mounting considerations

Gears can be mounted centrally (symmetrically), or non-centrally (non-symmetrically) on shafts supported at both ends. Alternatively, they can be mounted on overhanging shafts (cantilevers), or on ball or roller bearings. In all the cases, deflection of the shaft due to load, affects distribution of load. **Table 2.28** outlines the bending fatigue factors K_{fb}, for various mountings, tooth hardness, and width/pcd $\left(\dfrac{b}{d_p}\right)$ ratio for gears. The lesser the gear width, the lesser will be the bending fatigue. **Table 2.29** gives us the dynamic load factor K_d, for various precision grades, hardness, and peripheral speeds.

These factors can be taken into account by increasing the load proportionately, while determining the module. If under the mounting method, the torque is termed T_m and power kW_m, then,

$$T_m = K_{fb} K_d T; \text{ and } kW_m = K_{fb} K_d kW$$

These values of T_m should be used in Eqns. 2.40 and 2.41 for module, or while using the graphs in [**Figs. 2.18a and b**].

To smoothen the operation of the gears affected by mounting constraints, one of the gears, usually the bigger one, is made less hard than the mating pinion. The softer gear wears out gradually, at the contact points, to increase the contact area between the mating gears, with the passage of time and gradual wear.

Table 2.28: Bending fatigue factor K_{fb}, for various mounting styles

Schematic illustration	Gear arrangement with respect to bearings	Tooth surface hardness, BHN	$\psi_{hd} = b/d_p$						
			0.2	0.4	0.6	0.8	1.2	1.6
	On cantilevers, ball bearings	Up to 350	1.16	1.37	1.64	—	—	1.6
		Over 350	1.33	1.70	—	—	—	—
	On cantilevers, roller bearings	Up to 350	1.10	1.22	1.38	1.57	—	—
		Over 350	1.20	1.44	1.71	—	—	—
	Symmetrical	Up to 350	1.01	1.03	1.05	1.07	1.14	1.26
		Over 350	1.02	1.04	1.08	1.14	1.30	—
	Non-symmetrical	Up to 350	1.05	1.10	1.17	1.25	1.42	1.61
		Over 350	1.09	1.18	1.30	1.43	1.73	—

d_p = Pitch of pinion

Table 2.29: Dynamic load factors [K_d] for various accuracy grades and speeds

Degree of accuracy of gear drive	Tooth surface hardness, BHN	Peripheral speed, m/sec					
		1	2	4	6	8	10
7	up to 350	1.08/1.03	1.16/1.06	1.33/1.11	1.50/1.16	1.62/1.22	1.80/1.27
	over 350	1.04/1.01	1.06/1.02	1.09/1.03	1.13/1.05	1.17/1.07	1.22/1.08
8	up to 350	1.10/1.03	1.20/1.06	1.38/1.11	1.58/1.17	1.78/1.23	1.96/1.29
	over 350	1.04/1.01	1.06//1.02	1.12/1.03	1.16/1.05	1.21/1.05	1.26/1.08
9	up to 350	1.13/1.04	1.28/1.07	1.50/1.14	1.72/1.21	1.98/1.28	2.25/1.35
	over 350	1.04/1.01	1.07/1.02	1.14/1.04	1.21/1.06	1.27/1.08	1.34/1.09

Note: The figures in the numerators refer to spur gears, and those in the denominators, to helical gears.

Even with the simple swinging quadrant [**Fig. 2.12**], a wide range of feed rates can be obtained by replacing the gears on the main spindle I, intermediate shaft II, and feed shaft III.

The usage of two-gear trains permits transmission ratios, ranging from $\frac{1}{96}$ to 81.

2.10.5 Rapid change gear boxes

Sophisticated gear boxes can allow an instant, speed change-over through levers. Moving a lever in an appropriate position can give us 3 speeds. Operation of 2 levers can provide 9 speed choices. The desired speed can be obtained in a small fraction of a minute (6 seconds/lever, at most). The speed change levers generally have 2 or 3 positions (right or left, or up and down; or right, center, or left; or up, middle, or down). Most of the gear boxes use cluster gears.

2.10.5.1 Cluster gears

Figure 2.19 shows schematically, a cluster of two gears, A and B. This can be slid on a splined input shaft, to mesh with gear C on the left, or gear D on the right. The positions provide 2 speeds on the output shaft. **Figure 2.20** shows clusters of 3 gears. Each cluster and the 3 corresponding mating gears, give 3 speed choices. Thus, two, 3-position levers on the gear box in [**Fig. 2.22**], can provide 9 speed choices.

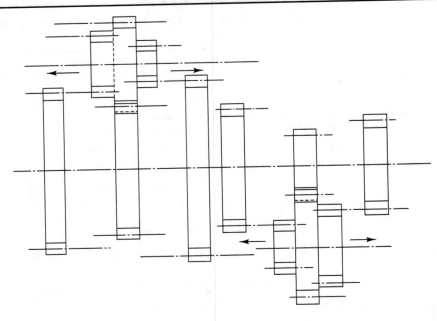

Fig. 2.20: Using 2 clusters of 3 gears and intermediate shaft for obtaining 9 speeds

To avoid making the gear boxes too wide or too long, transmission ratios for gears in the boxes, are limited to the ranges stated in the following table.

Table 2.30: Transmission ratios (i) for gear boxes

Sr. no.	Gear box type	Gear type	Transmission ratio
1	Main spindle	Straight spur	4–0.5
		Helical spur	4–0.4
2	Feed shaft and	Straight spur	5–0.36
	small high speed applications.	Helical spur	5–0.25

As 'i' = $\dfrac{N_{in}}{N_{out}}$; it is obvious that whereas it is possible to reduce speed to 1/4 to 1/5 of the input value, the possible increase in rotary speed in a gear train, is much lesser: only 2 to 2.8 times the input value.

$$\text{Speed range} = R = \frac{4}{0.5} = 8, \text{ for straight spur gears,}$$

and

$$\frac{5}{0.36} = 14, \text{ for helical spur gears.}$$

2.10.6 Design of gear box

This involves finding shaft rotary speeds (RPM), drawing ray diagrams to meet transmission ratio limitations (0.5–4), finding the number of teeth of individual and cluster gears, and computing torque and shaft sizes. Fine points about shaft design and choice of mounting bearings, will be dealt with in the chapter on 'Design of machine tool elements'. As rotary speeds, ray diagrams and transmission ratios have already been discussed, we will now discuss the various limitations in the choice of the number of teeth in the gear trains, used in the sliding cluster-type gear boxes.

2.10.7 Design of gear trains

In sliding cluster-type gear boxes, each shaft carries a number of gears, which mesh with the gears on other shafts. To ensure that all the gears on the two shafts mesh properly, the sum of the meshing gears on the two carrying shafts must be equal, if the gear modules are equal. At the same time, the gear trains should provide the designed transmission ratios. The number of teeth for the gear trains for various transmission ratios, can be computed as follows:

1. First, calculate all the required transmission ratios between the two shafts; say $i_h, i_m, i_l,$ ..., etc.
2. Convert the transmission ratios to fractions, with both numerator and denominator as integers.

$$\text{Say, } i_h = \frac{a_1}{b_1} \,;\, i_m = \frac{a_2}{b_2} \,;\, i_l = \frac{a_3}{b_3},$$

where, $a_1, a_2, a_3 \ldots b_1, b_2, b_3 \ldots$ are integers

3. Add the numerator and denominator,
Say $(a_1 + b_1); (a_2 + b_2); (a_3 + b_3)$ ..., etc.
4. Find the lowest common multiplier (L.C.M.) of the sums of the numerators and denominators, for all the transmission ratios, i.e. find LCM of $(a_1 + b_1); (a_2 + b_2); (a_3 + b_3)$.
5. The sum of the meshing gears on the two shafts, is an integer multiple of the L.C.M.
∴ Sum of the teeth of the meshing gears = $N \times$ L.C.M., where N is an integer. The value of N depends upon the minimum number of teeth, allowed on the smallest gear. This generally ranges from 18 to 22 teeth.
6. Find the minimum sum of the meshing gears (t_{sm}), from the minimum number of the teeth on the pinion, using the following formula:

$$t_{smin} = \frac{(a + b) \times t_{min}}{\text{lesser between } a \text{ and } b} \quad \text{(Eqn. 2.57a)}$$

a, b: Values of a, b for maximum transmission ratio i_1
t_{min}: Minimum no. of teeth in a pinion (18–22)
t_{smin}: Sum of the minimum number of teeth in the gear trains.

This [t_{smin}] should be increased to the nearest higher integer multiple of the L.C.M., to get the final sum of teeth of the meshing gears [t_s]

7. The sum can be divided in 2 parts according to the transmission ratio.

$$t_1 = t_s \frac{a_1}{a_1 + b_1}$$ (Eqn. 2.57b)

$$t_2 = t_s - t_1$$ (Eqn. 2.57c)

Example 2.15: Design a 2-stage, 6-speed gear box, for 125 to 710 R.P.M. spindle speeds. Draw a ray diagram for a 1,500 R.P.M. input motor, and calculate the no. of teeth for all the gears. Minimum no. of teeth = 20.

Solution: $\phi = \sqrt[z-1]{\frac{N_{max}}{N_{min}}} = \sqrt[(6-1)]{\frac{710}{125}}$

$= 1.415$

The rotary speeds will be: 125, 125 × 1.415, 125 × 1.415², 125 × 1.415³, 125 × 1.415⁴, and 125 × 1.415⁵, i.e. 125, 180, 250, 355, 500, and 710 R.P.M.

The motor speed 1,500 R.P.M. must be reduced to 2 speeds, convenient for the intermediate shaft II. The maximum possible reduction is determined by the maximum transmission ratio 4.0, for the straight spur gear.

For the intermediate shaft II,

$$N_{II min} = 1,500 \times \frac{1}{4} = 375 \text{ R.P.M.}$$

This must be reduced to minimum spindle R.P.M. 125, in the next stage.

Transmission ratio = $i_{II, III, 1} = \frac{375}{125} = 3$

The fourth speed, i.e. 355 R.P.M., will be obtained with the same transmission ratio $\left(\frac{1}{3}\right)$, from the other higher speed on the intermediate shaft II. It can be arrived at as below:

$$i_{II, III, 1} = 3 = \frac{N_{in}}{N_{out}} = \frac{N_{in}}{355}$$

∴ $N_{in} = 355 \times 3 = 1,065$ R.P.M.

The middle transmission ratio = $i_{II, III, m}$

$$i_{II, III, m} = \frac{N_{in}}{N_{out}} = \frac{375}{180} = 2.08$$

Also, $i_{II, III, m} = \frac{1,065}{500} = 2.13$

$$\therefore \quad i_{\text{II, III, m}} \simeq 2$$

$$i_{\text{II, III, h}} = \frac{N_{in}}{N_{out}} = \frac{1{,}065}{710} = 1.5 \simeq \frac{3}{2}$$

Also $\quad \dfrac{375}{250} = \dfrac{3}{2}$

We can now draw the ray diagram [**Fig. 2.21**]. The next stage is to find the number of teeth for the individual gears and clusters.

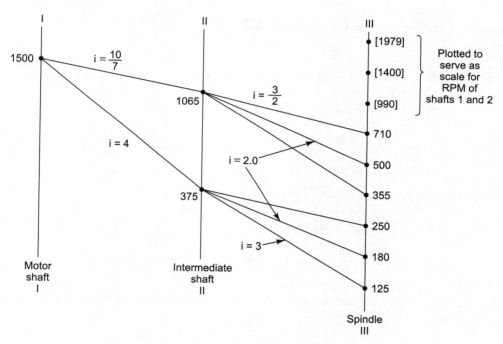

Fig. 2.21: Ray diagram for Example 2.15

The transmission ratio between shafts I (motor) and II (intermediate) are:

$$i_{\text{I, II, 1}} = \frac{N_{in}}{N_{out}} = \frac{1{,}500}{375} = 4 = \frac{4}{1}$$

$$i_{\text{I, II, h}} = \frac{N_{in}}{N_{out}} = \frac{1{,}500}{1{,}065} = 1.408 \simeq \frac{10}{7}$$

Adding numerators and denominators of the fractional numbers,

$$a_1 + b_1 = 1 + 4 = 5; \; a_2 + b_2 = 7 + 10 = 17$$

L.C.M. of 5 and 17 = $5 \times 17 = 85$

The sum of actual number of teeth must be an integer multiple of the L.C.M. = 85.

$i_{max} = 4$

$t_{min} = 20$ (given)

Sum of minimum no. of teeth = $s_{min} = \dfrac{(a+b)t_{min}}{\text{lesser between } a \text{ and } b}$

$$t_{smin} = \dfrac{(1+4) \times 20}{1} = 100$$

The nearest, higher integer multiple of L.C.M. = 85 is $85 \times 2 = 170$.
From Eqn. 2.10:

$$t_1 = \dfrac{1}{1+4} \times 170 = 34; \text{ and from Eqn. 2.11:}$$

$$t_2 = 170 - 34 = 136$$

$$t_3 = \dfrac{7}{17} \times 170 = 70$$

$$t_4 = 170 - 70 = 100$$

For transmission between shafts II and III,

$$i_{III, II, 1} = 3; \; i_{II, III, m} = 2; \; i_{II, III, h} = \dfrac{3}{2}$$

$a_3 + b_3 = 1 + 3 = 4; \; a_4 + b_4 = 1 + 2 = 3; \; a_5 + b_5 = 2 + 3 = 5$

L.C.M. of 4, 3, and $5 = 4 \times 3 \times 5 = 60$.

The sum of the teeth for meshing gears on shafts II and III, must be an integer multiple of the L.C.M. = 60.

Maximum transmission ratio = $i_{II, III, 1} = \dfrac{3}{1}$

$\therefore \quad t_{smin} = \dfrac{1+3}{1} \times 20 = 80$

The next integer multiple of L.C.M. = $60 \times 2 = 120$. It is necessary to check that this sum (120) gives adequate center distance between shafts II and III, and that shaft III will not obstruct the biggest gear on shaft II. The biggest gear on shaft II is no. 2 with 136 teeth.

$$\text{Outside radius of gear no. 2} = \dfrac{m(t_2 + 2)}{2}$$

$$= \dfrac{m(136+2)}{2} = 69 \, m$$

$m = $ module of gear

For $t_s = 120$, the center distance between shafts II and III $= m \dfrac{t_s}{2} = m \dfrac{120}{2} = 60$ m.

As 60 m is less than the outside radius of 69 m of gear no. 2, the center distance between shafts II and III must be increased, i.e. t_s must be increased. The next higher multiple of L.C.M. 60, is $60 \times 3 = 180$ m, and center distance between shafts II and III $= m \cdot \dfrac{180}{2} = 90$ m

As 90 m is much bigger than the outside radius of 69 m of gear no. 2, there is no risk of gear no. 2 obstructing shaft III.

For $t_s = 180$

$$t_5 = \dfrac{1}{1+3} \times 180 = 45; \text{ and } t_6 = 180 - 45 = 135$$

$$t_7 = \dfrac{1}{1+2} \times 180 = 60; \text{ and } t_8 = 180 - 60 = 120.$$

$$t_9 = \dfrac{2}{2+3} \times 180 = 72; \text{ and } t_{10} = 180 - 72 = 108$$

Figure 2.22 shows the layout of the gears in the gear box. The circled numbers show the gear, and the corresponding suffix used, in specifying the number of teeth i.e. the number of teeth in gear no. $7 = t_7$. Also note the following points: for gear box layout.

1. Clustered gears are mounted on splined shafts (I and III)
2. Axial travel of clusters is limited by stop collars, to facilitate blind-positioning, and prevent overshoot
3. In the 3-gear cluster, the biggest gear is placed in the center
4. There should be a difference of at least 4 teeth between adjacent gears in a cluster
5. Fixed, single gears on shaft II are provided with grub screw clamping, to prevent axial motion.
6. Center distance between adjacent shafts should be large enough to provide a positive clearance between the shaft and the biggest gear on the adjacent shaft. (Refer to Example 2.15)
7. For the same module, the sum of teeth (t_s) between the meshing gears on the two shafts should be equal

Example 2.16: Design a 9-speed, two-stage gear box for spindle speeds 30–500 R.P.M. Use a 1,500 R.P.M. motor. Minimum no of teeth in the pinion = 20. Use belt transmission for reducing motor speed to a level, suitable for the gear box input shaft.

Solution: $\phi = \sqrt[(9-1)]{\dfrac{500}{30}} = 1.421 \approx 1.414$ [standard]

Nearest standard step ratio = 1.414. The speeds will therefore be:
30, 30×1.414, 30×1.414^2 ... 30×1.414^8

$= 30, 42.45, 60.06, 85, 120.27, 170.18, 240.8, 340.73,$ and 482.14.

Range ratio = $\dfrac{482.14}{30} \approx \dfrac{48}{3} \approx 16$

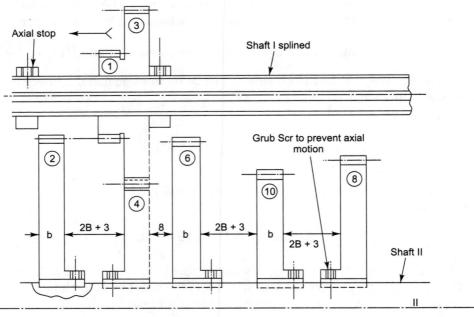
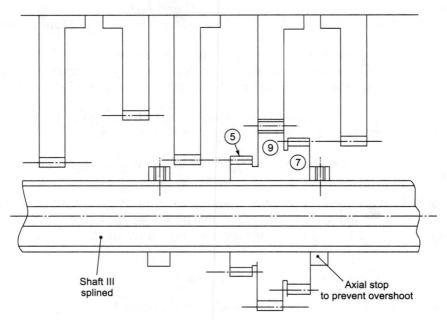

Fig. 2.22: Layout of gear box for Example 2.15

For drawing a ray diagram between spindle shaft III and the preceding shaft II, take $i_1 = 4$ max. Therefore, for the preceding intermediate shaft II, the minimum speed will be:

$$N_{II} \min = N_{III} \min \times i \max = 30 \times 4 = 120 \text{ R.P.M.}$$

For maximum speed 480 of shaft III minimum transmission ratio $i = \dfrac{1}{2}$ can be used

$$\dfrac{N_i}{N_6} = \dfrac{N_i}{180} = \dfrac{1}{2}; \; N_i = \dfrac{480}{2} = 240 : \text{max RPM of shaft II}$$

This speed will have transmission ratio (4), with the corresponding lower speed on the spindle. Connect the 240 R.P.M. point on intermediate shaft II with the 60 $\left(\dfrac{240}{4}\right)$ R.P.M. point on the spindle.

For finding the middle speed on intermediate shaft II, find the speeds corresponding with spindle speeds 42 and 340 R.P.M.

For maximum transmission ratio (4), the middle speed on the intermediate shaft $= 42 \times 4 = 168$ R.P.M.

For minimum transmission ratio $\left(\dfrac{1}{2}\right)$, the middle speed on the intermediate shaft $= 340 \times \dfrac{1}{2} = 170$.

Select 170 R.P.M. and mark the point on the intermediate shaft. Connect this point (170) with the points for 42 and 340 R.P.M. on the spindle shaft line III.

That still leaves 3 central spindle speeds: 85, 120, and 170 R.P.M., unconnected. Connect them to the points for 120, 170, and 240 R.P.M. respectively, on the intermediate shaft, to finish the ray diagram between the spindle and the intermediate shaft [**Fig. 2.23**].

For the ray diagram between the gear box input shaft I, and the intermediate shaft II, extend the left-most line connecting speeds 30 and 120, to meet the input shaft line I at the 480 R.P.M. [120 × 4] point. Connect the 480 R.P.M. point of the input speed to the speed points 170, and 240 R.P.M. on the intermediate shaft, to complete the ray diagram for the gear box.

The motor speed of 1,500 R.P.M. should be reduced to 480 R.P.M. by belt transmission. The motor will have a small (say 125) diameter pulley. The diameter of the pulley, on the input shaft of the gear box will be $= D = 125 \times \dfrac{1{,}500}{480} = 390.6$.

Figure 2.23 also shows the single ray for transmission between the motor and the gear box input shaft I.

Let us find the number of teeth for various gears.

First, we have to convert the transmission ratios into fractions, where both, the numerators and denominators are integers. Next, find the L.C.M. of the sums of the numerators and denominators of all the transmission ratios.

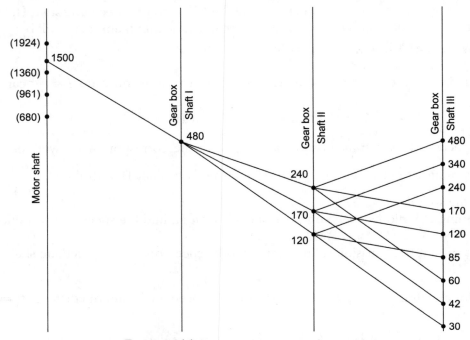

Fig. 2.23(a): Ray diagram for Example 2.16

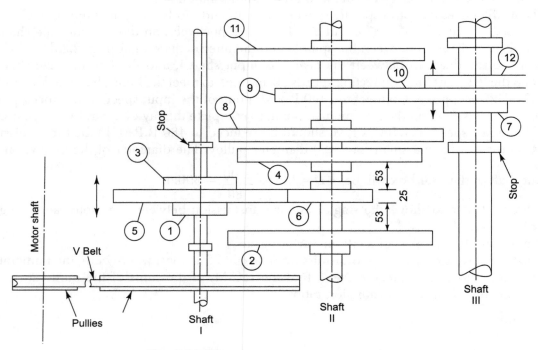

Fig. 2.23(b): Layout of drive for Example 2.16

$$i_{I, II, 1} = 4 = \frac{a_1}{b_1} \quad \therefore a_1 + b_1 = 1 + 4 = 5$$

$$i_{I, II, m} = \frac{485}{170} \approx \frac{48}{17} \quad \therefore a_2 + b_2 = 17 + 48 = 65$$

$$i_{I, II, h} = \frac{1}{2} = a_3 + b_3 = 2 + 1 = 3$$

As $65 = 13 \times 5$,

L.C.M. $= 13 \times 5 \times 3 = 195$

$t_{min} = 20$ (given)

$i_{max} = 4$

$$t_s \min = \frac{1+4}{1} \times 20 = 100$$

The sum of the minimum no. of meshing teeth [ts] will be the nearest, higher integer multiple of the L.C.M. found earlier, i.e. 195. Nearest higher integer multiple $= 195 \times 1 = 195$
Referring to [**Fig. 2.23b**],

$$t_1 = \frac{a_1}{a_1 + b_1} \times 195 = \frac{1}{1+4} \times 195 = 39$$

$$t_2 = 195 - 39 = 156$$

$$t_3 = \frac{a_2}{a_2 + b_2} \times ts = \frac{17}{17 + 48} \times 195 = 51$$

$$t_4 = 195 - 51 = 144$$

$$t_5 = \frac{a_3}{a_3 + b_3} \times 195 = \frac{2}{2+1} \times 195 = 130$$

$$t_6 = 195 - 130 = 65$$

For transmissions between shafts II and III,

$$i_{II, III, 1} = \frac{4}{1} = \frac{a_4}{b_4} \quad \therefore a_4 + b_4 = 5$$

$$i_{II, III, m} = \frac{125}{85} = \frac{25}{17} \quad \therefore a_5 + b_5 = 42 = 14 \times 3$$

$$i_{II, III, h} = \frac{1}{2} \quad \therefore a_6 + b_6 = 3$$

L.C.M. = 5 × 14 × 3 = 210

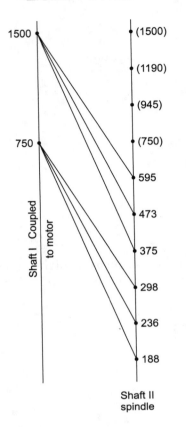

Fig. 2.24(a): Ray diagram for lower speeds from 1500/750 R.P.M. motor

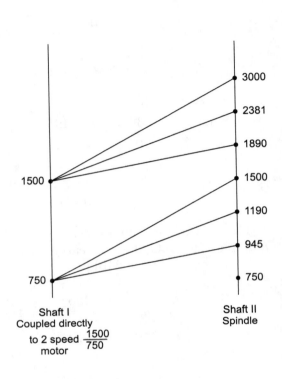

Fig. 2.24(b): Ray diagram for higher speeds from 1500/750 R.P.M. two-speed motor in a single tramsmission

$$t_{min} = 20;\ i_{max} = 4$$

$$\therefore\quad t_{s\,min} = \frac{1+4}{1} \times 20 = 100$$

Nearest higher integer multiple of L.C.M. 210 is: 210 × 1 = 210.

$$t_s = 210.$$

Let us check for interference between the biggest gear on shaft II (t_2) and shaft III, for the center distance corresponding with the sum of teeth (t_s) 210. Center distance between shafts II and III = $m\dfrac{t_s}{2} = 105\ m$.

Outside radius of the biggest gear 'No. 2' $= \dfrac{(t_2 + 2)}{2}$ m

$$= \dfrac{156 + 2}{2} \text{ m}$$

$$= 79 \text{ m}$$

As 79 m is less than the center distance of 105 m, shaft III would not interfere [obstruct] rotation [or even assy.] of gear t_2. We can now find the number of teeth for individual gears, nos. 7 to 12.

$$t_7 = \dfrac{a_4}{a_4 + b_4} \times t_5 = \dfrac{1}{5} \times 210 = 42;\ t_8 = 210 - 42 = 168;$$

$$t_9 = \dfrac{17}{42} \times 210 = 85;\ t_{10} = 210 - 85 = 125;\ t_{11} = \dfrac{2}{3} \times 210 = 140;\ t_{12} = 210 - 140 = 70.$$

For finding actual sizes of the gears, we will have to find the tooth module (size) and width of the gear, as explained earlier in **Gear design**. (Eqns. 2.40–2.43 and Tables 2.23–2.26).

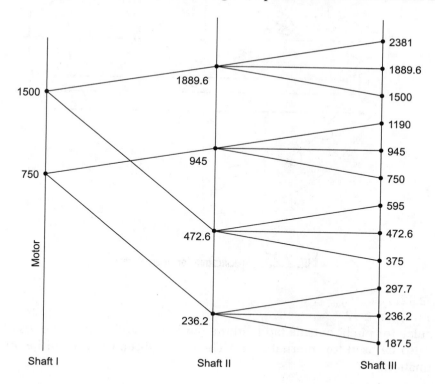

Fig. 2.24(c): Ray diagram for 12 speeds from 2 stage gear box and 2 speed (1500/750) motor

Internal gears

These are compact and strong, and run smoothly. In internal gears, the driver and the driven gears run in the same direction. The bigger internal gear meshes with a smaller external gear [**Fig. 2.25**]. The correction is usually applied radially outward for the internal gear, and considered negative. Also, $(x_1 + x_2) = 0$. The cutter used for cutting the internal teeth should have at least a dozen lesser teeth, than the internal gear.

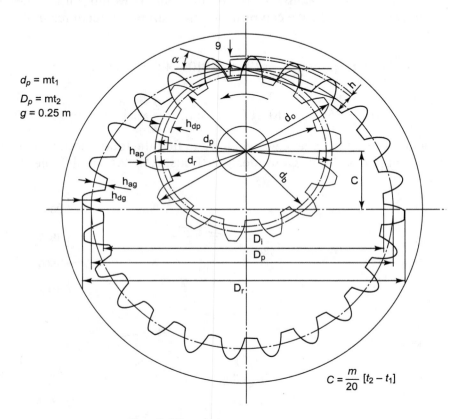

Fig. 2.25: Specification for internal gears

$t_2 - t_{cutter} \geq 12$; Also, $t_{cutter} \geq 16$
$t_1 \geq 14$; and $t_2 - t_1 \geq 12$

The formulae for pitch, outside (addendum), and base, and root diameters of gears stated earlier, can also be used for internal gears. The center distance [C_i] can be found from the following equation:

$$C_i = \frac{m(t_2 - t_1)}{20} \text{ cms} \qquad \textbf{(Eqn. 2.58)}$$

For the shaping cutter for an internal gear,
$$C = 0.05(t_2 - t_c)m$$
For changed center distance (C'),

$$\text{Working pressure angle} = \alpha'n = \cos^{-1}\left[\frac{C}{C'} \cdot \cos\alpha_n\right] \quad \textbf{(Eqn. 2.59)}$$

$$x_1 + x_2 = \left\{\frac{t_2 - t_1}{2\tan\alpha_n}\right\}(\text{Inv } \alpha' \ n = \text{Inv } \alpha n) \quad \textbf{(Eqn. 2.60)}$$

It will be useful to enlarge the inner (addendum) diameter (D_i) of the internal gear, and increase the outside diameter of the pinion, to reduce interference during operation. When $t_1 > 16$,

$$D_i = D_p - 1.2 \text{ m} = \{t_2 - 1.2\} \text{ m} \quad \textbf{(Eqn. 2.61)}$$
$$d_o = d_p + 2.5 \text{ m} = \{t_1 + 2.5\} \text{ m} \quad \textbf{(Eqn. 2.62)}$$

The tooth depth (h) is not changed, and the outside (dedendum) diameter of the internal gear (D_r) and, the root diameter of the pinion (d_r), are increased accordingly.

$$D_r = D_i + 4.5 \text{ m}$$
$$d_r = d_o - 4.5 \text{ m}$$

Example 2.17: Determine the important diameters of 1.5 module gears, having an internal gear and pinion with a transmission ratio of 1.75. Pressure angle = 20°. Enlarge the gear and pinion to reduce interference. The internal gear is shaped with a 16-teeth cutter.

Solution: Considering the constraints for the number of teeth.,

$$t_1 \geq 14; \ t_2 - t_1 \geq 12$$
$$t_2 = 1.75 \ t_1$$

If we take $t_1 = 20$,

$$t_2 = 1.75 \times 20 = 35$$
$$t_2 - t_1 = 35 - 20 = 15, \text{ i.e.} > 12$$
$$D_p = 35 \times 1.5 = 52.5$$
$$D_i = 52.5 - 1.2 \times 1.5 = 50.7$$
$$d_p = 20 \times 1.5 = 30$$
$$d_o = 30 + 2.5 \times 1.5 = 33.75$$
$$D_r = 50.7 + 4.5 \times 1.5 = 57.45$$
$$d_r = 33.75 - 4.5 \times 1.5 = 27$$
$$C = 0.5 \ (35 - 20) \times 1.5 = 11.25 \text{ mm}$$

Tip clearance at the root of the internal gear (g) can be obtained as follows:

$$g = \frac{D_r}{2} - \left(C + \frac{d_o}{2}\right) = \frac{57.45}{2} - \left(11.25 + \frac{33.75}{2}\right)$$

$$= 0.6 \approx 0.4 \text{ m}$$

Tip clearance at the root of the pinion can be obtained as follows:

$$g = \frac{D_i}{2} - \left(C + \frac{d_r}{2}\right) = \frac{50.7}{2} - \left(11.25 + \frac{27}{2}\right)$$

$$= 0.6 \approx 0.4 \text{ m}$$

2.10.7.1 Crossed helical gears

Helical gears can be mounted cross: on non-parallel, non-intersecting shafts. While mating, helical gears mounted on parallel shafts must have opposite helix angles, crossed helical gears can have the same hand helix angles. The helix angles can also have different values. If two crossed helical gears with helix angles β_1 and β_2 have to mesh, the angle between the mounting shafts (S) can be found using the following equation:

For same hand, helix shaft angle, $S_s = \beta_1 + \beta_2$
For opposite hand helix shaft angle, $S_o = \beta_1 - \beta_2$

If the angle between the shafts carrying crossed helical gears is small ($S < 10°$), it is abvisable to have helix angles of opposite hand.

All formulae used for helical gears can be used for crossed gears as well. It should however be remembered, that there is only a point contact between crossed helical gears. Increasing gear width (b) does not increase the transmission capacity for the crossed drive. The gear with bigger helix angle $[\beta]$ should be the driver.

$$\text{Working face width} = b' = \frac{2m_n \sin \beta}{\tan \alpha_n} \leq 3m_n$$

Helix angles generate tangential, axial, and separating forces.

$$\text{Tangential force} = F_t = \frac{9{,}55{,}000}{N_{\min}} \cdot \frac{2}{d_p} \cdot \text{kW}$$

$N_{\min}$ = Min R.P.M.; d_p = pinion pitch ϕ, kW = Kilowatt rating.

Axial force = $F_a = F_t \tan(\beta_1 - \phi)$

$$\phi = \tan^{-1}\left\{\frac{\mu}{\cos \alpha_n}\right\}; \quad \mu = \text{coefficient of friction}$$

Coefficient of friction (μ) depends upon the relative sliding velocity (V_r), and shafts angle [S]

$$V_r = \frac{V_1 \sin S}{\cos \beta_2} = \frac{V_2 \sin S}{\cos \beta_1}$$

V_1 = Pinion sliding velocity (m/sec)

$$= \frac{\Pi d_p N_1}{1{,}000 \times 60}, \text{ where } d_p = \text{pinion PCD } (mm)$$

and, N_1 = pinion R.P.M.

V_2 = Gear sliding velocity (m/sec)

$$= \frac{\Pi D_p N_2}{60{,}000}, \text{ where } D_p = \text{gear PCD (mm)}$$

N_2 = gear R.P.M.

Coefficient of friction (μ) can be found from Table 2.31.

The crossed helical gears drive can be rendered self-locking in a forward direction, by making the helix angle of the pinion (β_1) equal to angle ϕ, i.e.

$$\beta_1 = \phi$$

Separating force = $F_s = F_a \dfrac{\tan \alpha_n}{\sin \beta_1}$; Axial force = $F_t \tan [\beta - \phi]$

Resultant force = $F_r = \sqrt{F_t^2 + F_a^2 + F_s^2}$

For maximum efficiency,

$$\beta_1 = \frac{S + \phi}{2}; \quad \beta_2 = \frac{S - \phi}{2}$$

$$\eta_{max} = \frac{1 - 0.5 \tan \phi \tan (S/2)}{1 + 0.5 \tan \phi \tan (S/2)}$$

Table 2.31: Coefficient of friction (μ) for crossed helical gears

V_r (m/s)	0.05	0.1	0.15	0.2	0.25	0.3	0.35		
μ	0.1209	0.0993	0.0859	0.0764	0.0693	0.0637	0.0591		
V_r (m/s)	0.4	0.45	0.5	0.75	1.0	1.5	2	3	4
μ	0.0553	0.0522	0.0495	0.408	0.0365	0.033	0.0327	0.0349	0.0384
V_r (m/s)	5	7.5	10	15	20	30	40	50	
μ	0.042	0.0506	0.0582	0.0712	0.0822	0.1007	0.1163	0.13	

It may be noted that the friction coefficient (μ) is minimum [0.0327] when the relative velocity [V_r] is 2 m/sec.

2.10.7.2 Worm and wheel reduction gears

Used widely in high reduction gear boxes, worm and wheel transmit power between non-parallel non-intersecting shafts. A single enveloping worm gearing comprises a cylindrical, thread-like, worm; and a globoid wheel. Usually, the axes of the wheel and the worm, are square with each other. The tooth profile of the hob, used for machining the wheel teeth, resembles the mating worm.

Worm types

1. **ZA–Spiral worm profile** has rack-like straight sides in the axial section [**Fig. 2.26a**]. The pressure angle is measured in the axial plane. The name spiral, comes from the Archimedean spiral profile, in the transverse plane.

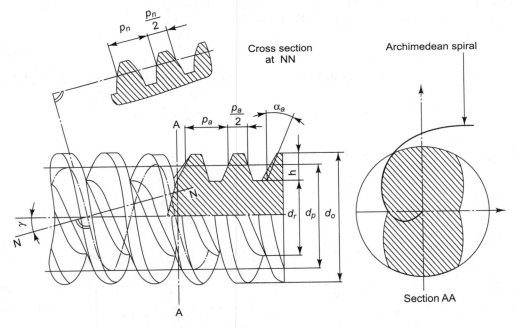

Fig. 2.26(a): ZA type worm

2. **ZN-Worm** too, has a rack-like profile, but in the normal plane square to the flanks [**Fig. 2.26b**], i.e. square to the lead angle. The standard pressure angle is in the normal plane.
3. **ZI [Fig. 2.26c] or involute helicoids worm** has an involute profile in the transverse section. It is, in fact, a helical gear with a very high helix angle, and only a few number of teeth (starts). The standard pressure angle is in the normal plane.

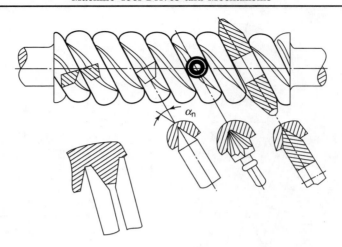

Fig. 2.26(b): ZN type worm

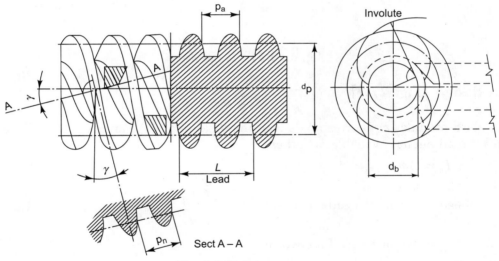

Fig. 2.26(c): ZI type worm

4. **ZK-Worm [Fig. 2.26d]** is produced by a rotary tool with straight sides. The tool is tilted to the lead angle and the pressure angle is measured in the normal plane.

   The wheel teeth profile depends upon the worm profile. For machining hob resembles the worm.

Precision and ease in manufacture

ZN : Simplest to manufacture. Simple tool geometry of a fly tool.
ZI : Higher precision possible, but the profile needs grinding.
ZK : Advantages of ZI precision, without the need for grinding.

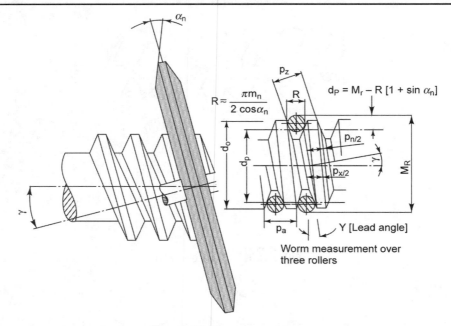

Fig. 2.26(d): ZK type worm

Worm design [Fig. 2.27]

The standard module (m_a) corresponds with the axial pitch (P_a), and is equal to one of the standard thread pitches available on lathes.

$$P_a = \Pi m_a$$

$$\text{Lead} = L = t_1 P_a; \quad \text{Lead angle} = Y = \tan^{-1}\left(\frac{t_1 m_a}{d_p}\right) \qquad \textbf{(Eqn. 2.63)}$$

t_1 = No. of starts of the worm

Pitch diameter of worm (d_p):

$$d_p = q m_a = \frac{t_1}{\tan Y} m_a \qquad \textbf{(Eqn. 2.64)}$$

$$q = \text{Diametrical quotient} = \frac{t_1}{\tan Y} = 6 - 20 \text{ [11 for initiating design]}$$

D_p = P.C.D. of wheel = $t_2 m_a$

$$\text{Efficiency} = \eta = \frac{(1 - \mu \tan Y)}{(1 + \mu \cot Y)} \qquad \textbf{(Eqn. 2.65)}$$

μ = coefficient of friction.

The efficiency increases with lead angle, till it reaches the highest value at 45. The lead angle however, is rarely kept above 6° per start, to minimize manufacturing problems.

Thus, for obtaining 24° lead angle, a 4 start worm is used. A higher pressure angle permits a higher lead angle.

Table 2.32(a): Recommended pressure angles [α], lead angles [γ] and No. of starts [t_1] for worm gears

Indexing	Power transmission			
Pressure angle [α]	14.5°	20°	25°	30°
Lead angle Y°	15° Max	25° Max	35° Max	45° Max
No. of starts [t_1]	1–2	1–2	3 or more	

Worm and wheel drive can be made irreversible or self-locking, by making $\tan^{-1} \mu \geq Y$ (Lead angle).

Generally, if the lead angle is less than 6°, the gearing becomes irreversible.

Important dimensions [Figs. 2.27a, b, c]

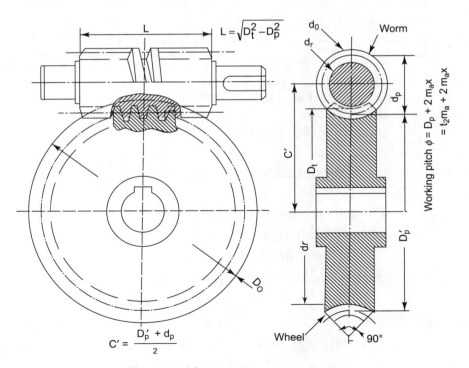

Fig. 2.27(a): Worm and wheel assembly

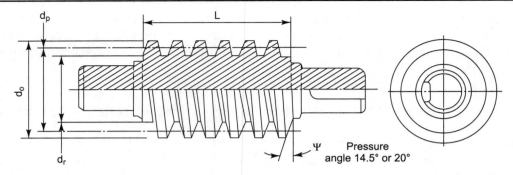

Fig. 2.27(b): Worm (pinion)

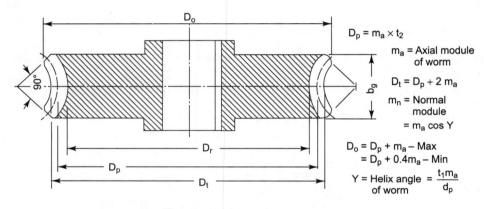

Fig. 2.27(c): Wheel for worm

Addendum for worm = $h_{ap} = m_a$

Worm outside $\phi = d_o = d_p + 2\ m_a = \left(\dfrac{t_1 m_a}{\tan Y}\right) + 2\ m_a$

Dedendum for worm = $h_{dp} = 1.2\ m_a$

Worm root $\phi = d_r = d_p - 2.4\ m_a$

Tip clearance = $g = 0.2\ m_a$

Addendum for wheel = $h_{ag} = (1 + x)\ m_a$

x = correction coefficient

Dedendum for wheel = $h_{dg} = (1.2 - x)\ m_a$

Working ϕ of wheel = $D'_p = D_p + 2x\ m_a$ **(Eqn. 2.66)**

x = Correction coefficient

Throat ϕ of wheel = $D_t = D_p + 2h_{ag}$ **(Eqn. 2.67)**

Root ϕ of wheel = $D_r = D_t - 2(h_{ag} + h_{dg}) = D_t - 4.4\, m_a$ (Eqn. 2.68)

External ϕ of wheel = $D_o = (D_p + m_a)$ Max

$= (D_p + 0.4\, m_a)$ Min

Width of wheel = $b_g = 2\, m_a \sqrt{q+1}$

$= 2\, m_a \sqrt{\dfrac{t_1}{\tan Y} + 1}$ Note Y = Lead angle

$= \tan \dfrac{[t_1\, ma]}{d_p}$

Throat radius of wheel = $r = \left[\dfrac{d_r}{2} + 0.2 m_a\right]$

Length of worm = $l = \sqrt{D_t^2 - D_p^2}$ (Eqn. 2.69)

Tooth correction

The minimum number of teeth required in wheel, to avoid undercutting, depends upon the axial pressure angle (α_a).

α_a	15°	20°	25°
t_2	36	22	15

When the standard pressure angle is in the normal plane (ZN, ZI, ZK worms), the axial pressure angle (α_a) increases with an increase in the lead angle (Y). This permits the use of a lesser number of teeth, than specified above. For $\alpha_a = 20°$ and correction (x) = +0.3, even a 16-teeth wheel can be used, instead of 22, as stated in the above table.

For ZA worms with $\alpha_n = \alpha_a$,

$$x = 1 - \dfrac{t_2}{2} \sin^2 \alpha_a$$

Only the wheel needs to be corrected. The correction coefficient (x) should not exceed 0.5. Also, the sum of the teeth of wheel (t_2) and worm (starts) should not be less than 40.

i.e., $\quad t_1 + t_2 \geq 40$

Worm root interference

In ZI worms, an increase in the lead angle (Y) decreases the transverse pressure angle (a_t)

$$\tan \alpha_t = \dfrac{\tan \alpha_n}{\sin Y}$$

A decrease in the transverse pressure angle (α_t) increases the worm base circle diameter (d_b). For,

$$d_b = d_p \cos \alpha_t,$$

Worm root ϕ (d_r) = $d_p \cos \alpha_t - 0.4\ m_a \cos Y$

Worm dedendum = $0.5\ m_a\ (q(1 - \cos \alpha_t) + 0.4 \cos Y)$

Worm-wheel addendum = $0.5\ m_a\ q\ (1 - \cos \alpha_t)$

Load capacity Capacity is dependent on 2 considerations: wear and strength. For worm,

$$\text{Wear torque} = T_{wp} = 1.9\ X_{wp}\ S_{wp}\ Y_t\ D_p^{1.8}\ m_a \quad \textbf{(Eqn. 2.70)}$$

For wheel

$$T_{wg} = 1.9\ X_{wg}\ S_{wg}\ Y_t\ D_p^{1.8}\ m_a \quad \textbf{(Eqn. 2.71)}$$

X_w = Speed Factor for wear **[Fig 2.28a]**

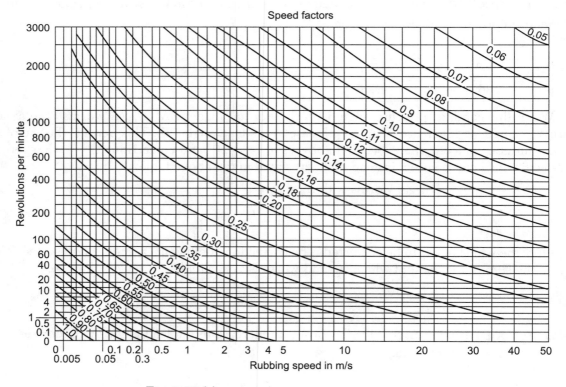

Fig. 2.28(a): Speed factor for worm gears for wear X_w

The lesser of the above two values should be used.

The suffixes p and g indicate worm [pinion] and wheel [gear], respectively.

Speed factor X_w depends upon the rubbing speed. It can be found from [**Fig. 2.28c**]

$$\text{Rubbing speed} = \frac{\Pi d_p N_1}{60,000 \cos Y}$$

Stress factor S_w can be found from Table 2.32b, while the zone factory Y_t can be found from Table 2.33.

Torque capacity for bending strength can be found from the following equations:

$$T_{sp} = 1.8\ X_{sp}\ S_{sp}\ m_a\ l_r\ D_p \cos Y \qquad \text{(Eqn. 2.72)}$$

l_r = Curved length of the radiused root of wheel

$$T_{sg} = 1.8\ X_{sg}\ S_{sg}\ m_a\ l_r\ D_p \cos Y \qquad \text{(Eqn. 2.73)}$$

S_s = Bending stress factor [Table 2.32]

$$l_r = (d_0 + 2g) \sin^{-1}\left(\frac{b}{d_o + 2g}\right) \qquad \text{(Eqn. 2.74)}$$

X_s = Strength speed factor [**Fig. 2.28b**]

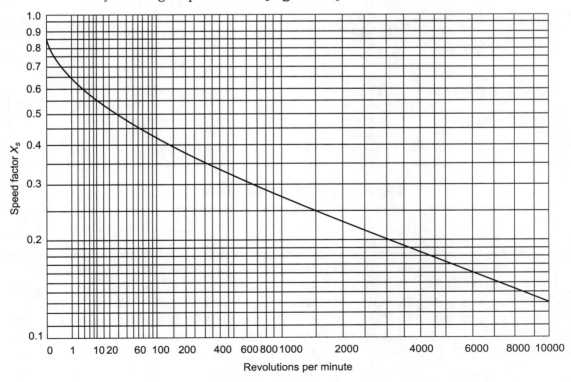

Fig. 2.28(b): Speed factor for worm gears for strength X_s

The lesser of the two values should be used. The least of the 4 values of torques, gives the design torque. These values give 26,000 hours life, and 0.003 d_p deflection of the worm shaft.

Table 2.34 gives the dimensions of IS standard worms, for which machining hobs are commercially, easily available.

Table 2.32(b): Stress factors for worm gears S_s and S_w (Extract from IS 7443)

	Materials	IS reference	Bending stress factor S_s	Surface Stress factor S_w when running with				
				A	B	C	D	E
A	Phosphor bronze centrifugally cast	IS 28-1958	7	—	0.85	0.85	0.92	1.55
	Phosphor bronze sand cast chilled	—	6.4	—	0.63	0.63	0.7	1.27
	Phosphor bronze sand cast	—	5	—	0.47	0.47	0.54	1.06
B	Grey cast iron	Grade 20 IS 210-1970	4.09	0.63	0.42	0.42	0.42	0.53
C	0.4% carbon steel normalized	C40 IS 1570-1961	14.1	1.1	0.7	—	—	—
D	0.55% carbon steel normalized	C55Mn75 IS 1570-1961	17.6	1.55	0.85	—	—	—
E	Carbon case hardening steel	C10, C14 IS 4432-1967	28.2	4.93	3.1	—	—	1.55
	3% nickel and nickel molybdenum case hardened steel	16Ni80Cr60 20Ni2Mo25 IS 4432-1967	33.11	5.41	3.1	—	—	1.55
	3 ½ nickel chromium	13Ni3Cr80 15Ni4Cr1 IS 1570-1961	35.22	6.19	3.1	—	—	1.55

Table 2.33: Worm gear zone factor Y_t

$q \rightarrow$	6	6.5	7	7.5	8	8.5	9	9.5	10	11	12	13	14	16	17	18	20
t_1							Y_t										
1	1.045	1.048	1.052	1.065	1.084	1.107	1.128	1.137	1.143	1.16	1.020	1.26	1.318	1.374	1.402	1.437	1.508
2	0.991	1.028	1.055	1.099	1.144	1.183	1.214	1.223	1.231	1.25	1.28	1.32	1.36	1.418	1.447	1.49	1.575
3	0.882	0.89	0.969	1.109	1.209	1.26	1.305	1.333	1.35	1.365	1.393	1.422	1.442	1.502	1.532	1.58	1.674
4	0.826	0.883	0.981	1.098	1.204	1.301	1.38	1.428	1.46	1.49	1.515	1.545	1.57	1.634	1.666	1.71	1.798
5	0.947	0.991	1.05	1.122	1.216	1.315	1.417	1.49	1.55	1.61	1.632	1.652	1.675	1.735	1.765	1.805	1.886
6	0.132	1.145	1.172	1.22	1.287	1.35	1.438	1.521	1.588	1.675	1.694	1.714	1.733	1.789	1.818	1.854	1.928
7			1.316	1.35	1.37	1.405	1.452	1.54	1.614	1.704	1.725	1.74	1.76	1.817	1.846	1.88	1.95
8					1.437	1.462	1.5	1.557	1.623	1.715	1.738	1.753	1.778	1.838	1.868	1.898	1.96

contd.

$q \rightarrow$	6	6.5	7	7.5	8	8.5	9	9.5	10	11	12	13	14	16	17	18	20
t_1									Y_t								
9							1.573	1.604	1.648	1.72	1.743	1.767	1.79	1.85	1.88	1.91	1.97
10								1.68	1.728	1.748	1.773	1.798	1.858	1.888	1.92	1.98
11									1.732	1.753	1.777	1.802	1.862	1.892	1.924	1.987
12										1.76	1.78	1.806	1.866	1.895	1.927	1.992
13											1.784	1.806	1.867	1.898	1.931	1.998
14												1.811	1.871	1.9	1.933	2

1. The values are based on $b = 2m_a \sqrt{q+1}$, symmetrical about the center plane of the wheel
2. For smaller face widths, the value of Y_t must be reduced proportionately
3. When it is necessary to obtain greater load capacity the width of the worm wheel face may be increased up to a maximum of $2.3\, m_a \sqrt{q+1}$, and the zone factor increased proportionately
4. The table applies to worm wheels with 30 teeth. Variations in the number of teeth produce negligible changes in the value of Y_t
5. Diametrical quotient $[q] = \dfrac{d_p}{m_a}$ [6] Max momentary torque $T_w = 3.85_w\, Y_t\, D_p^{1.8}\, m_a$ for wear
   For strength $T_s = 4\, S_s\, D_p\, lm_a\, \cos Y$

Table 2.34: Dimensions of standard worms (IS 3734)

Axial module m_a	Axial pitch p_a	No. of starts t_1	Dimetral quotient q	Reference dia. of worm d_p	Tip dia. of worm d_o	Root dia. of worm d_y	Lead angle γ
1	3.142	1	16	16	18	13.6	3°35′
1.5	4.712	1	16	24	27	20.4	3°35′
2	6.283	1	(11)	22	26	17.2	5°12′
2	6.283	2	(11)	22	26	17.2	10°18′
2.5	7.854	1	16	40	45	34	3°35′
2.5	7.854	2	(11)	27.5	32.5	21.5	10°18′
3	9.425	1	(11)	33	39	25.8	5°12′
3	9.425	2	(11)	33	39	25.8	10°18′
3	9.425	4	(11)	33	39	25.8	19°59′
4	12.566	1	10	40	48	30.4	5°43′
4	12.566	2	10	40	48	30.4	11°19′
4	12.566	4	10	40	48	30.4	21°48′
5	15.708	1	10	50	60	38	5°43′
5	15.708	2	10	50	60	38	11°19′
5	15.708	4	10	50	60	38	21°48′

contd....

Axial module m_a	Axial pitch p_a	No. of starts t_1	Dimetral quotient q	Reference dia. of worm d_p	Tip dia. of worm d_o	Root dia. of worm d_r	Lead angle γ
6	18.85	1	10	60	72	45.6	5°43
6	18.85	2	10	60	72	45.6	11°19′
6	18.85	4	10	60	72	45.6	21°48′
6	18.85	6	10	60	72	49.36	30°58′
8	25.133	1	10	80	96	60.8	5°43′
8	25.133	2	10	80	96	60.8	11°19′
8	25.133	4	10	80	96	60.8	21°48′
8	25.133	4	8	64	80	48.52	26°34′

Note: Bracketed values are non-preferred.

Example 2.18: Design a worm and wheel pair for $\frac{1}{10}$ reduction ratio and 2.2 kW power. $N_{\min} = 150$. Use C 40 and phosphor bronze as raw materials for worm and wheel. Take $q = 11$.

Solution: From torque equations,

$$T_{\max} = \frac{955{,}000 \times 2.2}{150} = 14{,}007 \text{ kg mm}$$

For $\frac{1}{10}$ reduction,

Wheel teeth $(t_2) = 10 \times t_1 = 10\, t_1$

As $t_1 + t_2 \geq 40$,

$t_1 + 10\, t_1 = 40$

$$t_1 = \frac{40}{11} \approx 4$$

$t_2 = 4 \times 10 = 40$

It is necessary to find the rubbing speed, for determining the speed factor for wear (X_w). Initially, it can be assumed to be an ideal 2 m/sec. From **[Fig. 2.28a]**, for 1,500 R.P.M. and 2 m/sec speed, $X_{wp} = 0.13$

From Table 2.32b, $S_{wp} = 1.1$

From Table 2.33 for $t_1 = 4$ and $q = 11$, wheel zone factor $(Y_t) = 1.49$, for zero correction.

$D_p = 40\, m_a$

Substituting these values,

Wear torque = $T_{wp} = 14{,}007 = 1.9\, X_{wp}\, S_{wp}\, Y_t\, D_p^{1.8}\, m_a$
$= 1.9 \times 0.13 \times 1.1 \times 1.49 \times (40 m_a)^{1.8}\, m_a = 14{,}007$

$$m_a = \sqrt[2.8]{\frac{14007}{0.4048 \times 765}} = 3.9 \simeq 4$$

$d_p = q m_a = 11 \times 4 = 44$

$q = 11 = \dfrac{t_1}{\tan Y} = \dfrac{4}{\tan Y}$

$Y = \tan^{-1} \dfrac{4}{11} = 19.98° \simeq 20°$

$D_p = t_2/m_a = 40 \times 4 = 160$

Let us check the torques using the other formulae.
First, for the wear:

$T_{wg} = 1.9 \times X_{wg}\, S_{wg}\, Y_t\, D_p^{1.8}\, m_a$

Rubbing speed $= \dfrac{\Pi d_p N_1}{60{,}000 \cos Y}$

$= \dfrac{\Pi \times 44 \times 1{,}500}{60{,}000 \times \cos 20°}$

$= 3.7$ m/sec

From **[Fig. 2.28a]**, for the wheel with 150 R.P.M. and 3.7 m/sec rubbing speed, $X_{wg} = 0.23$
From Table 2.32, for sandcast chilled the wheel of phosphor bronze and 0.4 C steel worm,

$S_{wg} = 0.63$
$T_{gw} = 1.9 \times 0.23 \times 0.63 \times 1.49 \times 160^{1.8} \times 4$
$= 15{,}222,$ i.e. $> 14{,}007$

Let us check the torques for strength,

$T_{ps} = 1.8\, X_{sp}\, S_{sp}\, m_a\, l_r\, D_p \cos Y$

From **[Fig. 2.29]** for 1500 R.P.M., $X_{sp} = 0.24$
From Table 2.32 for 0.4 C steel, $S_{sp} = 14.1$, and from Table 2.33 for $t_1 = 4$; $q = 11$, $Y_t = 1.49$

$T_{ds} = 1.8\, X_{sd}\, S_{sd}\, m_a\, l_r\, D_p \cos Y$

It is necessary to find wheel width 'b', and worm root curved length 'l_r' first.

Wheel width $(b) = 2\, m_a \sqrt{q+1}$

$$= 2 \times 4\sqrt{11+1} = 27.7$$

$$l_r = (d_o + 2_g) \sin^{-1}\left(\frac{b}{d_o + 2_g}\right) \qquad \text{...from Eqn 2.74}$$

$$d_o = d_p + 2\, m_a = 44 + 2 \times 4 = 52.0$$

Clearance $= g = 0.2\, m_a = 0.2 \times 4 = 0.8$

$$l_r = (52 + 2 \times 0.8) \sin^{-1}\left(\frac{27.7}{53.6}\right)$$

$$= 29.1$$

$$T_{ps} = 1.8\; X_{sp}\; S_{sp}\; m_a\; l_r\; D_p\; \cos Y$$

$$= 1.8 \times 0.24 \times 14.1 \times 4 \times 29 \times 160 \times 0.9397$$

$= 1,10,675$, i.e. $\geq 14,007$. The steel worm can transmit nearly 8 times the required torque.

For the wheel R.P.M. 150 from **[Fig. 2.28b]**, $X_{sg} = 0.39$

From Table 2.32b for Phosphor Bronze $S_{sg} = 5$

$$T_{gs} = 1.8 \times X_{sg}\; S_{sg}\; m_a\; l_r\; D_p\; \cos Y$$

$$= 1.8 \times 0.39 \times 5 \times 4 \times 29 \times 160 \times 0.9397$$

$= 61217$, i.e. $> 14,007$. The wheel can transmit 4.4 times the required torque.

2.10.7.3 Bevel gears

Shaped like cones, bevel gears are used for power transmission between intersecting axes **[Fig. 2.30a]**. The crown gear **[Fig. 2.30b]** is like the basic rack. It has a flat pitch circle plane, with teeth perpendicular to the axis of rotation. The crown is actually a bevel gear with a 90° cone angle (δ). The pitch plane passes through the point of intersection of the axes, of the crown gear and the mating bevel gear. As $2\delta = 180°$, the pitch cone surface of a crown gear is a flat surface.

Manufacturing and assembly errors in bevel gears lead to a shift in the contact area, excessive noise and wear. This can be reduced by barreling the tooth along with length **[Fig. 2.30c]**.

The face width 'b' is usually limited to 1/3 to 1/4 of the cone distance (R), to avoid excessive tooth thinning at the small end of the bevel gear. The module (m) is measured at the bigger end. The most popular octoid tooth form **[Fig. 2.30d]**, produces an 8-like line of contact between a mating pair.

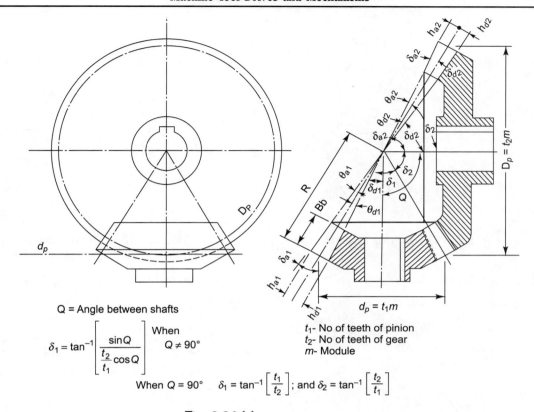

Q = Angle between shafts

$\delta_1 = \tan^{-1}\left[\dfrac{\sin Q}{\dfrac{t_2}{t_1}\cos Q}\right]$ When $Q \neq 90°$

When $Q = 90°$ $\quad \delta_1 = \tan^{-1}\left[\dfrac{t_1}{t_2}\right]$; and $\delta_2 = \tan^{-1}\left[\dfrac{t_2}{t_1}\right]$

t_1 - No of teeth of pinion
t_2 - No of teeth of gear
m - Module

Fig. 2.30(a): Bevel gears trains

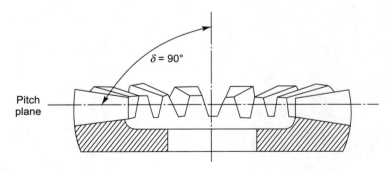

Fig. 2.30(b): Crown gear with 90° cone angle (δ)

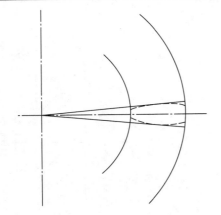

Fig. 2.30(c): Barreling bevel gear teeth

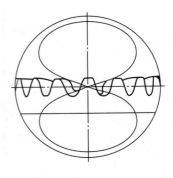

Fig. 2.30(d): Octoid profile bevel gears

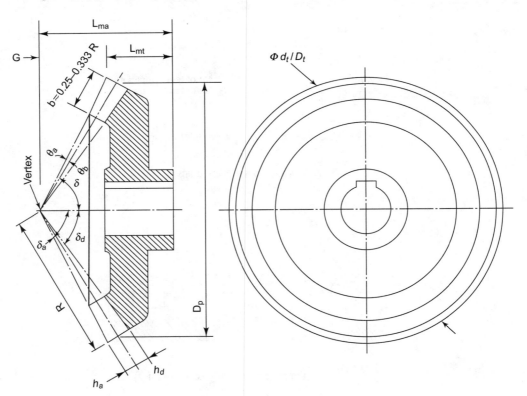

Fig. 2.30(e): Bevel gear specifications

An octoid form can be produced by using tools with a straight edge.

Bevel gear types:
  a. Straight tooth bevel gears
  b. Spiral bevel gears
  c. Zerol bevel gears
  d. Hypoid bevel gears

Straight tooth bevel gears The teeth taper in both, height and thickness. If extended beyond the face width 'b', the tooth flanks will intersect at the intersection point of the gears' axes, i.e. at the apex. The speed of the unground gears should be less than 3 m/sec. Grinding permits speeds up to 10 m/sec.

Bevel gears function without any correction, only for a transmission ratio of 1:1. The correction for other ratios can be found from **[Fig. 2.31a]**. The number of teeth on the beveled pinions are represented by curves. The intersection point of the pinion curve with the vertical line (ordinate) for the number of teeth on the bigger bevel gear, gives unit correction (**X**) on the horizontal axis. Refer to the following example for further details. Both the gears are corrected such that,

$$X_1 + X_2 = 0$$

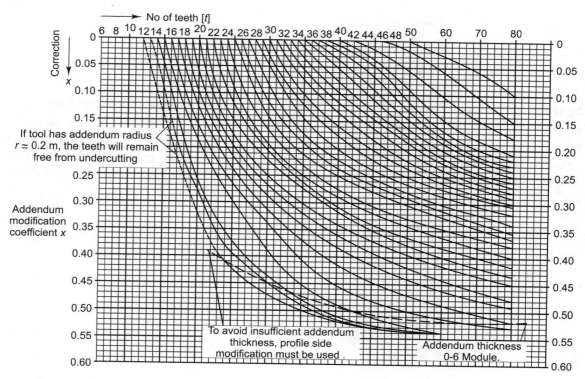

Fig. 2.31(a): Profile modifications for straight bevel gears with 90° shaft angle and 20° pressure angle
(Extract from *VDF News*, No. 34, 1968)

Load capacity can be found from the formulae for spur gears. In fact, the bevel gear is considered as a spur gear with a pitch diameter equal to the mid-width diameter of the pitch cone ('δ_m' in [**Fig. 2.31b**]).

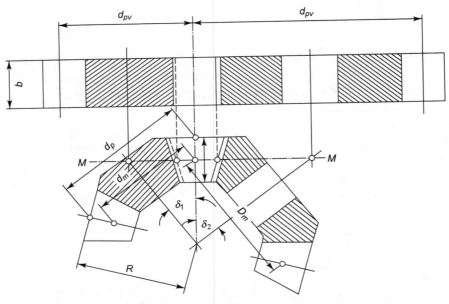

Fig. 2.31(b): Equivalent spur gears for bevel gears

Spiral bevel gears Usage of spiral angle, helps facilitate gradual engagement and continuous pitch line contact. This increases the load capacity, and enables usage of a higher transmission ratio (10:1), and faster speed (up to 11 m/sec for unground gears, and 35 m/sec for ground gears). The spiral angle varies along the gear width. It is measured mid-width [**Fig. 2.32a**], and usually measures 35°. Mating gears must have opposite hand spiral angles [**Fig. 2.32b**].

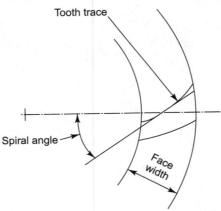

Fig. 2.32(a): Specifying spiral angle at mid-width of bevel gears

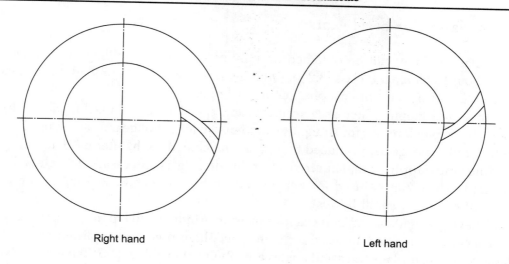

Fig. 2.32(b): Opposite hands of mating spiral bevel gears

Zerol bevel gears Although manufactured on spiral bevel gear generators, zerol gears have zero spiral angle. They have an excellent localized tooth contact [**Fig. 2.32c**]. Unground gears can be used up to a speed of 5 m/sec, while ground gears can be run up to a speed of 16 m/sec.

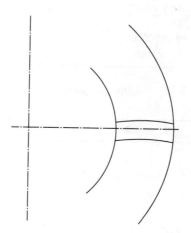

Fig. 2.32(c): Zerol spiral bevel gears

Precision classes for bevel gears

Accuracy class 9 alone, can be obtained only by machining: by milling or shaping. Used mostly without hardening, 9 class (3.2–6.3 μm) gears are used for light load, low-frequency, low-speed (up to 1 m/sec) applications, i.e. manual operation.

Hardening and lapping, increases accuracy and speed range of machined bevel gears. High speeds of up to 2 m/sec for straight teeth (5 m/sec for spiral gears) can be used for class 8 (1.6–302 μm) bevel gears, produced by lapping machined teeth, after hardening.

Grinding and lapping (or shaving) after hardening is necessary for class 7 gears (0.8–1.6 μm), which can be used for heavy loads at higher speeds: 2–5 m/sec for straight bevels (5–10 m/sec for spiral bevels).

Class 6 bevel gears (0.34–0.8 μm) call for precision grinding and lapping. These are among the most accurate gears and are used for high-speed, high-precision applications.

Tolerances on gear blank sizes and angles are tabulated in Table 2.35. Table 2.36 gives the permissible runouts for various classes. Table 2.37 states the limits for adjacent and cumulative pitch errors. Tooth thickness tolerances are tabulated in Table 2.38.

Table 2.35: Bevel gear blank tolerances [Fig. 2.30e]

	Module		
	0.3–0.5 [.0118″–.0197″]	0.6–1.25 [.0236″–.0492″]	1.5–10 [.059″–.3937″]
Tip $f(d_t)$	−0.075 [.0029″]	−0.100 [.0039″]	−0.125 [.0049″]
Tip to mounting face (L_{mt})	−0.025 [.001″]	0.050 [.002″]	−0.075 [.0029″]
Apex to mounting face (L_{ma})	−0.025	−0.050	−0.075
Tip angle (δ_a)	+0°, 0′, 30″	+0°, 0′, 15″	+0°, 0′, 8″
Back cone angle (δ_d)	±0°, 0′, 60″	±0°, 0′, 30″	±0°, 0′, 15″

Table 2.36: Runouts limits for bevel gears

	Accuracy class			
	6	7	8	9
Face and back cone runout mounting face runout	IT6	IT6	IT6	IT6
<50 ϕ [1.968″]	0.005 [.0002″]	0.008 [.0003″]	0.025 [.001″]	0.025 [.001″]
> 50 ϕ	0.01 [.0004″]	0.016 [.0006″]	0.04 [.0016″]	0.04 [.0016″]

Table 2.37(a): Permissible adjacent and cumulative pitch errors in bevel gears

	Errors in bevel gears Module					
	1–3		3.5–6		7–10	
	Adjacent pitch error	Cum. Pitch error	Adjacent pitch error	Cum. pitch error	Adjacent pitch error	Cum. Pitch error
Pitch circle $\phi < 50$ [1.968″]						
Class 6	0.016 [.0006″]	0.03 [.0012″]	0.017 [.0007″]	0.03 [.0012″]	0.021 [.0008″]	0.03 [.0012″]
7	0.022 [.0009″]	0.045 [.0018″]	0.024 [.0009]	0.045 [.0018]	0.030 [.0012″]	0.045 [.0018]
8	0.030 [.0012]	0.07 [.0028]	0.034 [.0013]	0.07 [.0028]	0.040 [.0016]	0.07 [.0028]
9	0.042 [.0016″]	0.11 [.0043]	0.48 [.0189]	0.11 [.0043]	0.058 [.0023]	0.11 [.0043]
Pitch circle ϕ 51–100 [2.008″–3.937″]						
Class 6	0.016 [.0006″]	0.036 [.0014]	0.018 [.0007]	0.036 [.0014]	0.022 [.0009]	0.036 [.0014]
7	0.022 [.0009″]	0.055 [.0022]	0.025 [.001]	0.055 [.0022]	0.030 [.0012″]	0.025 [.001]
8	0.032 [.0013″]	0.09 [.0035]	0.035 [.0014]	0.09 [.0035]	0.042 [.0016″]	0.09 [.0035]
9	0.044 [.0017″]	0.14 [.0055]	0.050 [.002]	0.14 [.0055]	0.060 [.0024]	0.14 [.0055]
Pitch circle ϕ 101–200 [3.976″–7.874″]						
Class 6	0.017 [.0007]	0.046 [.0018]	0.018 [.0007]	0.046 [.0018]	0.022 [.0009]	0.046 [.0018]
7	0.024 [.0009]	0.07 [.0028]	0.026 [.001]	0.07 [.0028]	0.032 [.0013″]	0.07 [.0028]
8	0.032 [.0013″]	0.112 [.0044]	0.036 [.0014]	0.112 [.0044]	0.044 [.0017″]	0.112 [.0044]
9	0.045 [.0017]	0.185 [.0073]	0.050 [.002]	0.185 [.0073]	0.060 [.0024]	0.185 [.0073]
Pitch circle ϕ 201–400 [7.913″–15.748″]						
Class 6	0.018 [.0007]	0.055 [.0022]	0.020 [.0008]	0.055 [.0022]	0.024 [.0009]	0.055 [.0022]
7	0.026 [.001]	0.090 [.0035]	0.028 [.0011]	0.090 [.0035]	0.032 [.0013″]	0.090 [.0035]
8	0.035 [.0014]	0.14 [.0055]	0.038 [.0015]	0.14 [.0055]	0.045 [.0018″]	0.14 [.0055]
9	0.05 [.002]	0.24 [.0009]	0.055 [.0022]	0.24 [.0009]	0.065 [.0026]	0.24 [.0094]
Pitch circle $\phi > 400$ [15.748″]						
Class 6	0.020 [.0008]	0.07 [.0028]	0.022 [.0009]	0.07 [.0028]	0.024 [.0009]	0.07 [.0028]
7	0.028 [.0011]	0.11 [.0043]	0.03 [.0012″]	0.11 [.0043]	0.035 [.0014]	0.11 [.0043]
8	0.038 [.0015]	0.18 [.0071]	0.042 [.0016″]	0.18 [.0071]	0.05 [.002]	0.18 [.0071]
9	0.055 [.0022]	0.28 [.0011]	0.06 [.0024]	0.28 [.0011]	0.07 [.0028]	0.28 [.0011]

Backlash

Although zero backlash bevel gears are also used, it is common to provide some intentional backlash in bevel gears, to simplify their manufacture. The standard backlash depends upon the cone generator line length (R).

Table 2.37(b): Backlash in terms of pitch cone generator length [R]

	0 50 [1.968"]	51 [2.008"] 80 [3.149]	81 [3.189] 120 [4.724]	121 [4.764] 200 [7.874]	201 [7.914] 320 [12.598]	321 [12.638] 500 [19.685]	501 [19.724] 800 [31.496]	801 [31.535] 1250 [49.213]
Std. backlash	0.085 [0.0033]	0.1 [0.0039]	0.13 [0.0051]	0.17 [0.0067]	0.21 [0.0083]	0.26 [0.0102]	0.34 [0.0134]	0.42 [0.0165]

Minimum tooth thinning, necessary for obtaining the standard backlash, depends upon the module, and the pitch circle ϕ (d_p or D_p)

Table 2.37(c): Std. minimum tooth thinning for std. backlash for accuracy classes 6–9 of bevel gears

Module	Pitch circle ϕ (d_p or D_p)					
	0 50 [1.968"]	51 [2.008"] 80 [3.15"]	81 [3.189"] 120 [4.724]	121 [4.764"] 200 [7.874]	201 [7.914"] 320 [12.598]	321 [12.638"] 500 [19.685"]
1–2.5 [0.0394"–.0984"]	0.05 [.002]	0.06 [.0024]	0.08 [.0031]	0.10 [.0039]	0.12 [.0047]	0.16 [.0063]
2.75–6 [.108–.236]	0.052 [.0021]	0.065 [.0026]	0.085 [.0033]	0.105 [.0041]	0.13 [.0051]	0.17 [.0067]
7–10 [.2756–.3937]	—	0.07 [.0027]	0.09 [.0035]	0.11 [.0043]	0.14 [.0055]	0.18 [.0071]

The permissible radial runout on the tooth flanks, depends upon the accuracy class of the bevel gears, pitch circle ϕ, and the cone angle (6).

Table 2.37(d): Radial runout for bevel gear tooth flanks

	Pitch Cone Angle			
Accuracy class	up to 20°	21° 30°	31° 40°	above 41°
6 Pitch circle ϕ				
Up to 50 [1.968"]	0.021 [.00083"]	0.024 [.0009"]	0.028 [.0011"]	0.03 [.0013"]
51–80 [2.008–3.15]	0.025 [.001]	0.028 [.0011]	0.032 [.0013]	0.036 [.0014]
81–120 [3.189–4.724]	0.03 [.0013]	0.034 [.0013]	0.038 [.0015]	0.042 [.0016]
121–200 [4.764–7.874]	0.036 [.0014]	0.040 [.0016]	0.045 [.0018]	0.05 [.002]
201–320 [7.914–12.598]	0.048 [.0019]	0.048 [.0019]	0.052 [.0021]	0.06 [.0024]
321–500 [12.638–19.685]	0.050 [.002]	0.055 [.0022]	0.06 [.0024]	0.07 [.0027]

Accuracy Class	Pitch Cone Angle			
	upto 20°	21° 30°	31° 40°	above 41°
7 P.C. φ				
Up to 50 [1.968"]	0.034 [.0013"]	0.036 [.0014"]	0.045 [.0018"]	0.048 [.0019"]
51–80 [2.008–3.15]	0.04 [.0016]	0.045 [.0018]	0.052 [.0021]	0.055 [.0022]
81–120 [3.189–4.724]	0.048 [.0019]	0.055 [.0022]	0.06 [.0024]	0.065 [.0026]
121–200 [4.764–7.874]	0.055 [.0022]	0.065 [.0026]	0.07 [.0027]	0.08 [.0031]
201–320 [7.914–12.598]	0.065 [.0026]	0.075 [.0029]	0.08 [.0031]	0.095 [.0037]
321–500 [12.638–19.68]	0.08 [.0031]	0.09 [.0035]	0.095 [.0037]	0.11 [.0043]
8 P.C. φ				
Up to 50 [1.968"]	0.052 [.0021]	0.06 [.0024]	0.07 [.0027]	0.075 [.0029]
51–80 [2.008–3.15]	0.063 [.0025]	0.07 [.0027]	0.08 [.0031]	0.09 [.0035]
81–120 [3.189–4.724]	0.075 [.0029]	0.09 [.0035]	0.095 [.0037]	0.105 [.0041]
121–200 [4.764–7.874]	0.09 [.0035]	0.105 [.0041]	0.12 [.0047]	0.125 [.0049]
201–320 [7.914–12.598]	0.105 [.0041]	0.120 [.0047]	0.125 [.0049]	0.15 [.0059]
321–500 [12.638–19.68]	0.125 [.0049]	0.14 [.0055]	0.15 [.0059]	0.18 [.0071]
9 P.C. φ				
Up to 50 [1.968"]	0.085 [.0033]	0.095 [.0037]	0.11 [.0043]	0.12 [.0047]
51–80 [2.008–3.15]	0.1 [.0039]	0.11 [.0043]	0.13 [.0051]	0.14 [.0055]
81–120 [3.189–4.724]	0.12 [.0047]	0.13 [.0051]	0.15 [.0059]	0.17 [.0067]
121–200 [4.764–7.874]	0.14 [.0055]	0.16 [.0063]	0.18 [.0071]	0.2 [.0079]
201–320 [7.914–12.598]	0.17 [.0067]	0.19 [.0075]	0.21 [.0083]	0.24 [.0094]
321–500 [12.638–19.68]	0.2 [.0079]	0.22 [.0087]	0.24 [.0094]	0.28 [.011]

Tolerances on tooth thickness depend upon the radial runout of the tooth flanks.

Table 2.38: Tooth thickness tolerances for bevel gears

Radial runout on tooth flanks		Tolerance on tooth thickness	Rad. runout on tooth flanks		Tol. on tooth thickness
μm	micro inches		μm	micro inches	
8–16	315–630	0.045 [.0018"]	61–80	2401–3150	0.11 [.0043"]
17–20	669–787	0.048 [.0019"]	81–100	3189–3937	0.13 [.0051]
21–25	828–984	0.052 [.0021]	101–120	3976–4724	0.15 [.0059]
26–32	1023–1260	0.06 [.0024]	121–160	4764–6299	0.19 [.00748]
33–40	1299–1575	0.07 [.0028]	161–200	6338–7874	0.24 [.00945]
41–50	1614–1968	0.08 [.0031]	201–250	7913–9842	0.28 [.011]
51–60	2008–2362	0.09 [.0035]	251–320	9881–12600	0.34 [.0134]

For smooth operation, bevel gears are thinned slightly.
The minimum thinning depends upon the backlash, pitch circle diameter, and the module.

Table 2.39: Minimum tooth thinning for bevel gears

Module	1–2.5		2.75 – 6		7 – 10	
Backlash	Zero *	Std. *	Zero *	Std.	Zero	Std.
Pitch φ						
Below 50 [1.968"]	0.012 [.0005"]	0.050 [.002"]	0.013 [.0005"]	0.052 [.0021"]	–	–
51–80 [2.008–3.15]	0.016 [.00063]	0.060 [.0024]	0.017 [.0007]	0.065 [.0026]	0.018 [.0007"]	0.070 [.0027"]
81–120 [3.19–4.27]	0.020 [.0008]	0.080 [.0031]	0.021 [.0008]	0.085 [.0033]	0.022 [.0009]	0.090 [.0035]
121–200 [4.76–7.87]	0.025 [.001]	0.100 [.0039]	0.026 [.001]	0.105 [.0041]	0.028 [.0011]	0.110 [.0043]
201–320 [7.91–12.6]	0.032 [.0013]	0.120 [.0047]	0.034 [.0013]	0.130 [.0051]	0.036 [.0014]	0.140 [.0055]
321–500 [12.64–19.7]	0.040 [.0016]	0.160 [.0063]	0.042 [.0016]	0.170 [.0067]	0.045 [.0018]	0.180 [.0071]

Note: Zero backlash bevel gears must be made to 6th or 7th accuracy classes only. Eight or ninth class gears have tolerances too wide to allow for a zero backlash.

As stated earlier, formulae for spur gears can be used for computing bevel gear pitch circle diameter (d_p or D_p). Other important dimensions can be calculated, using the following formulae:
For shaft angle 90°,

For pinion, Pitch cone angle = $\delta_1 = \tan^{-1}\left(\dfrac{t_1}{t_2}\right)$ (Eqn. 2.75a)

For bigger gear Pitch cone angle = $\delta_2 = \tan^{-1}\left(\dfrac{t_2}{t_1}\right)$ (Eqn. 2.75b)

$$\text{Cone distance} = R = \frac{d_p}{2\sin\delta_1} = \frac{D_p}{2\sin\delta_2}$$ (Eqn. 2.76)

$$\text{Virtual no. of teeth} = t_v = \frac{t}{\cos\delta}$$ (Eqn. 2.77)

$$t_{v1} = \frac{t_1}{\cos\delta_1}; \quad t_{v2} = \frac{t_2}{\cos\delta_2}$$

$$\text{Correction for pinion} = x = \frac{14 - t_{v1}}{17}$$

Figure 2.31b shows the equivalent spurs for bevel gears. Corrections from [**Fig. 2.31a**], give a balanced design.

If the above equation gives a negative value for x, then zero correction ($x = 0$) should be used.

$$\text{Addendum for pinion} = h_{a1} = m_n (1 + x) \quad \textbf{(Eqn. 2.78a)}$$

$$\text{Addendum for gear} = h_{a2} = m_n (1 - x) \quad \textbf{(Eqn. 2.78b)}$$

$$\text{Dedendum for pinion} = h_{d1} = m_n (1.188 - x) \text{ for Gleason system} \quad \textbf{(Eqn. 2.79a)}$$

$$\text{Dedendum for gear} = h_{d2} = m_n (1.188 + x) \text{ for Gleason system} \quad \textbf{(Eqn. 2.79b)}$$

$$\text{Addendum angle for pinion} = \theta_{a1} = \tan^{-1}\left(\frac{h_{a1}}{R}\right) \quad \textbf{(Eqn. 2.80a)}$$

$$\text{Addendum angle for gear} = \theta_{a2} = \tan^{-1}\left(\frac{h_{a2}}{R}\right) \quad \textbf{(Eqn. 2.80b)}$$

$$\text{Dedendum angle for pinion} = \theta_{d1} = \tan^{-1}\left(\frac{h_{d1}}{R}\right) \quad \textbf{(Eqn. 2.81a)}$$

$$\text{Dedendum angle for gear} = \theta_{d2} = \tan^{-1}\left(\frac{h_{d2}}{R}\right) \quad \textbf{(Eqn. 2.81b)}$$

$$\text{Pinion tip angle} = \delta a_1 = \delta_1 + \theta_{a1}$$

$$\text{Pinion root angle} = \delta_{d1} = \delta_1 - \theta_{d1}$$

$$\text{Gear tip angle} = \delta_{a2} = \delta_2 + \theta_{a2}$$

$$\text{Gear root angle} = \delta_{d2} = \delta_2 - \theta_{d2}$$

$$\text{Tip diameter of pinion} = d_{a1} = d_p + 2h_{a1} \cos \delta_1 \quad \textbf{(Eqn. 2.82a)}$$

$$\text{Tip diameter of gear} = Da_2 = D_p + 2h_{a2} \cos \delta_2 \quad \textbf{(Eqn. 2.82b)}$$

$$\text{Circular tooth thickness} = S = \frac{\Pi m_n}{2} + 2m_n \times \tan \alpha_n$$

$$\alpha_n = \text{Normal pressure angle}$$

$$\text{Pinion chordal tooth thickness} = \overline{S}_1 = S\left(1 - \frac{1}{6}\left(\frac{S}{d_p}\right)^2\right)$$

$$\text{Gear chordal tooth thickness} = \overline{S}_2 = S\left(1 - \frac{1}{6}\left(\frac{S}{D_p}\right)^2\right)$$

$$\text{Chordal tooth height} = \bar{h}_{a1} = h_{a1} + \frac{\bar{S}_1^2 \cos \delta_1}{4 d_p}$$

For pinion,

$$\text{Chordal tooth height} = \bar{h}_{a2} = \frac{h_{a2} + \bar{S}_2^2 \cos \delta_2}{4 D_p}$$

For gear,

Clearance = $g = 0.188\ m + 0.05$, for Gleason system.
$= 0.2\ m$, as per IS 5037

$$\text{Face width} = b = 0.25 - 0.33\ R \qquad \textbf{(Eqn. 2.83)}$$

$$\text{Transverse module} = m_t = m_n = \frac{b}{8-10} \qquad \textbf{(Eqn. 2.84)}$$

Tip to apex = G (**[Fig 2.30e]** page 198)

$$\text{For pinion } G_1 = \frac{d_P}{2} \cot \delta_1 - h_{a1} \sin \delta_1 \qquad \textbf{(Eqn. 2.85a)}$$

$$\text{For Gear } G_2 = \frac{D_P}{2} \cot \delta_2 - h_{a2} \sin \delta_2 \qquad \textbf{(Eqn. 2.85b)}$$

Usually, the distance from mounting face to apex (L), is selected to suit the application.

The distance between the mounting face and the reference plane passing through the tip ϕ (d_{a1} or D_{a2}), depends upon the application: the length required for accommodating suitable grub screw is for clamping drive key and the gear, onto the shaft. Refer to Example 2.19 for guidance.

The above formulae are applicable for straight tooth bevel gears, mounted on shafts, square with each other. If the angle between the shaft axes is not 90°, but different, i.e. ($Q \neq 90°$).

$$\text{Pitch cone angle for pinion } (\delta_1) = \tan^{-1}\left(\frac{\sin Q}{\frac{t_2}{t_1} \cos Q}\right) \qquad \textbf{(Eqn. 2.86a)}$$

$$\text{Pitch cone angle for gear } (\delta_2) = \tan^{-1}\left(\frac{\sin Q}{\frac{t_1}{t_2} \cos Q}\right) \qquad \textbf{(Eqn. 2.86b)}$$

When the shafts for the gears are at 90°, the equivalent straight spur gear dimension can be found from the following formulae (**[Fig 2.31b]** page 200). For virtual number of teeth [t_v],

$$t_v = \frac{t}{\cos \delta}, \text{ i.e. } t_{v1} = \frac{t_1}{\cos \delta_1}, \text{ and } t_{v2} = \frac{t_2}{\cos \delta_2}$$

Equivalent straight spur pinion pitch $\phi = d_{vp} = \dfrac{d_m}{\cos \delta_1}$

$$d_m = d_p - b \sin \delta_1 \quad \text{(Eqn. 2.87a)}$$

Equivalent St. spur gear pitch $\phi = D_{vp} = \dfrac{D_m}{\cos \delta_2}$

$$D_m = D_p - b \sin \delta_2 \quad \text{(Eqn. 2.87b)}$$

Peripheral velocity $= V = \dfrac{\Pi d_m N_1}{60{,}000}$ m/sec. \quad **(Eqn. 2.88)**

$$= \frac{\Pi D_m N_2}{60{,}000}$$

Example 2.19: Design a pair of mating bevel gears for transmitting 11 kW and 1:2 reduction. Assume that the driver pinion is running at 1,000 R.P.M. Use 25 grade C.I. for pinion, and 20 grade C.I. for the gear. The gear axes are square with each other. Use the Gleason system. [Clearance $[g] = 0.188$ m]

Solution:

$$\text{Torque on pinion} = T = \frac{95{,}500 \text{ KW}}{N}$$

$$= \frac{95{,}500 \times 11}{1{,}000} = 1{,}050.5 \text{ kgcm}$$

Taking tentatively, the number of the pinion teeth (t_1) as 20,

$$t_2 = 2 \times 20 = 40$$

From the graph in **[Fig. 2.31a]**, (p 199) following the curve for 20 teeth, we notice that it intersects the 40-teeth vertical line at the point, which gives us the correction coefficient (x) of 0.33, on the horizontal line from the intersection point. From Table 2.26 [p 152] for 20 teeth and 0.33 correction coefficient, form factor $Y = 0.456$.

Taking tentatively the 9th grade accuracy and $\psi_m = \dfrac{b}{m} = 6$, for unhardened unlapped gears. From Table 2.23 (p 151) for grade 20 C.I., $f_b = 9$ kg/mm² $= 900$ kg/cm².
For 25 grade C.I., $f_b = 13$ kg/mm² $= 1300$ kg/cm².
From Eqn. 2.40, (p 145) the pinion average module (m_{av}) can be found.

$$m_{av} = 12.6 \sqrt[3]{\frac{T}{Y f_b \psi_m t_1}} = 12.6 \sqrt[3]{\frac{1050.5}{0.456 \times 1300 \times 6 \times 20}}$$

$$= 3.09 \text{ mm} \approx 3.0 \text{ module}$$

Tip (Diameter) module (m_t) can be found from the following equation:

$$m_t = m_{av} + \frac{b}{t} \sin \delta \qquad \text{(Eqn. 2.89)}$$

It is necessary to first find m_t to find pitch cone angle δ

$$\text{Pinion pitch cone angle} = (\delta_1) = \tan^{-1}\left(\frac{t_1}{t_2}\right) = \tan^{-1}\left(\frac{20}{40}\right)$$

$$= 26.565°$$

$$\text{Gear pitch cone angle } (\delta_2) = \tan^{-1}\left(\frac{t_2}{t_1}\right) = \tan^{-1}\left(\frac{40}{20}\right)$$

$$= 63.435$$

As $\psi_m = 6$; $b = 6\, m_t$

Substituting for pinion in Eqn. 2.58,

$$m_t = m_{av} + \frac{6 m_t}{t} \sin \delta_1 = 3.09 + \frac{6 m_t}{20} \times 0.4472$$

$$= \frac{3.09}{1 - \frac{0.4472 \times 6}{20}} = 3.57 \approx 3.5 \text{ nearest std. module}$$

Addendum for pinion $(h_{a1}) = (1 + x)\, m = (1 + 0.33)\, 3.5$

$$= 4.655$$

Addendum for gear $(h_{a2}) = (1 - x)\, m = (1 - 0.33)\, 3.5$

$$= 2.345$$

Dedendum for pinion $(h_{d1}) = (1.188 - x)\, m = (1.188 - 0.33)\, 3.5$

$$= 3.003$$

Dedendum for gear $(h_{d2}) = (1.188 + x)\, m = (1.188 + 0.33)\, 3.5$

$$= 5.313$$

$$\text{Cone distance } (R) = \frac{m_t t_1}{2 \sin \delta_1} = \frac{m_t t_2}{2 \sin \delta_2}$$

$$= \frac{3.5 \times 20}{2 \sin 26.565} = \frac{3.5 \times 40}{2 \sin 63.435}$$
$$= 78.262$$
$$\text{Gear width} = 0.25 - 33 \ R = 0.25 - 0.33 \ (78.262)$$
$$= 19.56 - 25.82$$
$$\simeq 25$$

Let us find the addendum and dedendum angles for both the gears.

For pinion,
$$\theta a_1 = \tan^{-1}\left(\frac{h_{a1}}{R}\right) = \tan^{-1}\left(\frac{4.655}{78.262}\right) = 3.404°$$

For gear,
$$\theta_{a2} = \tan^{-1}\left(\frac{h_{a2}}{R}\right) = \tan^{-1}\left(\frac{2.345}{78.262}\right) = 1.716°$$

$$\text{Dedendum angle for pinion } (\theta_{d1}) = \tan^{-1}\left(\frac{h_{d1}}{R}\right) = \tan^{-1}\left(\frac{3.003}{78.262}\right)$$
$$= 2.197°$$

$$\text{Dedendum angle for gear } (\theta_{d2}) = \tan^{-1}\left(\frac{h_{d1}}{R}\right) = \tan^{-1}\left(\frac{5.313}{78.262}\right)$$
$$= 3.884°$$

$$\text{Pinion tip angle } (\delta_{a1}) = \delta_1 + \theta_{a1} = 26.565 + 3.404$$
$$= 29.969 \simeq 30°$$

From Table 2.35 for 3.5 module, tolerance on tip angle + 0° 0', 8"

$$\text{Pinion root angle } (\delta_{d1}) = \delta_1 - \theta_{d1} = 26.565 - 2.197$$
$$= 24.368°$$

$$\text{Gear tip angle } (\delta_{a2}) = \delta_2 + \theta_{a2} = 63.435 + 1.716°$$
$$= 65.151° - 0°$$

$$\text{Gear root angle } (\delta_{d2}) = \delta_2 - \theta_{d2} = 63.435 - 3.884$$
$$= 59.551°$$

$$\text{Pinion tip } \phi \ (d_{a1}) = d_p + 2h_{a1} \cos \delta_1$$
$$= m_t \ t_1 + 2ha_1 \cos \delta_1$$
$$= 3.5 \times 20 + 2 \times 4.655 \cos 26.565$$
$$= 78.327$$

From Table 2.35 (p 202) for 3.5 module, tolerance on $d_{a1} = -0.125$

$$\text{Gear tip } \phi(D_{a2}) = D_p + 2h_{a2}\cos\delta_2$$
$$= 3.5 \times 40 + 2 \times 2.345 \cos 63.435$$
$$= 142.097 + 0 \text{ or } 142.097 - 0.125$$

$$\text{Pinion tip to apex }(G_1) = \frac{d_p}{2}\cot\delta_1 - h_{a1}\sin\delta_1$$

$$= \frac{70}{2}\cot 26.565 - 4.655 \sin 26.565$$

$$= 67.918$$

$$\text{Gear tip to apex }(G_2) = \frac{D_p}{2}\cot\delta_2 - h_{a2}\sin\delta_2$$

$$= \frac{140}{2}\cot 63.435 - 2.345 \sin 63.435$$

$$= 32.902$$

The mounting face to apex distance (L_{ma}) is controlled, to control the tip to apex distance. For 3.5 module, Table 2.35 (p 202) gives tolerance of –0.075 for L_{ma}.

L_{ma} depends upon practical considerations such as space available, and the diameter of the mounting shaft. The shaft diameter depends upon the small-end diameter of the bevel gear.

$$\text{Small end } \phi = \text{Tip } \phi - 2\,b\sin\delta_a$$

For the pinion,

$$d_{as1} = d_{a1} - 2 \times 25 \times \sin d_{a1}$$
$$= 78.327 - 50 \sin 30° = 53.327$$

Assuming shaft $\phi(d_s)$ half of pinion small end $dia\,(d_{as1})$ shaft $\phi(\delta_s) = \frac{53.327}{2} = 26.77 \approx 25$.

Grub screw used for clamping the gear onto shaft is usually one fourth of the shaft ϕ. The grub screw should have margin equal to its diameter on both sides.

$$\therefore \quad \text{Gear boss length} = 2 \times \frac{25}{4} = 12.5 \approx 13$$

$$\text{Mounting face to tip} = 13 + 4.4m \sin\delta \text{ (tentatively)} = 26.8 \approx 27$$
$$\text{Mounting face to apex }(L_{ma}) = \text{Tip to apex }(G_1) + 27$$
$$= 67.92 + 27 \approx 95$$

From Table 2.35 tolerance on L_{ma} is –0.075 for 3.5 module. The reader can find the mounting face to apex distance for the gear as an exercise.

Circular tooth thickness $(S) = \dfrac{\Pi m_n}{2} + 2m_n \tan \alpha_n$

$$= \dfrac{\Pi \times 3.5}{2} + 2 \times 3.5 \tan 20°$$

$$= 8.0456$$

Pinion chordal thickness $(\overline{S}_1) = S\left(1 - \dfrac{1}{6}\left(\dfrac{S}{d_p}\right)^2\right)$

$$\overline{S}_1 = 8.0456\left(1 - \dfrac{1}{6}\left(\dfrac{8.0456}{70}\right)^2\right) = 8.0278$$

Chordal tooth addendum $= \overline{h}_{a1} = h_{a1} + \dfrac{\overline{S}_1^2 \cos \delta_1}{4 d_p}$

$$= 4.655 + \dfrac{8.0278^2 \cos 26.565}{4 \times 70}$$

$$= 4.861$$

Tolerance on the tooth thickness depends upon the permissible runout. For pitch cone angle between 21°–30° and pitch diameters 51–80, permissible runout is 0.028. From Table 2.38 (p 205) for 0.026–0.032 runout, tolerance on tooth thickness is 0.06.

Minimum tooth thinning necessary for 51–80 pitch ϕ and 3.5 module is 0.065 (for Std. backlash). So chordal tooth thickness for pinion $= (4.861 - 0.065) - \gamma$

$$= 4.796 - 0.06$$

The Std. backlash for R between 50 to 80 is 0.1. [Table 2.37b]

Bevel gear design can be expedited, by using graphs for pinion pitch diameter (d_p), and number of teeth (t_1). It is necessary to convert the power (kW_N) at working R.P.M. (N), into the power possible at 100 R.P.M. (kW_{100}), the unit used in **[Fig. 2.34]**.

$$kW_{100} = \dfrac{100\, kW_N}{N K_m}$$

$K_m =$ Material factor (Table 2.40)

Figure 2.35 plots the bevel pinion teeth number (t_1) against pinion pitch diameter, for various speed ratios. Knowing the approximate pinion pitch ϕ (d_p), and the number of teeth, the module can be found from the standard equation: $d_p = m t_1$.

Table 2.40(a): Material factor (k_m) for bevel gears durability for various material combinations of pinion and gear materials.

Material		and	Hardness	Km
Pinion		Gear		
Material	Hardness	Material	Hardness	
C.I	—	C.I.	—	0.30
Annealed steel	160–200 HB	C.I.	—	0.30
Case or flame hardened steel	RC 55	C.I.	—	0.40
C' hardened steel	50 RC 55	Heat treated steel	250–300 HB	0.50
Oil hrd steel	—	Oil hardened steel	—	0.65
Flame or cased hardened steel	RC 50–55	Flame or case hardened steel	RC 50	1.0

For Example 2.19, from Table 2.40,
For both, pinion and gear of cast iron, km = 0.30

$$kW_{100} = \frac{100 \times 11}{1000 \times 0.30} = 3.67$$

The graph in [**Fig. 2.34**], for 3.67 kW at 2:1 ratio, gives pinion ϕ of 115.
From graph in [**Fig 2.35**], for 115 ϕ pinion and 2:1 ratio, the no. of pinion teeth (t_1) is 22.5 ≈ 23.
The gear will have 23 × 2, i.e. 46 teeth.

$$d_p = \text{Pinion pitch } \phi = m \, t_1 = m \times 23 = 115$$

$$\therefore \quad m = \frac{115}{23} = 5$$

$$d_p = 23 \times 5 = 115$$

Pinion pitch cone angle = $\delta_1 = \tan^{-1}\left(\frac{23}{46}\right) = 26.565°$

Gear pitch cone angle = $\delta_2 = \tan^{-1}\left(\frac{46}{23}\right) = 63.435°$

$$R = \frac{d_p}{2 \sin \delta_1} = \frac{115}{2 \sin 26.565°} = 128.57$$

$D_p = 46 \times 5.0 = 230$

$b = 0.25 - 0.33 \, R = [0.25 - 0.33] \times 128.57 = 32 - 42.8$

$= 43$

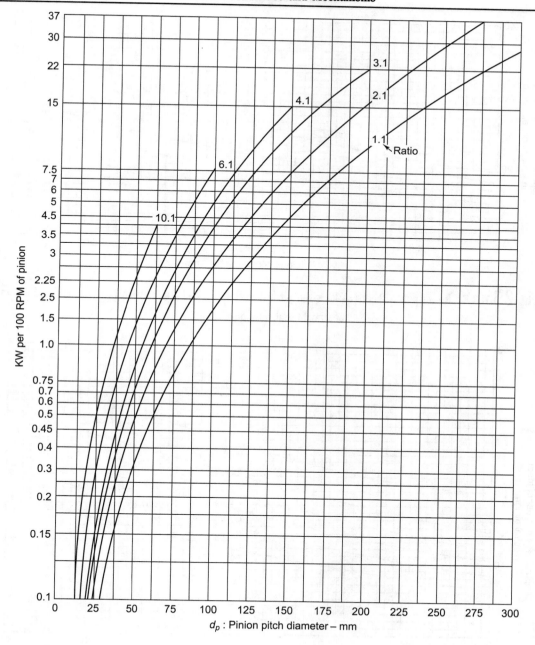

Fig. 2.34: Bevel gear pinion pitch diameters for 100 R.P.M. for various transmission ratios

In [**Fig. 2.31a**], (p 199) the curve for the 23 teeth pinion intersects the vertical line for the 46 teeth gear, at a point that gives a correction coefficient (x) of 0.32.

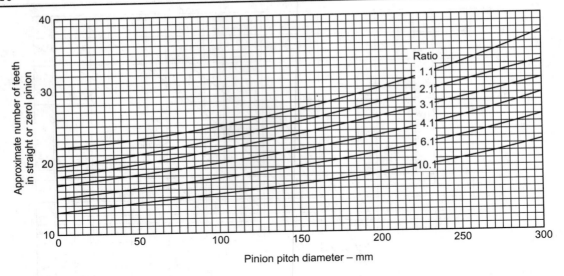

Fig. 2.35(a): No. of pinion teeth (t_1) for given pitch diameter (d_p) and transmission ratio for balanced design of straight and zerol bevel gears

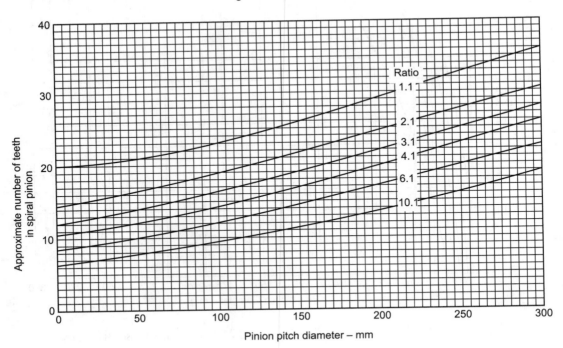

Fig. 2.35(b): No. of pinion teeth (t_1) for given transmission ratio and pitch diameter (d_p) for balanced spiral bevel gears

$$\therefore \text{Addendum for pinion } (h_{a1}) = m(1 + x) = 5.0(1 + 0.32)$$
$$= 6.6$$
$$\text{Addendum for gear } (h_{a2}) = m(1 - x) = 5.0(1 - 0.32)$$
$$= 3.4$$
$$\text{Dedendum for pinion } (h_{d1}) = m(1.188 - x)$$
$$= 5.0(1.188 - 0.32) = 4.34$$
$$\text{Dedendum for gear } (h_{d2}) = m(1.188 + x) = 7.54$$

$$\text{Pinion addendum angle } (\theta_{a1}) = \tan^{-1}\left(\frac{h_{a1}}{R}\right) = \tan^{-1}\left(\frac{6.6}{128.57}\right)$$
$$= 2.939°$$

$$\text{Gear addendum angle } (\theta_{a2}) = \tan^{-1}\left(\frac{h_{a2}}{R}\right) = \tan^{-1}\left(\frac{3.4}{128.57}\right)$$
$$= 1.515°$$

$$\text{Pinion tip angle } (\delta_{a1}) = \delta_1 + \theta_{a1} = 26.565 + 2.939$$
$$= 29.5$$

$$\text{Pinion root angle } (\delta_{d1}) = \delta_1 - \theta d_1 = 26.565 - \tan^{-1}\left(\frac{4.34}{128.57}\right)$$
$$= 24.632$$

$$\text{Gear tip angle} = (\delta_{a2}) = \delta_2 + \theta_{a2} = 63.435 + 1.515 = 64.95°$$

$$\text{Gear root angle} = (\delta_{d2}) = \delta_2 - \theta_{d2} = 63.435 - \tan^{-1}\left(\frac{7.54}{128.57}\right)$$
$$= 60.08°$$

As the module is still in the 1.5–10 range, the tolerances on the gear blank sizes and angles will be the same as the 3.5 module gears. (Table 2.35).

An increase in pitch diameter will allow more standard backlash and tooth thinning. Let us check the mean linear speed.

$$V = \frac{\Pi d_m N_1}{60,000} = \frac{\Pi(115 - 43 \times \sin 23.565)1,000}{60,000}$$
$$= 5.1 \text{ m/sec.}$$

As the speed exceeds 1 m/sec, class 9 gears cannot be used. The 2 m/sec limit also rules out the use of class 8 gears. It will therefore, be necessary to use class 7 accuracy, with a 5 m/sec speed limit.

2.10.7.4 Planetary [epicyclic] gears

Planetary gears are mounted on a carrier which rotates around the axis of an external gear called the sun-wheel. These equi-spaced planets (gears) connect the sun-wheel with another gear that is co-axial with the sun-wheel. In the simple planetary train [**Fig. 2.36a**], the planets mesh with the internal gear called 'annulus', co-axial with the sun-wheel.

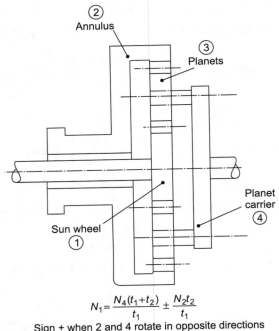

$$N_1 = \frac{N_4(t_1+t_2)}{t_1} \pm \frac{N_2 t_2}{t_1}$$

Sign + when 2 and 4 rotate in opposite directions
Sign − when 2 and 4 rotate in the same direction

Fig. 2.36(a): Simple planetary train

In a compound planetary train [**Fig. 2.36b**] the planet shaft has two co-axial gears. One engages with the sun-wheel, while the other meshes with the internal annulus. The planetary train, shown in [**Fig. 2.36c**], replaces the annulus with an external gear, while [**Fig. 2.36d**] shows the bevel gears and the sun-wheel, planets and the third gear, co-axial with the sun-wheel. In this case, as in the previous one, the planet gears move like planets around the two co-axial gears. Both these external co-axial gears are called sun-wheels.

If one of the three transmission elements (sun-wheel, planets carrier, or annulus) is fixed by locking its shaft, we get a fixed ratio transmission at the remaining 2 points. Either of the shafts can be used conveniently for input, or output.

It should be remembered that when two external gears mesh, and rotate in opposite directions, the transmission ratio is negative. Thus, when an external sun-wheel, and the planets carrier rotate in the same direction, the transmission ratio is positive. The signs are

Machine Tool Drives and Mechanisms

opposite when one of the meshing gears is internal. The transmission ratio is po[s?]— the planet carrier and the internal annulus rotate in opposite directions.

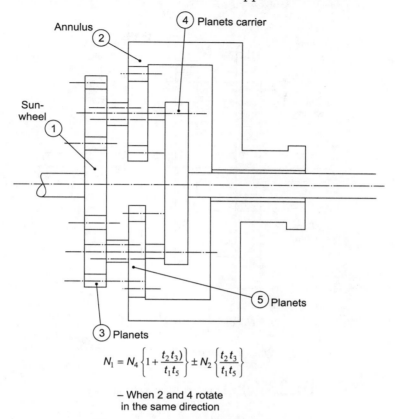

$$N_1 = N_4 \left\{ 1 + \frac{t_2 t_3}{t_1 t_5} \right\} \pm N_2 \left\{ \frac{t_2 t_3}{t_1 t_5} \right\}$$

– When 2 and 4 rotate in the same direction

Fig. 2.36(b): Compound planetary train

When one of the elements (sun-wheel, carrier, or the third gear) is fixed, its R.P.M. will be zero, and the train will have a fixed transmission ratio.

Compactness, high-transmission ratio, low weight, high efficiency, low rolling and sliding velocities, differential output torques, and speeds and silent running, make planetary gears appropriate for co-axial input and output applications. Load division between number of planetary pinions, reduces flywheel effect, vulnerability to shock loads, and balances the static forces. These gears do not generate any load on the shaft bearings, when straight spur gears are used for the constituent gears.

Example 2.20: In [Fig. 2.36a], the sun-wheel (1) has 40 teeth, the annulus (2) has 100 teeth, and the planets (3) have only 20 teeth. Find:

(a) Speed of the planet carrier (4) if the sun-wheel (1) is fixed, and annulus (2) is rotated at 150 R.P.M.

(b) Speed of the sun-wheel, when annulus is fixed, and the carrier (4) is rotated at 150 R.P.M.

speed of annulus (2), when the carrier (4) is fixed, and the sun-wheel is rotated at 150 R.P.M.

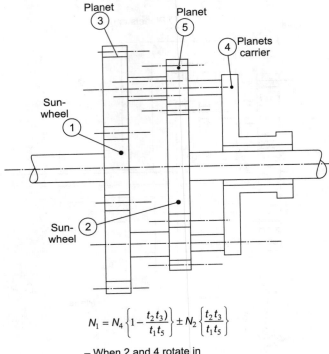

$$N_1 = N_4 \left\{1 - \frac{t_2 t_3}{t_1 t_5}\right\} \pm N_2 \left\{\frac{t_2 t_3}{t_1 t_5}\right\}$$

— When 2 and 4 rotate in opposite directions

Fig. 2.36(c): Planetary train W/2 sun-wheels

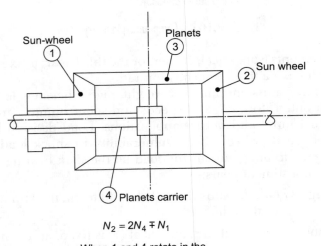

$$N_2 = 2N_4 \mp N_1$$

— When 1 and 4 rotate in the same direction

Fig. 2.36(d): Differential drive W/equal bevel gear planetary train

Solution: From [**Fig. 2.36a**], general equation for the drive speeds is:

$$N_1 = \frac{N_4(t_1 + t_2)}{t_1} \pm \frac{N_2 t_2}{t_1}$$

(a) The sun-wheel is fixed. Therefore $N_1 = 0$

$$0 = \frac{N_4(40 + 100)}{40} \pm \frac{150 \times 100}{40}$$

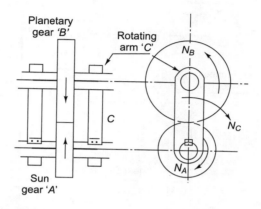

Fig. 2.36(e): Planetary (epicyclic) differential gears

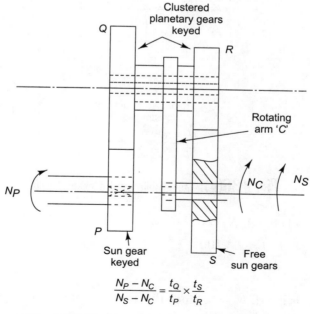

$$\frac{N_P - N_C}{N_S - N_C} = \frac{t_Q}{t_P} \times \frac{t_S}{t_R}$$

Fig. 2.36(f): Clustered planetary differential gears

The carrier will rotate in the same direction as the annulus (internal gear), so the sign will be negative.

$$N_4 = \frac{15{,}000}{40} \times \frac{40}{140} = 107.14$$

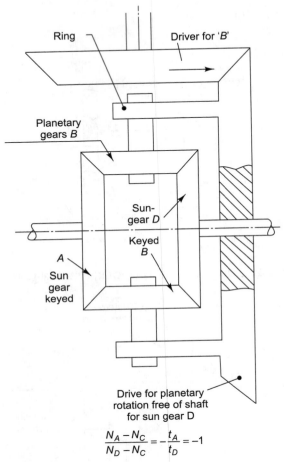

Fig. 2.36(g): Automobile bevel differential gear

(b) When the annulus is fixed, $N_2 = 0$

$$N_1 = 150 \times \frac{140}{40} \pm 0 \times \frac{100}{40}$$
$$= 525$$

(c) When the carrier is fixed, $N_4 = 0$. Since the annulus rotates in a direction opposite to the sun-wheel, the sign will be positive.

$$N_1 = 0 \times \frac{140}{40} + N_2 \times \frac{100}{40} = 150$$

$N_2 = 60$ R.P.M.

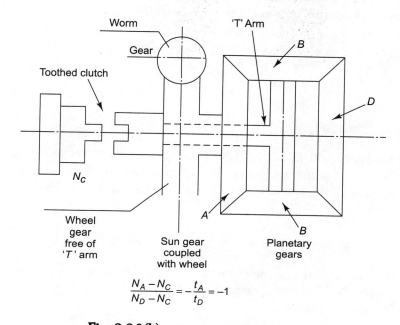

$$\frac{N_A - N_C}{N_D - N_C} = -\frac{t_A}{t_D} = -1$$

Fig. 2.36(h): Machine tool differential gear

If none of the shafts is locked, and all the three shafts are left to rotate freely, we will get a differential drive. In the case of a differential drive, any one of the three shafts can be used as a driver, to drive the other two.

2.10.7.5 Differential drives

These add (or subtract) two motions, by using planetary (epicyclic) gears. Gear trains can have spur, helical, or bevel gears.

The spur or helical gear differential [**Fig. 2.36e**] uses a stationary sun gear (A), and a planetary gear (B), mounted on an arm which rotates around the sun gear (A). Suppose gear A is rotating at 'N_A' revolutions per minute, and the arm for planetary gear (B) makes 'N_C' revolutions per minute around the axis of gear A, in the same direction (say clockwise), then the relative rotary speed of the sun gear (A), with respect to the planetary gear (B), will be $(N_A - N_C)$. If the planetary gear 'B' is rotating around its own axis at N_B R.P.M., the rotary speed of the planetary gear (B), with respect to the sun gear (A), will be $(N_B - N_C)$. The transmission ratio for external planetary gears is negative.

$$N_B = N_C \left(1 + \frac{t_A}{t_B}\right) - N_A \frac{t_A}{t_B}$$

2.10.7.6 Clustered planetary differential gears

Sometimes, a double cluster planetary gear is used. In this arrangement [Fig. 2.36f], a cluster of gears 'Q' and 'R' rotates around the common axis of gears 'P' and 'S' at 'N_C' revolution per minute. The relationship between the speeds of the spindles 'P' and 'S' is as per the equation below:

$$\frac{N_P - N_C}{N_S - N_C} = \frac{t_Q}{t_P} \times \frac{t_S}{t_R}$$

2.10.7.7 Bevel gear differentials

Automobile differential gears Figure 2.36g shows a bevel gear differential, used in automobiles. The pair of planetary gears 'B' is rotated around the common axis of gears 'A' and 'D' by a ring gear. If the speed of rotation of the planetary arm (ring for gear 'B') is N_C, then

$$\frac{N_A - N_C}{N_D - N_C} = \frac{t_A}{t_D} = -1 \text{ (as } t_A = t_D\text{)}$$

The minus sign indicates that if the planetary arm stops rotating, i.e. $N_C = 0$, gears 'A' and 'D' rotate in the opposite direction at equal speed $N_A = N_D$, and at all other times, when the planetary arm is rotating, then

$$N_A + N_D = 2 N_C$$

On a curve, the differential gear allows the inner and outer wheels of 4 wheeler vehicles to run at different speeds: the wheel near to the center of the radius of the curve runs slower, than the wheel farther from the center.

Example 2.21: Find the rotary speeds (R.P.M.) of the outer wheel and the planetary gears arm (ring), if the radius of turn for the inner wheel is 11 meters, and for the outer wheel, 13 meters. The inner wheel R.P.M. $N_A = 500$/min and wheel dia = 0.6 meter.

Solution: Linear speed of the inner wheel = ΠDN

$$= \Pi \times 0.6 \times 500 = 942.5 \text{ m/min}$$

$$\text{Linear speed of outer wheel} = \frac{13}{11} \times 942.48$$

$$= 1{,}113.9 \text{ m}$$

$$\text{R.P.M. of outer wheel} = \frac{1{,}113.9}{\Pi \times 0.6} \approx 591$$

$$N_C = \frac{N_A + N_D}{2} = \frac{500 + 591}{2} = 545.5$$

Differentials in machine tools [**Fig. 2.35h**] These use a 'T'-shaped arm for rotating the planetary gears 'B', around the common axis of the bevel gears 'A' and 'D'. The driving gear 'A' is rotated by a worm and wheel drive. The wheel, coupled with the driver 'A', is free to rotate on the 'T' shaft, rotating the planetary gear arm. Usually, a clutch is provided between the 'T' shaft and its driver to facilitate stoppage of the planetary arm rotation, even while free gear 'A' is rotating. Stoppage of the rotation of the planetary gear arm (around the common axis of gears 'A' and 'D'), reverses the direction of the rotation of gear 'D', to the opposite of gear 'A'.

2.10.7.8 Gears manufacture

Milling form cut gears The manufacturing method for gears depends upon the quantity required, and the type of the gear required. If only one gear is required as a replacement for a broken gear, it can be profile-milled by a fly tool, ground to the correct shape, or one can use a gear milling cutter of the required module, suitable for the number of teeth of the required gear, for the profile of the cutter depends upon the number of teeth. **Figure 2.37a** and **b**, give the profile of the cutter for up to 35, and more than 35 teeth. Table 2.40b gives the cutter nos. for various ranges of the number of teeth. Table 2.40 (c) gives the 'X' and 'Y' co-ordinates for the various points of the profile, for a 1.0 module gear. Needless to say, the co-ordinates for any module can be found, by multiplying the co-ordinates with the required module.

For hardened gears, the blanks are rough-machined before hardening. Finishing is done after hardening to RC 40–45. Finishing can be done by milling or grinding. If it is not possible to grind or finish-mill, gears can be made of oil hardening, non-shrinking steel (OHNS), and finished to size, by milling before hardening.

For a small batch production of similar gears, a mandrel accommodating a number of gears is used. The no. of blanks stacked on a mandrel can be machined in a single operation.

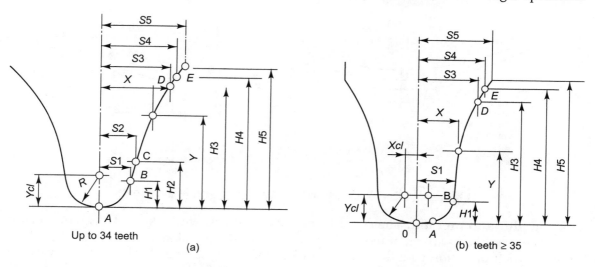

Fig. 2.37: Tooth profile for gear milling cutters

Table 2.40(b): Coefficients for determining coordinates of profile points B, C, D, and E [Fig. 12.6], of gear cutters for 1.0 module

Coordinates of non-involute elements of profile

Cutter no.	Number of gear teeth required	Point B		Points C		Center of circle		Point D		Point E	
		H_1	S_1	H_2	S_2	X_c	Y_c	H_3	S_3	H_4	S_4
1	12–13	0.5878	0.6412	0.8588	0.6652		0.6439	2.0351	1.5102		1.5861
2	14–18	0.5744	0.6228	0.7828	0.6451	0	0.6292	2.0649	1.4638	2.10	1.6059
3	17–20	0.5883	0.6093	0.6726	0.6793		0.6116	2.0942	1.4141	2.20	1.5087
4	21–25	0.5151	0.5853	0.5716	0.5925		0.5901	2.1186	1.3662		1.5123
5	26–34	0.4755	0.5642	0.4975	0.5680		0.5725	2.1369	1.3276		1.4442
6	35–54	0.4180	0.5365	–	–	0.0167	0.5321	2.1554	1.2842	–	1.3738
7	55–134	0.3375	0.4989	–	–	0.0578	0.4570	2.1731	1.2388	2.30	1.3040
8	135–rack	0.2550	0.4545	–	–	0.1034	0.3693	2.1897	1.1859		1.2333

A dividing head is used to divide equally, the gear periphery for the required number of teeth. The cutter must be set accurately above the center of the blank. One can carry out roughing almost to the full depth, leaving only a little (0.5–1 mm) material for the purpose of finishing. The gear blanks should be clamped rigidly by a nut.

Generation of teeth can be done by three methods:
(a) Sunderland method [**Fig. 2.37c**] with rack cutter.
(b) Gear shaper with pinion cutter.
(c) Hobbing.

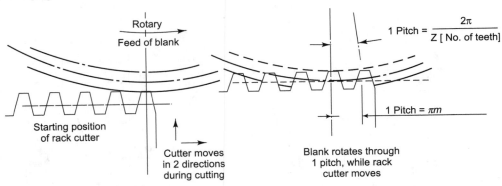

Fig. 2.37(c): Sunderland method of generating gears by a rack cutter

(a) **Sunderland method:** The cutter is shaped like a rack, with rake and clearance angles. The gear blanks are mounted on an arbor (mandrel) on a slide, which rotates the arbour in synchronization with the cutter's strokes and the cross travel. The cutter travels one pitch, while the gear blanks rotate by one pitch, before the cutter is withdrawn and

Table 2.40(c): Coordinates (mm) x and y of profile points on an involute section of gear cutters, for module 1.00 mm

Cutter 1		Cutter 2		Cutter 3		Cutter 4		Cutter 5		Cutter 6		Cutter 7		Cutter 8	
y	x	y	x	y	x	y	x	y	x	y	x	y	x	y	x
0.90	0.6778	0.80	0.6488	0.70	0.6295	0.60	0.5939	0.50	0.5684	0.50	0.5555	0.40	0.5160	0.30	0.4662
1.00	0.7138	0.90	0.6770	0.80	0.6464	0.70	0.6172	0.60	0.5888	0.60	0.5817	0.50	0.5547	0.40	0.5025
1.10	0.7586	1.00	0.7137	0.90	0.6766	0.80	0.6441	0.70	0.6141	0.70	0.6106	0.60	0.5718	0.50	0.5359
1.20	0.8108	1.10	0.7569	1.00	0.7134	0.90	0.6459	0.80	0.6420	0.80	0.6373	0.70	0.6077	0.60	0.5708
1.30	0.8697	1.20	0.8061	1.10	0.7551	1.00	0.7127	0.90	0.6764	0.90	0.6764	0.80	0.6414	0.70	0.6051
1.40	0.9352	1.30	0.8611	1.20	0.8017	1.10	0.7531	1.00	0.7131	1.00	0.7125	0.90	0.6764	0.80	0.6404
1.50	1.0077	1.40	0.9217	1.30	0.8529	1.20	0.7976	1.10	0.7527	1.10	0.7515	1.00	0.7129	0.90	0.6760
1.60	1.0860	1.50	0.9876	1.40	0.9084	1.30	0.8456	1.20	0.7952	1.20	0.7925	1.10	0.7505	1.00	0.7129
1.70	1.1711	1.60	1.0591	1.50	0.9685	1.40	0.8959	1.30	0.8408	1.30	0.8353	1.20	0.7898	1.10	0.7499
1.80	1.2635	1.70	1.1357	1.60	1.0325	1.50	0.9528	1.40	0.8891	1.40	0.8804	1.30	0.8301	1.20	0.7871
1.90	1.3631	1.80	1.2181	1.70	1.1010	1.60	1.0103	1.50	0.9408	1.50	0.9273	1.40	0.8720	1.30	0.8271
2.00	1.4704	1.90	1.3059	1.80	1.1710	1.70	1.0729	1.60	0.9941	1.60	0.9764	1.50	0.9149	1.40	0.8639
2.10	1.5861	2.00	1.3983	1.90	1.2507	1.80	1.1368	1.70	1.0505	1.70	1.0270	1.60	0.9594	1.50	0.9030
2.20	1.7106	2.10	1.4997	2.00	1.3324	1.90	1.2052	1.80	1.1095	1.80	1.0805	1.70	1.0050	1.60	0.9422
2.30	1.8436	2.20	1.6059	2.10	1.4181	2.00	1.2770	1.90	1.1711	1.90	1.1354	1.80	1.0520	1.70	0.9821
2.40	1.9884	2.30	1.7191	2.20	1.5087	2.10	1.3519	2.00	1.2354	2.00	1.1919	1.90	1.0999	1.80	1.0229
2.50	2.1464	2.40	1.8395	2.30	1.6041	2.20	1.4303	2.10	1.3024	2.10	1.2507	2.00	1.1493	1.90	1.0572
2.60	2.3173	2.50	1.9685	2.40	1.7040	2.30	1.5123	2.20	1.3720	2.20	1.3114	2.10	1.1996	2.00	1.1060
2.70	2.5046	2.60	2.1037	2.50	1.8090	2.40	1.5978	2.30	1.4442	2.30	1.3738	2.20	1.2509	2.10	1.1485
—	—	2.70	2.2435	2.60	1.9190	2.50	1.6887	2.40	1.5190	2.40	1.4380	2.30	1.3040	2.20	1.1907
—	—	—	—	2.70	2.0353	2.60	1.7796	2.50	1.5965	2.50	1.5042	2.40	1.3580	2.30	1.2333
—	—	—	—	—	—	2.70	1.8760	2.60	1.6767	2.60	1.5725	2.50	1.4127	2.40	1.2778
—	—	—	—	—	—	—	—	2.70	1.7597	2.70	1.6422	2.60	1.4694	2.50	1.3213
—	—	—	—	—	—	—	—	—	—	—	—	2.70	1.5269	2.60	1.3654
—	—	—	—	—	—	—	—	—	—	—	—	—	—	2.70	1.4112

moved back by one pitch. The arbor is indexed by one pitch to repeat the process, till all the teeth are generated. Productivity can be increased by using a machine with two arbors and two cutters. Machined gears from one arbor are replaced with fresh blanks, while the machine cuts teeth on the blanks, on the other arbor.

(b) **Gear shaper [Fig. 2.37d]:** It uses a pinion gear, with rake and clearance angles, to generate the required gear. Depending upon the shape of the required gear, the cutting stroke can be downward or upward. Both, the gear blanks and the shaping pinion cutter, are mounted on their arbors. Both the arbors are geared to each other by change gears.

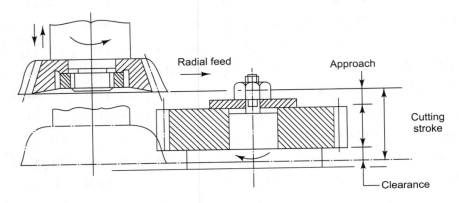

Fig. 2.37(d): Gear-shaping

Cutting commences with the gradual radial feed of the pinion cutter, to full-tooth depth in the blank, while the cutter reciprocates in cutting strokes. After cutting to full depth, the indexing feed is engaged. It rotates both the arbors. Gradually, the entire periphery of the blanks is shaped. Bigger gears may require 2 or 3 cuts, each cut after one full rotation of the blanks arbor. Needless to say, it is necessary to change gears on the arbors for different sizes of gears.

Internal gears can be shaped only on a gear shaper. To facilitate the manufacture of internal gears with minimum overshoot, the cutter is made saucer-shaped **[Fig. 2.37d,f]**. There is a counterbore for the arbor nut, to ensure that no part of the arbor assembly protrudes below the cutting teeth level. Table 2.40d below, gives the standard pitch diameters of the shaping cutters for the standard modules.

Table 2.40(d): Standard pitch diameters (D_p) for gear shapers.

Module	1	1–2	1–3	1–5	1–8	2–10	6–10	8–12
D_p (mm)	35	38	50	75	100	125	160	200

No. of teeth = $\dfrac{D_p}{m}$

The pitch diameter (D_p) is increased to provide material for resharpening. This is done by increasing the tip thickness (tt). The 'D_p' increase is however, limited by a permissible variation in the tip thickness of the required gear. The following table gives the range of permissible tooth thickness, in gears.

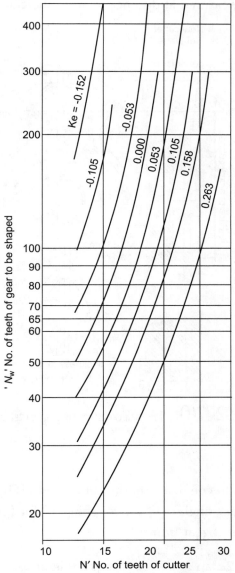

Fig. 2.37(e): Extension coefficient for gear shaping cutters

Table 2.40(e): Permissible tooth thickness (tt) in gears [Fig. 2.37f]

Module	1–1.5	1.75–2.75	3–4	4.25–6
'tt' range	0.41–0.5 m	0.31–0.4 m	0.25–0.3 m	0.20–0.25 m

A new cutter shapes the gear tip to the minimum permissible tip thickness. The cutter can be used till it wears out to generate a gear with the maximum permissible tip thickness.

The pitch diameter is measured at distance 'a' from the cutting face plane. [**Fig. 2.37f**]

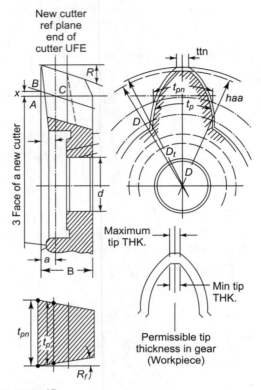

Fig. 2.37(f): Tooth thickness of shaping cutter and gear

$$a = \frac{K_e m}{\tan R}$$

K_e = Extension coefficient of the cutter (from [**Fig. 2.37e**] on p 229)

R = Clearance angle = 6° for roughing, 9° for finishing.

m = Module of gear (mm)

Cutter tip dia $(D_t) = D_p + 2a.\tan R + \dfrac{2.5 m}{\cos R}$ (to js 8 tolerance)

Rake angle $(\alpha) = 5°$

Bore ($H4$) to suit arbor diameter. Note that there is no keyway. The cutter drive is thru friction with the arbor.

Tolerances for gear shaping pinion cutters
Roundness within 0.005

Faces squareness within 0.005, with arbor bore.
Faces parallel within 0.008, with each other.
Surface finish 0.25 μ for cutting angles, and,
   0.63 μ for clamping face

Hobbing uses a rotating, screw-like cutter (hob), to generate gear teeth.

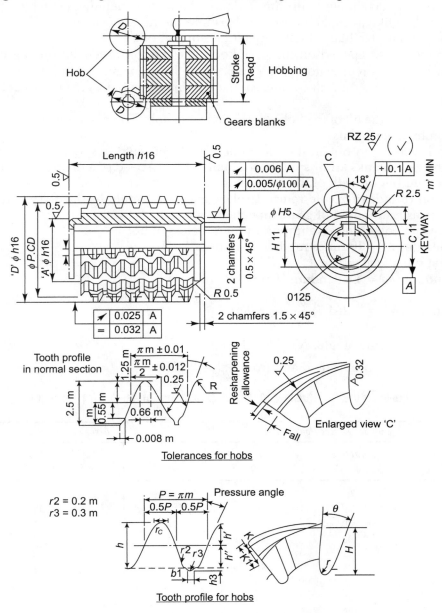

Fig. 2.37(g): Hobbing gear teeth

The hob [**Fig. 2.37g**] is a worm gear, slotted radially, to provide cutting teeth with the required rake and clearance angles, and the gullet space, for the cut-off chips. The pitch is equal to the circular pitch of the required gear. Rotation of the hob generates an action, similar to gradually pushing a rack in a direction that is square to the gear axis. It is like rotating the gear-blanks arbor by the hob-rack, a condition necessary for generating involute teeth. The condition is similar to the Sunderland method of generating teeth by a rack-shaped cutter. The hob is like a rotating rack. Furthermore the number of cutting teeth reduces the wear, per tooth, and facilitates higher material-removal rate.

The hob teeth are relieved on the periphery, like the form-relieved milling cutters. The relief is however, in the direction of the helix of the hob. For machining straight spur gears, the hob axis is tilted by the helix angle, so that the thread on the worm is parallel to the gear blanks axis.

For helical gears, the tilt of the hob should make the hob-thread, parallel to the helical teeth being cut on the gear blank.

The hob is rotated to obtain the recommended linear cutting speed. The hob spindle is geared to the gear-blanks arbor spindle, to rotate it at the speed of a mating wheel, with teeth equal to the number of teeth of the required gear.

Elimination of indexing, idle return strokes, and faster multi-teeth rotary cutting, render hobbing, the fastest method of gears manufacture.

The following equations give important parameters of hobs:
1. Hob length $(L) = 10$ times module $= 10$ m
2. Normal pitch $(pn) = \pi$ m
3. Addendum $(ha) =$ Dedendum $(hd) = 1.25$ m
4. Cam relief $(K) = (1.2–1.7) \dfrac{\pi D}{N} \tan R$;

$R =$ Relief angle, $N =$ No. of teeth,
$D =$ Hob outside diam. from Table 2.40(e)

Table 2.40(e): Outside ⌀, arbor ⌀, and no. of teeth (*N*) for hobs

Gear module	Grade 7 precision			Grade 8–10 precision		
	Outside ⌀ *'D'*	Arbor ⌀ *'A'*	No. of teeth *N*	Outside *D*	Arbor *A*	Teeth *N*
1–1.25	70	32	16	63	27	12
1.5–1.75	80	40	16	63	27	12
2–2.25	90	40	14	70	27	12
2.5–2.75	100	40	14	80	32	10
3–3.75	112	40	14	90	32	10
4–4.5	125	50	14	100	32	10
5–5.5	140	50	14	112	40	10
6–7	160	60	12	125	40	9
8	180	60	12	140	40	9
9	200	60	12	140	40	9

5. Corner radius $(rc) = 0.25 - 0.3$ m
6. Taking into account the reduction in outside dia (D) during resharpening,

$$D_p = D - 2.5 \text{ m} - 0.3 \frac{\pi D}{N} \tan R$$

7. Hob thread lead angle $(W) = \sin^{-1}\left(\frac{nm}{D_p}\right)$

   n = No. of starts of worm thread; m = module
   D_p = Pitch diameter from (6)

8. Axial pitch $(pa) = \dfrac{pn}{\cos W}$
9. Axial lead $= n\,(pa)$
10. Axial pressure angle $= PA = \cot^{-1}(\cot P \cos W)$

    P = Normal pressure angle $(14.5°/20°)$
    W = Lead angle from (7)

11. Flute (Gullet)
    a. Root radius $(r) = 0.2 - 0.3$ m
    b. Flute depth (H)

    $$H = 2.5 \text{ m}\left(1 + \frac{\pi D}{N}\tan R\right) + r$$

    c. Axial tooth thickness $(ta) = \dfrac{\pi m}{2} \cdot \dfrac{1}{\cos W}$
    d. Flutes lead $= \dfrac{\pi D_p}{\tan W}$
    e. Flute angle $\theta = 22° - 25°$

2.10.7.9 Bearing loads due to gears

1. Spur gears

Gears exert radial and tangential, loads and moments, on the mounting shafts, which pass them on to the bearings carrying the shafts. For straight spur gears,

$$\text{Tangential force} = F_t = \frac{9,55,000 \times 2 \text{ kW}}{Nd_p}$$

$$\text{Radial force} = F_r = F_t \tan \alpha' t$$

$$\cos \alpha'_t = \frac{C}{C'} \cos \alpha_t$$

d_p = Pitch ϕ of the gear $= mt_1$ [mm]

C = Theoretical center distance = $\dfrac{m(t_1 + t_2)}{20}$ [cms]

C' = Actual center distance [cms]

α_t = Transverse pressure angle
 = $\tan^{-1}(\tan \alpha_n \sec \beta)$, where β = Helix angle,

and, α_n = Normal pressure angle $[20°/14\tfrac{1}{2}°]$

$\alpha_n = \alpha$, for straight spur gears.

Helical gears also generate axial force (F_a) due to the helix angle. This causes an axial thrust on the bearings. The resultant radial forces, on the bearings at A and B [**Fig. 2.38**], can be found from the following equations:

$$F_{rA} = F_t \sqrt{\left(\dfrac{1_B}{1}\right)^2 + \left(\dfrac{1_B}{1}\tan\alpha'_t + \dfrac{d_p}{21}\tan\beta\right)^2}$$ (Eqn. 2.90a)

$$F_{rB} = F_t \sqrt{\left(\dfrac{1_A}{1}\right)^2 + \left(\dfrac{1_A}{1}\tan\alpha'_t - \dfrac{d_p}{21}\tan\beta\right)^2}$$ (Eqn. 2.90b)

Axial thrust = $F_a = F_t \tan\beta$

Note that the radial forces F_{rA} and F_{rB}, are specified for the same direction of rotation, and helix angle, as shown in [**Fig. 2.38**]. The plus (or minus) sign preceding the term $\left(\dfrac{d_p}{21}\tan\beta\right)$ will change to its opposite, if either the helix angle, or the direction of rotation is reversed.

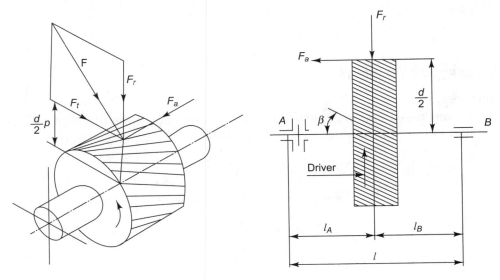

Fig. 2.38: Bearing loads due to spur gears

2. Bevel gears [Fig. 2.39]

Angular contact of the bevel gears tends to separate the meshing gears, and causes an axial thrust. Separating force $= F_s = F_t \left(\dfrac{\tan \alpha_n \cos \delta}{\cos \beta} \pm \tan \beta \sin \delta \right)$

The spiral direction is observed from the apex side. For a driver bevel, the sign preceding the term $\tan \beta \sin \delta$, is positive, when the direction of rotation is opposite to the direction of helix (as viewed from apex). For the driven gear, it is opposite: the sign for $\tan \beta \sin \delta$ is minus, when the direction of rotation is opposite to the direction of the spiral.

$$\text{Axial force} = F_a = F_t \left(\dfrac{\tan \alpha_n \sin \delta}{\cos \beta} \pm \tan \beta \cos \delta \right)$$

The sign preceding $\tan \beta \cos \delta$ is positive, when the driving bevel gear is rotating in the same direction as the spiral. In case of the driven gear, the sign is positive when the direction of rotation is opposite to the direction of the spiral as in [Fig. 2.39a]. For mountings in [Fig. 2.39b], for both types of mounting sign– for Fig p + for Fig. q.

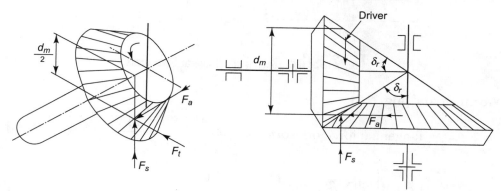

$d_n = (d_p - b \sin 2\delta)$
M = Tilting moment on shaft, caused by axial force
F_t is calculated as in spur and helical gears

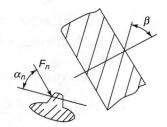

Fig. 2.39(a): Separating and axial forces generated by bevel gears

Total load at Brg A $= F_a = \sqrt{\left(\dfrac{F_t l_b}{1} \right)^2 + \left(\dfrac{F_s l_b}{1} - \dfrac{F_a d_m}{21} \right)^2}$ for both types of mounting

Total load at Brg B = $F_B = \sqrt{\left(\dfrac{F_t 1_a}{1}\right)^2 + \left(\dfrac{F_s 1_a}{1} \mp \dfrac{F_a d_m}{21}\right)^2}$ sign – for Fig. p
+ for Fig. q

$$d_m = d_p - b \sin 2\delta$$

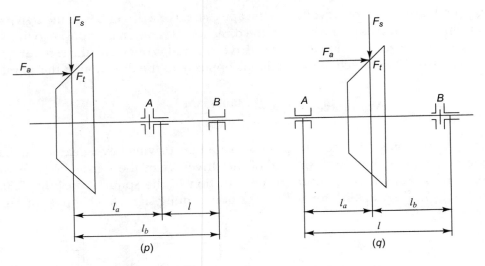

Fig. 2.39(b): Bearing loads due to bevel gears

3. Worm gears [Fig. 2.4]

$$\text{Tangential force on worm} = F_{t1} = \dfrac{9,55,000 \times 2\,\text{kW}}{N_1 d_p}$$

$$F_{a1} = F_t \cot\left[(90 - \beta) \pm \tan^{-1}\left(\dfrac{\mu}{\cos \alpha_n}\right)\right] \quad \text{(Eqn. 2.91)}$$

β = Helix angle of worm [Fig. 2.40]

μ = Coefficient of friction

α_n = Normal pressure angle $\left(20° \text{ or } 14\dfrac{1}{2}°\right)$

Positive sign, for worm as a driver, and negative sign, for wheel as a driver.

$$F_r = F_a \dfrac{\tan \alpha_n}{\cos(90 - \beta)} \quad \text{(Eqn. 2.92)}$$

$$F_A = \sqrt{\left(F_{t1} \dfrac{1_B}{1}\right)^2 + \left(\dfrac{F_{s1} 1_B}{1} \pm \dfrac{F_{a1} d_p}{21}\right)^2} \quad \text{(Eqn. 2.93a)}$$

Positive sign should be used for F_a for the direction of rotation, shown in **[Fig. 2.40]**. The sign will be negative for rotation in the opposite direction.

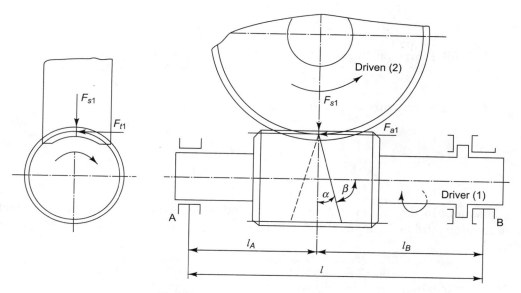

Fig. 2.40: Bearing loads due to worm and wheel

$$F_B = \sqrt{\left(F_{t1}\frac{1_A}{1}\right)^2 + \left(\frac{F_{s1}1_A}{1} \pm \frac{F_{a1}d_p}{21}\right)^2}$$ (Eqn. 2.93b)

We will use the minus sign for F_B for rotation direction as shown in the figure, and positive sign for the opposite direction.

Force on the wheel:
Tangential force = F_{a1}
Axial force = F_{t1}
Radial force = F_{s1}

Example 2.22: Find the forces F_A and F_B on bearings, in **[Fig. 2.38]**, if span $l = 160$; $1_A = 88$, $1_B = 72$, and the power to be transmitted is 2.2 kW at 74 R.P.M. The helix angle of the gear is 11°. It is leading in the same direction, as the direction of rotation, shown in the figure. The pitch diameter of the gear is 81.497, and $\alpha_n = 20$.

Solution:

$$\text{Tangential force } F_t = \frac{95{,}5000 \text{ kW}}{N} \times \frac{2}{d_p}$$

$$= \frac{95,5000 \times 2.2}{74} \times \frac{2}{81.497}$$

$$= 696.76 \text{ kg}$$

$$\text{Radial force on Brg A } (F_{rA}) = F_t \sqrt{\left(\frac{1_B}{1}\right)^2 + \left(\frac{1_B}{1} \tan \alpha'_t \pm \frac{d_p}{21} \tan \beta'\right)^2}$$

$$\text{Transverse pressure angle } (\alpha_t) = \tan^{-1}\left(\frac{\tan \alpha_n}{\cos \beta}\right)$$

$$= \tan^{-1}\left(\frac{\tan 20°}{\cos 11°}\right)$$

$$= 20.3439°$$

As no adjustment in the center distance is specified, $\beta' = \beta$ and $\alpha'_t = \alpha_t$. For the given direction of rotation, and helix angle, the sign before the last term will be: positive for F_{rA}, and negative for F_{rB}.

$$\therefore \quad F_{rA} = 696.8 \sqrt{\left(\frac{72}{160}\right)^2 + \left(\frac{72}{160} \tan 20.3439 + \frac{81.497}{2 \times 160} \tan 11°\right)^2}$$

$$= 347.91 \text{ kg}$$

$$F_{rB} = 696.8 \sqrt{\left(\frac{88}{160}\right)^2 + \left(\frac{88}{160} \tan 20.3439 - \frac{81.497}{320} \times 0.194\right)^2}$$

$$= 398.1 \text{ kg}$$

If the direction of the helix, or the rotation is changed, the signs before the term $\frac{d_p}{21} \tan \beta'$ will become the opposite.

Example 2.23: Find the loads on bearings A and B [**Fig. 2.39b/p**], for a pair of straight bevel gears, with 20° pressure angle (α_n), pinion cone angle (δ_1) of 26.565°, gear cone angle (δ_2) = 63.435°, driver pinion pitch ϕ (d_p) = 70, and width (b) = 25.
Driven gear pitch ϕ (D_p) = 140,
Tangential force (F_t) = 300 kg,
Distance between the bearings (l) = 200, and,
Overhang of the pinion (1_A) = 85
Refer to [**Figs. 2.39a and 2.39b/p**]

Solution: It is necessary to first calculate the radial or separating force (F_s), and axial force (F_a).

For the pinion,

$$F_s = F_t\left(\frac{\tan\alpha_n \cos\delta_1}{\cos\beta} \pm \tan\beta \sin\delta_1\right)$$

The sign is positive for the driving member, and negative for the driven gear. Thus, it will be positive for the pinion. Helix angle $\beta = 0$.

$$F_{sp} = 300\left(\frac{\tan 20° \cos 26.565°}{\cos 0} + \tan 0° \sin 26.565\right)$$

$$= 300\left(\frac{0.36397 \times 0.8944}{1} + 0 \times 0.447\right)$$

$$= 97.663 \text{ kg}$$

Axial force on pinion $(F_a) = F_t\left(\dfrac{\tan\alpha_n \sin\delta_1}{\cos\beta} \pm \tan°0 \cos 26.565\right)$

$$= 300\left(\frac{0.36397 \times 0.447}{\cos 0°} + \tan 0° \times 0.89\right)$$

$$= 48.83 \text{ kg}$$

From **[Fig. 2.39b]**, reactions at bearing A, due to radial (F_s), axial (F_a), and tangential (F_t) forces are

$\dfrac{F_s l_B}{1}, \dfrac{F_a d_m}{21}$ and $\dfrac{F_t l_B}{1}$, respectively.

Resultant radial force at $A = F_A$

$$F_A = \sqrt{\left(\frac{F_t l_B}{1}\right)^2 + \left(\frac{F_s l_b}{1} + \frac{F_a d_m}{21}\right)^2}$$

It is necessary to find d_m first.

$$d_m = d_p - b \sin 2\delta_1 = 70 - 25 \sin(2 \times 26.565)$$

$$= 70 - 25 \sin 53.13 = 50$$

$$F_A = \sqrt{\left(\frac{300 \times (85 + 200)}{200}\right)^2 + \left(\frac{97.663 \times 285}{200} - \frac{48.83 \times 50}{2 \times 200}\right)^2}$$

$$= 447.7 \text{ kg}; \text{ Similarly } F_B = 132.32$$

2.10.8 Multi-speed electrical motors

Two-speed motors, double the range of spindle speeds. Moreover, the speed can be changed even while the spindle is running, without any stoppage, as is otherwise required, during stepped change, by a quick action gear box.

A two-speed motor (3000/1500 or 1500/750 R.P.M.) can reduce one stage of gear transmission. In a six-speed gear box, the first stage of gears between shafts I and II, can be eliminated altogether, by using a two-speed electric motor. For the next stage, two choices are available: higher speed range, or lower speed range. **Figure 2.24a** (p 178) shows ray diagrams for the lower range (188–595). **Figure 2.24b** shows the ray diagram for the higher range (945–3,000 R.P.M.). The motor is coupled directly with shaft I.

Figure 2.24c shows the ray diagram for obtaining 12 speeds from a two-stage gear box, driven by a two-speed [1500/750 R.P.M.] motor.

Although two-speed motors cost much more, the advantage of changing speed without stoppage, and the possibility of doubling the number of speeds, justifies their usage in certain applications.

Smaller speed ratios (1.12) can also be considered for machines, calling for more speeds in a lesser range, or machines warranting the usage of a two-speed electric motor.

Feed gear boxes   Generally, the feed values are also arranged in a geometric progression. Heavy-duty engine lathes however, use an approximate arithmetic progression.

A wide range of applications, calling for different types of motions (continuous/intermittent), degrees of precision (thread cutting/turning), types of loads (heavy/light), frequency of change (after minutes/hours), and sources of power (drive from main spindle/separate motor), led to the development of a wide variety of feed gear boxes, suitable for different applications.

The classification of feed change methods is done on the lines of that in the speed change, with the same criteria: frequency of change, and proportion of running (cutting) time (minutes/hours/shifts) to the setting time (minutes/hours). As explained earlier, under para 2.121a (p 134), under sub-heading 'Change gears', for long-run mass production machines, replaceable change gears can be used economically to change feed. We have already seen in **[Fig. 2.12a]**, how a pivoting quadrant can be conveniently used for mounting and setting (meshing), the change gears in center lathes.

In bar autos, the fixed centers of the feed shaft, and the machine spindle, are generally close enough to mesh the gears, mounted on them. No intermediate idler gear is required. The center distance between the spindle and the feed shaft, and the module of the gears used, determines the range of the possible transmission ratios.

Example 2.24: Find the ideal center distance between the spindle and feed shafts, to avail the maximum transmission ratio range, of helical spur gears. Module of gears = 2.5, min no. of teeth on pinion = 20.

Solution: From Table 2.7(p 102), we know that feed drive with helical spur gears, can have a wide transmission ratio range: $0.2 \left(\dfrac{1}{5}\right) - 4$.

If t_1 = No. of teeth on pinion on spindle
= 20 (min. given),
t_2 = No. teeth on feed shaft, then,

Transmission ratio $[i] = \left[\dfrac{N\ out}{N\ in}\right] = \dfrac{t_1}{t_2} = \dfrac{1}{5}$

$t_2 = \dfrac{t_1}{1/5} = 5\ t_1 = 5 \times 20 = 100$

Center distance $= \dfrac{m(t_1 + t_2)}{20}$

$= \dfrac{2.5\,(20 + 100)}{20}$

$= 15.0$ cms

Figure 2.41a shows the arrangement of the feed gears.

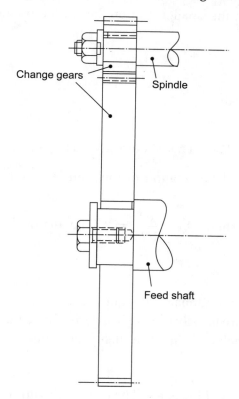

Fig. 2.41(a): Replaceable gears for changing feed

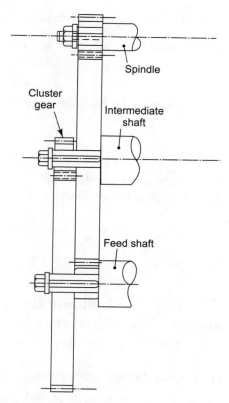

Fig. 2.41(b): Using intermediate shaft to increase speed range

The introduction of an intermediate shaft, can reduce the speed of the feed shaft to $\frac{1}{25}$ of the spindle speed.

Example 2.25: Find the minimum number of teeth on the spindle gear, intermediate cluster gear, and feed shaft gear, to avail the maximum transmission ratio, range for straight spur gears. Minimum number of teeth on the pinion = 18. Find the pitch of feed lead screw if the minimum feed = 0.1/Rev.

Solution: The feed shaft will run the slowest, and the spindle, the fastest.

∴ Smallest gear on spindle will have 18 (t_1) teeth. The mating gear on the intermediate shaft will have 't_2' teeth.

$$i = \frac{N\text{ out}}{N\text{ in}} = \frac{t_1}{t_2} = \frac{1}{5}$$

∴ $\quad t_2 = 5\, t_1 = 5 \times 18 = 95$

To avail the same minimum transmission ratio $\left(\frac{1}{5}\right)$, the smaller gear on the cluster, on the intermediate shaft will have a minimum of 18 teeth.

$$t_3 = 18$$

The gear on the feed shaft will have $\frac{18}{1/5}$, i.e. 95 teeth.

Now, the feed shaft runs at $\frac{1}{25}$ of the spindle R.P.M. This should give us a 0.1 axial movement, through the lead screw. In other words, 1/25 rotation of the feed lead screw, should give us 0.1 axial motion.

∴ For 1 R.P.M. of the feed shaft, the axial motion should be $= \frac{0.1}{1/25} = 2.5$, i.e. the pitch of the screw (by definition).

Gear boxes with sliding cluster gears

These are similar to the quick-change, spindle speed gear boxes, discussed earlier. Such gear boxes are usually used in heavy-duty machines, with a separate motor for powering feed. Milling machines often use such gear boxes.

Feed gear boxes with cone gears [Fig. 2.42]:

These are similar to cone pullies and have two sets of meshing gears. The driving [input] shaft (I) has one set of cone gears keyed rigidly to it. However, the driven gears on shaft II, are not keyed to their shaft. They rotate freely on shaft II, which is provided with slot for an axially sliding key. This key is pivoted in a keyholder, which can be moved axially, along the driven shaft. II. The key therefore, can be slid axially, to engage with any of the gears on shaft II. The driven shaft rotates at the speed of the gear engaged with the sliding key. A flat spring, attached to the sliding key, holds the key, that is engaged with the gear keyway. The spring also allows the key to swing out of the gear keyway, to enable an axial shift. The spacers between the driven gears, prevent risky, simultaneous engagement of the sliding key, with the two adjacent gears.

The weakening of the driven shaft II, by the slot, the limited width of cone gears, and the slenderness of long and small sliding keys, renders this arrangement unsuitable for high-torque, high-speed applications. Still, they are quite convenient to use, especially in small machines.

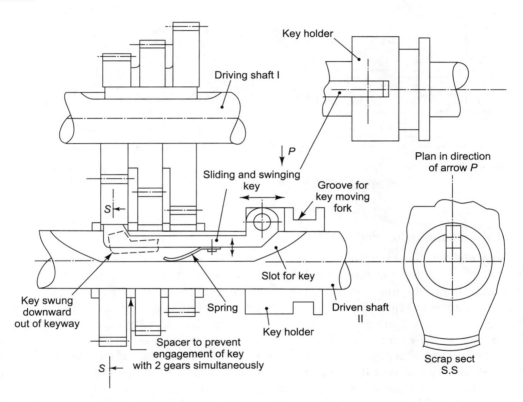

Fig. 2.42: Cone gears with sliding key for feed gear boxes

Norton's tumbler gear [Fig. 2.43]

This has only one cone gear which is mounted on the input shaft I. Transmission is accomplished by using a pair of meshed gears. The smaller one is mounted on the output (driven) shaft (II) and can be slid axially, on shaft II. This gear meshes permanently, with an intermediate gear, mounted on an arm that is swivelable, around the output shaft (II). The swivel is used to engage (mesh) and disengage (unmesh) the intermediate idler gear, with any of the cone gears mounted on the input (driving) shaft. While pivoting away from the axis of the input shaft, unmeshes the intermediate gear, pivoting towards it, meshes the intermediate gear. Thus, axial adjustment of the driven gear, and swiveling adjustment of the intermediate gear, allow the intermediate gear to mesh with any of the cone gears, and transmit motion from driving shaft I to driven shaft II. The transmission ratio can be changed suitably, by enmeshing the intermediate gear with a suitable gear on the cone.

Although location pin and mating holes are provided to hold the meshed gears in position, wear of pin/hole, reduces the meshing accuracy. Also, position-holding by a small pin reduces rigidity. Tumbler gears are used widely in thread-cutting applications on lathes. They are more suitable for arithmetic progression, used on engine lathes.

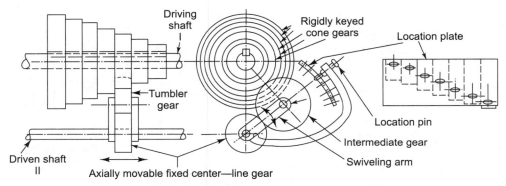

Fig. 2.43: Norton's tumbler gear

Meander's mechanism [Fig. 2.44]

Meander's mechanism combines the speed range of a sliding clusters gear box, and easy engagement of Norton's tumbler gear. Like in the case of a 2-stage gear box there are 3 shafts. Driving shaft I, and the intermediate shaft II are mounted with a number of similar, two-gear clusters, with an equal number of teeth for the smaller and bigger gears.

Only one cluster, on the extreme left, on the input shaft I, is keyed to its shaft. No other cluster is keyed. So except for the cluster with gear nos. 1 and 2 [**Fig. 2.44a**], all other clusters are free to rotate, independent of the shafts on which they are mounted. The keyed cluster with gears ½, drives the cluster with gears 3/4, and 5/6. The cluster with gears 5/6, drives the cluster with gears 7/8, which transmits the drive through gear 8, to cluster gears 9/10, 11/12, and 13/14. So gears 1 to 14 rotate continuously, as long as the input shaft I is running.

Feed shaft III gets its drive from the gears on the intermediate shaft II. The splined feed shaft is mounted with a gear, which can slide axially onto shaft III. The feed shaft gear is permanently meshed with an intermediate idler gear, mounted on an arm, which can swivel around the feed shaft. Like Norton's tumbler gear, the idler is swung towards shaft II, to mesh with any of the chosen gears on the intermediate shaft. The swiveling arm is swung away from shaft II, to unmesh the gears.

The smaller, as well as the bigger gears on the clusters, act as the driver, or driven. Hence, the transmission ratio (i) cannot exceed the maximum limit of 2.0 for straight spur (feed) gears (Table 2.7 p 103), and 2.5 for helical spur (feed) gears.

Machine Tool Drives and Mechanisms

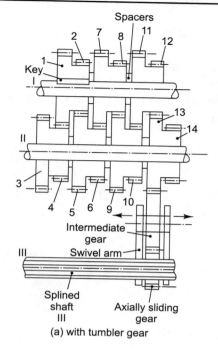
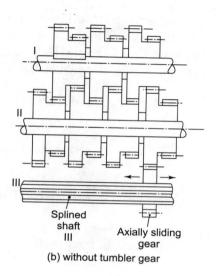

Fig. 2.44: Meander's mechanism

Example 2.26: Find the rotary speeds for feed shaft I in [**Fig. 2.44a**], if the speed of the input shaft I is 750 R.P.M. The smaller gear of the clusters has 20 teeth (t_1), and the bigger one has 40. The gear on the feed shaft has 25 teeth (t_f).

Solution: For transmission between gear nos. 1 and 4,

$$i = \frac{N_{\text{out}}}{N_{\text{in}}} = \frac{t_1}{t_2} = \frac{40}{20} = 2$$

∴ R.P.M. of gear nos. 3/4 $= i \times N_{\text{in}} = 2 \times 750 = 1,500$

R.P.M. of gear nos. 5/6 $= N_2 \times \dfrac{t_2}{t_5} = 750 \times \dfrac{20}{40} = 375$

R.P.M. of gear nos. 7/8 $= N_6 \times \dfrac{t_6}{t_7} = 375 \times \dfrac{20}{40} = 187.5$

R.P.M. of gear nos. 9/10 $= N_8 \times \dfrac{t_8}{t_9} = 187.5 \times \dfrac{20}{40} = 93.75$

R.P.M. of gear nos. 11/12 $= N_{10} \times \dfrac{t_{10}}{t_{11}} = 93.75 \times \dfrac{20}{40} = 46.9$

R.P.M. of gear nos. 13/14 $= N_{12} \times \dfrac{t_{12}}{t_{13}} = 46.9 \times \dfrac{20}{40} = 23.44$

The number of teeth on the swivelable idler gear is insignificant. It just transmits torque from the cluster gears to the feed shaft gear (t_f). The idler gear does not affect the transmission ratio between shafts II and III. The transmission ratio depends upon the number of teeth of the gear with which the swivelable idler gear is engaged.

With gear 3, feed shaft III R.P.M. $= N_3 \times \dfrac{t_3}{t_f} = 1{,}500 \times \dfrac{40}{25} = 2{,}400$

With gear 4, feed shaft III R.P.M. $= N_4 \times \dfrac{t_4}{t_f} = 1500 \times \dfrac{20}{25} = 1200$

With gear 5, feed shaft III R.P.M. $= N_5 \times \dfrac{t_5}{t_f} = 375 \times \dfrac{40}{25} = 600$

With gear 6, feed shaft III R.P.M. $= 375 \times \dfrac{20}{25} = 300$

With gear 9, feed shaft III R.P.M. $= 93.75 \times \dfrac{40}{25} = 150$

With gear 10, feed shaft III R.P.M. $= 93.75 \times \dfrac{20}{25} = 75$

With gear 13, feed shaft III R.P.M. $= 23.44 \times \dfrac{40}{25} = 37.5$

Thus, we get a geometric progression of speed, with $\phi = 1/2$ (or 2).

In heavy-duty machines, the problems resulting from the wear of the holding pin in the Norton tumbler gear, are eliminated by removing the swiveling idler gear, altogether. The sliding gear on the feed shaft III is meshed directly with the bigger gears, in the clusters on the intermediate shaft II [**Fig. 2.44b**]. This enhances the rigidity of the mechanism. But the three intermediate choices of speeds (1200, 300, and 75 R.P.M.) are lost, as the sliding gear on the feed shaft (t_f) cannot be engaged with the smaller gears in the clusters.

2.10.9 Infinitely variable stepless drives

Although costlier than stepped speed changes, stepless drives offer certain advantages.
1. *Optimizing cutting speed:* particularly during facing, cutting off, and turning irregular contours, and stepped workpieces. This leads to higher tool life, more uniform surface finish, and enhanced productivity and profit.
2. *Speed change:* even while the spindle is running. But the change can only be gradual. You cannot double or halve the cutting speed instantly, as is possible in a 2-speed electric motor.
3. *Smooth operation:* with less noise. No jerks due to worn out gears.

Machine Tool Drives and Mechanisms

Types of stepless drives

2.10.9a. *Mechanical friction drives:* are subject to frictional losses, distortion of contact surfaces, and slip. Still, the overall efficiency is rarely less than 95%. The speed can be changed only while the spindle is running.

2.10.9b. *Hydraulic drives:* are vulnerable to leakages and temperature changes. They are unsuitable at very low speeds (below 15 mm/min). But the speeds can be changed, and even reversed instantly. They are self-lubricating and there is almost no risk of damage due to overload. They are suitable for automation.

2.10.9c. *Electrical stepless drives:* use D.C. motors with shunt adjustment, to obtain a 10–15 speed range ratio. Ward Leonard system can have a range ratio up to 1,000. Automation in this case is simplified.

2.10.9d. *Mechanical friction drives:* are classified according to the shape of the transmission elements such as face plate variators, cone variators, and spheroidal variators.

2.10.10 Face plate variator [Fig. 2.45a]

In this arrangement, an axially adjustable roller determines the contact diameter (D), on the friction disk (face plate). By moving the roller along the friction disk, the contact diameter 'D', and the transmission ratio 'i' can be varied. Both, the disk as well as the roller, can be used as driver (input) or driven (output). If the friction disk is the driver, and the roller the driven element, then,

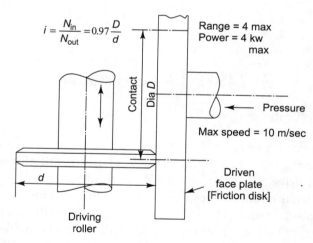

Fig. 2.45(a): Single plate face variator

$$\text{Transmission ratio} = \frac{N_{in}}{N_{out}} = \frac{\text{Driven Dia}}{\text{Driving Dia}} = \frac{D}{d}$$

Taking into consideration the inevitable slip due to the small contact area,

Effective transmission ratio = $i_E = 0.97 \dfrac{D}{d}$

Maximum speed variation range = R = 4
Maximum linear speed = 10 m/sec.
Maximum transmittable power = 4 kW
Overall efficiency (η) = 0.8 (80%)

In another arrangement [**Fig. 2.45b**], two face plates are used, to almost entirely eliminate roller deflection. In this arrangement, the face plates remain stationary. The roller is under the pressure of the friction disks from opposite sides that neutralizes deflection of the roller. The output speed [N_{out}] can be varied, by axially adjusting the central roller. This changes the contact diameters 'D_{in}' and 'D_{out}', and the transmission ratio 'i'. As there are 2 frictional transmissions, there is a double slip, and the transmission ratio is less.

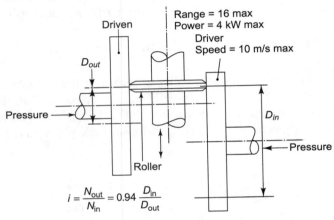

Fig. 2.45(b): Face variator with two plates

Effective transmission ratio = $i_E = 0.94 \dfrac{N_{in}}{N_{out}} = 0.94 \dfrac{D_{out}}{D_{in}}$

Maximum speed variation range = R = 16.

The maximum linear speed, transmittable power, and overall efficiency are the same as the single friction disk drive, i.e. 10 m/sec, 4 kW, and 80%, respectively.

Simple and cheap, friction disk drives are used in cutting-off machine spindles, and camshaft drives in multi-spindle autos. The rollers are usually made of cast iron, and coated with either leather or plastic, to increase friction. The face plates can be made of steel, or cast iron.

The pressing force exerted by the springs, is increased by 20 to 50%, to ensure positive contact at frictional face.

Spring force = $Q \geq 1.2 - 1.5 \dfrac{P}{\mu}$ (**Eqn. 2.94**)

P = The peripheral force (kg) at contact dia (D_{in} or D_{out})
μ = Coefficient of friction.

2.10.11 Cone variators [Figs. 2.46 a, b, c, d, e]

These provide better linear speed distribution over the contact area. The cone axis can be angular [**Fig. 2.46a**], or parallel with the friction disk axis [**Fig. 2.46b**]. If the driving cone has an angular axis, the cone contact edge is kept parallel with the friction disk face. By moving the cone in a direction parallel to the friction disk face [**Fig. 2.46a**], the contact radii 'R' on the friction disk, and 'r' on the cone, can be varied to change the transmission ratio (i).

$$i = \frac{N_{in}}{N_{out}} = \frac{R}{r}$$

Max linear speed range = 4

Max transmittable power = 2.5 kW

Max linear speed = 15 m/sec.

$\eta = 90\%$

Alternatively, the cone and the friction disk axes can be kept parallel, and the face of the friction disk, as well as the cone adjustment direction, can be made angular: parallel with the cone edge [**Fig. 2.46b**].

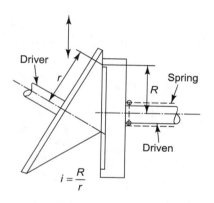

Fig. 2.46(a): Speed variator with angular cone axis

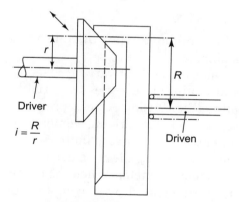

Fig. 2.46(b): Speed variator with cone axis parallel to plate axis

Mounting a friction disk on the arm of a planetary gear [**Fig. 2.46c**], increases the linear speed range $R(5)$, and the transmittable power (7.5 kW), as there is a self-tightening action between the cone and the disk, even under large torques. The speed of the disk is varied by moving the cone or the disk at cone angle.

In another arrangement, the cone is mated with a swivelable spherical friction disk [**Fig. 2.46d**].

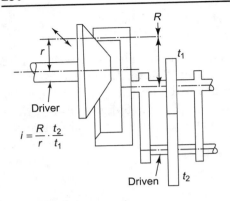
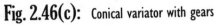

Fig. 2.46(c): Conical variator with gears

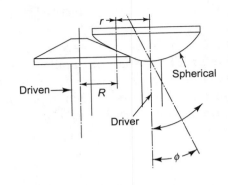

Fig. 2.46(d): Conical variator W/spherical driver

The cone is the driven element. The swivel of the disk, changes the contact radius 'R' on the cone, and 'r' on the disk.

Max. linear speed range = 8

Max. transmittable power = 3 kW

Max. linear speed = 15 m/sec.

$\eta = 85\%$.

All the above types of cone variators are used widely, in small and medium drilling machines.

Other methods of varying speed, use pairs of axially adjustable cones as variable diameter pullies [**Fig. 2.46e**]. A belt, a ring, or a chain is used as a separate driving element. Bringing two conical halves closer, increases the contact diameter. Moving the cone halves apart, reduces the contact diameter. When the diameter of one pulley is increased by moving the cones closer, the halves of the other pulley automatically move apart, by a suitable distance. This arrangement can be used for increasing as well as reducing speed.

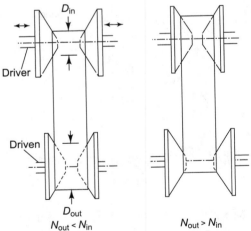

Fig. 2.46(e): Axially adjustable variable contact dia conical pullies

The driving element can be any of the following:

1. **V Belts:** Narrow V belts, with higher speed capacity, but lesser range (R) and power capacity (P), are used widely for light lathes and drilling machines. Wider V belts are necessary for heavier machines, which can do with lesser speed capacity, but call for a wider speed range.

Table 2.41: Comparison of narrow and broad V belts

Belt type	Speed		Power limit	Efficiency
	Range R	Limit V m/s	P kW	η %
Narrow 'V'	9	18	2.5	92
Wider 'V'	12	15	15	92

2. **Steel rings** of hardened bearing steel (En 31/103 Cr1) run in oil. The pullies must also be hardened. The speed range is wider ($R = 16$), but the velocity distribution in the area of contact is unfavorable. This reduces the efficiency ($\eta = 85\%$). Despite the higher cost, the effective wear-resistance, makes rigid steel ring drives, convenient for medium-sized (up to 10 kW) drilling and grinding machines, and automats, subjected to higher wear.
3. **Nylon belts** are usually cogged, to prevent slip, in spite of the elasticity of the material.
4. **Chains** provided with rollers on the inside contact surface, reduce friction. In spite of a low speed limit (10 m/sec), and range ($R = 7$), they are suitable for heavy machines ($P = 15$ kW max).

Chains are also used in the positive, infinitely variable drive [**Fig. 2.47a**]. The cone pulley too has flutes, similar to bevel gears. The flutes guide thin, laterally-movable laminas, on the chain. The laminas are made of high-strength steel, while the pullies are case-hardened, after manufacture from suitable steel.

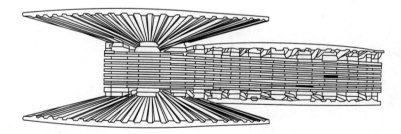

Fig. 2.47(a): Axially adjustable variable diameter pullies with chain drive

Maximum linear speed range = $R = 10$
Maximum transmittable power = 14 kW
Efficiency = $\eta = 90\%$

Epicyclic speed variation [Fig. 2.47b]

This drive combines axially adjustable, variable speed cone pulley drive, with epicyclic (planetary) differential gears train.

In this arrangement, the same motor is used for driving the planetary gear arm, as well as gear '3'. The cluster of gears '2' and '3' is free to rotate on its shaft. This shaft is coupled with the drive motor, and is keyed to the planetary arm. So, the planetary arm rotates around gear '3' axis, at the R.P.M. of the motor (N_M). The R.P.M. of gear '3' is varied by using train of gears '1' and '2', and stepless speed drive, similar to the one in [Fig. 2.46e]. The speed of clustered gears '2' and '3' is the same.

$$N_3 = N_2 = \frac{t_1}{t_2} \times N_1$$

Using gear nos. (in [Fig. 2.47b]) as suffixes, for referring to their speeds (N), and no. of teeth (t), for N_M rotations per minute, of the planetary arm around the common axis of gears 2, 3 and 6,

$$\frac{N_3 - N_m}{N_6 - N_m} = \frac{t_6}{t_5} \times \frac{t_4}{t_3}$$

The above equation can be used for computing the output speed (N_6) of the driven gear '6'.

Example 2.27: Find the range of speeds for gear '6', if for [Fig. 2.47b], the speed of rotation of the planetary arm motor (N_m) is 750 R.P.M. The range of the variable speed drive transmission ratio $\frac{N \text{ out}}{N \text{ in}}$ is $\frac{1}{2}$ to 3. The no. of teeth for the various gears are stated below:

$t_1 = 40$; $t_2 = 20$; $t_3 = 80$
$t_4 = 30$; $t_5 = 40$; $t_6 = 70$

Solution: Max. transmission ratio for the variable speed drive is 3.

∴
$$N_1 \max = 750 \times 3 = 2{,}250$$

$$N_3 = N_2 = \frac{t_1}{t_2} N_1$$

∴
$$N_3 \max = \frac{40}{20} \times 2{,}250 = 4{,}500$$

$$\frac{N_3 - N_m}{N_6 - N_m} = \frac{t_6}{t_5} \times \frac{t_4}{t_3} = \frac{70}{40} \times \frac{30}{80} = \frac{21}{28}$$

$$\frac{N_3 \max - 750}{N_6 \max - 750} = \frac{21}{28}$$

$$\frac{4{,}500 - 750}{N_6 \max - 750} = \frac{21}{28}$$

$$N_6 \max = \frac{28}{21}(4{,}500 - 750) + 750$$
$$= 5000 + 750 = 5750 \text{ R.P.M.}$$

Min. transmission ratio for the variable speed drive is $\frac{1}{2}$

$$N_1 \min = 750 \times \frac{1}{2} = 375$$

$$N_3 \min = \frac{40}{20} \times 375 = 750; \text{ and}$$

$$N_6 \min = \frac{28}{21}(750 - 750) + 750$$
$$= 750$$

Speed Range $= \frac{5750}{750} \approx \frac{23}{3} = 7.67$

Due to very low efficiency (5–33%), at higher speeds, epicyclic variators are used for low-speed drives, where efficiency of transmission is of little consequence.

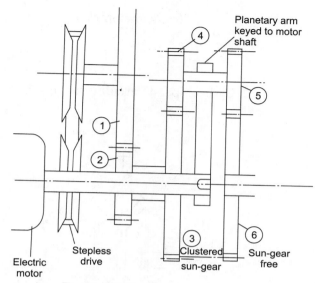

Fig. 2.47(b): Epicyclic speed variator

2.10.12 Three-member spheroid and cone variators

(a) **Svetozarov's variator [Fig. 2.48a]** uses two tekstolite swiveling disks for changing the transmission ratio (i), between two hardened steel (RC 60–65) spheroids, mounted on the driving (I) and driven (II) shafts.

$$i = 0.94 \frac{D_{out}}{D_{in}}$$

Max. linear speed range $(R) = 8$
Max. linear speed $(V) = 20$ m/sec
Max. transmittable power $= 25$ kW
Efficiency $(\eta) = 85\%$

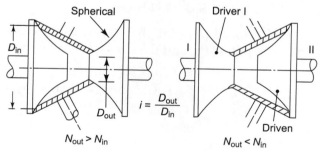

Fig. 2.48(a): Svetozarov's spheroid variator with swiveling discs

(b) **Wulfel-Kopp variator [Fig. 2.48b]** uses two spheres as drive elements. The driving shaft (I) which is mounted with a conical driver, drives the sphere which transmits power to another conical member, mounted on the driven shaft (II). The spheres are mounted on separate shafts, to enable tilting them when required. The tilt of the sphere shafts determines the contact radii $r1$ and $r2$ on the sphere. Other factors contributing to the transmission ratio (i), are the contact diameters d_1 and d_2, on the driving (I) and driven (II) shaft cones.

Transmission ratio $(i) = \dfrac{N_i}{N_o} = \dfrac{2r}{d_1} \cdot \dfrac{d_2}{2R}$

$$i = \dfrac{d_2}{d_1} \cdot \dfrac{r}{R}$$

Max. speed variation range = 9

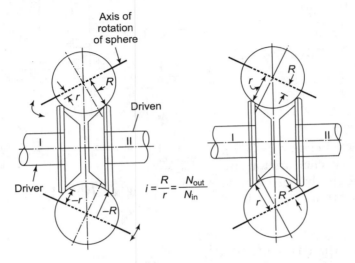

Fig. 2.48(b): Wulfel–Kopp variator

2.10.12.1 Electrical stepless drives

The rotary speed [N] of 3 phase synchronous motor depends upon A.C. frequency [f], the number of poles [P], and the slip [s] between the rotor and the rotating magnetic field.

$$\text{A.C. motor R.P.M.} = N_m = \dfrac{60f(1-s)}{p} \qquad \textbf{(Eqn. 2.95)}$$

f = frequency (In India, it is 50 cycles/sec.)

p = number of poles

In two-speed motors, the number of poles is halved by electrically disconnecting half the number of poles, to double the rotary speed (R.P.M.) and vice versa.

As there can be only an even number of poles, increasing or decreasing the number of poles can result only in a stepped speed change.

Frequency (f) of commercial electricity supply is constant for the entire country. For obtaining supply with changeable frequency, it is necessary to use costly, special A.C. generators and amplifiers.

Slip quantum can be changed by introducing variable active resistance in the rotor winding. However, in addition to causing a lot of energy loss, the resistance also affects the motor characteristics beyond permissible limits.

It is much more economical and convenient to use Direct Current (D.C.) motors. The stator electromagnets are activated by an independent source of power, or a separate excitation circuit.

$$\text{Motor R.P.M.} = N = \frac{V - Ia(Ra + Rv)}{Ce\,\phi} \quad \text{(Eqn. 2.96)}$$

$V=$ voltage of supply; $Ia=$ armature current; $Ce = \dfrac{PN_a}{60\,a}$

$Ra, Rv =$ Armature resistance, and variable resistance.

$P = \dfrac{1}{2}$ No. of poles, $N_a =$ Coils in armature,

$a =$ No. of parallel segments in armature winding

$\phi =$ Magnetic flow [flux] of each pole.

Stepless variation of D.C. motor speed can be accomplished through the following means:
1. Variable resistance in series with armature, causes large heat losses due to the passage of the variable resistance current, through the armature. This limits its usage to small D.C. motors only
2. Variable resistance in parallel (shunt) with armature winding [Fig. 2.49]. Increase in variable resistance decreases the shunt current, and increases the rotor R.P.M. Although the power remains constant, the torque decreases, as the R.P.M. increases. Moreover, weakening of the magnetic field causes an overload on the armature. Limitations on the permissible armature overload, limits the speed increase by this method, to only 20%.

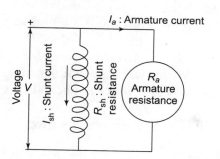

Fig. 2.49: Speed variation by varying shunt resistance in D.C. motor

3. Varying input voltage (Ward–Leonard system) uses 4 electrical machines [Fig. 2.50]. An A.C. motor and D.C. generator are coupled together. The D.C. generator drives a D.C. motor. The A.C. motor is also coupled with a smaller D.C. generator, called 'exciter'. The exciter provides independent excitation to the D.C. motor, and the bigger D.C. generator. A rheostat (Rg) facilitates varying resistance, and through it, the voltage of excitation for the D.C. generator. With its two contacts, the rheostat 'Rg' serves as a potentiometer. Furthermore, interchanging the contacts of the rheostat, reverses the direction of rotation of the D.C. motor instantly. Rheostat 'Rg' provides a speed variation

range up to 8. The range can be doubled by using another variable resistance (Rm), to vary the excitation of the D.C. motor. Thus, the two rheostats together, provide a total speed range of 16.

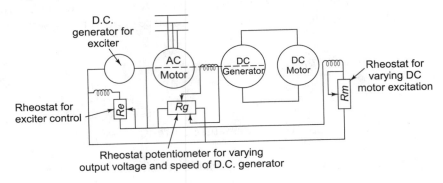

Fig. 2.50: Ward-Leonard system of stepless speed control

4. **Amplidyne with Voltage Feedback.**
   This method uses sophisticated Synchro Transmitters and Selsyn Receivers, to boost the speed variation range up to 4,000. Used mainly for feed motion, this exorbitantly costly system can provide for rapid traverses, as well as much slower, setting movements.
5. Magnetic amplifiers can give us a speed range of up to 100, but their efficiency is only 50%. They can be found in applications, in the feed systems of milling and grinding machines.
6. Electronic methods, such as the use of thyristors, and ionic devices, are developing fast. With a wide speed range of up to 200, and a power range of up to 10 kW, these methods will find wider applicability, in this century.

2.10.12.2 Hydraulic stepless drives

Hydraulic drives use pressurized oil as a medium for transmission of power. These drives are eminently suitable for stepless variation of speed, as well as force/torque. While speed can be altered, by varying the flow of the oil, force or torque can be varied by varying the oil pressure.

Hydraulic system comprises the following elements:

1. Hydraulic oil, stored in a tank of suitable capacity: usually 3 times the oil flow, per minute
2. Pump for pressurising oil
3. Pressure relief valve for controling the maximum pressure
4. Direction control valve for controlling the direction of rotary (clockwise/anticlockwise), or linear (right to left/left to right) motion
5. Flow control valve is used only if speed adjustment is necessary
6. Other manipulation devices, such as sequence valve, pressure reducing valve, etc. These too are used only if required

Figure 2.51 (p 259) shows a schematic arrangement of a typical hydraulic system, for obtaining rotary motion, through a hydraulic motor.

Advantages of hydraulic power

1. Compactness: high power in small spaces
2. Long life: due to continuous self-lubrication
3. Low friction: $\mu = 0.01$ for oil and pipe, but for metal to metal contact, $\mu = 0.05\text{--}0.1$
4. Easy transmission: through flexible hose pipes, which can bend around an obstruction
5. Flexibility: Force, as well as speed can be varied steplessly, by varying the pressure and flow of oil. Speed range ratio $(R) = 50\text{--}60$
6. Easy automation: through multiple operations of many actuators, through a single-direction control valve. Ease in synchronization and sequencing the operations of various actuators
7. Overload protection: of hydraulic and mechanical elements, through control of the maximum pressure, by using a pressure relief valve. Rise of pressure above the set value, directly bypasses pressurized oil to the tank, stopping the operations of the hydraulic actuators altogether
8. Smooth operation: no jerky action, as happens in worn gearboxes. Moreover, once set, the set speed remain stable and immune to fluctuations in load

Disadvantages

1. Vulnerable to the dirt in hydraulic oil, which can jam the valves
2. Leakages at higher pressure can reduce rigidity, but only marginally, as the pump replenishes oil continuously
3. Vulnerable to temperature rise, which considerably reduces the viscosity of the oil. This also reduces the maximum working pressure
4. Liquid friction, though lower compared to metals, can cause considerable losses when high velocities and long pipes are used

Requisites of hydraulic oil

1. **Viscosity**, or the resistance to flow, helps in pressurization of oil. Temperature rise reduces viscosity. The measure of resistance to change, in viscosity, due to a change in the temperature 'Viscosity Index', is high for good hydraulic oils. The lubricity is however, inversely proportional to the viscosity. Higher the viscosity, lesser the lubricity.

Table 2.42: Recommended viscosity for various oil pressure (kg/cm^2)

Pressure kg/cm^2	Up to 70	Above 175
Viscosity cm^2/sec.	0.35	1.0
	0.65	2.0

2. **Pour point,** the lowest temperature at which oil becomes fluid, should be at least 11°C below the likely minimum temperature of the system.
3. **Demulsibility** is the resistance to emulsification with water. Demulsibility prevents water from sneaking in, and circulating through the hydraulic system, and causing corrosion.
4. **Oxidation resistance,** to minimize acidity, sluggishness, and formation of insoluble sludge, causing malfunction of valves. Addition of a oxygen inhibiters, use of a heat exchanger to keep the temperature below 56°C, and preventing water and vegetation contamination, minimizes oxidation.
5. **Low-inflamability in applications,** subjected to high temperature. Fire-resistant fluids, such as water-in oil emulsions, water-glycol solutions, phosphate esters or silicon fluids used in high temperature applications, call for special, costly viton seals.

Fluid power generation

In all modern hydraulic systems, fluid energy is generated by a pump. Pumps are specified by the maximum pressure and volume of the fluid they can deliver, at that pressure.

Pumps can be classified according to their functional characteristics:

1. **Constant volume pumps:** discharge a certain invariable volume per rotation (or stroke). Rise in No. of rotations (or strokes), per minute, increases the delivery per minute.
2. **Variable discharge pumps:** have facilities for obtaining two pressures and two discharges from a single pump. Adjustment of pump discharge according to demand, helps in reducing the power requirement (kW) of the hydraulic drive.

Choice of pressure

Although high-working pressure provides steady flow of pressurized oil, the high pressure pumps are costlier. Consequently, most of the hydraulic systems use cheaper, low-pressure pumps, wherever possible, often along with a high-pressure pump. Low-pressure pumps cost only $1/8^{th}$ to $1/3^{rd}$ the cost of high pressure pumps. A two-pump system generally uses a cheap, high-volume, low-pressure pump for a low-load, fast approach; and a costlier high-pressure, low-volume pump for the slow, short, peak load working stroke.

A single-variable discharge pump can replace the combination of a low-pressure and high-pressure pump, for, the delivery of the pump is high when the pressure is low, under lesser load. The discharge decreases automatically, and progressively, as the load and the fluid pressure increases.

Conditioning hydraulic oil involves

1. **Filtering** by strainers to 150 microns (0.15 mm), before entry into the suction pipe. Providing a 30 micron filter between pump and valves, and return-line filter, before draining into the tank. Filters can be of a mechanical type, with baffles to trap particles in

the long tortuous path, or of an absorbent type, with porous elements of paper, wood pulp, cotton, cellulose, etc. to absorb the dirt. The return-lines are provided with full-flow filters, which bypass the flow, when the filtering element gets clogged. Pressure lines use proportional flow filters, with partial filtration.

2. **Temperature control** with air coolers (finned tubes), or water coolers with oil pipes, surrounded by circulating water.
3. **Chemical stabilization** by addition of oxygen inhibitors, and control of temperature, as explained above.

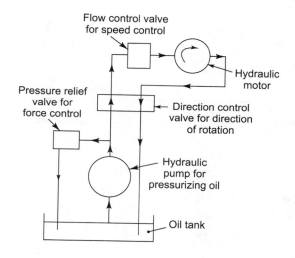

Fig. 2.51(a): Hydraulic system for stepless rotary drive in clockwise direction

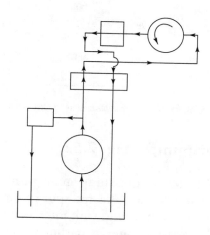

Fig. 2.51(b): Hydraulic system arrangement for rotation in opposite direction (anti-clockwise)

2.11 PUMPS

Hydraulic pumps draw (suck) oil from the tank and pressurize it, to deliver it to the hydraulic actuators, for generating the required force and speed.

2.11.1 Types of Pumps

The following are the types of contemporary pumps, with their pressure ranges, and power requirements.

Table 2.43: Comparison of hydraulic pumps

Type	Pressure 'P' kg/cm² (bar)		Efficiency η %	Power Reqd. (kW) for V lits/min at pressure 'p'	Remarks
	Normal	Max *			
Gear pumps	30 100	204	70 80	$\dfrac{pV}{150}$	Noisy, sturdy, compact vulnerable to wear, leakages
Vane pumps	70 100	160	75 85	$\dfrac{pV}{187.5}$	Good life
Piston pumps	100 300	600*	85 95	$\dfrac{pV}{255}$	Long life, less pulsations, large size.

*These pumps, for the highest possible pressure, are not available commercially. They are usually manufactured as 1 off, at a higher cost, against an order.

Gear pumps [Fig. 2.52]

A simple, single-stage gear pump comprises two identical, hardened-Nickel Chrome spur gears, meshing with each other in a close fit cast housing, with intersecting counter-bores. The rotation of the gears creates a partial vacuum, through which the oil is sucked, entrapped between the teeth, and pressed out, at diametrically opposite ends.

Sturdy and cheap gear pumps are suitable for low-pressure, high-volume applications. They are usually used for increasing the speed for a rapid approach, along with a high-pressure, low-volume pump, for a short-duration peak load.

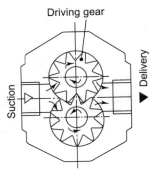

Fig. 2.52: Gear pump

2.11.1.1 Vane pumps

In its simplest form, a vane pump comprises a hardened steel rotor, mounted in an eccentric stator ring, of high-grade cast iron [Fig. 2.53].

The rotor is slotted radially, to house hardened steel vanes, which are pushed outwards by the centrifugal force, caused by the rotary motion of the rotor. The vanes press against the stator ring and provide the radial sealing between the pockets, created between the adjacent vanes. During rotation from A to B, the space between vanes gradually increases, and the oil is sucked through the kidney-shaped inlet slot I, in the port plate. Between B and A, the volume between the vanes decreases and the oil is pushed through the outlet slot 'II', in the port plate. The placement of the inlet and outlet ports diagonally opposite each other, causes unbalances in forces, which limits the pressure of oil in such pumps, to 100 kg/cm².

2.11.1.2 Dual vane pumps

The unbalance in forces, resulting from the diagonally opposite placement of suction and delivery ports, can be corrected by using an oval cam track ring instead of an eccentric stator ring. The oval cam track acts like two eccentric stator rings [**Fig. 2.54**].

There are two sets of ports: two suction ports, placed diagonally opposite each other, and two delivery ports, at right angles to the suction ports. Positioning similar ports, opposite each other, balances the forces and oil can be pressurized up to 160 kg/cm^2.

Although costlier than gear pumps, vane pumps cost lesser than piston pumps. They are less noisy and more durable than gear pumps, and are used extensively in medium-pressure [100 kg–160 kg/cm^2] applications.

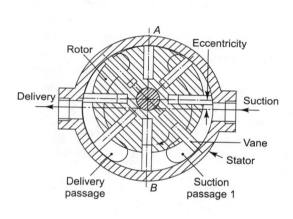

Fig. 2.53: Eccentric rotor vane pump

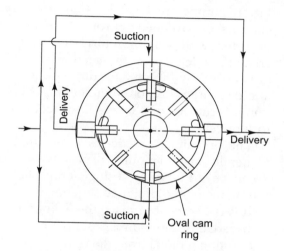

Fig. 2.54: Oval track dual vane pump

Variable discharge vane pumps

Discharge of a vane pump is proportional to the eccentricity of the stator ring, with respect to the axis of the rotor. Consequently, discharge of a pump can be altered, by varying the eccentricity of the stator ring.

In a pressure compensated variable discharge pump, the maximum eccentricity and through it, the maximum displacement of the pump, can be adjusted with the flow-adjustment screw [**Fig. 2.55**]. The stator ring is held, pressed against the flow-adjustment screw, by a spring which can be adjusted for pressure, with a screw. This provides

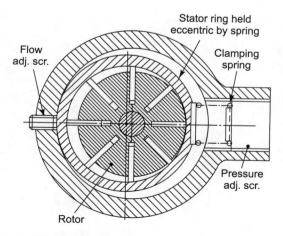

Fig. 2.55: Variable discharge vane pump

an adjustable clamping force. The adjustment facilitates the control of pressure in the pump. When the oil pumped by the vanes, reaches a certain set pressure, the hydraulic force on the stator ring compresses the clamping spring, and reduces the eccentricity, and also the discharge of the pump. Higher oil pressure compresses the spring more, and reduces the eccentricity further. At the set pressure, the stator ring becomes concentric with the rotor ring.

When eccentricity is zero, and the pump discharge negligible. This makes the variable discharge vane pump eminently suitable for clamping mechanisms, with a rapid approach and automatic slowdown after the clamp touches the workpiece.

2.11.1.3 Piston pumps

Piston pumps are ideal for high-pressure, high-load applications. A single piston pump comprises a hardened or bronze-coated steel plunger, reciprocating in a bronze-lined cylinder, furnished with two non-return valves for suction and delivery. During in-stroke, a vacuum is created, which opens the suction valve to admit the fluid into the cylinder. The fluid is pressurized by the plunger during out-stroke. The pressurized oil opens the delivery valve, and discharges it into the outlet line. As there is no delivery during in-stroke, a single piston pump is subjected to severe flow fluctuations in discharge. Hence, its use is confined mainly to lubricating systems.

The prevailing practice is to use multi-plunger pumps, to reduce the flow fluctuations. Generally, an odd number of pistons are chosen to prevent a cyclic ripple effect. If the cylinders are arranged in a line, a single crankshaft, with an appropriate number of cranks, can drive the plungers in a designed sequence.

It is convenient to place the cylinders in a circular configuration; radially or axially.

In radial piston pumps, a set of an odd number of identical cylinders are placed with their axes running radial from the center line of the housing. In the pumps illustrated in **[Fig. 2.56]**, three sets of identical cylinder sub-assemblies are mounted radially in the housing. The rotation of the central eccentric, reciprocates the pistons in a cyclic succession.

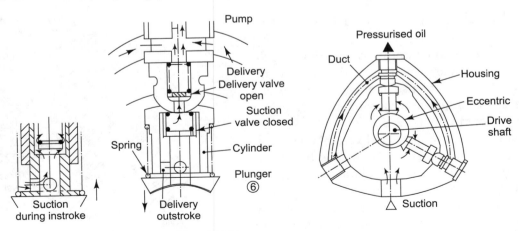

Fig. 2.56: Radial 3-piston pump

Machine Tool Drives and Mechanisms

The fluid enters the housing through port *S*. During the stroke of the piston towards the center, the suction port in the cylinder is opened. The oil enters and fills the cylinder. During the stroke, away from the center, the filled oil is pressurized and delivered through the spring-loaded delivery [non-return] valve. The ducts in the housing, convey the pressurized oil to the delivery port *P* of the pump. These pumps can develop up to 600 kg/cm^2 of pressure. They are rotated between 1,000 and 2,000 R.P.M., with the delivery increasing as the R.P.M. increases.

Axial piston pumps For high delivery, high pressure source of hydraulic oil axial piston pumps are suitable.

In the inclined axis pump, illustrated in **[Fig. 2.57]**, the drive shaft on the left, rotates the piston plate carrying axial pistons, with pinless ball joints. There is a spherical bearing surface between the bronze-coated valve plate and the cylinder housing. In the uppermost position during rotation, the plunger is at the farthest end of the out-stroke, whereas it reaches the end of the in-stroke at the lower-most position of the rotation. Consequently, oil is sucked from the kidney-shaped suction port '*S*' in the valve plate, during half rotation, when the plunger is ascending in an out-stroke. This oil is pumped into pressure port '*P*' in the valve plate, when the plunger descends into an in-stroke, during the remaining half rotation. Thus, the numerous pistons in the pump, suck oil from '*S*' and pump it into '*P*' in quick succession.

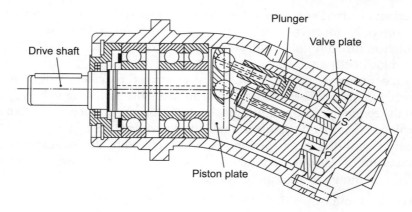

Fig. 2.57: Inclined axis multi-piston pump

In the swash plate pump **[Fig. 2.58]**, the swash plate effects suction, pressurization, and delivery. The pump discharge can be changed by changing the swash plate angle.

2.12 HYDRAULIC VALVES

Hydraulic valves are used to control the pressure and flow of the hydraulic fluid. They are classified according to function, construction, and method of actuation.

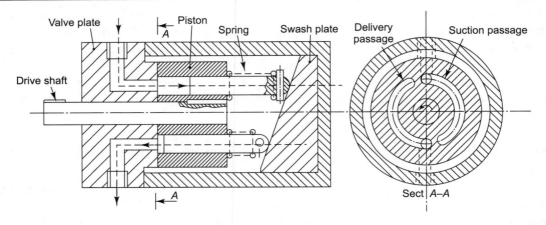

Fig. 2.58: Swash plate multi-piston pump

2.12.1 Classification of valves

1. Functional classification

- **(1a) Flow control valves** control the quantum of flow, including total stoppage or zero flow.
- **(1b) Direction control valves** are used to direct the hydraulic fluid to various actuators, and more often for reversing the direction of motion, of the actuator, e.g. reversal of the rotation direction of a hydraulic motor (clockwise/anti-clockwise), or a cylinder (out-stroke/in-stroke).
- **(1c) Pressure control valves** are used to control the system pressure (relief valve), pressure in a part of the hydraulic system (pressure reducing valves), or for ensuring a certain sequence in operation of the actuators. Sequence valve can ensure that a workpiece is clamped, before metal-cutting commences.

Comprehensive symbols have been developed to schematically indicate the functions and methods of actuation of various hydraulic valves. **Figure 2.60** shows the symbols used, to represent various elements in hydraulic systems.

2. Constructional classification

- **(2a) Sliding spool valves** with a spool, which can be slid axially to connect the valve ports in different ways **[Fig. 2.59a]**.

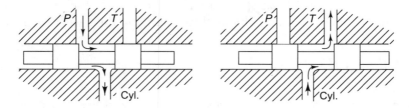

Fig. 2.59(a): 2-way sliding spool direction control valve

(2b) **Rotating spool valves** use a spool, which can be rotated about its axis to change the manner in which various ports are connected [**Fig. 2.59b**].

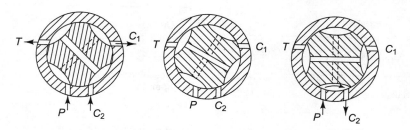

Fig. 2.59(b): Rotary spool 4 away D.C. valve

(2c) **Poppet valves** use poppets to keep the valve ports closed. Lifting different poppets, connects the ports in different ways, through the built-in-passages, in the valve body [Fig 2.59c].

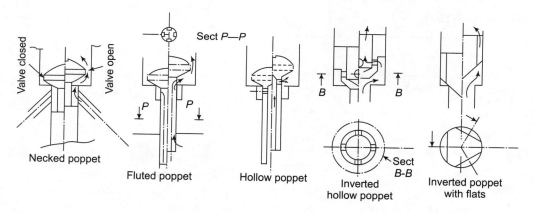

Fig. 2.59(c): Poppet valves

3. Actuational classification [Fig. 2.60]

(3a) **Manually operated valves** with levers, knobs, etc. for actuation.

(3b) **Solenoid operated valves** with electrical coils, which become temporary electro-magnets to the slide valve spool, or lift valve poppet, for electrical operation. They are very convenient for automation.

(3c) **Pilot operated valves** use hydraulic pressure to move the spool/poppet, by piston action. Most of the pressure control valves, some direction control valves, and flow controls, use pilot operation for automation. The following pages describe a few typical valves, illustrating the types cited above.

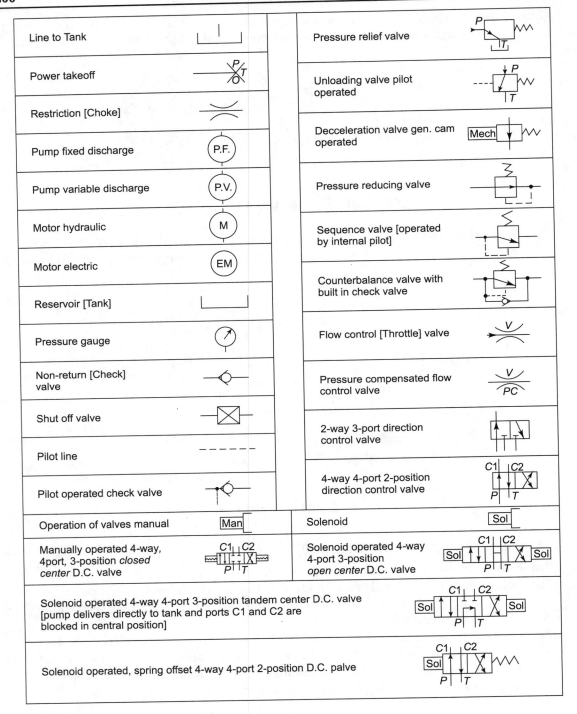

Fig. 2.60: Symbols for hydraulic circuits

(1a) Flow control valves

i. **Throttle valves** are like ordinary water taps/cocks. The area of the fluid passage can be increased to a fully-open, maximum-flow position, or the area can be decreased to reduce the flow, to the required level. Being vulnerable to load changes, these cheap valves are used in applications where the marginal change in speed of actuator is insignificant, e.g. for controlling the idle return stroke speed of a linear actuator (cylinder).

ii. **Pressure compensated (P.C.) flow control valve** [Fig. 2.61] maintains a constant rate of flow, even if the load and the resulting pressure differential, causing the flow, changes. An eccentric groove in the throttle spool 'T' [Fig. 2.61] can be positioned according to the dial setting, which determines the areas of the flow passage, and the flow rate. This rate will remain constant, as the spring-loaded compensator spool 'P' positions itself automatically, according to the load/pressure variations, to automatically adjust the flow, to the eccentric spool 'T'. When the pressure rises, the spool moves towards the left, against the spring, to decrease the area of flow to the throttle 'T'. This automatically compensates for the increase in the flow, due to increases in pressure/load.

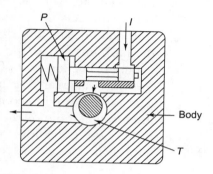

Fig. 2.61: Pressure compensated flow control valve

2.12.2 Location of flow control valve

P) **Meter-in system** [Fig. 2.62a] is suitable for positive load.
Q) **Meter-out system** [Fig. 2.62b] is used for negative load.
R) **Bleed-off system** [Fig 2.62c] is not as accurate as the Meter-in or Meter-out systems, under variable load conditions. Bleed-off is usually used for control of the speed of the hydraulic motors.

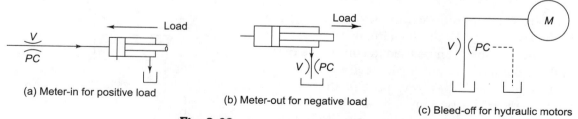

(a) Meter-in for positive load
(b) Meter-out for negative load
(c) Bleed-off for hydraulic motors

Fig. 2.62: Placement of flow control valve

(1b) **Direction control valves** are like traffic signals for hydraulic actuators. There are 2 types of valves: two-way valves and four-way valves. In each type, a number of varieties, with different dispositions and actuators are available.

b (i) Two-way 3-port valves provide two paths of flow. It is like a junction of 3 roads. Arriving along one road, you can choose either of the other two roads. A two-way valve is used for operating a single-action hydraulic cylinder. In a single-action cylinder pressurized oil is used for moving the piston and the attached rod (ram), in only one direction. Load, dead-weight, or a spring is used for the return stroke, in the opposite direction.

For example, when a vertically mounted cylinder is used for lifting load, the 2-way valve connects the pressurized oil to the lower port of the cylinder, during lifting **[Fig. 2.63a]**. For lowering the load, the valve connects the lower port of cylinder to the tank line. Load/spring pushes the piston downward, to drive the oil out of the cylinder. The top port of the cylinder is left open, to allow the entry or exit of air.

Two-way valves are also used as full-flow stop valves, by blocking one port **[Fig. 2.63b]**. The valve is used to connect the pressurized oil line with the cylinder, or disconnect it, entrapping the oil already present in the cylinder. This method is widely used to instantly stop the flow from a high-volume, low-pressure pump; at the end of a rapid approach. This cut-off reduces the actuator speed to a slower feed-speed.

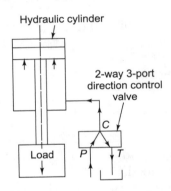

Fig. 2.63(a): Using 2-way valve for vertical lifting cylinder

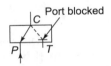

Fig. 2.63(b): Using 2- way valve as stop valve by blocking 1 port

There is more information on such arrangements, in the section on 'Hydraulic circuits'.

Two-way valves can be used to control the direction of the motion of a double-action cylinder **[Fig. 2.63c]**. In this method, when one port of the cylinder is connected to the pressurized oil, the other port is connected to the tank, and vice versa. Both the valves must be operated simultaneously. This can be accomplished by using solenoid actuation for the valves. Alternatively, one can combine the two valves into a single, composite valve, i.e. a four-way valve.

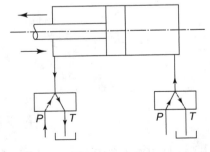

Fig. 2.63(c): Using 2-way valve to operate double action cylinder

b (ii) Four-way 4-port valves provide four flow paths **[Fig. 2.64a]**. They can be used for double-action actuators: hydraulically reversible motor, or cylinder. Each port of the actuator is connected alternately, with the pressurized oil, and tank. When one port is connected to the pressurized oil, the other is connected to the tank, and vice versa. Four-way valves are further subdivided into two-position and three-position valves.

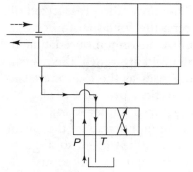

Fig. 2.64(a): Using 4-way 2-position D.C. valve for double action cylinder

In 2-position valves, the actuator cannot be stopped midway. The hydraulic cylinder rod will keep on moving out-stroke (or in-stroke), till it reaches the stroke end. Changing the position of the valve mid-stroke, reverses the motion, but cannot stop it. If we want to stop the actuator mid-stroke, we will have to use a 3-position valve.

In 3-position valves, there is an additional neutral position of the valve. Three different versions of the neutral position are available.

In the closed center version, all the four ports of the valve are blocked in the central position **[Fig. 2.64b]**. The actuator will remain stationary, wherever it is, locked in that position, when the valve spool is shifted to the neutral (central) position. This type of valve is used when we have to use more than one actuator; one cylinder for clamping, and another for moving the machine table or spindle. When a closed center valve is used for operating a clamping cylinder, bringing the valve in the neutral (central) position, it will hold the piston in the clamped position. Also, the clamping cylinder will be isolated (disconnected) from the hydraulic system.

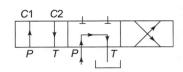

Fig. 2.64(b): 4-way 3-position closed center valve for operating two or more actuators independently

Fig. 2.64(c): Tandem center valve for avoiding oil heating when piston is held mid- stroke in a single cylinder

A closed center valve should not be used for a single actuator. Although the actuator can be locked in any intermediate position, both pressure line as well as the tank line also get blocked. The pressurized oil has no other way, except to pass through the pressure relief valve, into the tank. The relief valve will not open till the oil is pressurized to the value set, on the relief valve. Pressurization to such a high level will cause a detrimental heating of oil. Furthermore, energy is wasted in pressurizing oil, without extracting any work, i.e. when it is simply drained into the tank, through the relief valve. For operating a single cylinder, a tandem center valve is suitable.

In a tandem center valve, the cylinder ports are blocked in a central position **[Fig. 2.64c]**. But the pressure port (P) of the valve is connected to the tank port. Therefore, the pressurized oil is drained into the tank at a low pressure, just enough to overcome the frictional forces.

It must be remembered that the oil pressure is proportional to the load it encounters. During rapid approach, the pressure is just enough to overcome the frictional resistance. The pressure rises as the load increases. In the hypothetical case of zero load, and zero friction, the theoretical pressure is also zero. Hence, hydraulic systems aim to pressurize oil only when required (under load), and only to the extent (pressure) necessary.

In some applications, the piston stopped at an intermediate position, should be free to move, and adjust itself, to align with some external locator. Under such circumstances, an open center valve is necessary.

In an open center valve in a neutral position, all the four ports of the valve are interconnected [**Fig. 2.64d**]. As the cylinder ports are interconnected, the piston is free to move out-stroke or in-stroke. The piston moves the cylinder rod approximately, to the required position. For locking it in this position, additional means are necessary. Needless to say, open center valves are used only for horizontally mounted cylinders, which are not affected by gravitational force, caused by the weight of the piston, its rod, and the connected load.

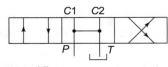

Fig. 2.64(d): Open center valve to permit piston movement in central position

2.12.2.1 Pressure control valves

Pressure control valves are used to limit the maximum pressure of the whole or a part of the system, to protect the system against overload, or to ensure a desired sequence in the operation of two or more actuators.

Relief valve [**Fig. 2.65**] is used to provide protection against overload. Spring holds the valve-closing ball in position. When the inlet pressure exceeds the spring load, the ball is lifted off its seat, and the oil flows into the tank. When the pressure drops below the spring load, the ball resumes its seat, thereby closing down the passage.

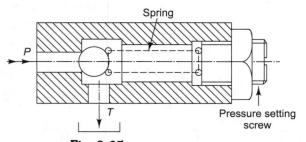

Fig. 2.65: Pressure relief valve

2.12.2.2 Pressure reducing valve [Fig. 2.66]

This is used to limit the pressure in a part of a system. The setting of this valve must always be lower than the system pressure, set on the pressure relief valve.

The pressure at the outlet port is transmitted to spool 'A' through passage 'P'. The small aperture at the center of 'A', equalizes the pressures on both sides of spool 'A'. It also connects the outlet to the spring-loaded conical poppet 'B', which can be set for the desired pressure, using screw 'S'.

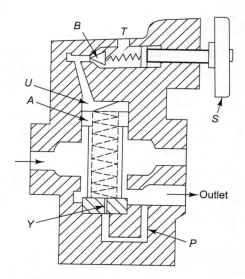

Fig. 2.66: Pressure reducing valve

When the pressure at the outlet rises above the set intensity, the conical poppet 'B' is lifted off its seat that connects passage 'P' with the drain (T). The resulting pressure difference between chamber 'U' and passage 'P', lifts the spool 'A' against the spring. The lift in 'A' is proportional to the difference between the pressure at the outlet, and the pressure set by the screw 'S'. The lift reduces the area of the outlet port, restricting the flow through the valve, to the extent necessary, to prevent further build-up of pressure.

2.12.2.3 Sequence valve [Fig. 2.67]

This is used when it is necessary to ensure that there is sufficient pressure in one part of the circuit, before oil is admitted into another part of the circuit. A typical example is when, a part must be clamped securely, before machining begins.

The spring-loaded spool usually keeps port 2, closed. It will open only after the pressure in the line connected

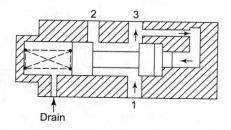

Fig. 2.67: Sequence valve

to port 3, rises to an intensity, necessary to move the spool against the spring. Thus, the actuator connected to port 3 operates positively, before the actuator connected to port 2.

2.12.3 Hydraulic actuators

Hydraulic actuators convert fluid energy into linear or rotory mechanical energy. Rotary motion can be obtained from motors, whereas for linear motion, cylinders are used.

2.12.4 Hydraulic motors

Generally, most of the pumps used for pressurizing oil, can be used as motors. However, a hydraulic motor **receives** pressurized oil which rotates it, whereas a hydraulic pump ***delivers*** pressurized oil. Thus, a pump converts electrical energy into fluid power, whereas the motor converts fluid power into mechanical energy.

The salient feature of hydraulic motors is their constant torque, which does not change even if the speed changes. The torque remains almost constant for a given pressure and flow. Nevertheless, at very low speeds, such as 1 R.P.M., used for feed drives in copying, and numerical control m/cs, it is necessary to use special motors such as roll vane motors, and multi-lobe ball and piston motors. These are more compact and can be used for a wide range of speeds, from 1 to 2500 R.P.M. Easy speed regulation, reversibility, and constant torque-speed, make hydraulic motors suitable for feed drives in copying, countouring, die-sinking, and spark erosion m/cs.

Oscillating motors cannot rotate fully. They can only turn through a partial arc, which can be up to 270°. These motors are particularly useful for indexing, feeding, transferring, and clamping mechanisms.

2.12.5 Hydraulic cylinders

A hydraulic cylinder consists of a piston, which slides under hydraulic pressure, in a tube (cylinder) fixed with end covers **[Fig. 2.68]**. The motion and force of the piston is transmitted to load, by the rod (or ram) fixed to the piston. In a double-ended cylinder, the piston has two rods protruding from both ends of the piston. The piston rod slides through the central hole in the end cover fitted with seals, to prevent the hydraulic fluid from leaking out of the cylinder. The external, free-end of the piston rod is generally threaded, to facilitate its coupling with the mechanism dealing with load.

There are two ports at both ends, for the passage of the hydraulic fluid. For cylinder operation, pressurized oil is admitted through the port at one end, while the port at other end is used for draining out the unpressurized fluid on the other side of the piston. For example, for out-stroke of the cylinder shown in **[Fig. 2.68]**, pressurized oil is admitted through port 'C1', at the closed end of the cylinder. The hydraulic pressure on the piston face, pushes it towards the rod end of the cylinder. The moving piston drives out the unpressurized oil on the rod side through port C2. For in-stroke in the opposite direction, the pressurized oil is admitted through port C2 to the rod side of the piston, while unpressurized oil emerges through port C1, at the closed end.

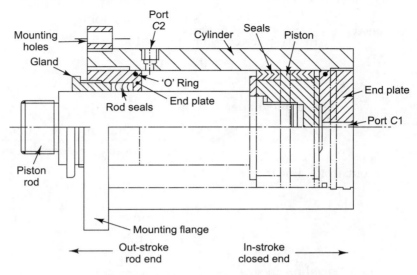

Fig. 2.68: Heavy duty hydraulic cylinder

The pressure bearing areas of the piston, on the closed side and the rod side of the cylinder, differ considerably. The area on the rod side is usually half to one-third the area on the closed side. Consequently, the hydraulic force during the return, or in-stroke is 1/2 to 1/3 of the force available during the out-stroke. Also, the linear speed during the in-stroke is faster: 2 to 3 times the out-stroke speed.

Example 2.28: Find the out-stroke and in-stroke forces and speeds of a hydraulic cylinder, if cylinder bore = 100 mm; piston rod dia = 70 mm; hydraulic pressure = 100 kg/cm^2; and pump discharge = 5 litres/min.

Solution:

$$\text{Circular area at closed end } [Ac] = 0.7854 \times 10^2$$
$$= 78.54 \text{ cm}^2$$

Annular area at rod end [Ar] = Area of closed end − Area (cross-sect) of piston rod
$$= 78.54 \text{ cm}^2 - 0.7854 \times 7^2$$
$$= 39.68 \text{ cm}^2$$

$$\frac{Ar}{Ac} = \frac{39.68}{78.54} = 0.51$$

For hydraulic pressure of 100 kg/cm^2,

$$\text{Out-stroke force} = Fc = Ac \times \text{pressure}$$
$$= 78.54 \times 100$$
$$= 7854 \text{ kg} \approx 7.8 \text{ Tonnes}$$
$$\text{In-stroke force} = Fr = Ar \times 100$$
$$= 0.51 \times Fc$$
$$= 3968 \text{ kgs.} \approx 3.9 \text{ Tonnes}$$
$$\text{Speed of cylinder} = S$$
$$S = \frac{\text{Volume of fluid entering cylinder}}{\text{Cross-sectional area of cylinder}}$$

For outstroke, linear speed = Sc

$$Sc = \frac{\text{Volume}}{\text{Closed end area }[Ac]} = \frac{5 \times 1000}{78.54}$$
$$= 63.6 \text{ cms/min.}$$

$$\text{Speed for in-stroke} = S_r = \frac{\text{Volume}}{\text{Rod End area }[A_r]}$$
$$S_r = \frac{5000}{39.68} = 124.6 \text{ cms/min.}$$

2.12.5.1 Cushioned cylinders

Sometimes, it is necessary to reduce the speed of the piston rod near the end of the stroke, to reduce impact and shock. In such applications, the cylinders are provided with cushions at ends. Cushioning is accomplished, by gradually reducing the flow area of the oil coming out of the cylinder, near the end of the stroke [Fig. 2.69]. The area can be reduced automatically, by providing a tapered pilot on piston. As the pilot enters the flow passage, the taper gradually reduces the flow area. The same effect can be achieved by providing an angular groove on a straight pilot. For adjustable cushioning, it is necessary to use a needle valve to adjust the flow

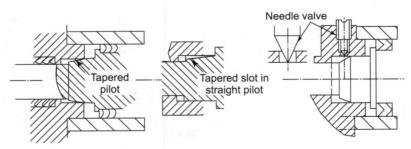

Fig. 2.69: Cushioning of hydraulic cylinders

area. When a needle valve is used, the pilot completely blocks the central flow passage. The oil is thus, compelled to flow through the holes in the end cover, through the needle valve, to the central counter-bore in the end plate.

The end plates of a cylinder are generally screwed into the cylinder tube [**Fig. 2.69**]. In short stroke cylinders, the closed end plate is made integral to the cylinder [**Fig. 2.70a**]. In small diameter cylinders, end plates are usually assembled with four or more tie rods [**Fig. 2.70b**]. The rod side end plate generally has a gland which facilitates the replacement and tightening of the piston rod seals [**Fig 2.68**].

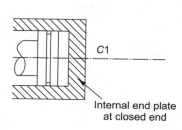

Fig. 2.70(a): Short stroke cylinder with integral end plate

Fig. 2.70(b): Small cylinder with tie rods

The choice of seals depends mostly on the working pressure. Synthetic rubber 'O' rings can be used in up to 160 kg/cm^2 pressure in dynamic applications, whereas their range for static application may be as high as 500 kg/cm^2. For high-pressure dynamic application up to 400 kg/cm^2, chevron 'V' and 'Y' packings with canvas reinforcement are more suitable. Generally, only piston and rod seals need to be 'V' or 'Y' shaped. Static elements can be sealed more economically, by 'O' rings.

Cylinders are classified according to their functions and mounting styles.

Functional classification

(1) **Action**

   (a) **Single-action cylinder**

      With only one port, it is connected to pressure to effect an out-stroke and tank for an in-stroke. The in-stroke is effected by a spring or dead load, which pushes the piston rod and piston to drive out the oil.

   (b) **Double-action cylinder**

      With 2 ports, which are connected alternately, to pressurized oil and tank line, for operating out-stroke and in-stroke.

(2) **Piston rod**

   (a) Single-end rod cylinder has piston, with a rod at one end only.

   (b) Double-end rod cylinder has piston rods on both sides. This is convenient for operating the slides of some machine tools such as surface grinder.

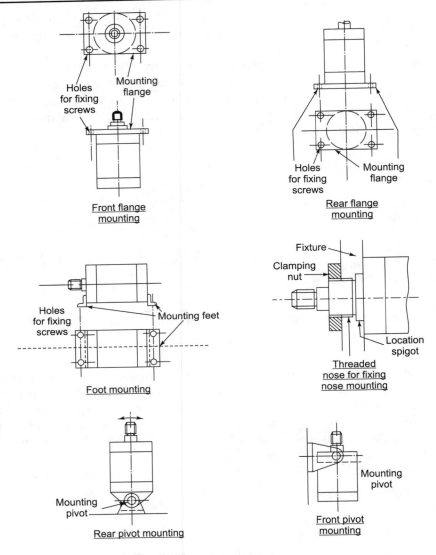

Fig. 2.71: Cylinder mounting styles

(3) **Mounting styles**

  (a) **Flanged mounted cylinders [Fig. 2.71]:** These have a mounting flange at the front (rod end) or rear [closed] end. The flange is often provided with a location spigot for centralizing the cylinder. The flange also has holes for the fixing screws.

  (b) **Pivot or trunnion mounted cylinders:** These use pivot pins for mounting. The cylinder is provided with a trunnion with pivot holes at the front (rod end) or rear (closed end), or anywhere in between.

Trunnion mounted cylinders are particularly suitable for applications where the cylinder must be allowed to swing about a pivot, for automatic adjustment and positioning, to suit the application.

(c) **Foot mounted cylinders:** These are provided with mounting feet with fixing holes, for assembly on a face, parallel to the axis of the piston. As both ends of the cylinder are secured, this mounting is convenient for firm assembly of long cylinders.

(d) **Nose mounting**: In small, light-weight cylinders, the rod end flange is provided with a threaded nose and a clamping nut. The cylinder flange and nut, clamp onto the fixture (mounting plate), from opposite sides.

2.13 PIPING

Hydraulic elements are connected with oil pipes. The pipes can be broadly classified as below:

(a) **Rigid pipes** are used for connecting components, which remain fixed, with respect to each other, during operation of a hydraulic system. When there is no relative motion between two members, they can be connected by rigid steel pipes, suitably bent.

(b) **Flexible piping** is necessary when there is relative motion, between the members to be connected. Rubber hose pipes, braided with steel wires are common for pliable flexible piping.

2.13.1 Selection of pipes

(a) **Flow:** Oil velocity in pipe bore should not exceed 6 meters/sec.

$$\frac{\text{Velocity}}{\text{(M/Sec)}} = S = \frac{\text{Discharge CC / Min}}{\text{Bore area cm}^2 \times 100 \times 60}$$

Pipe bore size can be computed from the above equation.

(b) **Pressure:** Pipe should be able to withstand the system working pressure, without bursting. Bursting pressure in kg/cm^2 = P_b

$$P_b = \frac{2 \times \text{Ultimate stength kg/cm}^2 \times \text{Thickness cms}}{\text{Outside diameter cms}}$$

The bursting pressure should be 4 to 5 times the maximum working pressure in hydraulic systems.

2.13.1.1 Steel pipes [Fig. 2.72]

Hydraulic steel pipes are available in various grades such as standard, extra heavy, double extra heavy grades, with working pressures 250 kg/cm^2, 400 kg/cm^2, and 800 kg/cm^2, respectively. Steel pipes are connected to hydraulic elements using compression fittings [Fig. 2.73]. Pipes are cut to length, and ends are filed square to the axis.

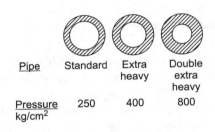

Fig. 2.72: Steel pipes

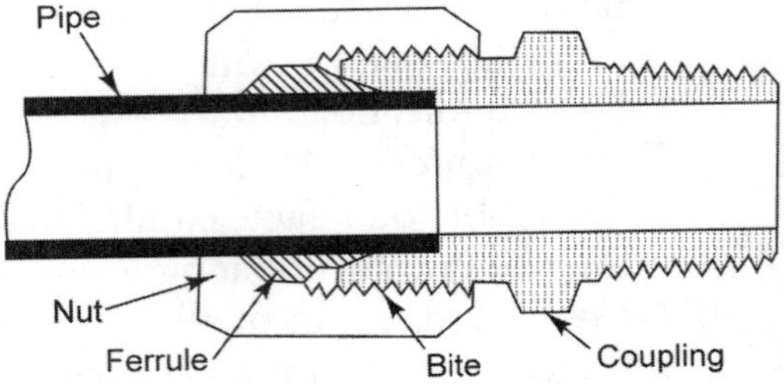

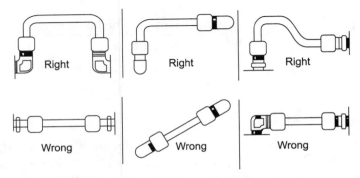

Fig. 2.73: Compression fitting

Hydraulic elements are generally provided with tapped holes, with pipe threads. The coupling of the compression fitting is screwed into the hole. Ferrule and nut are slid on to the cut pipe. The pipe is inserted in the coupling, and the nut is tightened. This presses the ferrule into the conical seat in the coupling. Furthermore, the ferrule bites into the pipe for a permanent sub-assembly. The joint can be dissembled by unscrewing the nut, and assembled again, by screwing it. Compression fittings are very popular in the industry, as the pipes need not be threaded or flared for assembly.

In a flared fitting, the tube is flared to a 37° angle. The flared portion is clamped between a sleeve and a coupling conical seat, by a nut, to provide a leakage-free joint. The tubes should have been flared before the fittings can be assembled.

2.13.1.2 Hose pipes

Couplings used for hose pipes are provided with a *conical* seat on the bore, which mates with the conical seat provided on the fittings, at the end of the hydraulic hose pipes. The pipe can be assembled by a nut, which presses the conical seats against each other, to provide a leakage-free joint. The joint can be assembled and dissembled any number of times, with the nut.

Hose pipes should be *extra* in *length* to ensure positive sag. For long durability pipes should not be bent sharply, or twisted in assembly **[Fig. 2.74]**.

Rubber, nylon and various other synthetic materials are used for hose pipes. High pressure pipes are *braided* with steel wires.

Fig. 2.74: Recommended pipe configurations

2.13.2 Hydraulic system design

This involves determining the sizes of pumps, actuators, valves, piping, etc. to meet the requirements of force and speed in a machine tool. The most important parameters in any hydraulic system are, pressure and flow. A hydraulic system resembles an electrical system in some respects. The pressure is like voltage, flow like the current, and the load like electrical resistance. Most of the hydraulic elements are specified by, maximum pressure and maximum flow possible.

The velocity of oil is maximum in the connecting pipes. The pipe bore should be big enough to avoid turbulence, occurring at a very high velocity. On the other hand, very low velocity increases the pipe bore size and cost, considerably. The following table gives the recommended oil velocities for connecting pipes.

Table 2.44: Recommended maximum velocities for hydraulic oils

Sr. No.	Line type	Maximum velocity (meters/sec.)
1	Pressure lines	6 m/s
2	Tank return lines	3
3	Pump suction lines	2

Hydraulic system design involves the following steps:

(1) Find the maximum size of the hydraulic actuator, which can be accommodated in the available space, i.e. area. The cylinder bore can be taken about half of the available height/width, whichever is minimum
(2) Determine the working pressure for the actuator size. This depends upon the hydraulic force required, which in turn, depends upon the maximum load on the actuator
(3) Determine the maximum flow, necessary to generate the maximum linear speed required
(4) Find pump motor power
(5) Determine the pipe bore to avoid turbulence
(6) Determine the maximum pressure and maximum flow capacities, necessary for the pump and valves
(7) Select the nearest higher standard size of the pump and valves
(8) Draw a hydraulic circuit

The following example illustrates this procedure.

Example 2.29: A broaching machine has to provide maximum 5 tonnes cutting force, at speeds ranging from 3 to 18 meters/min. Design a hydraulic system, with a vane pump as the source of pressurized oil.

$$\text{Max. pressure} = 100 \text{ kg/cm}^2.$$

Solution: For the hydraulic cylinder,

$$\text{Hydraulic force} = \text{Cylinder area} \times \text{Pressure}$$
$$(\text{kg}) \qquad\qquad (\text{cm}^2) \qquad (\text{kg/cm}^2)$$

$$\therefore \quad \text{Cylinder area} = A_c = \frac{5 \times 1000}{100} = 50 \text{ cm}^2$$

$$\text{Cylinder diameter} = D_c = \sqrt{\frac{50}{0.7854}} = 7.98 \text{ cms}$$

$$\approx 8 \text{ cms}$$

$$\text{Maximum speed} = 18 \text{ m/min}$$
$$= 18 \times 100 \text{ cms/min}$$

$$\text{Required flow (cc)} = \text{Cyl. area} \times \text{Max speed (cms/min)}$$
$$= 50 \times 1800 = 90,000 \text{ cc/min}$$
$$= 90 \text{ lits/min}$$

From Table 2.10, for Vane Pumps,

$$\text{Motor power (kW)} = \frac{PV}{150} = \frac{100 \times 90}{150}$$
$$= 60 \text{ kW}$$

The nearest, higher standard motor rating is 75 kW.

As both, pressure and tank ports of valves, are usually of the same size, the tank return line speed of 3 meters/sec (Table 2.12) is the criterion for determining the sizes of pipes and valves.

$$\text{Flow through pipes} = 90 \text{ lits/min} = \frac{90}{60} \text{ lits/sec}$$

$$= \frac{90}{60} \times 1,000 \text{ cc/sec.}$$

$$= 1,500 \text{ cc/sec.}$$

$$\text{Flow (cc/sec)} = \text{Pipe area (cm}^2) \times \text{Velocity (cm/sec)}$$

$$\therefore \quad \text{Pipe area} = \frac{1,500}{3 \times 100} = 5 \text{ cm}^2$$

$$\text{Pipe bore (cms)} = \sqrt{\frac{5}{0.7854}} = 2.52 \text{ cms}$$

$$\approx 1 \text{ inch}$$

A standard commercial pipe of 250 kg/cm² working pressure, can be used. Let us list the specifications of all the hydraulic elements.

Sr. No.	Description	Max pressure kg/cm²	Max flow lits/min	Size
1.	Vane pump	100	90	—
2.	Pressure relief valve	100	90	1"
3.	Direction control valve	100	90	1"
4.	Flow control valve	100	90	1"
5.	Piping	250	90	1"

Figure 2.75 shows the schematic hydraulic circuit for the hydraulic system. Items 1 to 4 are usually mounted on the oil tank, and connected according to the required configuration. The integral unit is called the hydraulic power pack. The power pack is connected to the hydraulic cylinder, by pipes of suitable lengths.

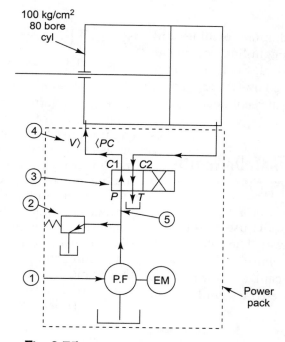

Fig. 2.75: Hydraulic circuit (for Example 2.28)

The D.C. valve 3 can be operated manully (by handle) or electrically (by push button).

CHAPTER 3

RECTILINEAR OR TRANSLATORY DRIVES

Machine tools such as shaping, planing, and slotting machines use straight-line, rectilinear motions for metal removal. Such machines use a reciprocating mechanism for cutting stroke, and an intermittent rectilinear motion for lateral, cross feed.

3.1 CONVERTING ROTARY DRIVE TO TRANSLATORY MOTION

Most of the machine tools use an electric motor as a source of power. The rotary drive is transformed to translatory (rectilinear) motion by using crank, cam, a screw and nut combination, or a geared rack and pinion combination.

3.1.1 The crank and slider mechanism

The crank and slider mechanism is widely used for converting the reciprocating motion of an engine piston, to the rotary driver of the engine flywheel. The same mechanism can be used to convert a rotary motion into a rectilinear reciprocating drive [**Fig. 3.1**]. The rotating end of the crank is pinned to one end of the connecting rod, while the other end is pinned to a slider, constrained by guides, to enable its movement in a straight line. During the half-rotation towards the slider, the connecting rod pushes the slider away from the crank, in a forward-cutting stroke. The remaining half-rotation away from the slider, pulls the slider towards the crank, in a return

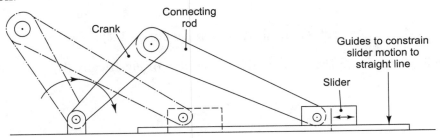

Fig. 3.1: Crank and slider mechanism

stroke. As both cutting and return strokes are effected in half-rotations with equal duration, the forward and return strokes run at equal speeds.

The speed of the forward-cutting stroke must suit the combination of the workpiece and cutting tool materials. But the idle, non-cutting return stroke can, and should be faster to save machining time, and increase productivity. Therefore, the crank and slider arrangement can be used only for small machine tools, with a stroke less than 300, so that the time and productivity loss, due to the slow return speed becomes negligible.

3.1.2 The crank and rocker mechanism [Fig. 3.2]

This is more productive than the crank and slider mechanism. In this mechanism, the connecting rod is replaced by a rocker arm which swings to and fro, like a pendulum, about a fixed pivot. The slot in the rocker arm guides a block, pinned to the rotating end of a crank. While the rocking end of the arm is pinned to one end of a connecting link, the other end of the link is pinned to a tool holder/workpiece slide. Needless to say, the slide must be guided, to constrain its motion to a straight line.

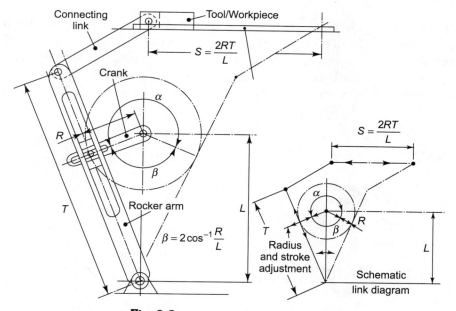

Fig. 3.2: Crank and rocker mechanism

Metal removal is carried out during the slow power stroke, which converts more than half the crank rotation (α) into a power stroke. The return stroke occupies less than half the rotation (β). The relatively lesser duration of the return stroke makes it much faster than the cutting stroke.

As in the case of the crank and slider motion, the stroke length is double the crank radius. The ratio of the cutting and return stroke speeds, depends upon angles α and β, which in turn depend upon the distance 'L', between the crank and rocker arm fulcrums, and the adjustable crank radius 'R'. For,

$$\angle\beta = 2\cos^{-1}\frac{R}{L} \qquad \text{(Eqn. 3.1)}$$

$$\text{and } \angle\alpha = 360 - \beta \qquad \text{(Eqn. 3.2)}$$

$$\frac{\text{Return speed}}{\text{Cutting speed}} = \frac{\alpha}{\beta} \qquad \text{(Eqn. 3.3)}$$

If the rocker arm length is 'T',

$$\text{Stroke length} = S = \frac{2RT}{L} \qquad \text{(Eqn. 3.4)}$$

Usually, the stroke is made adjustable by making the crank length R, adjustable. This can be done by providing a slot in the crank end.

The crank and rocker arrangement is widely used in long stroke machines.

3.1.3 Cam and follower [Fig. 3.3]

Cams can be broadly divided into plate-type, and drum-type. In a plate-type cam [Fig. 3.3a], the variation of radius 'R', from the pivot of rotation, moves the tool-holder slide radially, towards or away from the pivot. An increase in radius results in a cutting out-stroke, and a decrease causes the idle return stroke. Usually, the return stroke can be effected by gravity, or a spring. For a positive, precise return stroke, a slotted or a face-type cam [Fig. 3.3b] is used.

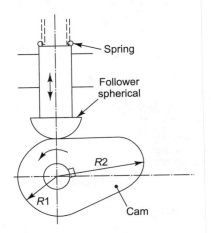
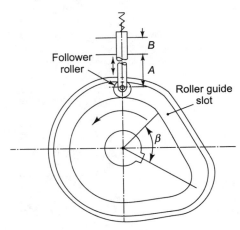

Fig. 3.3(a): Plate cam with spherical follower **Fig. 3.3(b):** Face cam with slot for roller follower

The speed of the cutting out-stroke, and return in-stroke can be varied, without changing the speed of the rotation of the cam. The speed depends upon the cam profile. It can be made suitable for the requirements of the metal removal cycle. In bar automatic lathes, the outstroke can be divided into rapid approach, and slow-cutting parts, in any desired ratio. Even the return in-stroke can be controlled as necessary, by controlling the cam profile. The average stroke speed (S) between any two points on a cam, can be found from the following equation:

$$\text{Speed} = S = \frac{R_2 - R_1}{\beta} \times 360 \frac{N}{1,000} \quad \text{(Eqn. 3.5a)}$$

S = Speed (m/min)
R_2 = Radius at point 2 (mm)
R_1 = Radius at point 1 (mm)
N = Rotary speed (Revolutions/min)
β = Angle for rise ($R_2 - R_1$) in degrees

Normal force [F_n] between the cam and the follower depends upon the pressure angle [α], the coefficient of friction [μ] between the cam and the follower, and unguided {A} and guided [B] lengths of the cam holder **[Fig. 3.3b]**. The maximum pressure angle should be less than the locking angle α_L, which will jam the follower in the guide.

$$\alpha_L = \tan^{-1}\left[\frac{B}{\mu\{2A + B\}}\right] \quad \text{(Eqn. 3.5b)}$$

α_m = Max permissible angle
     = 30° for translating rollers
     = 45° for oscillating rollers

The pressure angle [α] is the angle between the line joining the contact point to the roller center (normal), and the line joining the centers of the roller and the cam rotation axis **[Fig. 3.3c]**.

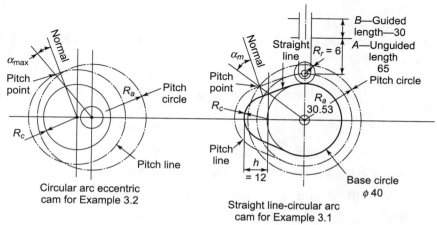

Fig. 3.3(c): Circular arc cams for Examples 3.1 and 3.2

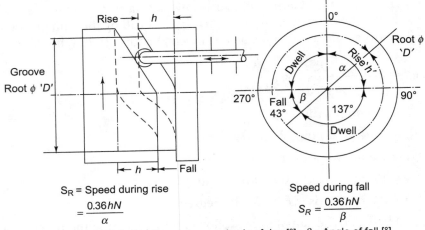

Fig. 3.3(d): Cylindrical (drum) cam

$$F_n = \frac{F}{\cos \alpha - \mu \left[\dfrac{2A + B}{B}\right] \sin \alpha} \qquad \text{(Eqn. 3.5c)}$$

F = Force on the cam [kg]
A = Unguided length of roller-holder [mm]
B = Guided length of roller-holder [mm]

F_n will be maximum when pressure angle α is maximum.

$$\text{Contact stress} = f_c = 0.591 \sqrt{\frac{2 F_n\, E_1 E_2}{b\, [E_1 + E_2]} \left[\frac{1}{R_c} \pm \frac{1}{R_r}\right]} \qquad \text{(Eqn. 3.5d)}$$

sign is + when the roller is in contact with convex (external) radius and – when the contact radius is concave.

$\dfrac{2 E_1 E_2}{E_1 + E_2}$ = Equivalent elastic modulus of cam and roller materials

$\qquad = 2.15 \times 10^4$ kg/mm² for Stl. cam and Stl. roller
$\qquad = 1.1 \times 10^4$ kg/mm² for C.I. cam and Stl. roller
$\qquad = 1.2 \times 10^4$ kg/mm² for Bronze cam and Stl. roller
$\qquad = 7 \times 10^2$ kg/mm² for Nylon cam and Stl. roller

b = cam width (mm); R_r = Follower roller radius (mm)
R_c = cam profile radius (mm) = h in circular arc, and circular arc-straight line cams

Rectilinear or Translatory Drives

A roller is usually made up of steel, and hardened. A cam can be made of cast iron, carbon alloy steel, case hardening steel, bronze, or nylon.

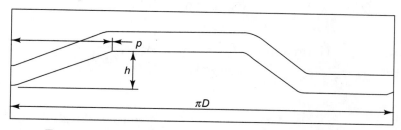

Fig. 3.3(e): Development of cylindrical cam in [**Fig. 3.3d**].

Cam material	C.I.				Steel		
Hardness HB	150	210	240	280	285	500	540
Permissible f_c (kg/mm^2)	36	45	54	53	110	113	122

In the simplest *Straight Line-Circular* Arc form, the cam radius is joined tangentially to the base circle, by a straight line. In the *circular arc* form, the base circle and cam radius are joined by a circular arc, tangential to both, i.e. the cam and base circle radii.

In both the forms, value of cam factor $\left[\dfrac{R_a \beta}{h}\right]$ depends upon the maximum pressure angle $[\alpha_{max}]$. Note: h = cam rise [throw]

R_a = Radius of cam at the point having maximum pressure angle $[\alpha_m]$

For straight line-circular arc form, $\dfrac{R_a \beta}{h} = 2 \tan\left[\dfrac{\alpha_m}{2}\right] + \cot \alpha_m$

For circular arc form, $\dfrac{R_a \beta}{h} = \cot\left[\dfrac{\alpha_m}{2}\right]$

Example 3.1: Design a straight line-circular arc type cam, for base circle of 40Φ, rise[h] = 12, angle for rise/fall [β] = 90°, max pressure angle [α_{max}] = 15°, guide length [B] = 65 and unguided length [A] = 30, cam load [F] = 50 kg; μ = 0.1, b = 10 Roller radius [R_r] = 6; $\dfrac{2E_1 E_2}{E_1 + E_2} = 2.15 \times 10^4$/mm^2. Ref [**Fig 3.3c**]

Solution:

For straight line-circular arc form $\dfrac{R_a \beta}{h} = 2 \tan\left[\dfrac{\alpha_m}{2}\right] + \cot \alpha_m$

$$\dfrac{R_a \dfrac{\pi}{2}}{12} = 2 \tan \dfrac{15}{2} + \cot 15$$

$$= 2 \times 0.13165 + 3.7321 = 3.995$$

$$R_a = 3.995 \times \frac{12}{1.57} = 30.53$$

$$\text{Normal force } [F_n] = \frac{50}{\cos 15 - \mu \left[\dfrac{2 \times 30 + 65}{65}\right] \sin 15}$$

$$= \frac{50}{0.96593 - 0.1\{1.923\}\, 0.25882}$$

$$= \frac{50}{0.916} = 54.585$$

$$\text{Cam radius } [R_c] = h = 12 \quad [\text{Ref } \textbf{Fig. 3.3c}]$$

From Eqn. 3.5d,

$$f_c = 0.591 \sqrt{\frac{2F_n\, E_1 E_2}{b(E_1 + E_2)} \left[\frac{1}{R_c} \pm \frac{1}{R_r}\right]}$$

$$= 0.591 \sqrt{\frac{54.585 \times 2.15 \times 10^4}{10} \left[\frac{1}{12} \pm \frac{1}{6}\right]} = 101.23$$

Checking materials specified below Eqn. 3.5d, it can be observed that the cam must be made of steel. With 285 Brinnel hardness, it can have a permissible stress level of up to 110 kg/mm², i.e. 10% higher than the computed stress of 101.23 kg/mm².

Example 3.2: Find the contact stress, if the profile of the cam in the above example is changed to 'eccentric', with a circular arc form.
Solution:

In an eccentric cam, the rise 'h' is obtained in half rotation, i.e. $\beta = \pi$

$$\frac{R_a \beta}{h} = \cot\left[\frac{\alpha_m}{2}\right]$$

$$R_a \times \frac{\pi}{12} = \cot\left[\frac{15}{2}\right] = 7.5958$$

$$R_a = \frac{12}{\pi} \times 7.5958 = 29.01$$

The cam radius [R_c] is 20 ($\frac{1}{2}$ base θ 40) in this cam.

$$f_c = 0.591 \sqrt{\frac{54.585 \times 2.15 \times 10^4}{10} \left[\frac{1}{20} \pm \frac{1}{6}\right]}$$

$$= 94.24 \text{ kg/mm}^2$$

A variety of curves (such as parabola) are used for cam profile.

Rectilinear or Translatory Drives

In cylindrical or drum cams [**Fig. 3.3d**], the follower roller moves in a peripheral groove, on a rotating cylinder (drum). The helix-like groove moves the follower roller and the attached tool holder/workpiece slide axially, to cause an out-stroke and in-stroke. Consequently, the speed (S) of the stroke depends upon the axial movement of the roller, and the peripheral distance in which the axial movement takes place: the angle traversed by the follower during the axial movement.

$$\text{Stroke speed} = S = \frac{0.36\,hN}{\alpha} \quad \text{(Eqn. 3.6)}$$

α = Angle causing the axial motion (degrees)
h = Axial movement (cam rise in mm)
N = Rotary speed of cam (Revs/min)
S = Speed (meters/min)

If the cam is developed (unrolled) into a flat plane [**Fig 3.3e**]

$$S = \frac{h}{p} \times \frac{\pi DN}{1{,}000} \quad \text{(Eqn. 3.7)}$$

h = Axial movement of the follower (mm)
p = Peripheral distance moved (mm)
D = Root diameter of the cam groove (mm)

$$p = \frac{\alpha D}{2}, \text{ when } \alpha \text{ is in radians}$$

$$= \frac{\pi \alpha D}{360}, \text{ when } \alpha \text{ is in degrees}$$

Each workpiece-machining cycle requires an exclusive cam. As precision cams are expensive, they are only used in mass production machines, such as automatic lathes. Though costly, cams are practically tamper-proof.

3.1.4 Nut and screw combination

This uses a screw which can rotate around its axis, but cannot move axially. The screw is provided with a collar which is held, pressed against some fixed face in housing, by a retainer screwed to the housing [**Fig. 3.4a**]. When the screw is rotated one turn (360°), the engaged nut moves axially by a distance equal to the lead (L) of the screw.

The axial speed of the nut depends upon the lead, and the rotary speed (N) of the screw.

$$S = \frac{LN}{1000} \text{ (m/min)} \quad \text{(Eqn. 3.8a)}$$

$$\text{Axial force } (F_a) = p \times \pi\, d_p\, hn \quad \text{(Eqn. 3.8b)}$$

d_p = Pitch f[mm]; h = thread engagement [mm] i.e. working thread depth for square threads

n = No. of mating threads = $\dfrac{\text{Nut length}}{\text{Pitch}}$

p = permissible bearing pressure (kg/mm^2)

= 0.8 kg for steel screw and cast iron nut

= 0.3 kg for precision feeding with steel screw and bronze nut

= 1.2 kg for milling m/c feed, with steel screw and bronze nut

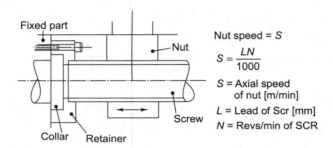

Fig. 3.4(a): Screw and nut mechanism for rectilinear motion

Compactness, ease in manufacture, and high load capacity makes the screw and nut combination very popular for auxiliary and feed motions. However, the axial speed cannot be changed without changing the rotary speed (R.P.M.). In sophisticated machine tools such as bar autos, the axial speed is increased during non-cutting approach and idle return stroke, by using a two-speed motor and running it at a high speed during the idle motions. The direction of an axial motion can be reversed by simply reversing the direction of rotation.

High frictional losses limit the usage of the simple screw and nut combination, for feed and auxiliary motions.

These frictional losses can be reduced by using a recirculating ball screw [**Fig. 3.4b**]. The screw, as well as the nut, are provided with precise concave grooves. The balls rolling in these grooves, propel the nut. As the sliding friction, is replaced by rolling friction the losses decrease considerably. A return passage for the balls is necessary, to facilitate their re-circulation. Although the efficiency of transmission of a ball screw is more than double the simple screw and nut combination, the cost is much higher, due to the high-precision concave grooves, which call for profiled thread grinding. Ball screws are hence, used only in high-precision machines such as grinders, jig boring machines, and numerically-controlled machines which must be free from backlash.

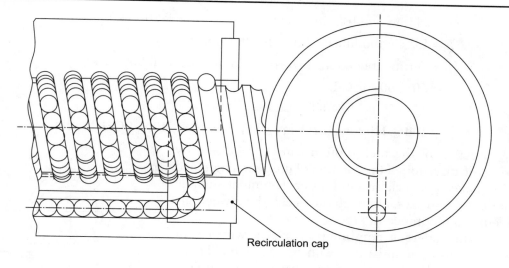

Fig. 3.4(b): Anti-friction ball screw with ball recirculation cap

Fig. 3.4(c): Form for ball screw threads

Fig. 3.4(d): Using packing shims to remove backlash

3.1.5 Rack and pinion combination

In this arrangement [**Fig. 3.5**], a rotating gear (pinion) drives a rack, guided to move in a straight line. The speed (S) of the rack depends upon the pitch diameter (D_p) of the gear, and the rotary speed (N) R.P.M.

$$S = \frac{\pi D_p N}{1000} \text{ meters/min} \qquad \textbf{(Eqn. 3.9a)}$$

$$\text{Permissible Force [F]} = \frac{1.4 f_c^2 \, D_p \, b \sin 2\alpha}{E} \qquad \textbf{(Eqn. 3.9b)}$$

D_p = Pinion pitch $\phi = tm$

b = Pinion width (mm) ; α = Pressure angle = 20°/14.5°

E = Elastic modulus = 2.1×10^4 kg/mm^2

f_c = Permissible contact stress (kg/mm^2) (from Table 2.23d)

= 50 – 60 for C.I.

= 45 – 50 for C 40/45 (En 8)

= 100 – 110 for Hrd Alloy stl. [30N; 4Cr1, 40N; 2Cr1 Mo28].

The direction of the rack motion can be reversed, by reversing the direction of the rotation of the gear. High load capacity makes the rack and pinion arrangement suitable even for cutting strokes, although the backlash between the gear teeth is a hindrance in precise motions.

The worm and rack combination uses a worm gear instead of a spur gear, and a rack with curved teeth, like the wheel usually used with a worm gear. This method combines screw and nut, and rack and pinion arrangements. Although the rectilinear motion is much smoother than the rack and pinion combination, the manufacturing cost is higher, and the efficiency lower than the worm and wheel combination.

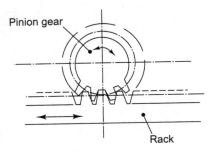

Fig. 3.5: Rack and pinion drive

3.2 GENERATING INTERMITTENT PERIODIC MOTION

In machines using reciprocating motion for metal cutting, the tool must be moved laterally, and fed to a new position, before commencing the next cutting stroke. This is accomplished by incorporating a periodic motion, usually propelled by the metal cutting drive.

Periodic motion is also used for indexing tool-holding turrets/drums, in mass production turning machines such as capstans, turrets, and automatic lathes.

3.2.1 Ratchet and pawl mechanism [Fig. 3.6]

In this arrangement, a spring-loaded pawl is used to turn a ratchet wheel, keyed to the lead screw for the lateral feed.

The pawl is swung clockwise and anti-clockwise through the set angle, around a ratchet wheel, with uni-directional teeth. A slotted plate is keyed to the drive for metal cutting [**Fig 3.6b**]. The pawl swing angle can be varied by adjusting the radial position of the drive rod end on the slotted plate. This facilitates adjustment of the lateral feed.

During the anti-clockwise swing around the ratchet axis [**Fig 3.6b**], the spring-loaded pawl climbs the slope of the uni-directional teeth. At the end of the swing, the spring pushes the pawl to the root of the nearest ratchet tooth.

During the clockwise swing of the pawl, ratchet wheel moves clockwise. This rotates the keyed lead screw, and pushes the engaged nut, and attached table, laterally, to effect the feed.

The feed is effected during the tool return stroke, while during the cutting stroke, the pawl moves anti-clockwise, around the ratchet climbing the slopes of the uni-directional teeth, swinging around the ratchet, without rotating it.

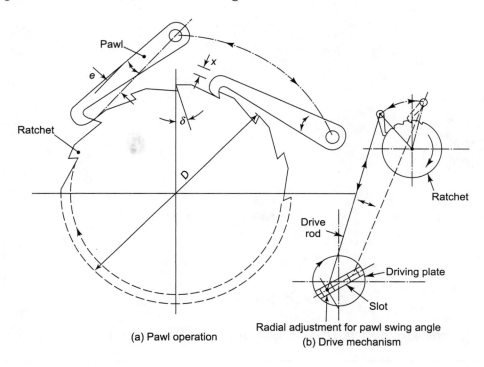

Fig. 3.6 (a) and (b): Ratchet and pawl mechanism

Ratchet and pawl mechanisms are used widely, for cross feed of shaping and planing machines. Immunity from reversal under load, makes them convenient for hoists using worm gears, and lifting jacks with rack and pinion gears.

Table 3.1: Recommended number of teeth [t] for ratchets

Transmission system	Rack and pinion	Worm and wheel	Arrestors
Number of teeth [t]	6–8	12–20	16–25

The permissible pressure on the ratchet tooth depends upon the materials used for ratchet and pawl. The tooth angle [δ] depends on the friction coefficient between ratchet and pawl.

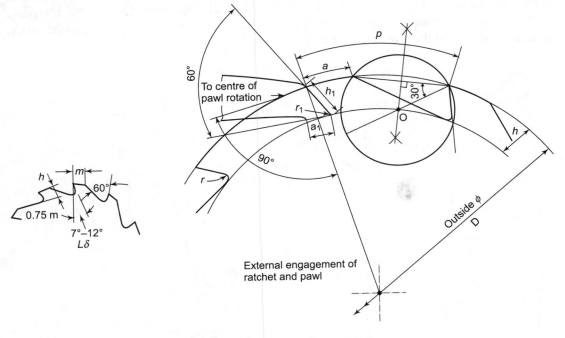

Fig. 3.6(c): Tooth profile of ratchet wheel.

Table 3.2: Permissible pressure [P] and tooth angle [δ] for common ratchet and pawl materials

Materials	C.I. ratchet–Steel pawl	Steel ratchet–Steel pawl
P [kg/mm²]	0.5–1	1.5–3
δ [minimum]	7°	11° 30′

Like gears, ratchets are specified by modules. Only pitch diameter is replaced by the outside diameter [D]. Refer to **[Fig. 3.6 b,c]**.

$$m = D/t; \quad h = 0.75\,m; \quad \text{tooth land} = a = m$$

$$m = 2 \sqrt[3]{\frac{T}{t.b.f}} \qquad \text{(Eqn. 3.10)}$$

T = Torque [kgmm]

t = Number of teeth of ratchet

b = Breadth of ratchet = 1.5 – 3 module (m) i.e. $\dfrac{b}{m}$ = 1.5 – 3

m = Module [mm]

f = Permissible stress [kg/mm²] = 3–5 kg/mm² for C45

The stress on pawl depends upon the eccentricity [e] of the pawl, when it reaches the tooth tip during a crossover [**Fig. 3.6a**].

	m	6	8	10	12	14	16	18	20	22	24	26	30
Ratchet wheel	Z	\multicolumn{12}{c}{from 6 to 30 inclusive}												
	P	18.85	25.13	31.42	37.70	43.98	50.27	56.55	62.83	69.12	75.4	81.68	94.25
	h	4.5	6	7.5	9	10.5	12	13.5	15	16.5	18	19.5	22.5
	a	6.0	8	10	12	14	16	18	20	22	24	26	30
	r	1.5	1.5	1.5	1.5	1.5	1.5	1.5	1.5	1.5	1.5	1.5	1.5
Pawl	h_1	6	8	10	12	14	14	16	18	20	20	22	25
	a_1	4	4	6	6	8	8	12	12	14	14	14	16
	r_1	2	2	2	2	2	2	2	2	2	2	2	2

Fig. 3.6(d): Tooth dimensions for standard ratchets and pawls. [Ref **Fig 3.6c**]

$$\text{Stress } [f] = \frac{6 F_t e}{b x^2} + \frac{F_t}{xb} \qquad \text{(Eqn. 3.11)}$$

$$F_t = \frac{2T}{tm}$$

$e =$ Eccentricity [mm]

$b =$ Pawl width = Ratchet width = $1.5 - 3\ m$

$x =$ Depends upon the shape of pawl [Ref **Fig 3.6a**]

The diameter of the pawl pin [d] depends upon the stress [f] and the distance [g] between the pawl-holder and ratchet.

$$d = 2.71 \sqrt[3]{\frac{F_t [0.5b + g]}{2f}} \qquad \text{(Eqn. 3.12)}$$

The ratchet tooth angle [δ] should be more than the friction angle Φ [$\Phi = \cos^{-1} \mu$, where $\mu =$ coefficient of friction]. And $g =$ Thickness of collar on pawl pin

Example 3.3: Design Ratchet and Pawl for transmitting 250 kgmm torque, for 20 teeth ratchet, if $b = 3$ m; $f = 1.5$ kg/mm^2; tooth land $[a] = m$, and $x = 3$; $g = 1.5$ and $e = 2.2$. The ratchet is made of C.I., while the pawl is made of steel.

Solution:

$$m = 2\sqrt[3]{\frac{250}{20 \times 3m/m \times 1.5}} = 2.81 \approx 3$$

$$F_t = \frac{2 \times 250}{20 \times 3} = 8.33 \text{ kg}; \quad b = 3\ m = 9$$

From Eqn. 3.11,

$$f = \frac{6 \times 8.33 \times 2.2}{9 \times 9} + \frac{8.33}{3 \times 9}$$

$$= 1.35 + 0.31$$

$$= 1.66$$

$$\text{Pawl pin } \Phi = d = 2.71\sqrt[3]{\frac{8.33}{2 \times 1.66}[0.5 \times 9 + 1.5]}$$

$$= 6.69$$

$$\approx 7$$

From Table 3.2, the tooth face angle $[\delta]$ should be more than 7°, say 8°.

Figure 3.6d gives us the dimensions of some standard ratchets and pawls for ready use.

3.2.2 Geneva mechanism [Fig. 3.7]

Used widely for indexing turrets, the Geneva mechanism comprises a rotating driver, and a pivoted indexing wheel, with equally-spaced radial slots. During operation, a roller on the rotating driver enters a radial slot on the indexing wheel. The rotation of the driver moves the roller in a circular path. After entering one of the radial slots of the indexing wheel, the engaged roller rotates the indexing wheel through an angle equal to the spacing between the radial slots: the indexing angle (2A).

Needless to say, for a given combination of driver and indexing wheel, the indexing angle (2A) is constant and cannot be varied. For a smooth and impact-free operation, the roller center line should be square with the radial slot at the time of entry into, and exit from, the slot [Fig. 3.7].

$$\angle A + \angle B = 90° = \pi/2 \qquad \textbf{(Eqn. 3.13)}$$

$$\frac{r}{e} = \cos B = \sin A = \sin\frac{\pi}{Z} \qquad \textbf{(Eqn. 3.14)}$$

Rectilinear or Translatory Drives

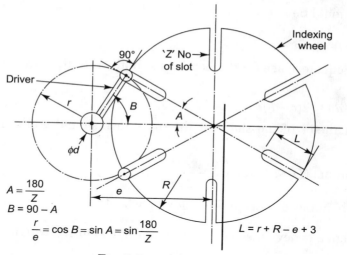

$$A = \frac{180}{Z}$$
$$B = 90 - A$$
$$\frac{r}{e} = \cos B = \sin A = \sin \frac{180}{Z}$$
$$L = r + R - e + 3$$

Fig. 3.7: Geneva mechanism

r = Radius of rotation of the driver roller (mm)
e = Center distance between drive disk and indexing wheel (mm)
z = No. of slots in the indexing wheel

The maximum indexing torque [T_{imax}] occurs at the angle ϕ_{max}, which depends upon the number of slots [z].

Table 3.3: Various constants for different nos. of indexing slots [z]

z	$\dfrac{\omega_{g\,max}}{\omega}$	$\dfrac{\alpha_g}{\omega^2}$ Initial	$\dfrac{\alpha_g}{\omega^2}$ Max	Φ_{max}
3	6.46	1.732	31.44	4°46′
4	2.41	1.00	5.409	11°24′
5	1.43	0.7265	2.299	17°34′
6	1.0	0.5774	1.350	22°54′
7	0.766	0.4816	0.9284	27°33′
8	0.620	0.4142	0.6998	31°38′
9	0.520	0.364	0.5591	35°16′
10	0.447	0.3249	0.4648	38°30′

The acceleration is much higher for a lesser number of slots: 23 times the 6-slot drive, for a 3-slot indexing wheel. Hence, it is convenient to have more number of slots, and use double indexing for bigger angles. This also reduces overload on the motor at the beginning of indexing. A 6-slot Geneva mechanism can be run with a standard motor, with a 20% overload capacity.

The slot length should be at least 3 mm more than the value computed from the following equation:
$$L = r + R - e \quad \text{(Eqn. 3.15)}$$

For providing adequate space for the bearings at both ends, the driver should conform to the following equation:

Driver shaft dia $(d) < 2(e - R)$ \quad **(Eqn. 3.16)**

Similarly, the diameter of the shaft for the drive disk should conform to the following equation:

Index wheel shaft dia $(D) < 4e \sin^2\left(\dfrac{\pi}{4} - \dfrac{\pi}{2z}\right)$ \quad **(Eqn. 3.17)**

where $\quad z =$ Number of slots in the indexing wheel

Example 3.4: Design a geneva mechanism for a hexagonal turret, with indexing angle = 60°. Center distance = 200 (approx).

Solution:
$$2A = 60$$
From Eqn. 3.13,
$$B = 90 - A = 90° - 30° = 60°$$
Draw **[Fig. 3.7]**
$$r = 200 \sin 30 = 100$$
$$R = 200 \cos 30 = 173.20$$
From Eqn. 3.15,
$$L = r + R - e = 100 + 173.2 - 200 = 73.2$$
Slot length should be slightly more than L, say 76 mm.
From Eqn. 3.16,
$$d < 2(e - R), \text{ i.e. } < 2(200 - 173.2)$$
The driver diameter (d) should be less than 53.6, say 50.
From Eqn. 3.17,
$$D < 4\, e \sin^2\left(\dfrac{\pi}{4} - \dfrac{\pi}{2z}\right)$$
$$< 4 \times 200 \times \sin^2\left(\dfrac{\pi}{4} - \dfrac{\pi}{2 \times 6}\right) \quad \text{(Eqn. 3.18)}$$
$$< 200$$

The actual size of the wheel shaft can be much smaller: suitable for the torque, dependent on the mass to be indexed, and the speed of indexing [Ref. Example 3.5].

The theoretical outside the radius of wheel [R] should be increased, taking into account the roller pin radius [r_p].

$$\text{Corrected wheel rad } [R'] = R\sqrt{1 + \left(\frac{r_p}{R}\right)^2} \quad \text{(Eqn. 3.19)}$$

The Geneva wheel is subjected to two types of torques: Frictional torque [T_{gf}], which is constant, and Inertia torque [T_{gi}], which varies during indexing.

$$T_{gi\,max} = I\alpha_{g\,max} = I\left[\frac{\alpha_{g\,max}}{\omega^2}\right]\left(\frac{\pi N}{30}\right)^2$$

$$= 0.011\left[\frac{\alpha_{g\,max}}{\omega^2}\right] I N^2$$

The torques on the driver [T_f, T_i] are much higher.

$$\text{Driver max frictional torque } [T_{f\,max}] = \frac{\alpha_{g\,max}}{\omega^2} \cdot \frac{T_{gf}}{\eta} \quad \text{(Eqn. 3.20)}$$

$$\text{Driver max indexing torque } [T_{i\,max}] = \frac{\omega_{g\,max}}{\omega}\alpha_{g\,max}\frac{I}{\eta} \quad \text{(Eqn. 3.21)}$$

$$T_{max} = (T_{f\,max} + T_{i\,max})$$

(Note that $T_{f\,max} < T_{i\,max}$)

$$\text{Power [kW]} = \frac{T_{max} \times \omega}{102{,}000} \quad \text{(Eqn. 3.22)}$$

$$\omega = \text{Angular velocity (radians/sec)} = \frac{\pi N}{30}$$

Example 3.5: Find power requirement for a 6-slot Geneva mechanism, if $r = 100$, $R = 173.2$, $e = 200$, and Geneva wheel thickness = 35. The driver rotates at 6 R.P.M. and $\eta = 0.85$. Frictional torque for wheel [T_{gf}] = 120 kgm. Wheel $I = 400$ kgm^2

Solution:

$$T_{gi\,max} = 0.011\left[\frac{\alpha_{g\,max}}{\omega^2}\right] I N^2$$

From Table 3.3, for $z = 6$; $\left[\frac{\alpha_g}{\omega_{max}}\right] = 1.35$; $N = 6$ (given)

$$T_{gi\,max} = 0.011 \times 1.35 \times 400 \times 6^2 = 213 \text{ kgm}$$

The driver torques will be as follows:

Driver friction torque
$$[T_f]_{max} = \frac{\omega_{g\,max}}{\omega} \cdot \frac{T_{gf}}{\eta}$$

From Table 3.3, for 6 slots, $\dfrac{\omega_{g\,max}}{\omega} = 1.0$

$$T_{gf} = 120 \text{ kgm (given)}$$

$$T_{f\,max} = 1 \cdot \frac{120}{0.85} = 141.18 \text{ kgm}$$

From Eqn. 3.21,
$$T_{i\,max} = \frac{\omega_{g\,max}}{\omega} \cdot \alpha_{g\,max} \cdot \frac{1}{\eta}$$

From Table 3.3,
$$\frac{\alpha_g}{\omega^2} = 1.35$$

$$\omega = \frac{\pi N}{30} = \frac{\pi \times 6}{30} = 0.6286, \text{ and } \omega^2 = (0.6286)^2 = 0.395$$

$$\alpha_g = 1.35 \times 0.395 = 0.5334$$

$$T_{i\,max} = 1 \times 0.5334 \left[\frac{400}{0.85}\right]$$

$$= 251.02 \text{ kgm}$$

$$T_{max} = T_{f\,max} + T_{i\,max} = 141.18 + 251.02$$
$$= 392.2 \text{ kg m}$$
$$= 392,200 \text{ kg mm}$$

From Eqn. 3.22,
$$kW = \frac{T_{max}\,\omega}{102,000} = \frac{392,200 \times 0.6286}{102,000}$$
$$= 2.417$$

Driving tangential force on roller pin $[F] = \dfrac{T_{max}}{r} = \dfrac{392,200}{100}$
$$= 3922 \text{ kg}$$

If, permissible shear stress on alloy steel pin $= 11 \text{ kg/mm}^2$,

$$\text{Pin } \Phi = \sqrt{\frac{3,922}{0.7854 \times 11}} = 21.28 \approx 22$$

Rectilinear or Translatory Drives

Square force on the wheel slot $= F_g$

$$F_g = \frac{T_g}{\text{Minimum distance between roller and wheel center}}$$

$$= \frac{T_g}{e-r} = \frac{T_{gi\,max} + T_{gf}}{e-r} = \frac{(213+120)1{,}000}{200-100}$$

$$= 3{,}330 \text{ kg}$$

As $F_g < F$ the shear stress on the pin will be less than 11 kg/mm^2: the permissible stress

$$\text{Linear pressure} = \frac{3{,}922}{35[\text{wheel thickness}]}$$

$$= 12.06 \text{ kg/mm}$$

For $\Phi\, 22 \times \Phi\, 39 \times 30$ needle roller bearing,

$$\text{Roller } \Phi = r_p = 39$$

From Eqn. 3.19,

$$\text{Corrected wheel radius } [R'] = R\sqrt{1+\left(\frac{r_p}{R}\right)^2}$$

$$= 173.2\sqrt{1+\left(\frac{39}{173.2}\right)^2}$$

$$= 182$$

$$\text{Wheel } \Phi = 2R' = 364$$

If the pin for the roller is fixed in a fork with a 38 span, it will simply act as a supported beam, with a uniformly distributed load of 3,922 kg.

$$I = \frac{\pi}{64}d^4 = \frac{\pi}{64}22^4 = 11{,}503.6 \text{ mm}^4;$$

$$\text{Section modulus} = \frac{11503.4}{11} = 1{,}046 \text{ mm}^3$$

$$\text{B.M.} = \frac{3{,}922 \times 38}{8} = 18{,}582 \text{ mm};$$

$$\text{Stress} = \frac{18{,}582}{1{,}046} = 17.76 \text{ kg/mm}^2.$$

The alloy steel pin should be hardened and tempered, to increase the strength to 120 kg/mm^2.

$$\text{Deflection } [\delta] = \frac{5}{384} \frac{Wl^3}{EI} = \frac{5 \times 3{,}922 \times 38^3}{384 \times 2.1 \times 10^4 \times 11{,}503.6}$$

$$= \frac{10.733 \times 10^8}{9.27 \times 10^{10}} = 0.0115$$

For using the driver shaft simply as a supported beam, i.e. using the bearings at both the ends [Eqn. 3.16],
$d < 2\{e - R\}$, i.e. $< 2\{200 - 182\}$, i.e. < 36
Reducing shaft to ϕ 35 to avail standard bearings,

$$\text{Shear stress } f_s = \frac{16 T_{max}}{\pi 35^3} = \frac{16 \times 392{,}200}{\pi \times 42875} = 46.6 \text{ kg/mm}^2$$

It will be necessary to use an alloy like 40 Ni 2 Cr 1 Mo 28 [En 24], which can be tempered to enhance the tensile strength to 110–125 kg/mm^2. Alternatively, a bigger shaft like a cantilever, with a bearing at one end only, can also be used.

The Geneva wheel is subjected to an equally high torque, i.e. 392,200 kgmm. The wheel shaft diameter (D) should be less than $\left[4e \sin^2 \left(\frac{\pi}{4} - \frac{\pi}{2z} \right) \right]$

$$4e \sin^2 \left(\frac{\pi}{4} - \frac{\pi}{2z} \right) = 4 \times 200 \sin^2 \left(\frac{\pi}{4} - \frac{\pi}{2 \times 6} \right) = 200$$

For a stress of 3 kg/mm^2,

$$D = \sqrt[3]{\frac{T}{\frac{\pi}{32} \times 3}} = \sqrt[3]{\frac{392{,}200}{0.2946}} = \sqrt[3]{1{,}327{,}902} = 109.9 \approx 110$$

3.2.2.1 Materials and heat treatment

(1) Rollers should have RC 59/63 hardness. They can be made of bearing steel [103 Cr 1/En 31], any carborizing steel [15 Cr 65, 15 Ni Cr 1 Mo 15/En 354, or 17 Mn 1 Cr 95/En 207] and case hardened to 59–63 RC range.
(2) Geneva wheels being much bigger than the roller, are made up of full-hardening steels like 40 Cr 1 and tempered to RC 45–50 hardness. Alternatively, the wheel can be lined with the hardened liner on the wear surface of the slot. In such a case, the wheel plate need not be hardened.

CHAPTER 4

Drive transmission and manipulation

The generated drive is transmitted to various machine elements by using fixed connections like couplings, or quick-engaging and disengagable clutches. These transmit rotary power from one shaft to the other. For example, an electric motor shaft is coupled with a reduction gearing input shaft by a coupling, which once mounted, is removed only for maintenance or replacement. Clutches on the other hand, can be disengaged instantly, to disconnect the driving and driven shafts, to cease power transmission and the resultant motion. Similarly, the engagement of clutches can commence transmission instantly.

4.1 COUPLINGS

Couplings can be rigid or flexible. Rigid couplings call for perfect alignment of the coupled shafts. Misalignment can cause bending of shafts, uneven wear of bearings, and even jamming, especially in case of excessive misalignment. Flexible couplings have a pliable, intermediate element which can endure a certain degree misalignment of the shafts. The flexible transmitting element distorts itself to compensate for the misalignment.

4.1.1 Rigid couplings

Muff or sleeve couplings are sleeves [**Fig. 4.1**] bored and keywayed, to suit the coupled shafts. A wide undercut at the center, provides clearance for overshoot of the tool while machining the keyways. Radial set screws enable clamping of the shafts, to prevent an axial motion of the coupling during operation.

In a **half lap muff coupling [Fig. 4.2]**, the shafts are provided with an angular overlap, to prevent their separation during operation.

In a **split muff coupling [Fig. 4.3]**, the sleeve is split into two halves which can be bolted together on the shafts. This coupling can be dissembled and reassembled, without moving the shafts and the associated equipment. So the risk of misalignment during re-assembly is eliminated altogether. The split muff coupling is very convenient for heavy assemblies.

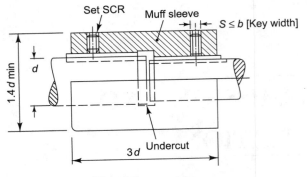

Fig. 4.1: Muff coupling

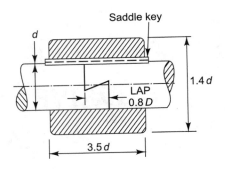

Fig. 4.2: Half lap muff coupling

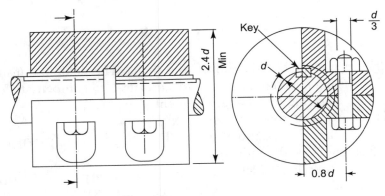

Fig. 4.3: Split muff coupling

Solid flanged couplings [Fig 4.4] are very convenient for shafts with integral (cast, forged, or welded) flanges. Flanges of the shafts are assembled together by using threaded, tapered pins. The flanges are machined with male and female-precision running fit spigots to ensure the requisite alignment of the shafts. The threaded, tapered pins can be replaced by close-fit bolts, and nuts, with washers.

Flanges can be made detachable **[Fig. 4.5]** by using keys for anchoring flanges on the shafts.

In all bolted assemblies, it is safer to provide counterbores for bolt-heads and nuts, to prevent risky projections.

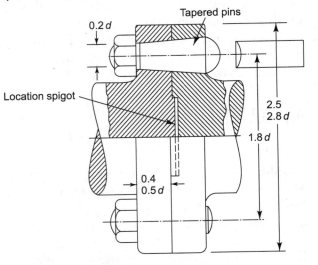

Fig. 4.4: Solid flange coupling

Sometimes, the flange body is provided with a recess [**Fig. 4.6**] to prevent risky projections of bolt-heads and nuts projections.

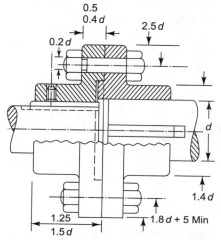

Fig. 4.5: Coupling with keyed detachable flanges

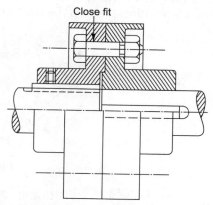

Fig. 4.6: Recessing coupling flanges to prevent bolt-head/nut projections

4.1.2 Flexible couplings

4.1.2.1 Bushed pin/bolt flange coupling

In a bushed pin/bolt flange coupling [**Fig. 4.7**], one of the flanges has rubber bushes or leather washers around the flange bolts. This permits a limited tilting of the bolts, and takes care of any marginal misalignment. While Table 4.1 gives the dimensions of standard bushed flexible couplings, Table 4.2 gives the service factors for various driver/driven machines.

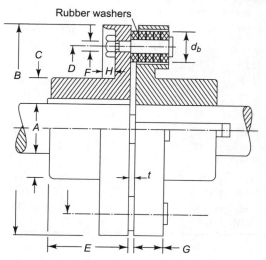

Fig. 4.7: Pin-type flexible coupling

4.1.2.2 The Oldham coupling

The Oldham coupling [**Fig. 4.8**] uses an intermediate double-slider with two integral keys, on opposite faces. These keys (projections) are square with each other. Each shaft is mounted with a flange, with a diametrical keyway on its face. These slots allow the double-slider to assume a position that will facilitate trouble-free transmission between the misaligned shafts.

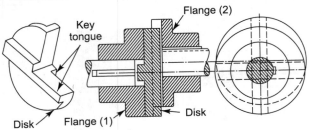

Fig. 4.8: Oldham's coupling

Table 4.1: Dimensions of standard bushed flexible couplings [Fig. 4.7]

Shaft Φ A (inch)	B min (inch)	C min (inch)	D (inch)	E min (inch)	F (inch)	G (inch)	H min (inch)	d_b * (inch)	t (inch)	No. of bolts	Torque kgm (inch)
12–16 [.47–.63]	80 [3.15]	25 [.98]	53 [2.09]	28 [1.1]	8 [.31]	18 [.71]	10 [.39]	20 [.79]	2 [.08]	3	4 [.16]
17–22 [.67–.87]	100 [3.94]	30 [1.18]	63 [2.48]	30 [1.18]	10 [.39]	20 [.79]	12 [.47]	22 [.63]	2	3	6 [.24]
23–30 [.9–1.18]	112 [4.41]	42 [1.65]	73 [2.87]	32 [1.26]	10	22 [.63]	12	22	2	3	8 [.31]
31–45 [1.22–1.77]	132 [5.2]	64 [2.52]	90 [3.54]	40 [1.57]	12 [.47]	30 [1.18]	15 [.59]	25 [.98]	4 [.16]	4	25 [.98]
46–56 [1.81–2.2]	170 [6.69]	80 [3.15]	120 [4.72]	45 [1.77]	12	35 [1.38]	15	25	4	4	40 [1.57]
57–75 [2.24–2.95]	200 [7.87]	100 [3.94]	150 [5.9]	56 [2.2]	12	40 [1.57]	15	30 [1.18]	4	4	60 [2.36]
76–85 [2.99–3.35]	250 [9.84]	140 [5.51]	190 [7.48]	63 [2.48]	16 [.63]	45 [1.77]	22 [.63]	40 [1.5]	5 [.2]	6	160 [6.3]
86–110 [3.39–3.98]	315 [12.4]	180 [7.09]	250 [9.84]	80 [3.15]	16	50 [1.97]	22	40	5	6	250 [9.84]
111–130 [4.37–5.12]	400 [15.75]	212 [8.35]	315 [12.4]	90 [3.54]	18 [.71]	56 [2.2]	28 [1.1]	45 [1.77]	6 [.24]	8	520 [20.47]
131–150 [5.16–5.9]	500 [19.68]	280 [11.02]	400 [15.75]	100 [3.94]	18	60 [2.36]	28	45	6	8	740 [29.13]

* Rubber bushes hardness = 80° IRH

Permissible alignment error between coupled shafts = 0.25/100 of Φ B

Table 4.2: Service factors for coupling

Driven Machine	Driver					
	Motor or Turbine	Engine				
		Petrol		Oil		Gas
		Up to 4 cyl	Above 4 cyl	Up to 4 cyl	Above 4 cyl	
Generators, Printing machines, Pumps	1.5	3	2.5	5	3.5	2
Compressors, Conveyors Wood/Metal cutting machines, Mixers Elevators	2	3.5	3	5.5	4	2.5
High speed crushers, compressors and Fans, Metal Forming machines Low speed compressors, Lifting Gears, Planing machines; Brick, Tile and	2.5	4	3.5	6	4.5	3
Rubber machines, Welding Generators	3.5	5	4.5	7	5.5	4

Design torque $\{T_d\}$ = Service factor × Required torque

4.1.2.3 Universal coupling or hook's joint

Universal coupling or hook's joint is very convenient for transmission between parallel or intersecting shafts. A single universal joint [**Fig. 4.9a**] cannot give uniform speed in the driven shaft. That limits its application to the drives, where variation in speed, and jerky motion is permissible. For uniform speed, it is necessary to use a double universal joint [**Fig. 4.9b**], or an intermediate shaft with two univeral joints. The axes of the pivot holes in the forks at both ends of the intermediate shaft, must be in the same plane [**Fig. 4.9c**].

Universal joints are generally fitted to shafts with taper pins. Table 4.3 and [**Fig. 4.9b and e**] give the dimensions of universal joints for ϕ 6–50 shafts, along with the size and position of the taper pin for each size of the shaft to be coupled. Table 4.3 also gives the torque capacity of the joints made from steel, having a tensile strength of 60 kg/mm², and fitted with bush bearings.

The torque capacity decreases considerably, when the angle between the coupled shafts, or the angle between the intermediate shaft and the driver/driven shaft axis exceeds 10°. **Figure 4.9f** gives the angle factor (η) for angles up to 45°.

Shocks also reduce the torque capacity of universal joints considerably. Shock factor (S) ranges from 1 to 3.

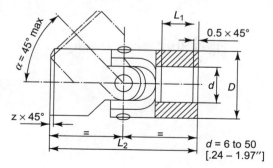

Fig. 4.9(a): Single universal joint

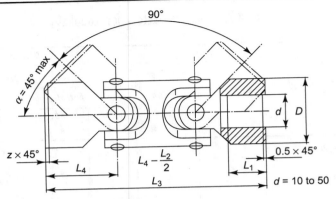

Fig. 4.9(b): Double universal joint

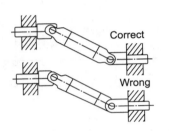

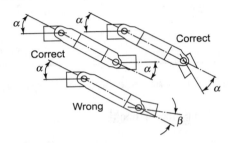

Fig. 4.9(c): Alignment of both ends

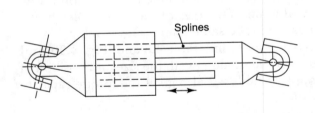

Fig. 4.9(d): Telescopic universal joint

Fig. 4.9(e): The size [t] of and margin [L_5] for taper pins used for assembly of universal joints on mating shafts

D	6 [.236″]	8 [.315]	10 [.39]	12 [.47]	16 [.63]	20 [.79]	25 [.98]	32 [1.26]	40 [1.57]	50 [1.97]
t		2 [.08]	3 [.12]	4 [.16]	5 [.2]	6 [.24]	8 [.31]	10 [.39]	12 [.47]	14 [0.55]	16 [.63]
L_5		4.5 [.18]	5 [.2]	6 [.24]	7.5 [.29]	9 [.35]	11 [.43]	15 [.2]	18 [.71]	22 [0.87]	27 [1.06]

Notes: 1. The shear strength of the taper pin should be between 16 to 25 kgf/mm² in order to transmit the given torque.
2. The length of the taper pin should conform to dia D indicated in Table 4.3

$$\text{Design torque } (T_d) = \frac{\text{Required torque } (T) \times \text{Shock factor } (S)}{\text{Angle factor } (\eta)}$$

$$T_d = \frac{TS}{\eta} \qquad \text{(Eqn. 4.1)}$$

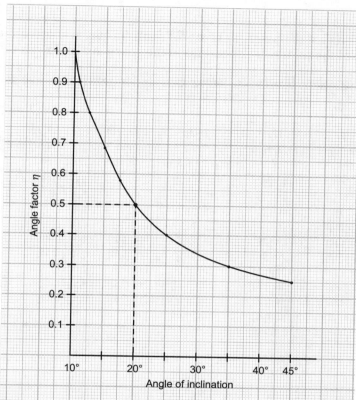

Fig. 4.9(f): Angle factors for universal joints

The usage of needle bearings significantly enhances the joints' life. The joint life is specified in millions (10^6) of revolutions. Table 4.3 gives the torque capacity of various joint sizes.

The power capacity (in kW) can be found easily from the following equation:

$$\text{kW} = \frac{T_d N (\text{RPM})}{1000} \qquad \text{(Eqn. 4.2)}$$

[1HP = 0.746 kW] [1 pound-foot = .1356 kgm]

Example 4.1: Find universal coupling for trasmitting 1.1 kW at 700 RPM, if the angle of inclination of the intermediate shaft is 20°. Shock factor = 2.

Solution: From Eqn. 4.2

$$T = \frac{\text{kW} \times 1000}{\text{RPM}} = \frac{1.1 \times 1000}{700} = 1.57 \text{ kgm}$$

From [**Fig. 4.9d**] for 20° inclination, $\eta = 0.5$

From Eqn. 4.1

$$T = \frac{TS}{\eta} = \frac{1.57 \times 2}{0.5} = 6.28 \text{ kgm}$$

From Table 4.3, shaft ϕ diameter 40 gives 5.8 kgm torque, while ϕ 50 joint gives 9 kgm torque. The choice of the joint will depend upon the space available.

Table 4.3: Universal joints: Dimensions, torque capacity for shafts' angle (up to 10°) [Fig 4.9a]

[1mm = 0.3937"]   [1 kgm = 7.376 Lb Ft]

Shaft dia (d) H7	Outside dia (D) K11	Shaft engagement L_1	Joint single L_2 + 1[.04]	Length double L_3 ± 1[.04]	Torque capacity kgm	Taper pin ϕt	L_4
6 [.236"]	16 [.63"]	9 [.35"]	34 [1.34"]	—	0.25 [1.844 Lb Ft]	2 [.08"]	4.5 [.18"]
8 [.315]	16	11 [.43]	40 [1.57]	—	0.33 [2.43]	3 [.12]	5 [.2]
10 [.394]	20 [.787]	15 [.59]	52 [2.05]	74 [2.91"]	0.48 [3.54]	4	6 [.24]
12 [.472]	25 [.984]	15	56 [2.2]	86 [3.39]	0.58 [4.28]	5 [.2]	7.5 [.29]
16 [.63]	32 [1.26]	19 [.75]	68 [2.68]	104 [4.09]	1.00 [7.38]	6 [.24]	9 [.35]
20 [.787]	40 [1.575]	23 [.9]	82 [3.23]	128 [5.04]	1.70 [12.54]	8 [.31]	11 [.43]
25 [.984]	50 [1.968]	29 [1.14]	105 [4.13]	160 [6.3]	2.7 [19.91]	10 [.39]	15 [.59]
32 [1.26]	63 [2.48]	36 [1.42]	130 [5.12]	198 [7.83]	4.5 [33.19]	12 [.47]	18 [.71]
40 [1.575]	75 [2.953]	44 [1.73]	160 [6.3]	245 [9.65]	5.8 [42.78]	14z [.55]	22 [.87]
50 [1.968]	90 [3.543]	54 [2.13]	190 [7.48]	290 [11.42]	9 [66.38]	16 [.63]	27 [1.06]

4.2 CLUTCHES

4.2.1 Positive action jaw clutches

Positive action jaw clutches [**Fig. 4.10a**] are totally free from slip and have teeth called jaws, on the mating faces. The clutch half on the driving shaft remains (axially) stationary, while the other half on the driven shaft can be axially slid on a splined (or keyed) shaft to mesh its teeth with the jaws on the stationary half. Jaw teeth can be made fine, for quick engagement or disengagement. With their numbers ranging from 11 to 165, fine teeth [**Fig. 4.10b**] have 60°–90° tooth angle for a bi-directional operation. Even uni-directional teeth have a 45°–60° angle on one side, and a 0°–5° angle on the other.

The number of coarse teeth ranges from 2–9. The straight (flat) sided teeth [**Fig. 4.10a**] must have positive clearance. Keeping their number at 2, 3, or 6 simplifies manufacture and inspection. Trapezoidal teeth [**Fig. 4.10b**] used for high torque, have sides inclined at 5° to 45°. The number of teeth is 5, 7, or 9, and the angular face engagement eliminates the need for clearance. Chamfering the mating side-face to 30°, facilitates easy engagement.

Drive Transmission and Manipulation

Clutches are engaged or disengaged with the help of a shifting lever. The lever must have a spring detent to hold it in an engaged or disengaged position. There must be a minimum of 1 mm clearance between the clutch faces, in the disengaged position. The bearing length of the sliding half of the clutch must be 2 to 2.5 times the diameter of the shaft to minimize the wear.

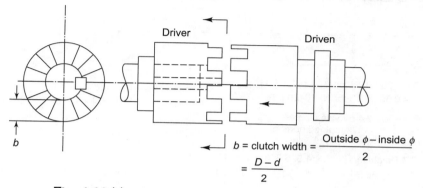

b = clutch width = $\dfrac{\text{Outside } \phi - \text{inside } \phi}{2} = \dfrac{D-d}{2}$

Fig. 4.10(a): Positive toothed (jaw) clutch with straight teeth

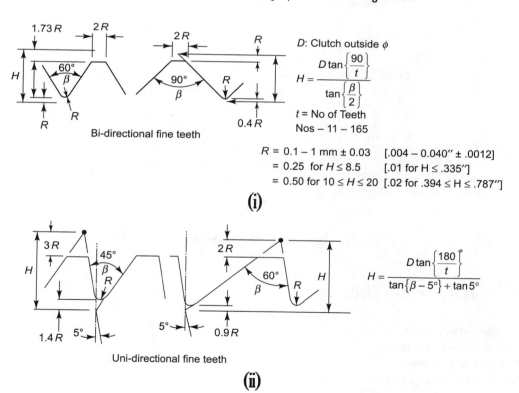

D: Clutch outside ϕ

$$H = \dfrac{D \tan\left\{\dfrac{90}{t}\right\}}{\tan\left\{\dfrac{\beta}{2}\right\}}$$

t = No of Teeth

Nos – 11 – 165

$R = 0.1 - 1$ mm ± 0.03 [.004 – 0.040″ ± .0012]
  = 0.25 for $H \le 8.5$ [.01 for $H \le .335″$]
  = 0.50 for $10 \le H \le 20$ [.02 for .394 ≤ H ≤ .787″]

(i)

$$H = \dfrac{D \tan\left\{\dfrac{180}{t}\right\}}{\tan\{\beta - 5°\} + \tan 5°}$$

Uni-directional fine teeth

(ii)

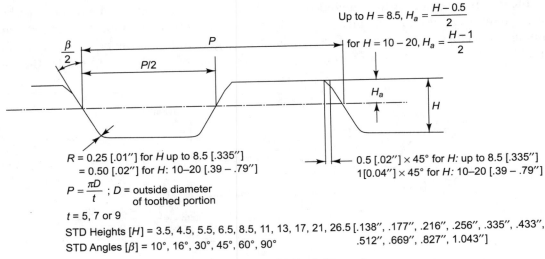

Bi-directional trapezoidal teeth [Coarse]

(iii)

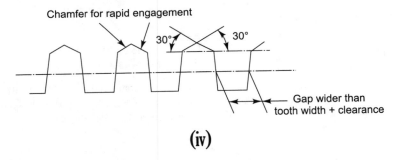

(iv)

Fig. 4.10b: Clutch teeth forms

4.2.2 Safety or slip clutches

These disengage under overload. Safety clutches are of three types:
  Toothed safety clutches
  Ball safety clutches
  Frictional safety clutches

Toothed safety clutches The tooth angle (β) in these clutches ranges from 90° to 120° in bi-directional clutches [**Fig. 4.11a**]. A bigger angle facilitates slip. As a 90° angle requires only 60% of the spring force required for a 120° angle, it is the most popular. In uni-directional clutches, the angle (β) ranges from 45° to 60°.

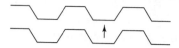

Fig. 4.11(a): Developed profile of bi-directional teeth of slip clutch

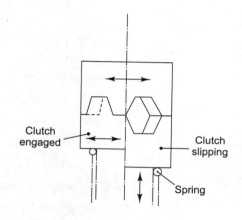

Fig. 4.11(b): Slip clutch operation

The safety clutches are held engaged by a spring. The holding force can be adjusted by a nut [**Fig. 4.11b**]. When the torque exceeds the set level, the movable half of the clutch climbs the slope of the fixed half, compressing the spring to disengage (slip). As soon as the torque decreases to the set value, the spring pushes the moving half forward, to engage the clutch automatically.

Uni-directional slip clutches have a bigger inclination angle only in the direction of slip [**Fig. 4.11c**].

Before discussing the other types of clutches, let us have a look at the force and strength requisites of toothed clutches. The clutch engagement and disengagement forces depend upon the frictional forces. The clutch tooth angle: $\frac{\beta}{2}$, for bi-directional clutches, and β for uni-directional clutches.

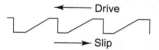

Fig. 4.11(c): Uni-directional slip teeth

$$\text{Engagement force } (F_{eb}) \text{ for bi-directional clches} = 2\ T_d \left(\frac{\mu_1}{d} + \frac{\tan\left(\frac{\beta}{2} + \phi_2\right)}{D_m} \right) \quad \textbf{(Eqn. 4.3)}$$

[1 mm = 0.3937″] [1 kgm = 7.376 Lb Ft]

For uni-directional clutches,

$$F_{eu} = 2\,T_d \left(\frac{\mu_1}{d} + \frac{\tan(\beta + \phi_2)}{D_m} \right) \quad \text{(Eqn. 4.4)}$$

Similarly, the disengagement forces for bi-directional and uni-directional clutches respectively, will be:

$$F_{db} = 2\,T_d \left(\frac{\mu_1}{d} - \frac{\tan\left(\frac{\beta}{2} - \phi_2\right)}{D_m} \right) \quad \text{(Eqn. 4.5)}$$

$$\text{and } F_{du} = 2\,T_d \left(\frac{\mu_1}{d} - \frac{\tan(\beta - \phi_2)}{D_m} \right) \quad \text{(Eqn. 4.6)}$$

T_d = Design torque (kgmm); [1 kg mm = .007376 Lb Ft]
μ = Coefficient of friction between shaft and clutch
   = 0.05 – 0.1 (with lubrication)
   = 0.15 – 0.3 (without lubrication)
μ_1 = Coefficient of friction between shaft and clutch
μ_2 = Coefficient of friction between clutch teeth
d = shaft ϕ (mm) [1 mm = 0.3937″]
$\phi_2 = \tan^{-1} \mu_2$
D_m = Mean ϕ of clutch teeth $(D-b)$, mm
D = Clutch outside ϕ (mm) = 1.8 – 2.5 d
b = Clutch tooth width (mm) [Ref. **Fig. 4.10a**]
   = 0.125 – 0.2 D
Clutch length = 1.3 – 1.7 d for fixed driver half
   = 2 – 2.5 d for sliding driven half

Bearing pressure between teeth $(p_b) = \dfrac{2T_d}{b\,D_m\,h_m\,t}$ \quad **(Eqn. 4.7)**

   = 2 – 3 kg/mm² [2840 – 4270 Lbs/Inch²] for running engagement
   = 8 – 12 kg/mm² [11380 – 17060 Lbs/Inch²] for stationary engagement
h_m = Tooth height at mean dia D_m (mm) = 0.6 – 1 b
t = No. of teeth.

Bending stress on teeth $(f_b) = \dfrac{2T_d\,h_m\,K}{D_m\,t\,Z}$ \quad [1 kg/mm² = 1422 psi] \quad **(Eqn. 4.8)**

K = Manufacturing errors constant = 2 – 3
Z = Sectional modulus at tooth root (mm^3)

$$= \frac{bP^2}{24}$$

P = Tooth circular pitch (mm) (see **[Fig. 4.10b]**) [1 mm = 0.3937″]

$$= \frac{\pi D}{t} \quad \text{where } D = \text{outside diameter of toothed toothed portion [mm]}$$

For making clutch self-locking,

For a bi-directional clutch, $\left(\dfrac{\beta}{2}\right) \leq \tan^{-1}\left[\mu_2\left(1+\left[\dfrac{\mu_1}{\mu_2} \times \dfrac{D_m}{d}\right]\right)\right]$ (Eqn. 4.9)

For a uni-directional clutch $\beta \leq \tan^{-1}\left[\mu_2\left(1+\left[\dfrac{\mu_1}{\mu_2} \times \dfrac{D_m}{d}\right]\right)\right]$ (Eqn. 4.10)

Example 4.2: Design a self-locking bidirectional clutch for transmitting 2.2 kW at 200 RPM, if shaft dia $(d) = 24$; $\mu_1 = 0.05$, and $\mu_2 = 0.15$; $D = 50$. Find the bearing pressure, if bending stress on the teeth = 8 kg/mm^2; $K = 2$. Also find the clutch engagement and disengagement forces.

Solution: For self-locking in a bi-directional clutch,

$$\frac{\beta}{2} = \tan^{-1}\mu_2\left(1+\left[\frac{\mu_1}{\mu_2} \times \frac{D_m}{d}\right]\right)$$

Taking $b = 0.2\ D \approx 0.2 \times 50 = 10$

$D_m = D - b = 50 - 10 = 40$

$\therefore \quad \tan\dfrac{\beta}{2} \leq 0.15\left(1 + \dfrac{0.05}{0.15} \cdot \dfrac{40}{24}\right) = 0.0833$

$$\frac{\beta}{2} \leq 4° \ 45.5;\ \beta \leq 9° \ 31' \approx 9°$$

$$\text{Torque} = \frac{955\,\text{kW}}{N} = \frac{955 \times 2.2}{200} = 10.505 \text{ kgm}$$

$$= 10{,}505 \text{ kgmm}$$

Taking design torque as double of the required torque,

$$T_d = 10{,}505 \times 2 = 21{,}010 \text{ kgmm}$$
$$h_m = 0.6 - 1 \quad b = 6 - 10, \text{ say } 8 \text{ mm}$$

If no. of teeth $(t) = 5$

$$P = \frac{\pi \times 50}{5} = 31.42$$

$$Z = \frac{bP^2}{24} = \frac{10 \times 31.42^2}{24} = 411.34 \text{ mm}^3$$

$$f_b = \frac{2T_d h_m K}{D_m t Z} = \frac{2 \times 21{,}010 \times 8 \times 2}{40 \times 5 \times 411.34}$$
$$= 8.17 \text{ kg/mm}^2$$

The stress will be lesser, if the number of teeth are reduced to 3.

$$\text{Bearing pressure } (p_b) = \frac{2T_d}{b D_m h_m t} = \frac{2 \times 21{,}010}{10 \times 40 \times 8 \times 5}$$
$$= 2.6 \text{ kg/mm}^2, \text{ i.e. between 2 and 3.}$$

Therefore, the clutch can be engaged while the shaft is running.

$$\text{Engagement force } (F_{eb}) = 2 T_d \left(\frac{\mu_1}{b} + \frac{\tan\left(\frac{\beta}{2} + \phi_2\right)}{D_m} \right)$$

$$\phi_2 = \tan^{-1} 0.15 = 8° 32'$$

$$F_{eb} = 2 \times 21{,}010 \left(\frac{0.05}{10} + \frac{\tan(4.5 + 8.53)}{40} \right)$$
$$= 453 \text{ kg}$$

$$\text{Disengagement force } (F_{db}) = 2 \times 21{,}010 \left(0.005 - \frac{\tan(4.5 - 8.53)}{40} \right)$$
$$= 284 \text{ kg}$$

Slip torque (T_S) for safety clutches

$$T_S = F_d \cdot \frac{D_m}{2} \left(\tan\left(\frac{\beta}{2} - \phi_2\right) - \frac{D_m \mu_1}{d} \right) \quad \textbf{(Eqn. 4.11)}$$

As lubrication quantum, machining precision, alignment (co-axiality) error, and length/diameter ratio for the clutch can affect slip torque considerably, it is necessary to provide adjustment for the spring force in the form of a nut **[Fig. 4.12]**.

Drive Transmission and Manipulation

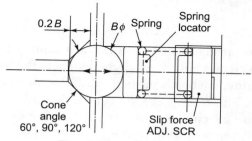

Fig. 4.12: Ball slip clutch

Efficiency (η) of the clutch depends upon the flank angle ($\beta/2$ for bi-directional clutch), and the coefficients of friction μ_1 and μ_2, for shaft and clutch teeth.

$$\eta = \frac{\tan\left(\frac{\beta}{2} - \phi_2\right) - D_m \frac{\mu_1}{d}}{\tan\frac{\beta}{2}}$$

If $\beta = 90°$ and μ_1 and $\mu_2 = 0.1$,
$\phi_2 = \tan^{-1} 0.1 = 5°\,42.5'$

$$\eta = \frac{\tan\left(\frac{90}{2} - 5°42.5'\right) - D_m \frac{0.1}{d}}{\tan\left(\frac{90}{2}\right)} \quad \textbf{(Eqn. 4.12)}$$

In Example 4.2, $d = 24$; $D_m = 40$. So for a 45° flank angle,

$$\eta = \frac{\tan 39°17.5' - 0.1\frac{40}{24}}{\tan 45} = \frac{0.8182 - 0.1666}{1}$$

$$= 65\%$$

Also, for the disengagement force (F_{db}) of 284 kg in Example 4.2, if β is 90°,

$$\text{Slip torque } (T_S) = F_d \frac{D_m}{2} \left(\tan\left(\frac{\beta}{2} - \phi_2\right) - \mu_1 \frac{D_m}{d} \right)$$

$$= 284 \, \frac{40}{2} \left(\tan(45 - 5°\,42.5') - 0.1 \frac{40}{24} \right)$$

$$= 284 \times 20 \,(0.8182 - 0.1666)$$

$$= 3701 \text{ kgmm}$$

Note that increasing angle β from 9° to 90° has reduced the torque capacity to nearly one-third of the clutch, with a 4.5° flank angle $\left(\frac{\beta}{2}\right)$.

4.2.3 Ball clutches

Ball clutches [**Fig. 4.12**] use spring-loaded balls for transmitting light torques. The spring is located by a sliding pad which prevents its entanglement with the ball. When the load exceeds the force exerted by the spring, the ball is pushed out of the conical seat against the spring, by the cam action of the cone. This disengages the clutch, providing slip and safety beyond the set slip torque (T_S). Cheap ball clutches are economical for light load and low operation frequency. But they are very vulnerable to heavy loads, generating high contact stresses. The balls can be placed axially, i.e. on the face [**Fig. 4.13a**], or radially [**Fig. 4.13c**]. A minimum of 2 balls are used. Their number depends upon the space available, and the slip torque required. In bigger clutches (315 mm PCD), even 40–45 balls can be used. The balls are placed equi-spaced, on their pitch circle diameter.

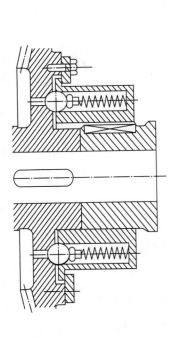

Fig. 4.13(a): Using ball on face

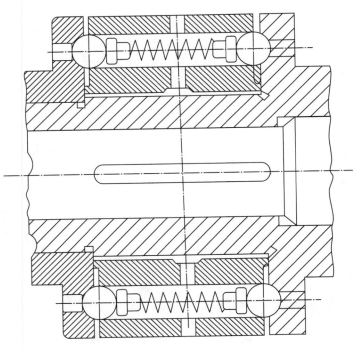

Fig. 4.13(b): Using two balls on two faces

The cone angle of the ball seat can be made to measure 60°, 90°, or 120° so that the seat can be machined conveniently with a standard center drill, counter-sinking cutter, or twist drill.

$$\text{Slip torque } (T_S) = R_p \, N F_t \qquad \textbf{(Eqn. 4.13)}$$

$$R_p = \text{Pitch radius of the balls (mm)} = \frac{\text{Pitch diameter}}{2}$$

N = No. of balls

F_t = Slip force per ball (kg)

$$= \frac{F_a}{C} = \frac{\text{Axial/Radial force exerted by spring}}{\text{Cone angle factor}}$$

C = 0.35 for 60° cone angle

    = 0.72 for 90° cone angle

    = 1.3 for 120° cone angle

The standard spring forces (F_a) are: 2, 4, 8, and 14.5 kg per spring. While the first two (2, 4) are used for ϕ 9.52 (3/8″) balls, 8, and 14.5 kg are used for ϕ 15.87 (5/8″) balls.

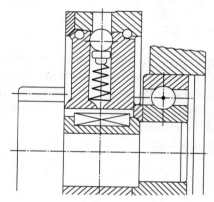

Fig. 4.13(c): Using balls on diameter

The slip torques range from 100 kgmm for ϕ 9.52 balls on 40 P.C.D., to 140,000 kgmm for ϕ 15.87 balls on 315 P.C.D. The balls should sink to about one-fifth of the diameter in the conical seat, i.e. 9.52 (3/8″) and 15.87 (5/8″) balls should sink to 2 mm and 3.2 mm respectively. The gap between the driving and the driven elements ranges from 0.5 to 0.8.

Example 4.3: Design a ball-type safety clutch for 800 kgmm slip torque, if the ball pitch diameter = 80 and cone angle = 120°.

Solution: From Eqn. 4.13,

$$T_s = R_P N F_t$$

For 120° cone angle C = 1.3

$$F_t = \frac{F_a}{1.3}; \text{ or } F_a = \text{Spring force} = 1.3 F_t$$

If we use a 15.87 ϕ ball, and spring with force F_a = 14.5

$$F_t = \frac{14.5}{1.3} = 11.15 \text{ kg}$$

$\therefore$
$$T_s = 800 = R_p \, NF_t = \frac{80}{2} \cdot N \cdot 11.15$$

$$N = \frac{800}{40 \times 11.15} = 1.79 \approx 2$$

$$\text{Ball radius } (r) = \frac{15.87}{2} = 7.935$$

$$\text{Cone depth} = \frac{r}{\cos 30} - (r - 3.2) = \frac{7.935}{0.866} - (7.935 - 3.2)$$

$$= 4.43 \approx 4.4$$

4.2.4 Friction clutches

Friction clutches use axial pressure on the friction disks for torque transmission. The engagement is shock-free. The torque capacity depends upon the mean radius of the tangential frictional force, dependent on the co-efficient of friction, and the normal pressure between clutch materials. The clutch should be able to dissipate the heat, resulting from the slip. Friction clutches are also used as brakes.

Friction clutches can be classified under two categories given below.
1. Disk clutches
2. Cone clutches

4.2.4.1 Disk clutches

Disk clutches provide greater contact area in a small space. There is uniform pressure, and no centrifugal effect. In a multi-plate disk clutch **[Fig. 4.14a]** there are a number of friction disks. The driving shaft is mounted with a housing, with a slotted or splined bore. The driven shaft is mounted with a sleeve having external spines or keyways, and a groove for operating the bell-crank lever. Alternately placed bigger friction disks have external projections which engage with the internal splines of the housing. The inter-spaced, smaller metal disks have internal splines or keyways which engage with the external splines or keyways of the sleeve. Both, the housing and the sleeve, remain axially stationary.

To engage the clutch, the operating sleeve is moved towards the left. The chamfer on the operating lever acts as a cam. It radially pushes the operating lever inward. This in turn pushes the friction disks towards the right. The pressure on the disk faces, and the ensuing frictional force, transmits the torque from the housing to the externally-splined friction disks, which transmit it to the internally-splined metal disks on the sleeve. As both, the housing and the sleeve are keyed to their shafts, both shafts begin to rotate together. Usually, the externally-splined, bigger disks are made of solid asbestos, or metal plates lined with asbestos. The smaller disks are made of metal. Both types of disks can slide freely, axially. The adjusting nut provides an axial adjustment of the spring force, and wear compensation for the disks.

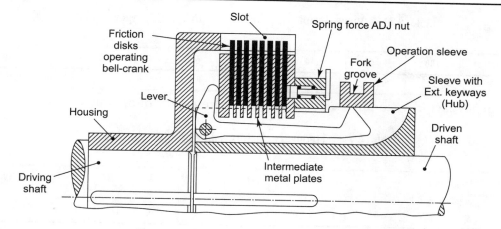

Fig. 4.14(a): Multi-plate friction disk clutch

The capacity of the clutch depends upon the size [dia] and the number of disks. These clutches are often operated by electromagnets in automated machine tools, and pneumatic and hydraulic operation through cylinders, and a direction control valve are also common.

Design of disk clutch

Driving and driven systems must be taken into account while computing the design torque (T_d). Electric motors have a higher starting characteristic than turbines. Using a directly-coupled or geared drive gives a better performance than a belt drive or intermediate transmissions between the motor and the clutch shaft. These considerations are reckoned by the factor K_1.

Table 4.4:   Driver factor [K_1] for clutches

Driver	Electric motor	Turbine or line transmission
Direct coupling	0.5	0.33
Belt transmission	0.33	0.33

When the starting torque is higher than the nominal or running torque, factor K_2 must be used.

Table 4.5:   Driven factor K_2 for clutches

M/C	K_2
1. Non-reversing machine tools without impact, piston compressors, centrifugal pumps	1.25
2. Reversible M/C tools, turbo compressors, cranes, heavy duty machines	1.6
3. Machines with impact forging, presses, piston pumps, paper machinery	2.5

Table 4.6: Speed factor [K_3] for disk clutch rotary speed (RPM)

RPM	100	160	240	400	620	1000	1400	1800
K_3	0.1	0.13	0.16	0.2	0.25	0.32	0.38	0.43

Frequency factor K_4 compensates for the higher frequency of operation.

Table 4.7: Frequency factors [K_4] for disk clutches

Frequency of operation in 8 hrs	1	8	16	32	48	96	240	480
K_4	0	0.20	0.55	0.75	0.90	1.2	1.8	2.0

All the above factors can be summed up in a single operating factor 'K'.

$$K = K_1 + K_2 + K_3 + K_4$$
$$T_d = KT_r = T_r(K_1 + K_2 + K_3 + K_4) \quad \text{(Eqn. 4.14)}$$

$$T_r = \text{Required (motor) torque} = \frac{97,400 \text{ kW}}{\text{RPM}(N)}$$

The number of friction surfaces required depends upon the friction coefficient (μ), permissible pressure (p), the pressure (contact) area ($2\pi r_m (R-r)$), and the mean radius of the pressure area $\left(r_m = \frac{R+r}{2}\right)$.

$$\text{No. of friction surfaces } (i) = \frac{T_d}{\mu p \, 2\pi (R-r) \frac{(R+r)}{2} \cdot \frac{(R+r)}{2}}$$

$$\therefore \quad i = \frac{2T_d}{\mu p \pi (R^2 - r^2)(R+r)} \quad \text{(Eqn. 4.15)}$$

$$\text{No. of driving (housing) disks} = \frac{i}{2}$$

$$\text{No. of driven (shaft) disks} = \frac{i}{2} + 1$$

Axial force for engaging clutch $(F_a) = p\pi(R^2 - r^2)$ \quad **(Eqn. 4.16)**

The permissible pressure (p) and coefficients of friction for various mating materials can be found from the following table.

Table 4.8: Permissible pressure (p) and friction coefficients (μ) for common clutch materials

Mating materials	Bearing pressure (p) kg/mm² With lubrication	Bearing pressure (p) kg/mm² Without lubrication	Coefficient of friction (μ) With lubrication	Coefficient of friction (μ) Without lubrication
Hrd Stl/Hrd Stl.	0.06–0.08 [85–114 psi]	0.025–0.03 [36–43 psi]	0.05–0.08	0.18
Stl/Sintered Bronze	0.05–0.2 [71–284]	0.05–0.08 [71–114]	0.06–0.08	0.12–0.18
Stl/Ferrodo	–	0.02–0.025 [28–36]	–	0.25–0.45
Stl/Fiber	–	0.035–0.04 [50–57]	–	0.2
C.I./C.I.	0.025–0.03 [36–43]	0.06–0.08 [85–114]	0.05	0.15

* If the linear speed at the mean radius (r_m) exceeds 2.5 m/sec, or the frequency of clutch operation exceeds 400/8 hours, the above-mentioned bearing pressures must be reduced to suit the conditions.

Table 4.9: Reduction factor S for high linear speeds and frequency of operation

Peripheral speed m/sec	Reduction factor S Frequency of engagement Below 400/8 hrs	Reduction factor S Frequency of engagement Above 400/8 hrs
Up to 2.5	1	0.75
2.6–5	0.8	0.6
5–10	0.65	0.49
10–15	0.55	0.41

Design pressure $= p_d = Sp$.

Metal disks are mounted on the sleeve of the driven shaft. Their thickness is 3 times that of the friction material (ferrodo, fibre, singered bronze), whose thickness ranges from 1 to 3 mm.

I.D. of driven disks $= 2$ shaft $\phi = 2d$

Outside of driven disks $= 2.5 - 3.6\ d$

$$\text{Shaft Diameter}\ (d) = \sqrt[3]{\frac{495 \times 10^4 \times \text{kW} \times K}{N f_t}}$$

$K = K_1 + K_2 + K_3 + K_4$

$N = $ R.P.M. (Revs/min)

$f_t = $ Permissible torsion stress

$\quad = 2.8 - 5$ kg/mm², for steels with tensile strength $50 - 80$ kg/mm²

Example 4.4: Design a disk clutch for transmitting 1.1 kW from a belt-driven shaft to a reversible machine tool, running at 240 RPM, without any impact load. The frequency of operation is about 60 times/hour. Use ferrodo friction disks, with steel plates in between.

Solution:

For belt drive $K_1 = 0.33$

For reversible m/c tool without impact, $K_2 = 1.6$

For 240 RPM, $K_3 = 0.16$

For 60 times/hr, i.e. 480 times/8 hrs, $K_4 = 2.0$

$$K = K_1 + K_2 + K_3 + K_4 = 0.33 + 1.6 + 0.16 + 2 = 4.09$$

$$\text{Shaft diameter } (d) = \sqrt[3]{\frac{495 \times 10^4 \times 1.1 \times 4.09}{240 \times 2.8}} = \sqrt[3]{33139}$$

$$\approx 32$$

∴ I.D. of driven disk $= 2d = 64$

∴ $$r = \frac{64}{2} = 32$$

Outside of driven disk $= 2.5 - 3.6\ d$

$$= 80 - 115.5$$

$$\approx 100$$

∴ $$R = \frac{100}{2} = 50$$

$$\text{Mean radius } (r_m) = \frac{r + R}{2} = \frac{32 + 50}{2} = 41$$

$$\text{Peripheral speed} = \frac{2\pi r_m N}{60} \times \frac{1}{1,000} \text{ m/sec}$$

$$= \frac{2\pi \times 41 \times 240}{60} \times \frac{1}{1,000}$$

$$= 1.03 \text{ m/sec, i.e.} < 2.5$$

Brg. pressure reduction factor for frequency above 400/8 hrs, and linear speed less than 2.5 m/sec $(S) = 0.75$.

Design of operating bell-crank lever [Fig. 4.14b]

The length (L) of the lever depends upon the number of disks and their thickness. The cross-section of the lever $(b \times h)$ depends upon the axial force (F_a), permissible bending stress (f_b), radial length (a) of the lever, and the number (n) of levers used.

$$h = \sqrt[3]{\frac{6 F_a \cdot a}{\psi f_b n}} \qquad \text{(Eqn. 4.17)}$$

$$\psi = \frac{b}{h} = 0.8 - 1$$

The deflection of the actuating arm (L) of the lever should be 0.05–0.15 mm. The deflection (δ) can be found from the following equation:

$$\delta = \frac{6 F_a \cdot a \cdot L^2}{n b h^3 E} \leq 0.05 - 0.15 \qquad \text{(Eqn. 4.18a)}$$

Bending stress on the lever (f_{b1}) can be found from the following equation:

$$f_{b1} = \frac{6 F_a \cdot a}{n b h^2} \qquad \text{(Eqn. 4.18b)}$$

Even the diameter (d) of the pivot pin should be computed, as it can help contain the shear stress (f_s) and bearing stress (f_c). Refer to **[Fig. 4.14c]**.

For limiting the shear stress,

$$d_s = \frac{F_a \sqrt{a^2 + L^2}}{1.57 \, nL \, f_s} \qquad \text{(Eqn. 4.19a)}$$

For limiting the bearing (compressive) stress (f_c),

$$d_c = \frac{F_a \sqrt{a^2 + L^2}}{nL \, bf_c} \qquad \text{(Eqn. 4.19b)}$$

The bearing length of the pin in the hub (b') should not be less than half of the lever thickness, i.e. $b' \geq 0.5 b$.

∴ For a ferrodo and stl combination, $p = 0.02 - 0.025$ kg/mm^2; and $\mu = 0.25 - 0.45$

Taking the minimum values, $p_d = Sp = 0.75 \times 0.02 = 0.015$ for Example 4.4

From Eqn. 4.15,

$$i = \frac{2 T_d}{\mu \, p_d \, \pi \, (R^2 - r^2)(R + r)}$$

$$T_d = K \, T_r = 4.09 \times \frac{974{,}000 \times 1.1}{240} = 18{,}258 \text{ kg/mm}$$

$$\therefore \quad i = \frac{2 \times 18{,}258}{0.25 \times 0.015 \times \pi (50^2 - 32^2)(50 + 32)}$$

$$= \frac{36516}{1426} = 25.6 \approx 26 \text{ i.e. very high.}$$

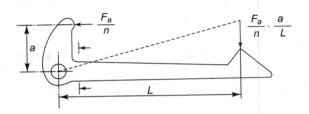

Fig. 4.14(b): Operating bell-crank lever

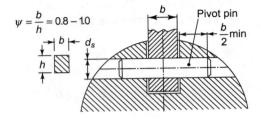

Fig. 4.14(c): Pivot pin design

Let us increase the outside diameter of the metal driven disks, i.e. R
Taking the higher value $115.6 \approx 116$,

$$R = \frac{116}{2} = 58$$

$$i = \frac{36{,}516}{0.011786 (58^2 - 32^2)(58 + 32)} = \frac{36{,}516}{2{,}481}$$

$$14.72 \approx 15$$

$$\text{Ferrodo driving disks reqd} = \frac{i}{2} = \frac{15}{2} = 7.5 \approx 8$$

$$\text{Metal driven disks reqd} = \frac{i}{2} + 1 = 8 + 1$$

$$= 9$$

Thin 1 mm ferrodo, and 3 mm thick metal disks can be used to minimize the clutch length. The bearing pressure between the disks and the hub (f_{cd}) at splines (or keyways) should not be more than 2–2.5 kg/mm² **[Fig. 4.14d]**.

$$f_{cd} = \frac{4 T_d}{i N_s t e D_m} \leq 2 - 2.5 \qquad \textbf{(Eqn. 4.19c)}$$

As already stated,

T_d = Design torque (kgmm); i = No. of friction faces = 8×2 for Example 4.4
N_s = No. of splines (lugs/keyways); t = Disc thickness (mm)
e = Height of bearing area; D_m = Mean Diameter (mm) = $[R + r]$

Drive Transmission and Manipulation

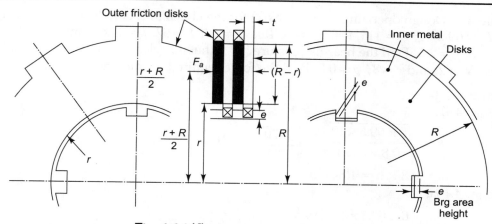

Fig. 4.14(d): Bearing pressure area [$t \times e$]

Force on the fork, used for moving the engaging ring, can be found from the following equation:

$$F_f = \frac{F_a a \mu}{L} \quad \text{(Eqn. 4.20)}$$

$\mu = 0.18$, for machined steel/steel (without lubrication).

Figure 4.14e shows a multi-disk clutch, operated by hydraulic (or pneumatic) power. Note that bell crank levers are replaced by small hydraulic/pneumatic cylinders, equispaced on their pitch diameter. The operating force can be easily altered by varying the pneumatic/hydraulic pressure. The springs are necessary to disengage the clutch.

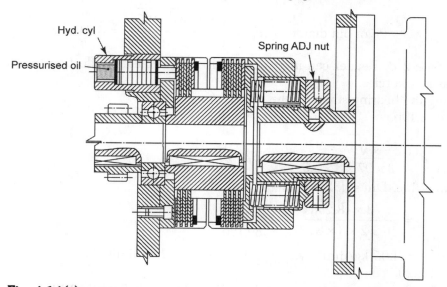

Fig. 4.14(e): Hydraulic operation of multi-disk clutch (with courtesy of Tata McGraw-Hill)

Example 4.5: Design operating levers (3 Nos.) for the clutch in Example 4.4, i.e. $T_d = 18,258$; $i = 16$; $F_a = 69.5$ kg; $D = 116$; $t = 3$; $n = 3$. Take $a = 10$; $L = 97$; $f_b = 8$ kg/mm². $E = 2.1 \times 10^4$/mm², $f_s = 3$, $f_c = 2.5$. Find the bearing pressure (f_c) if the hub (sleeve) on the driven shaft has 2 keyways: $N_s = 2$; and $e = 8$; $\psi = 0.8$. Find the force required for moving the fork.

Solution:
From Eqn. 4.17,
$$h = \sqrt[3]{\frac{6 \times 69.5 \times 10}{0.8 \times 8 \times 3}} = \sqrt[3]{\frac{4170}{19.2}} = 6.01 \approx 6$$
$$b = 0.8 \times 6 = 4.8 \approx 5$$

From Eqn. 4.18a,
$$\delta = \frac{6 \times 69.5 \times 10 \times 97^2}{3 \times 5 \times 8^3 \times 2.1 \times 10^4} = \frac{39,235,530}{1.613 \times 10^8} = 0.24 > 0.15$$

Increasing $\psi = \dfrac{b}{h}$ to 1.0, i.e. $b = 1 \times 8 = 8$

$$\delta = \frac{39,235,530}{3 \times 8 \times 8^3 \times 2.1 \times 10^4} = 0.152 \approx 0.15, \text{ i.e. maximum limit of deflection}$$

From Eqn. 4.18b, Dm = [32 + 58] = 90
$$f_{b1} = \frac{6 \times 69.5 \times 10}{3 \times 8 \times 8^2} = \frac{4,170}{1,536} = 2.71 \text{ kg/mm}^2$$

From Eqn. 4.19a, for shear stress (f_s) = 3 kg/mm²

$$\text{Pivot pin diameter } (d_s) = \frac{69.5 \sqrt{10^2 + 97^2}}{1.57 \times 3 \times 97 \times 3} = 4.94 \approx 5$$

Generally $h \geq 2 \, d_s$. Reader can increase h to 10 and calculate deflection 'δ' and bending stress [f_{b1}] as an exercise.

For limiting the bearing stress on the lever,
From Eqn. 4.19b,
$$d_c = \frac{69.5 \sqrt{10^2 + 97^2}}{3 \times 97 \times 8 \times 2.5} = 1.16 < 5$$

From Eqn. 4.19c, Dm = [32 + 58] = 90
$$f_{cd} = \frac{4 \times 18,258}{16 \times 2 \times 3 \times 8 \times 90} = \frac{73,032}{69,120}$$
$$= 1.05 \text{ kg/mm}^2, \text{ i.e. much less than 2.5 (the limit)}$$

From Eqn. 4.20,
$$\text{Force reqd. for moving fork } (F_f) = \frac{F_a \cdot a \cdot \mu}{L} = \frac{69.5 \times 10 \times 0.18}{97} = 1.3 \text{ kg}$$

4.2.4.2. Cone clutches

Cone clutches [Fig. 4.15] use a conical surface for frictional transmission.

The wedge action of the cone increases the normal force and friction, and hence, the torque capacity. Referring to [Fig. 4.15],

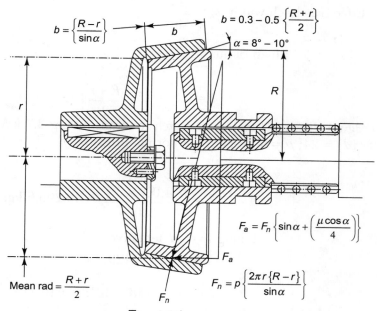

Fig. 4.15: Cone clutch

$$T = \frac{\mu (R + r) F_a}{2 K \sin \alpha} \qquad \text{(Eqn. 4.21)}$$

T = Torque (kgmm); R = Contact radius big end (mm)

r = Contact radius small end (mm); μ = Friction coefficient.

K = Constant = 1.3 – 1.5, for metal cutting m/cs

α = Cone angle = 8° – 10°, for metal to metal contact

= 12° – 15°, for asbestos-lined cones

Generally, $\tan \alpha = \mu$ [1 kgmm = .007376 LbFt] [1mm = .3937″]   (Eqn. 4.22)

If $\tan \alpha > \mu$, the cones disengage easily. It is necessary to provide a spring exerting axial for (F_a), for keeping the cones engaged.

If $\tan \alpha < \mu$, the clutch will not disengage by itself. It will consequently be necessary to use force to disengage the cones.

$$\text{Normal force } (F_n) = \frac{2 \pi r (R - r) p}{\sin \alpha} \quad [1 \text{ kg} = 2.2046 \text{ Lb}] \textbf{(Eqn. 4.23)}$$

where p = Permissible pressure (Table 4.8) for friction clutches (kg/mm²)

$$\text{Axial force } (F_a) = F_n \left(\sin \alpha + \frac{\mu \cos \alpha}{4} \right) \quad \text{(Eqn. 4.24)}$$

$$\text{Effective clutch width } (b) = \frac{R - r}{\sin \alpha}$$

$$\text{Mean radius} = \frac{R + r}{2}$$

$$b = 0.3 - 0.5 \left(\frac{R + r}{2} \right) \quad \text{(Eqn. 4.25)}$$

Example 4.6: Design a dry steel/steel cone clutch for transmitting 3.7 kW at 350 R.P.M. The shaft diameter $(d) = 24$; $\mu = 0.18$. Find the normal pressure (p).

Solution:

$$T = \frac{974{,}000 \times 3.7}{350} = 10{,}296.6 \text{ kgmm}$$

Taking $r = 5\, d$, tentatively,

$$r = 5 \times 24 = 120$$

From Eqn. 4.22,

$$\alpha = \tan^{-1} 0.18 = 10°12' \approx 10°$$

Taking mean value of 0.4 in Eqn. 4.25,

$$b = 0.4 \left(\frac{R + 120}{2} \right) = \frac{R - 120}{\sin 10°} = \frac{R - 120}{0.17{,}365}$$

$$\therefore \quad 0.03473 \, (R + 120) = R - 120.$$

$$R = \frac{120 + 4.17}{1 - 0.03473} = 128.6 \approx 129$$

Taking $K = 1.5$ in Eqn. 4.21,

$$T = \frac{\mu (R + r) F_a}{2 K \sin \alpha} = \frac{0.18 \, (129 + 120) F_a}{2 \times 1.5 \times 0.17365}$$

$$= 10{,}296.6$$

$$F_a = 10{,}296.6 \times \frac{0.52095}{44.82} = 119.7 \text{ kg}$$

$\therefore$ From Eqn. 4.23

$$F_n = \frac{2 \pi r (R - r) p}{\sin \alpha} = \frac{2\pi \times 120 \, (129 - 120) \, p}{0.17365}$$

$$= 39{,}077.7 \, p$$

From Eqn. 4.24,

$$F_n = \frac{F_a}{\sin\alpha + \dfrac{\mu\cos\alpha}{4}} = \frac{119.7}{0.17365 + \left(\dfrac{0.18 \times 0.9848}{4}\right)}$$

$$= 549.2 \text{ kg} = 39077.7\,p$$

$$\therefore \quad p = \frac{549.2}{39,077.7} = 0.014 \text{ kg/mm}^2$$

The normal pressure is much lesser than the range 0.025–0.03 recommended in Table 4.8. So the clutch cone size can be reduced.

If we take $\quad r = 4d = 4 \times 24 = 96$

$R = 103$ and $p = 0.0282$ kg/mm^2

As a thumb rule, $\quad r = 4 \times$ shaft $\phi = 4d$ **(Eqn. 4.26)**

4.2.4.3 Uni-directional roller clutches

Roller clutches are less noisy than toothed clutches. There is lesser engaging shock during engagement. Roller clutches can be operated at much higher speeds (up to 4 m/sec) than toothed clutches (0.7 m/s). Their simple design and ease in manufacture make them convenient for use in uni-directional operation, despite a bigger size and rapid wear. The wear is minimized by using hardened inserts at the contact faces [**Fig. 4.16a**].

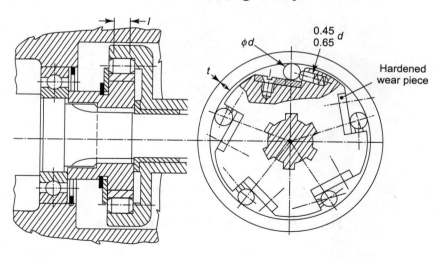

Fig. 4.16(a): Rooler clutch assy

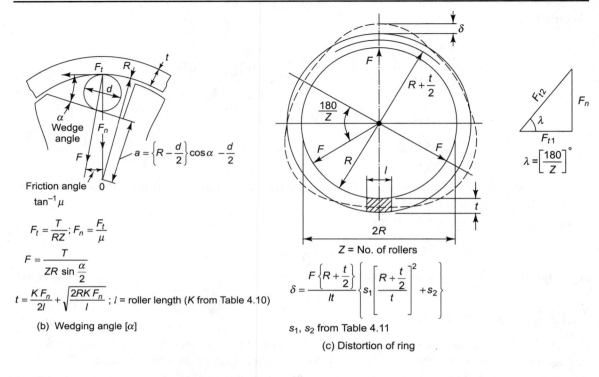

(b) Wedging angle [α]

(c) Distortion of ring

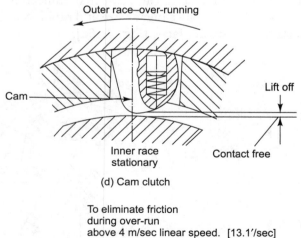

(d) Cam clutch

To eliminate friction during over-run above 4 m/sec linear speed. [13.1′/sec]

Fig. 4.16: Uni-directional roller and cam clutches

Roller clutches use a wedging action. The rollers are wedged between two inclined surfaces for transmitting the torque [**Fig. 4.16b**]. The wedging angle (α) depends upon the coefficient of friction (μ), which depends upon the surface finish of the mating faces and the lubrication status. For finish of $R_a = 0.4$, possible by grinding hardened steel, the coefficient of friction (μ)

is about 0.07. This gives a friction angle ($\tan^{-1}\mu$) of 4°. For a slip-free transmission, the wedging angle (α) should be less than twice the friction angle.

$$\text{Wedging angle } (\alpha) \leq 2 \tan^{-1} \mu \qquad \text{(Eqn. 4.27)}$$

μ = Coefficient of friction

α = Wedging angle $\leq 8°$, when $\mu = 0.07$

= 6° – 7°, for clutches with a high frequency operation ($\mu = 0.05 - 0.055$)

Hardened steel surfaces can be polished to reduce μ to 0.03, when frequency of operation is high and reduction in wear and increased life justifies the extra polishing expenses. The torque capacity depends upon the number of rollers and the inside radius of the ring.

$$\text{Torque } (T) = F_t R Z \qquad \text{(Eqn. 4.28)}$$

R = Inside radius of the ring (mm); Z = No. of rollers.

$$\text{Normal force } (F_n) = \frac{F_t}{\mu} \qquad \text{(Eqn. 4.29)}$$

$$\text{Resultant force } (F) = \frac{T}{ZR \sin\left(\dfrac{\alpha}{2}\right)} \qquad \text{(Eqn. 4.30)}$$

The resultant force (F) also depends upon the pressure (p) on the projected area of the roller, i.e. diameter (d) and length (l).

$$F = pdl \qquad \text{(Eqn. 4.31)}$$

= 5 dl for steel hardened to RC 63–64

d = roller diameter = 0.2 – 0.33 R

l = roller length = 1.4 – 2d = outer ring width

The thickness of the outer ring can be calculated from the following equation:

$$t = \frac{K F_n}{2 l} + \sqrt{\frac{2 R K F_n}{l}} \qquad \text{(Eqn. 4.32)}$$

Note that the width of the outer ring is only slightly (0.2 – 0.5) more than the roller length (l). The constant 'K' (Table 4.10) depends upon the number of rollers (Z), and bending stress (f_b), which is generally limited to 20 kg/mm^2.

Table 4.10: Constant K for no. of rollers at bending stress (f_b) of 20 kg/mm^2 [28440 psi]

No. of rollers	3	4	5	6	8	10	12
K	2.85 × 10^{-2}	2.05 × 10^{-2}	1.61 × 10^{-2}	1.34 × 10^{-2}	0.99 × 10^{-2}	0.79 × 10^{-2}	0.66 × 10^{-2}

The bending moment (M_o) at the outer diameter of the ring can be found from the following equation:

$$M_o = \left(R + \frac{t}{2}\right)\left(\frac{Z}{2\pi} F_n - \frac{F_n}{2\tan\left(\frac{180}{Z}\right)}\right) \quad \text{(Eqn. 4.33)}$$

The B.M. at the inner diameter of the ring (M_i) can be found from Eqn. 4.34.

$$M_i = \left(R + \frac{t}{2}\right)\left(\frac{Z F_n}{2\pi} - \frac{F_n}{2\sin\left(\frac{180}{Z}\right)}\right) \quad \text{(Eqn. 4.34)}$$

As *tan* of any angle is more than *sin*, it is obvious that $M_o > M_i$. The distortion (deflection) of the outer ring **[Fig. 4.16c]** resulting from the bending moment, can be calculated from the following equation:

$$\text{Deflection } (\delta) = \frac{F\left(R + \frac{t}{2}\right)}{l\, t}\left(S_1\left[\frac{R + \frac{t}{2}}{t}\right]^2 + S_2\right) \quad \text{(Eqn. 4.35)}$$

F = Resultant force from Eqn. 4.30 (kg)
R = Ring inside radium (mm); t = Ring thickness (mm)
l = Ring (and roller) width (mm)
S_1, S_2 = Constants dependent on the number of rollers (from the table below).

Table 4.11: Constants (S_1, S_2) for Eqn 4.35

Z	S_1	S_2
3	9.1×10^{-6}	4.84×10^{-5}
4	3.45×10^{-6}	4.91×10^{-5}
5	1.7×10^{-6}	5.18×10^{-5}
6	0.96×10^{-6}	5.62×10^{-5}
8	0.4×10^{-6}	6.87×10^{-5}
10	0.25×10^{-6}	8.20×10^{-5}
12	0.11×10^{-6}	9.55×10^{-5}

The resultant stress at the outside periphery (f_{ro}) is the sum of tensile and bending stresses.

$$f_{ro} = \frac{F_n}{2\, lt \cdot \tan\left(\frac{180}{Z}\right)} + \frac{6 M_o}{l\, t^2} \quad \text{(Eqn. 4.36)}$$

Similarly, the resultant stress at the inside periphery (f_{ri}) can be found as below:

$$f_{ri} = \frac{F_n}{2\, lt \cdot \sin\left(\frac{180}{Z}\right)} + \frac{6\, M_i}{lt^2} \quad \text{(Eqn. 4.37)}$$

[1 kg/mm² = 1422 psi]

The contact stress (f_c) can be found from Eqn. 4.38.

$$f_c = 0.59 \sqrt{\frac{F_n E}{ld}} \quad \text{(Eqn. 4.38)}$$

$E = 2.1 \times 10^4/\text{mm}^2$, $f_c \leq 150$ kg/mm² [2,13,300 psi] for life of 5×10^6 cycles

$f_c \leq 120$ kg/mm² [1,70,650 psi] for high precision applications

$$\text{Wedge face distance } (a) = \left(R - \frac{d}{2}\right)\cos\alpha - \frac{d}{2} \quad \text{(Eqn. 4.39)}$$

Example 4.7: Design a roller clutch for transmitting 4,000 kgmm torque. Coefficient of friction (μ) = 0.1; No. of rollers (Z) = 4; preferred inside ϕ = 70. Calculate the deflection, resultant stresses on the ring, and contact stress.

Solution:

From Eqn. 4.28,

$$T = F_t RZ = F_t \times \frac{70}{2} \times 4 = 4000$$

$$\therefore \quad F_t = \frac{4{,}000}{35 \times 4} = 28.57 \text{ kg} \approx 28.6 \text{ kg}$$

Friction angle = $\tan^{-1}\mu = \tan^{-1} 0.1$
$= 5° 43'$

From Eqn. 4.27, $\alpha \leq 2 \times 5°43'$
$\leq 11° 26'$

Taking $\quad \alpha = 11°$, from Eqn. 4.29,

$$F_n = \frac{F_t}{\mu} = \frac{28.6}{0.1} = 286 \text{ kg}$$

From Eqn. 4.30, Resultant force $(F) = \dfrac{T}{ZR \sin\dfrac{11}{2}}$

$$F = \frac{4{,}000}{4 \cdot \dfrac{70}{2} \sin 5.5} = 298.1 \text{ kg}$$

From Eqn. 4.31,
$$F = 5\, dl; \text{ Taking } l = 1.5\, d$$
$$5 \times d \times 1.5d = 7.5 d^2 = 298.1$$
$$d = \sqrt{\frac{298.1}{7.5}} = 6.3 \approx 6.5$$

$$\frac{d}{R} = \frac{6.5}{35} = 0.186, \text{ i.e. less than the recommended range of } 0.2\text{--}0.33\ R$$
$$d = 0.2\, R = 0.2 \times 35 = 7 \approx 8, \text{ i.e. the nearest standard size. [page 337]}$$
$$l = 1.5\, d = 1.5 \times 8 = 12$$

From Table 4.10, for K for 4 rollers, $K = 2.05 \times 10^{-2}$

So from Eqn. 4.32, $t = \dfrac{K F_n}{2l} + \sqrt{\dfrac{2\, R K F_n}{1}}$

$$= \frac{2.05 \times 10^{-2} \times 286}{2 \times 12} + \sqrt{\frac{2 \times 35 \times 2.05 \times 10^{-2} \times 286}{12}}$$
$$= 0.24 + 5.85 = 6.09 \approx 6.5$$

From Eqn. 4.35,
$$\delta = \frac{F\left(R + \dfrac{t}{2}\right)}{lt} \left(S_1 \left(\frac{R + \dfrac{t}{2}}{t}\right)^2 + S_2 \right)$$

From Table 4.11, $S_1 = 3.45 \times 10^{-6}$ and $S_2 = 4.91 \times 10^{-5}$

$$\therefore \quad \delta = \frac{298.1\left(35 + \dfrac{6.5}{2}\right)}{12 \times 6.5} \left(3.45 \times 10^{-6} \left(\frac{35 + \dfrac{6.5}{2}}{6.5}\right)^2 + 4.91 \times 10^{-5} \right)$$

$$= 146.18\, (11.95 \times 10^{-5} + 4.91 \times 10^{-5})$$
$$= 0.0246$$

From Eqn. 4.33,
$$M_o = \left(R + \frac{t}{2}\right)\left(\frac{Z}{2\pi} F_n - \frac{F_n}{2 \tan\left(\dfrac{180}{Z}\right)}\right)$$

$$= \left(35 + \frac{6.5}{2}\right)\left(\frac{4}{2\pi} \times 286 - \frac{286}{2 \tan\dfrac{180}{4}}\right)$$
$$= 38.25\, (182.05 - 143) = 1{,}493.67 \text{ kgmm}$$

Drive Transmission and Manipulation

From Eqn. 4.36,

$$f_{ro} = \frac{F_n}{2\, lt \cdot \tan\left(\frac{180}{Z}\right)} + \frac{6\, M_o}{lt^2}$$

$$= \frac{286}{2 \times 12 \times 6.5 \times \tan 45°} + \frac{6 \times 1{,}493.67}{12 \times 6.5^2}$$

$$= 1.833 + 17.67 = 19.51 \text{ kg/mm}^2: \text{ less than limit 20}$$

$$M_i = 38.25\left(182.05 - \frac{286}{2 \sin 45°}\right) = -772.16 \text{ kgmm}$$

From Eqn. 4.37,

$$f_{ri} = \frac{286}{2 \times 12 \times 6.5 \times \sin 45°} - \frac{6 \times 772.16}{12 \times 6.5^2}$$

$$= -6.5 \text{ kg/mm}^2$$

The minus sign indicates compressive stress. Usually, it is enough to calculate the resultant stress on the outside (f_{ro}), which is always more than the inside stress (f_{ri}).

From Eqn. 4.38,

$$\text{Contact stress } (f_c) = 0.59 \sqrt{\frac{F_n E}{ld}} = 0.59 \sqrt{\frac{286 \times 2.1 \times 10^4}{12 \times 8}}$$

$$= 147.6 \text{ kg/mm}^2, \text{ i.e. less than the limit 150, for life of 5 million cycles.}$$

From Eqn. 4.39,

$$a = \left(35 - \frac{8}{2}\right) \cos 5.5 - \frac{8}{2} = 26.86$$

Standard rollers:

$\phi 5 \times 8$, $\phi 6.5 \times 9$, $\phi 8 \times 12$, $\phi 9 \times 14$, $\phi 11 \times 15$, $\phi 12 \times 18$, $\phi 14 \times 20$, $\phi 16 \times 24$, $\phi 19 \times 28$, $\phi 22 \times 34$. [ϕ .197" × .315"], [ϕ .259" × .354"], [ϕ .315" × .472"], [ϕ .354 × .551"], [ϕ .433 × .59], [ϕ .472 × .709], [ϕ .63 × .945], [ϕ .748 × 1.1], [ϕ .866 × 1.339]

Standard rings (ID × O.D):

$\phi 40 \times \phi 55$, $\phi 50 \times \phi 65$, $\phi 56 \times \phi 70$, $\phi 70 \times \phi 85$, $\phi 90 \times \phi 110$, $\phi 100 \times \phi 120$, $\phi 120 \times \phi 140$, $\phi 140 \times \phi 165$, $\phi 160 \times \phi 190$. [$\phi 1.575" \times \phi 2.165"$], [$\phi 1.968 \times 2.559$], [$\phi 2.2" \times 2.756$], [$\phi 2.756 \times \phi 3.346$], [$\phi 3.543 \times \phi 4.331$], [$\phi 3.937 \times 4.724$], [$\phi 4.724 \times \phi 5.512$], [$\phi 5.512 \times \phi 6.496$], [$\phi 6.299 \times \phi 7.48$]

Preferred no. of rollers

Ring inside Dia. ($2R$)	40–50 [1.575" – 1.968"]	50–70 [1.968" × 2.756]	70–100 [2.756" × 3.937"]	120–160 [4.724" – 6.299]
No. of rollers Z	3	5	6	8

The spring-loaded pusher pin for rollers ranges from 0.43 d to 0.67 d (roller ϕ), but need not be more than 8 ϕ [.315"] for rollers above 16 ϕ. [.63"] The standard pin diameters are: ϕ 3, 4, 5, 6, and 8. [.118", .157", .236", .315"]

Springs

Each roller must have a separate spring placed at its length-wise center. The spring should be strong enough to push the roller weight, reckoning friction. The spring-actuated force should be comparatively higher in low speed indexing devices than high speed applications for indexing clutches, to take care of the low rotary speed.

Keying

The inner hub is usually keyed to shaft while facial slots and keys are more convenient for drive of the outer ring.

Lubrication

For linear speeds below 2 m/sec [6.56'/s], grease lubrication is satisfactory. For higher speeds (2 m–4 m/sec), the lubricating oil should have 20–37 Cst viscosity (at 50°C). The lubricant should be free from water, to eliminate the possibility of corrosion. It should not be acidic, lest it causes chemical erosion.

Clutch overrun causes slip, leading to wear. The roller clutches are suitable for speeds up to 4m/sec only. Above 4 m/sec [13.12'/s] linear (sliding) speed (at bore of the ring), it is necessary to replace the rollers by costlier cams **[Fig. 4.16d]**. The cams provide a contactless overrun.

Operation frequency and slip

The limit for the frequency of operation and the permissible slip depends upon the nominal diameter (inside dia of ring: 2R). The following table gives the values for a clutch with a 6° wedging angle (α).

Table 4.12: Max operation frequency (nos./min) and permissible slip (degrees) for roller clutches with 6° wedging angle (α)

Nominal Diameter (2R)	32 [1.26"]	50 [1.968"]	80 [3.15"]	125 [4.921"]	160 [6.299"]	200 [7.874"]
Max. permissible frequency (nos/min)	250	160	100	65	50	40
Max. permissible slip (degrees)	3°	2°30'	1°30'	0°, 45'	0°, 45'	0°, 30'

4.2.5 Electro-magnetic clutches and brakes

Quick-action electro-magnetic clutches [**Fig. 4.17a**] facilitate automation: pre-selection and programming. In these, the electro-magnetic coil replaces the clutch spring or hydraulic cylinder. Instead of a lever or direction control valve, an electrical switch is used for engagement. The clutch part is like a disk clutch discussed earlier: alternate outer and inner disks with slidefit projections and keyway slots, mating with the driving ring and the driven shaft, respectively. The magnetic force for pressing the disks together is generated by energizing the electro-magnetic coil placed strategically between the outer ring and the inner hub. On energization, the coil attracts the armature plate, which pushes the outer and inners disks against each other, to transmit the rotary drive, from the driving ring to the driven shaft. As both halves of the clutch rotate, slip rings are necessary to supply current to the coil. Furthermore, insulation is necessary between the driver and the driven, for the electro-magnetic field closes its circuit through outer plates [disks], armature, and the inner plates.

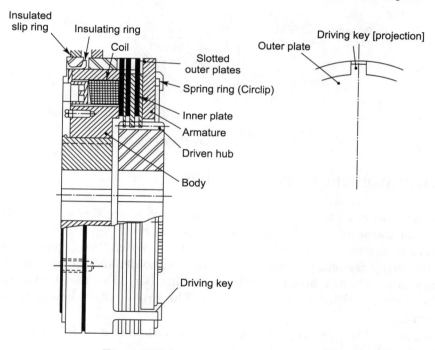

Fig. 4.17(a): Electro-magnetic disk clutch

Moderate lubrication through oil splash or mist, is necessary. Under no circumstances should slips ring be flooded with oil.

High pulsations of alternating current [AC], and ensuing high wear, make AC unsuitable for electro-magnetic clutches.

Brake clutches [**Fig. 4.17b**] are electro-magnetic clutches used for quick stoppage of rotating shafts or gear wheels. However, the housing halve is mounted on a stationary part of the machine, while the armature plate, the inner friction disks, and the sleeve are mounted on the shaft to be braked. As the housing does not rotate, slip rings are not required.

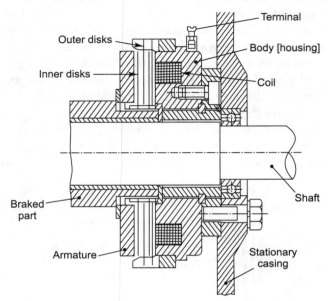

Fig. 4.17(b): Electro-magnetic brake

4.2.6 Ferro-magnetic powder clutches

Ferro-magnetic powder clutches use ferro magnetic carbonil powder with 0.7– 0.8% spherical grain carbon for transmission. The powder is mixed with oil.

The shaft to be driven is mounted with exciting coils. When the current is passed through these coils, the ferro-magnetic powder between the coil-holder and the housing is magnetized. The resulting magnetic flux holds the clutch housing [keyed to the driving shaft] and the exciter coils together, with 0.1 to 0.3 adhesion. The driven shaft, keyed to the coilholder begins to rotate.

Ferro-magnetic clutches are used in feed mechanisms and for precise motions in dynamometers and braking systems.

4.3 MOTION REVERSAL MECHANISMS

Some operations such as threading by taps, call for quick reversal of the direction of rotation of the machine spindle. Reversal can be accomplished mechanically, hydraulically, or electrically.

4.3.1 Mechanical reversal

Mechanical reversal is effected by using spur, helical, or bevel gears.

4.3.1.1. Spur gear reversal mechanism

Spur gear reversal mechanism comprises five gears [**Fig. 4.18a**]. Gears 'A' and 'D' are keyed onto the driver shaft I. Gear 'B', an idler, can rotate freely on intermediate shaft II. The driven shaft III is mounted with gears 'C' and 'E', which are free to rotate on the shaft, but cannot move axially. The gears on the driven shaft are fixed with toothed halves of a double jaw clutch. The central part of the special jaw clutch has teeth at both ends. Its internal splines mesh with the splines of the driven shaft. It can be slid axially. When the driving shaft I is rotating, all the gears rotate. Sliding the central part of the jaw clutch towards the left to mesh with the teeth of the half attached to gear 'C', rotates the driven shaft in the same direction as the driver. Meshing the central clutch half with the toothed half on gear 'E', rotates the driven shaft in the opposite direction. Thus, axial motion of the central clutch half reverses the direction of rotation of the driven shaft.

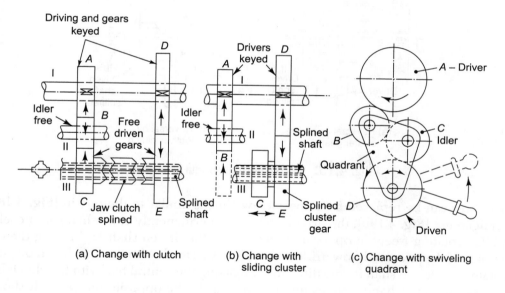

(a) Change with clutch   (b) Change with sliding cluster   (c) Change with swiveling quadrant

Fig. 4.18: Reversal mechanisms with idler gear

The jaw clutch can be eliminated altogether by clustering driven gears 'C' and 'E' together [**Fig. 4.18b**]. Sliding the splined cluster towards the left, meshes gear 'C' with idler 'B', to rotate the driven shaft in the same direction as the driver. Sliding the cluster towards the right to mesh gears '*E*' and '*D*', reverses the direction of rotation of driven shaft III.

Another method of reversing rotation direction uses a swiveling quadrant, incorporating an additional idler gear [**Fig. 4.18c**]. The swiveling quadrant enables engaging gears 'B' or 'C' with the driving gear 'A'. When gear 'C' is meshed with driver 'A', the gear 'D' and the keyed shaft are driven in the same direction as the driver 'A'. Swiveling the quadrant to mesh gear 'B' with driver 'A' however, rotates the driven gear 'D' and its shaft in the opposite direction.

As engagement of gear 'B' or 'C' with gear 'A' is radial, and driver 'A' is rotating during the engaging motion, all the gears in this arrangement usually have helical teeth. Mating gears have helix angles of the opposite hands.

Bevel gears reversal mechanism [**Fig. 4.19a**] comprises three gears. The driving shaft I is square with the driven shaft II. The driven shaft is mounted with a cluster of two bevel gears 'B' and 'C'. Splined internally, the cluster can be slid axially on the driven shaft. Engaging bevel gear 'B' of the cluster with driving gear 'A', rotates the driven shaft in one direction. Sliding the cluster towards the right meshes gears 'A' and 'C' to rotate the driven shaft in the opposite direction.

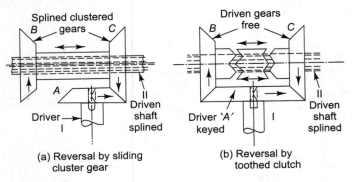

Fig. 4.19: Reversal mechanisms with bevel gears

The cluster can be replaced by a special jaw clutch, similar to the one in [**Fig. 4.18a**]. In this arrangement [**Fig. 4.19b**], the driving gear 'A' is permanently enmeshed with bevel gears 'B' and 'C', rotating freely in opposite directions on the driven shaft II. Moving the central splined part of the jaw clutch towards the left meshes its teeth with the clutch half on gear 'B'. This rotates the driven shaft in one direction. Engaging the central half with the clutch half on gear 'C', on the other hand, rotates the driven shaft in the opposite direction. It should be remembered that neither driven gear 'B' nor 'C' are keyed [or splined] to the driven shaft II. They rotate free of shaft II, in opposite directions.

CHAPTER 5

MACHINE TOOL ELEMENTS

Machine tools comprise a variety of elements: spindles, bearing supports, saddles (carriages) and guideways for feed motions, workpiece tables, and composite housings with columns, arms, beds, etc.

5.1 SPINDLES

Rotating elements which mainly bear bending loads, with little or no torsion, are called **axles**. Elements which transmit considerable torque are called **shafts**. They are usually mounted with torque transmitters like gears, belt pullies, and chain sprockets. **Spindles** are end-output shafts which carry tools or work-piece holders.

5.1.1 Functions of a spindle

1. Holding workpiece/cutting tool
2. Positioning (centering) workpiece/cutting tool
3. Imparting rotary motion (torque) to workpiece/cutting tool.

To fulfill these functions satisfactorily, a spindle should fulfill the following requisites.

5.1.1.1 Requisites of a good spindle

1. Precise mounting with minimum rotary and axial runout (for precision machining)
2. Adequate static and dynamic stiffness, to minimize distortions resulting from cutting forces
3. Wear resistance through hardening, or replaceable anti-friction bearings
4. Heat dissipation through selection of adequate bearings
5. Precision mounting surfaces for locating workpiece grippers (chucks, collets), or cutting tools (drills, cutters)
6. Long life against fatigue, by limiting the bending stresses.

5.1.2 Spindle materials

Distortion depends upon the modulus of elasticity (E). As 'E' is not significantly higher for costly high-tensile steels, they do not offer any marked advantage above the moderately-priced structural and alloy steels.

The choice of the material depends upon the following considerations:

5.1.2.1 Stress

This depends upon the power transmitted (kW/HP) and the bending stress, dependent upon the forces on the mounted transmission elements. The resultant bending stress can be computed from the following equation:

Resultant bending stress = f_b

$$f_b = \frac{\sqrt{M^2 + 0.45T^2}}{Z} \qquad \text{(Eqn. 5.1)}$$

M = Maximum bending moment (kgfmm) = $\sqrt{M_x^2 + M_y^2}$

M_x, M_y = Bending moments in planes x and y, square with each other

T = Torque (kgfmm)

$\quad = \dfrac{6000 P}{2 \pi N}$, where P = Power transmitted (kW)

N = Spindle Revs/Min

Z = Modulus of spindle section (mm^3)

$\quad = 0.0982\, D^3$, where D = Spindle outside ϕ (mm)

The permissible bending stress f_b depends upon the type of loading, and the ultimate tensile strength of the material used.

Table 5.1: Permissible bending stress for spindles

Sr. No.	Type of load	Permissible bending stress f_b in kgf/mm^2 [7.0307 × 10^{-4} × Lbs/Inch^2]			
		C45/55 En 8/9 RC 30 95 kg/mm^2 [135120 psi]	40Ni2 Cr1/En 24 RC 50 160 kg/mm^2 [227570 psi]	15Ni2 Cr1 En 354 Case Hrd 65 kg/mm^2 [92450 psi]	Spheroidal Graphite Cast Iron 37 kg/mm^2 [52630]
1	Steady load light bending moment	8 [11378]	12 [17068]	5 [7112]	2.8 [3982]
2	Fluctuating load light bending moment	6 [8534]	9.2 [13085]	3.7 [5263]	2.1 [2990]
3	Fluctuating load considerable bending moment	4.5 [6400]	7 [9956]	2.85 [4054]	1.6 [2276]

The cast spindles are used only in applications requiring hollow cavities or odd shapes, which usually take a lot of time for machining from bar-stock.

5.1.2.2 Stiffness or rigidity

Stiffness or rigidity is the ratio between the load (W) and deflection (δ). The overall stiffness of the spindle can be found from the following equation:

$$S = \frac{W}{\delta} \text{ (kgf/mm)} \qquad \text{(Eqn. 5.2)}$$

The cutting load varies according to the workpiece material, cut, feed, etc. The resultant (total) defection of the shaft [Fig. 5.1] comprises the deflection of the spindle at the overhanging end (δ_R), and the yield of the bearing, closer to the overhung end (δ_Q).

Deflection of the machine spindle must be limited, by providing for adequate rigidity.

$$\text{Max. deflection} = \delta_{max} \leq 0.0002\ L, \qquad \text{(Eqn. 5.3)}$$

where L = Span between the bearings (mm)

Even the deflection at the drive gears must be limited.

$$\text{Deflection at gear} = \delta_g \leq 0.01\ m \qquad \text{(Eqn. 5.4)}$$

m = module of the gear. [1 Diametral Pitch = 25.4 module]

The overall deflection (δ) can be computed from the following equation. Referring to [Fig. 5.1]:

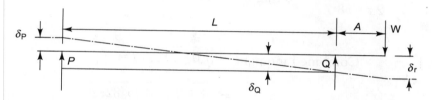

Fig. 5.1: Spindle with single overhanging load

$$\delta = W \left(\frac{A^2}{3E} \left(\frac{L}{I_L} + \frac{A}{I_A} \right) + \delta_Q \left(1 + \frac{A}{L}\right)^2 + \delta_P \frac{A^2}{L^2} \right) \qquad \text{(Eqn. 5.5)}$$

I_L = Moment of inertia of spindle, in the portion between bearings P and Q
I_A = Moment of inertia of spindle in the overhanging length A.
δ_Q = Deflection/kg of Bearing Q
δ_P = Deflection/kg of Bearing P.

Deflection can be minimized by optimizing the span between the bearings. Differentiating Eq. 5.5 for minimum deflection,

$$\text{Optimum span } (L_o) = \sqrt[3]{6EI_L \left(\delta_P + \delta_Q + \left(\frac{\delta_Q \cdot R}{A} \right) \right)} \qquad \text{(Eqn. 5.6)}$$

The value of R can initially be taken to be 4 times overhang A. The computed value of L_o can be substituted for R to find L_o by iteration.

The actual span can be up to 20% higher than the calculated value of L_o.

The values of δ_P and δ_Q can be found from the size and the type of bearing chosen. Equations 5.44 and 5.45, Tables 5.13 to 5.16, and **[Fig. 5.21]** in the later section on **Elastic Deformation**, give the details of the method for determining the deflection per kg, of various types of bearings.

$$\text{Inertia of shaft} = \frac{\pi}{64} D^4 = 0.0491 D^4 \quad [\text{Inch}^4 = 41.623 \times 10^4 \text{ mm}^4]$$

Example 5.1: Find the spindle size and optimum span of bearings, and deflection, if the spindle overhang (A) is 80. The cutting force at the overhung end is 110 kg. Powered by a 5 kW motor, the spindle runs in a 58–470 R.P.M. range. Use a roller bearing near the overhung end, and a ball bearing at the farther end.

Solution: The bending moment on the overhanging portion is a multiple of the cutting force and the overhang.

$$M = 110 \times 80 = 8800 \text{ kgmm.}$$

$$T = \frac{6000 \times 5}{2\pi \times 58} = 82.32 \text{ kg m} = 82320 \text{ kgmm.} \qquad [T_{\max} \text{ when RPM [N] is minimum}]$$

From Eqn 5.1, $\quad$ Combined moment $= \sqrt{M^2 + 0.45\, T^2}$

$$= \sqrt{8800^2 + 0.45\,(82320)^2}$$

$$= 55918.7 \text{ kgmm.}$$

Referring to Table 5.1, for $C\,45$ and fluctuating load $f_b = 6 \text{ kg/mm}^2$.
From Eqn. 5.1,

$$Z = 0.0982\, D^3 = \frac{\sqrt{M^2 + 0.45\, T^2}}{f_b} = \frac{55918.7}{6} = 9319.8$$

$$D = \sqrt[3]{\frac{9319.8}{0.0982}} = \sqrt[3]{94906} = 45.63$$

For a 45 $\phi \times$ 100 $\phi \times$ 36 roller bearing,

$$d_r = 0.25\,(100 - 45) = 13.75 \approx 14$$

Example 5.18a gives the details of the calculations for finding deflection of $\phi 45 \times \phi 100 \times 36$ roller bearing, and the calculated value: $\delta_q = 0.000137$.

A slightly smaller, say 40 ϕ ball bearing, will be used at the farther end. Choosing $\phi\, 40 \times \phi\, 68 \times 15$ Brg. with $C_o = 980$.
Example 5.17 computes the value of δ_P. It is 0.00045/kg.

$$I_L = I_A = 0.0491 \times 45^4 = 201340.7 \text{ mm}^4$$

From Eqn. 5.6, $L_o = \sqrt[3]{6EI_L \left(\delta_P + \delta_Q + \left(\frac{\delta_Q \cdot R}{A} \right) \right)}$

Taking $\quad R = 4A = 4 \times 80 = 320,$

$$L_o = \sqrt[3]{6 \times 2.1 \times 10^4 \times 201340.7(0.00045 + 0.000137 + 0.000137 \times 4)}$$

$$= \sqrt[3]{253.698 \times 10^8 (0.000587 + 0.000548)}$$

$$= \sqrt[3]{28.79 \times 10^6} = 306.5$$

Substituting this value for R,

$$L_o = \sqrt[3]{253.698 \times 10^8 \left(0.000587 + \left(0.000137 \times \frac{306.5}{80} \right) \right)}$$

$$= \sqrt[3]{28.19 \times 10^6} = 304.326$$

Again substituting 304.4 for R,

$$L_o = \sqrt[3]{253.698 \times 10^8 \left(0.000587 + \left(0.000137 \times \frac{304.326}{80} \right) \right)}$$

$$= \sqrt[3]{28.11 \times 10^6} = 304.06$$

This figure is close enough to preceding figure stop the iteration. The span can be 20% longer, i.e. between 304 and 365 say 335.

From Eqn. 5.5, deflection of the shaft can be found for a span of 305.

$$\delta = W \left(\frac{A^2}{3E} \left(\frac{L}{I_L} + \frac{A}{I_A} \right) + \delta_Q \left(1 + \frac{A}{L} \right)^2 + \delta_P \frac{A^2}{L^2} \right)$$

$$= 110 \left(\frac{80^2}{3 \times 2.1 \times 10^4} \left(\frac{335 + 80}{201340.7} \right) \right) + 0.000137 \left(1 + \frac{80}{335} \right)^2 + 0.00045 \frac{80^2}{335^2}$$

$$= 110 \, (00.000209 + 0.00021 + 0.0000257) = 0.05$$

From Eqn. 5.3,

Max. permissible deflection = δ_{max}

$$\delta_{max} = 0.0002 \, L = 0.0002 \times 335 = 0.067.$$

This deflection is excessive. The overall deflection should not exceed half the value of the permissible runout (usually 0.02). In fact, the preferable value is only one-third of the permissible runout. Generally, the maximum deflection is limited to 0.006 mm, i.e. 6 microns.

Even the slope at the edge of the bearing, closer to the load (Q in [Fig. 5.1]), must be limited to 0.001 Radians. Actual slope can be found from Eqn. 5.6.

$$Q = \frac{WAL}{3EI_A} \qquad \text{(Eqn. 5.7)}$$

Maximum permissible slope (misalignment) depends upon the type of bearing used.

Table 5.2: Permissible slopes for machine spindles

Brg type	Ball Brgs			Thrust	Roller Brgs	
	Single row	Double row	Angular contact		Cylindrical	Spherical
θ_{max}	0.001 Rad.	0.0002	0.003	0.0002	0.0003	0.009

Self-aligning, double row ball bearings can have a misalignment of up to 0.04 radians.

Example 5.2: Find the slopes for the shafts from the following data:

Shaft ϕ, mm	I_a mm^4	L_o mm	A mm	W kg
45	201340.7	305	80	110
60	636336	269	80	110
70	1178891	315	80	110

Solution: From Eqn. 5.7, $\theta = \dfrac{WAL}{3EI_A}$

$$\theta_{45} = \frac{110 \times 80 \times 305}{3 \times 2.1 \times 10^4 \times 201340.7} = 0.000211 \text{ radians}$$

$$\theta_{60} = \frac{8800 \times 269}{6.3 \times 10^4 \times 636336} = 0.00006 \text{ radians}$$

$$\theta_{70} = 0.1396 \; \frac{315}{1178891} = 0.000037 \text{ radians.}$$

All the slopes are within the permissible limits, specified in Table 5.2.

When there is a drive gear, on the part between the bearings, the overall deflection (δ) can be found from the following equation [Fig. 5.2]:

$$\delta = \frac{WA^2(A+L) - 0.5\, W_g ABL\left(1 - \dfrac{B^2}{L^2}\right) - M_r AL}{3\, E\, I} \quad \text{(Eqn. 5.8)}$$

W_g = Load due to gear (kgf); B = Distance between gear & rear bearing

M_r = Reactive bending moment (kgfmm) ≤ 0.35 WA

I = Average value of moment of inertia of the sections of the spindle. The slope at the edge of the bearing at Q can be found from the following equation:

$$\theta = \frac{1}{3EI}\left(WAL - 0.5\, W_g AL\left(1 - \frac{B^2}{L^2}\right) - M_r L\right) \quad \text{(Eqn. 5.9)}$$

Forces due to gears can be found from formulae 2.89 to 2.92 given in Chapter 2.

Example 5.3: Find the spindle diameters for the overhanging part, and the part between the anti-friction bearings, for the spindle in **[Fig. 5.2]**. The spindle is powered by a 2.5 kW motor, running at an R.P.M., ranging from 400 to 2400. The spindle is mounted with a gear at 60 mm from the rear bearing (P). It has 80 teeth, of 1.5 module. The gear generates a radial separating force of 101 kgf. The spindle is subjected to fluctuating load. Determine the deflections at the gear, and the overhanging end. Check the slopes for the limit of 0.001 radians. The spindle runout should not exceed 0.02 TIR [Total Indicated Runout]. The reactive bending moment can be taken as 10% of the bending moment, due to the overhanging load (M_Q).

Solution: Referring to **[Fig. 5.2]**,

$$M_Q = 70 \times 40 = 2800 \text{ kgfmm}$$

$$\text{Torque } (T) = \frac{6000 P}{2\pi N} = \frac{6000 \times 2.5}{2\pi \times 400} = 5.97 \text{ kgfm}$$

$$= 5970 \text{ kgfmm}$$

Combined (Equivalent) moment = M_c

$$M_c = \sqrt{2800^2 + 0.45 \times 5970^2} = 4886.55$$

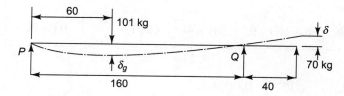

Fig. 5.2: Spindle for Example 5.3

From Table 5.1, for fluctuating and considerable bending load, permissible bending stress $(fb) = 4.5$ kgf/mm^2.

$$\therefore \quad D = \sqrt[3]{\frac{4886.55}{4.5 \times 0.0982}} = 22.28$$

≈ 25 the next bigger std. bearing size.
Permissible deflection $= \delta_{max}$
$\delta_{max} \le 0.0002\ L \le 0.0002 \times 160$
≤ 0.032

But the permissible runout is only 0.02. So the maximum deflection should be 1/3 of 0.02, i.e. 0.0067.

From Eqn. 5.8,

$$\delta = \frac{WA^2(A+L) - 0.5\ Wg\ ABL\left(1 - \frac{B^2}{L^2}\right) - 0.01\ WA^2 L}{3\ EI}$$

$$= \frac{70 \times 40^2 (40 + 160) - 0.5 \times 101 \times 40 \times 60 \times 160 \left(1 - \frac{60^2}{160^2}\right)}{}$$

$$\frac{- 0.1 \times 70 \times 40^2 \times 160}{3 \times 2.1 \times 10^4 \times 0.0491 \times 25^4}$$

$$= \frac{2.24 \times 10^7 - 1.666 \times 10^7 - 1.79 \times 10^6}{1.208 \times 10^9} = 0.0033$$

Slope at the edge of bearing $Q = \theta_Q$

$$\theta_Q = \frac{WAL - 0.5\ WgAL\left(1 - \frac{B^2}{L^2}\right) - 0.1\ WAL}{3\ EI}$$

$$= \frac{70 \times 40 \times 160 - 0.5 \times 101 \times 40 \times 160\left(1 - \frac{60^2}{160^2}\right) - 0.1 \times 70 \times 40 \times 160}{3 \times 2.1 \times 10^4 \times 19179.7}$$

$$= 1.063 \times 10^{-4} = 0.0001$$

Both the deflection (0.0033) and the slope (0.0001), are well within the permissible limits of 0.006 and 0.001, respectively. From Eqn. 5.4; $\delta_g = 0.01$ Gear module

Machine Tool Elements 351

$$\therefore \quad \delta_g = 0.01 \times 1.5 = 0.015$$

B.M. at gear $(M_g) = \dfrac{101 \times 60 \times 100}{160}$

$$= 3787.5 \text{ kgfmm}$$

$$\delta_g = \dfrac{M_g}{6EI}(2 \times 160(160-60) - 100^2 - (160-60)^2)$$

$$= \dfrac{3787.5}{6 \times 2.1 \times 10^4 \times 19179.7}(12000) = 0.0376$$

As δ_g is more than the permissible deflection of 0.015, a further increase in the shaft diameter is necessary.

For $\delta_g = 0.015$

$$I = \dfrac{3787.5 \times 12000}{0.015 \times 6 \times 2.1 \times 10^4} = 0.0491\, D_g^4$$

$$\therefore \quad D_g = \sqrt[4]{489768.2} = 26.4 \approx 30 \text{ (the next bigger standard bearing size)}.$$

Even the overhanging part of the spindle can be increased to $\phi 30$. This will further reduce the deflection and slope. Readers can find the deflection and slope for a $\phi 30$ spindle, as an exercise.

When a number of gears are mounted on the spindle, it is convenient to find their resultant, force and maximum bending moment to determine the maximum deflection in the portion between the bearings.

Example 5.4: Find the diameters in the overhanging part and between the bearings, for the spindle in [**Fig. 5.3**]. The reactive bending moment for Brg at Q can be taken as 15% of the moment, due to the overhanging load. The gears 1 and 2, mounted between the bearings, are of 2 module. The runout of the spindle should be limited to 0.03 TIR. The spindle is running in the range of 350 to 2500 R.P.M., and transmitting 1 kW. The load is fluctuating considerably.

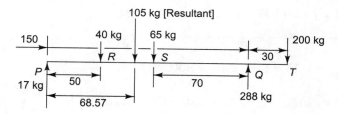

Fig. 5.3: Figure for Example 5.4

Solution:

$$M = 200 \times 30 = 6000 \text{ kgfmm}$$

$$T = \frac{6000 \times 1}{2\pi \times 350} = 2.728 \text{ kgfm}$$

$$= 2728 \text{ kgfmm}$$

$$M_c = \sqrt{6000^2 + 0.45 \times 2728^2} = 6272.87$$

From Table 5.1, for heavy bending moment and fluctuating load, permissible stress for C 45 [medium carbon steel] is 4.5 kgf/mm².

$$\therefore \quad D = \sqrt[3]{\frac{6272.9}{4.5 \times 0.0982}} = 24.21 \simeq 25$$

As the bending moment (6000) is more than double the torsional moment (2728), the spindle should be designed for bending strength. Refer to **[Fig. 5.3]**. Take moments about point P.

$$W_g \times B = W_{g1} \times B_1 + W_{g2} \times B_2$$

W_g = Resultant of loads $(W_{g1} + W_{g2})$ kgf.
W_{g1} = Radial load due to Gear 1 (kgf)
W_{g2} = Radial load due to Gear 2 (kgf)
B_1 = Distance of gear W_{g1} from left support P (mm)
B_2 = Distance of gear W_{g2} from left support P (mm)
B = Distance of load resulting from W_{g1}, W_{g2}

$$B = \frac{40 \times 50 + 65(150 - 70)}{40 + 65} = 68.57$$

Maximum permissible runout = 0.03 TIR

Maximum permissible deflection = $\frac{0.03}{3} = 0.01$

From Eqn. 5.7,

$$\delta = \frac{WA^2(A+L) - 0.5\, W_g ABL\left(1 - \frac{B^2}{L^2}\right) - M_r AL}{3\, EI}$$

$M_r = 0.15\, WA = 0.15 \times 200 \times 30 = 900$ kgf.

$$\delta = \frac{200 \times 30^2 \times 180 - 0.5 \times 105 \times 30 \times 68.57 \times 150 \times \left(1 - \frac{68.57^2}{150^2}\right) - 900 \times 30 \times 150}{3 \times 2.1 \times 10^4 \times 0.0491\, D^4}$$

$$= 0.01$$

$$\therefore \quad D = \sqrt[4]{\frac{3.24 \times 10^7 - 0.338 \times 10^7 - 0.4 \times 10^7}{0.01 \times 0.309 \times 10^4}}$$

$= 29.97 \approx 30$ (The nearest bigger Std. Brg. Size.)

For the portion between the bearings, i.e. for resultant force W_g, at distance B,

$$M_g = \frac{W_g B[L-B]}{L} = \frac{105 \times 68.57 \times [150 - 67.57]}{150} = 3908.6$$

$$\delta_g = \frac{M_g}{6EI} \{2L(L-B) - (L-B)^2 - (L-B)^2\}$$

$$= \frac{3908.6}{6 \times 2.1 \times 10^4 \times 0.0491 \times 30^4} \{2 \times 150(150 - 68.57) - 2(150 - 68.57)^2\}$$

$$= 7.8 \times 10^{-7}(24429 - 2 \times 6647.14)$$

$$= 0.0087$$

The module of the gears is 2.

$\therefore \quad \delta_g = 0.01 \times 2$ maximum permissible.

As the deflection 0.0087 is less than permissible 0.02, a ϕ 30 shaft would do.

Slope of the spindle at the outer edge of the bearing Q, can be checked from Eqn. 5.9,

$$\theta_Q = \frac{WAL - 0.5 W_g AL\left(1 - \frac{B^2}{L^2}\right) - M_r L}{3EI}$$

$$= \frac{200 \times 30 \times 150 - 0.5 \times 105 \times 30 \times 150\left(1 - \frac{68.57^2}{150^2}\right) - 900 \times 150}{3 \times 2.1 \times 10^4 \times 39771}$$

$= 0.00023$, i.e. much lesser than the permissible maximum of 0.001 radians.

Deflections can also be found graphically, by the **B.M. Area Moment** method [**Fig. 5.4**]. It is necessary to draw a bending moment diagram first. Consider the length of the beam as X axis. Then the difference in deflections at any two points, is equal to the moment of the bending moment area, between the two points around the Y axis, through one of the two points divided by EI.

[1 Lb. Ft. = 0.1356 kgm]

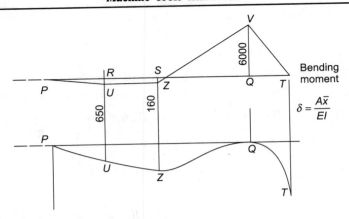

Fig. 5.4: Bending moment and deflection for Example 5.4

Referring to the B.M. diagram in [**Fig. 5.4**], the deflection at point R with respect to point P, can be found as below:

$$R = \frac{\text{Moment of B.M. area between } P \text{ and } R \text{ about } Y \text{ axis, through } P}{EI}$$

$$= \frac{\triangle PRU \text{ Area} \times \text{Distance of its centroid from } P}{2.1 \times 10^4 \times 0.0491 D^4}$$

$$= \frac{\frac{1}{2} RU \times PR \times \frac{2}{3} PR}{2.1 \times 10^4 \times 0.0491 \times 30^4}$$

$$= \frac{\frac{1}{2} \times 650 \times 50 \times \frac{2}{3} \times 50}{2.1 \times 10^4 \times 39771} = \frac{541669.4}{83519 \times 10^4}$$

$$= 0.00065$$

Similarly, for deflection at point 'S',

$$\delta_s = \frac{541669.4 + \text{Moment of Trapezoid RUSZ about } P}{83519 \times 10^4}$$

$$= \frac{541669.4 + \left(\frac{1}{2}(650 + 160)(150 - (70 + 50))\right) \times \left\{50 + \frac{30}{3}\left(\frac{2 \times 650 + 160}{650 + 160}\right)\right\}}{83519 \times 10^4}$$

$$= \frac{541669.4 + 826500}{83519 \times 10^4} = 0.0016$$

Deflection at T, with respect to Q

$$\delta_T = \frac{\text{Moment of Area VQT around } Y \text{ axis, through } Q}{83519 \times 10^4}$$

$$= \frac{\left(\frac{1}{2} \times 6000 \times 30\right) \times \frac{2}{3} \times 30}{83519 \times 10^4} = \frac{1,800,000}{83519 \times 10^4}$$

$$= 0.00216$$

It will be necessary to use a very high scale, i.e. $0.001 = 10$ mm, or $10,000 : 1$ for deflection (Y axis), to make it perceptible in **[Fig. 5.4]**.

Fig. 5.5(a): Centrifugal force due to deflection

Academicians in quest of subjects for post-graduate research, endeavor to find mathematically-ideal conditions. It has been found that the runout of a spindle, due to play between the anti-friction bearings, can be minimized, by keeping the $\frac{L}{A}$ ratio above 2.5.

$$\frac{L}{A} \geq 2.5 \qquad \text{(Eqn. 5.10)}$$

The diameter of the spindle between the bearings (D_L), should have stiffness (S_L) exceeding 35 kgf/micron. This can be ensured by satisfying the following condition:

$$L \leq \frac{D_L^{1.333}}{\lambda^{0.333}} \qquad \text{(Eqn. 5.11)}$$

$\lambda = 0.05$ for average accuracy, and,

$\quad = 0.1$ for high-precision machines.

For hydro-static bush bearings, the maximum deflection in the spindle portion between the bearings, should be limited to $\frac{L}{10^4}$.

$$\delta_L \leq \frac{L}{10^4} \qquad \text{(Eqn. 5.12)}$$

Example 5.5: The spindle of a high-precision machine is subjected to 200 kgf overhung load at 60, from the nearest bearing. It is subjected to 50 kgf load at mid-span, between the anti-friction bearings. Find the spindle diameters in the overhung portion (D_A), and between the bearings (D_L), to optimize the span (L). The spindle is transmitting 5 kW at 110 R.P.M. It is subjected to a fluctuating load. The deflection between the bearings should be limited to 0.03.

Solution:

$$M = WA = 200 \times 60 = 12000 \text{ kgfmm}$$

$$T = \frac{6000\,P}{2\pi N} = \frac{6000 \times 5}{2\pi \times 110} = 43.4 \text{ kgfm}$$

$$= 43400 \text{ kgfmm}$$

From Eqn 5.1, for a fluctuating load and light bending moment for a medium carbon steel (C45) spindle,

$f_b = 6$ kgf/mm^2. From Eqn. 5.1,

$$f_b = 6 = \frac{\sqrt{12000^2 + 0.45 \times 43400^2}}{0.0982\,D_A^3}$$

$\therefore \quad D_A^3 = \dfrac{\sqrt{9.91832 \times 10^8}}{6 \times 0.0982} = 53451$

$D_A = 37.67 \approx 40$ (Next bigger Std. Brg size).

For the portion supported between the bearings for central load,

$$\delta_L = \frac{WL^3}{48\,EI} = \frac{50\,L^3}{48 \times 2.1 \times 10^4 \times 0.0491\,D_L^4} = 0.001\,\frac{L^3}{D_L^4}$$

From Eqn. 5.11; $L \leq \dfrac{D_L^{1.333}}{\lambda^{0.333}}$

For a precision machine, $\lambda = 0.1$

$\therefore$ For a borderline case,

$$L = \frac{D_L^{1.333}}{0.1^{0.333}} = 2.154\,D_L^{1.333}$$

If $\quad D_L = D_A = 40$, then $L = 2.154 \times 40^{1.333} = 294.3$,

And $\quad \delta_L = 0.001\left(\dfrac{294.3^3}{40^4}\right) = 0.001\left(\dfrac{25490304}{2560000}\right)$

$$= 0.00996 \approx 0.01$$

From Eqn. 5.10; $\dfrac{L}{A} \geq 2.5$ or $L \geq 2.5 \times 60 \geq 150$. The optimum span of 294.3 satisfies Eqn. 5.10.

In actual practice, the span of the bearings is determined by practical considerations such as ease in assembly, and ease in maintenance of drive and control elements. The diameters of various portions of spindle are then determined, using various formulae cited for the design of the spindle, since the spindle is usually provided with steps for clamping the inner races of the bearings.

Machine Tool Elements

Usually, the span of the spindle bearings is made suitable for elements such as the gears and pullies, mounted on it. After working out a tentative design, its strength (stiffness), deflection, and slopes are checked for conformity with the permissible limits.

It is also advisable to check the twist (Angular deflection) angle (α). The angle depends upon the spindle length, and the power transmitted (kW).

Table 5.3: Angular deflection (α) per meter and spindle ϕ for solid and hollow spindles

Shaft type	Max. angular deflection $\alpha°$/metre	Spindle ϕ mm [Divide by 25.4 for Inch equivalent]
Solid	0.25°	$D = 128 \sqrt[4]{\dfrac{kW}{RPM}}$
	0.35°	$D = 118 \sqrt[4]{\dfrac{kW}{RPM}}$
Hollow	0.25°	$D^4 - d^4 \geq 2.742 \times 10^8 \dfrac{kW}{RPM}$
	0.35°	$D^4 - d^4 \geq 1.959 \times 10^8 \dfrac{kW}{RPM}$

D = Outside ϕ of shaft – (mm)
d = Inside ϕ of shaft – (mm)
kW = Power in kilowatts; R.P.M. = Revs/min

Example 5.6: Find the diameters for:
1. Solid shafts for transmitting 5 kW at 200 R.P.M., with angular deflection of 0.25° and 0.35°.
2. Hollow shafts, if inside $\phi = 30$

Solution:

1.
$$\text{For } 0.25°, D = 128 \sqrt[4]{\dfrac{5}{200}} = 50.9 \approx 51$$

$$\text{For } 0.35°, D = 118 \sqrt[4]{\dfrac{5}{200}} = 46.92 \approx 47$$

2.
$$d^4 = 30^4 = 810000$$

$$\text{For } 0.35°, D^4 = 1.959 \times 10^8 \times \dfrac{5}{200} + 810000$$

$$D = \sqrt[4]{570.75 \times 10^4} = 48.90$$

For 0.25°,
$$D^4 = 2.742 \times 10^8 \times \frac{5}{200} + 810000$$

$$D = \sqrt[4]{766.5 \times 10^4} = 52.6$$

5.1.3 Critical speed

Deflections resulting from external loads and the weight of the spindle itself, generate centrifugal forces. Their values depend upon the radius of rotation (deflection) and the angular velocity (ω).

$$\text{Centrifugal force} = \text{Weight} \times \omega^2 \delta$$
$$= \text{Area} \times \text{Density} \times \omega^2 \delta$$

At very high speed, the centrifugal force momentarily causes very detrimental, high-vibration resonance. This speed is called the critical speed for the particular spindle.

Usually, the spindle has a number of steps where the cross-sectional area is different. Under such circumstances, each step of a uniform cross-section is converted to a concentrated, weight-acting mid-length. Thus, theoretically it becomes a weightless shaft, with a number of concentrated loads at various points. There will be a number of critical speeds. But we have to consider only the first critical speed, to keep the highest spindle speed at less than 20% the first critical speed. **Figure 5.5b** gives the formulae for finding critical speeds under various load dispositions, and spindle supporting systems. If the length of the bearing (l) is longer than 0.3 of the bearing bore (d), the bearing is considered long and stiff, i.e., a rigid fixed point. Bearings, shorter in length, are considered simply-supported points. Self-aligning bearings are also considered simply-supported points.

Brg. length < 0.3 d .... Simply supported beam (short length bearing)
Brg. length > 0.3 d .... Fixed beam (Stiff, rigid bearing)

Example 5.7: Find the critical speed for a 50 ϕ spindle, carrying 100 kgf overhung load, at 40 mm from the nearest bearing. State the constraints posed by the critical speed, if the spindle is supported by rigid bearings, which can be considered fixed points.

Solution:

Referring to [**Fig. 5.5b**],

$$NC = 5.26 \times 10^4 \, \frac{d^2}{1\sqrt{W1}}$$

$$= 5.26 \times 10^4 \times \frac{50^2}{40\sqrt{100 \times 40}}$$

$$= 5.26 \times 10^4 \times \frac{2500}{2529.8}$$

$$= 51980 \text{ Revs/min}$$

Spindle speed should either be more than 62376 (1.2×51980) or less than 41584 (0.8×51980) RPM.

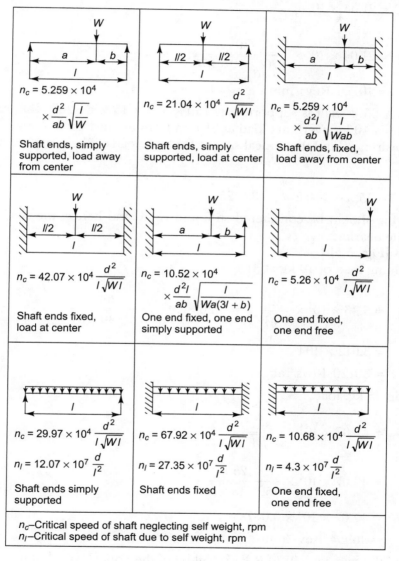

n_c–Critical speed of shaft neglecting self weight, rpm
n_l–Critical speed of shaft due to self weight, rpm

Fig. 5.5(b): Critical speeds

[$1'' = 25.4$ mm]   [1 Lb = 0.45 kg]

Example 5.8: Find the critical speed for a 70 ϕ spindle, if it carries a 100 kg load at the center of an 870 mm span. The shaft can be considered as simply supported.

Solution:

Referring to [**Fig. 5.5b**],

$$NC = 21.04 \times 10^4 \, \frac{d^2}{l\sqrt{Wl}}$$

$$= 21.04 \times 10^4 \times \frac{70^2}{870 \sqrt{100 \times 870}}$$

$$= 4017.6 \text{ Revs/min}$$

Example 5.9: A 25 ϕ shaft is supported in $25\phi \times 52\phi \times 15$ wide 25RNO2 roller bearings. The shaft carries a 40 kg overhung load at 31 mm from the nearest bearing, and 15 kg load mid-span the bearings. Find the critical speeds for both loads, if bearings are 150 apart.

Solution:

$$l = 15, \text{ i.e. } > 0.3 \, d \, (0.3 \times 25)$$

∴ The bearing should be considered stiff, and rigid, and hence, the shaft is considered fixed at both the bearings.

Referring to [**Fig. 5.5b**],
For an overhung load of 40 kg at 31,

$$NC_{40} = 5.26 \times 10^4 \times \frac{d^2}{l\sqrt{Wl}} = 5.26 \times 10^4 \times \frac{25^2}{31\sqrt{40 \times 31}}$$

$$= 3.012 \times 10^4$$

$$= 30120 \text{ Revs/min}$$

For mid-span 15 kg. load,

$$NC_{15} = 42.07 \times 10^4 \times \frac{d^2}{l\sqrt{Wl}}$$

$$= 42.07 \times 10^4 \times \frac{25^2}{150\sqrt{15 \times 150}}$$

$$= 42.07 \times 10^4 \times 0.0878$$

$$= 36954 \text{ Revs/min}$$

The lesser of the two, i.e. 30120 R.P.M. will be of the critical speed of the spindle.

Raleigh method is useful when there are a number of loads at different points, and/or the cross-section of the spindle is different, at different places. In the Raleigh method, the deflections at the loads are calculated, reckoning the spindle cross-section at the load point.

Then Critical speed (NC) = $946.48\sqrt{\dfrac{\Sigma W\delta}{\Sigma W\delta^2}}$ (Eqn. 5.13)

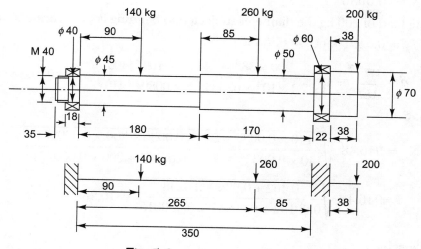

Fig. 5.6: Shaft for Example 5.10

Example 5.10: Find the critical speed for the shaft in [**Fig. 5.6**].

Solution:

As the bearing lengths at both the supports exceed $0.3D$, the shaft can be considered fixed at the bearings. It is necessary to compute the moments of inertia at the load points, to find the resulting deflection. At the point of load of 140 kg, the spindle ϕ is 45.

$$I_{45} = 0.0491 \times 45^4 = 201381.7 \text{ mm}^4$$

For the $\phi 50$ portion, $I_{50} = 0.0491 \times 50^4 = 306875 \text{ mm}^4$

For a beam fixed at both ends, for load W at distance x from L.H.,

$$\delta = \frac{Wx^3(l-x)^3}{3\,EI\,l^3}\text{ ; for 140 kg load, } x = 90,\ l = 350,\text{ and } I = 201381.7$$

$$\delta = \frac{140 \times 90^3 \times (350-90)^3}{3 \times 2.1 \times 10^4 \times 201381.7 \times 350^3}$$

$$= \frac{140 \times 729000 \times 17576000}{6.3 \times 201381.7 \times 42875000 \times 10^4}$$

$$= 0.0033$$

At 260 kg load, $x = 265$, $I = 306875$

$$\delta = \frac{260 \times 265^3 \times (350-265)^3}{6.3 \times 10^4 \times 306875 \times 42875000}$$

$$= \frac{260 \times 18609625 \times 614125}{6.3 \times 306875 \times 42875000 \times 10^4}$$

$$= 0.0036$$

At overhung load of 200 kg the diameter subjected to bending is the bearing mounting $\phi : 60$

$$I_{60} = 0.0491 \times 60^4 = 636336$$

$$\delta = \frac{Wl^3}{3EI} = \frac{200 \times 38^3}{3 \times 2.1 \times 10^4 \times 636336} = \frac{10974400}{4.00891 \times 10^{10}} = 0.000274$$

From Eqn. 5.12

$$N_C = 946.48 \sqrt{\frac{140 \times 0.0033 + 260 \times 0.0036 + 200 \times 0.000274}{140 \times 0.0033^2 + 260 \times 0.0036^2 + 200 \times 0.000274^2}}$$

$$= 946.48 \sqrt{\frac{0.462 + 0.936 + 0.0548}{0.00152 + 0.00337 + 0.000015}} = 16,289$$

5.1.3.1 Spindle noses

Machine spindles serve three purposes:
1. Locating workholder or tool
2. Securing workholder or tool
3. Driving workholder or tool
1. Location is accomplished by providing:
   (1) precisely machined spigot/bore
   (2) tapered bore/spigot with standard self-holding or quick-release taper.
2. Mounting workpiece-holder/tool can be done by providing:
   (1) tapped holes or clear holes for holding studs/screws
   (2) quick-action cam locks
   (3) bayonet system
3. Driving arrangement involves:
   (1) a screwed drive button (lathes), or
   (2) face keyways and keys (milling arbors)
   (3) a tang slot (drilling m/cs)

5.1.3.2 Lathes

Figure 5.7a shows A 1-type flanged spindle nose for lathes. The inner tapped holes are used for mounting scroll chucks of smaller sizes, secured by Allen screws. The outer tapped holes are used for securing face plates or fixtures. There is a precisely-machined counterbore for the drive button. Note the thru hole at the center of the spindles. The hole allows for the usage of the long bar as a raw material for the workpiece. Usually, the nose end of the spindle hole is precisely tapered to Morse/Metric standards, to provide a precision locating surface.

Machine Tool Elements

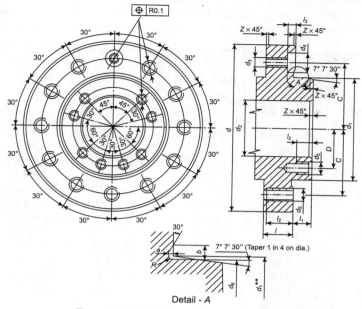

Fig. 5.7(a): Lathe spindle noses type A_1

Size No.	5	6	8	11
d	133 [5.236″]	165 [6.496″]	210 [8.268″]	280 [11.024″]
d_1	82.563 [3.25″]	106.375	139.719	196.869
	$\binom{+0.01}{0}$ [0.004]	$\binom{+0.01}{0}$ [0.0004″]	$\binom{+0.012}{0}$ [0.0005″]	$\binom{+0.014}{0}$ [0.00055″]
d_2 Max.	40 or Morse 5	56 or Morse 6	80 or Metric 80	125 or Metric 120
d_3	M6	M8	M8	M10
d_4 H8	15.9 [0.626″]	19.05 [0.75″]	23.8 [0.937″]	28.6 [1.126″]
d_5	M10	M12	M16	M18
d_6	82 [3.228″]	106 [4.173″]	139 [5.472″]	196 [7.716″]
a	1 [0.039″]	1 [0.039″]	1 [0.039″]	1 [0.039″]
b	6 [0.236″]	8 [0.315″]	8 [0.315″]	8 [0.315″]
C	52.4 [2.063″]	66.7 [2.62″]	85.7 [3.375″]	117.5 [4.625″]
D	30.95 [2.625″]	41.3 [1.625″]	55.55	82.55 [2.187″]
$l_1 \begin{bmatrix} 0 \\ -0.025 \end{bmatrix}$ [0.001″]	14.288 [0.562″]	15.875 [0.625″]	17.462 [0.687″]	19.05 [0.75″]
l_2	19 [0.75″]	22 [0.875″]	25 [1″]	32 [1.25″]
l_3	6 [0.236″]	8 [0.315″]	10 [0.394″]	12 [0.472″]
l	22 [0.875″]	25 [1″]	28 [1.125″]	35 [1.375″]
R	1 [0.039″]	1 [0.039″]	1 [0.039″]	1 [0.039″]
Z	1 [0.039″]	1.6 [0.063″]	1.6 [0.063″]	1.6 [0.063″]

1** Dimension d_1 is taken at the theoretical point of intersection between the generating line of the cone and the face of the flange.
2 Tolerance on un-toleranced dimensions: ±0.4. [3] Bracketed figures give inch equivalents

Figure 5.7b shows A2-type flanged nose. In this type the central bore size is bigger. So there

Size No.	3	4	5	6	8	11
d	92 [3.62"]	108 [4.17"]	133 [5.24"]	165 [6.5"]	210 [8.27"]	280 [11"]
d_1	53.975 [2.125"]	63.513 [2.5"]	82.563 [3.25"]	106.375 [4.187]	139.719 [5.5]	196.869 [7.75]
	$\binom{+0.008}{0}$ [0.0003]	$\binom{+0.008}{0}$ [0.0003]	$\binom{+0.01}{0}$ [0.0004]	$\binom{+0.01}{0}$ [0.0004]	$\binom{+0.012}{0}$ [0.0005]	$\binom{+0.014}{0}$ [0.00055]
d_2 Max.	32 or Morse 4	40 or Morse 4	50 or Morse 5	71 or Metric 80	100 or Metric 100	150 or Metric 160
d_3	—	M6	M6	M8	M8	M10
d_4 H8	—	14.25 [0.561]	15.9 [0.625]	19.05 [0.75]	23.8 [0.937]	28.6 [1.125]
d_5	M10	M10	M10	M12	M16	M18
d_6	53.5 [2.106]	63 [2.48]	82 [3.228]	106 [4.173]	139 [5.472]	196 [7.716]
a	1 [0.039]	1 [0.039]	1 [0.039]	1 [0.039]	1 [0.039]	1 [0.039]
b	5 [0.197]	6 [0.236]	6 [0.236]	8 [0.315]	8 [0.315]	8 [0.315]
c	35.3 [1.625]	41.3 [1.625]	52.4 [2.063]	66.7 [2.62]	85.7 [3.375]	117.5 [4.625]
l_1	11 [1.433]	11 [0.433]	13 [0.512]	14 [0.551]	16 [0.63]	18 [0.709]
l_2	14 [0.551]	17 [0.669]	19 [0.748]	22 [0.875]	25 [0.984]	32 [1.25]
l_3	—	5 [0.197]	6 [0.236]	8 [0.315]	10 [0.394]	12 [0.472]
l	16 [0.63]	20 [0.787]	22 [0.875]	25 [0.984]	28 [1.102]	35 [1.375]
R	1 [0.039]	1	1	1	1	1
Z	1	1	1	1.6 [0.063]	1.6	1.6

1 **Dimension d_1 is taken at the theoretical point of intersection between the generating line of the cone and the face of the flange.
2 Tolerance on un-toleranced dimensions: ±0.4.
3 For sizes 15–28, refer IS 2582

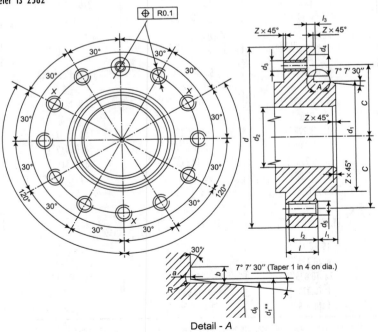

Fig. 5.7(b): Lathe spindle noses type A_2

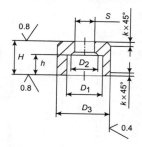

Fig. 5.7(c): Drive buttons for lathe spindles

Nominal size	4	5	6	8	11	15	20
$D_3 h_8$	14.25 [0.561]	15.9 [0.625]	19.05 [0.75]	23.8 [0.937]	28.6 [1.125]	34.9 [1.374]	41.3 [1.625]
$D_1 H_{12}$	10.4 [0.409]	10.4	13.5 [0.531]	13.5 [0.531]	16.5 [0.649]	18.5 [0.728]	18.5
D_2	—	—	—	—	—	16	16
$S H_{12}$	6.4 [0.25]	6.4	8.4 [0.33]	8.4	10.5 [0.413]	13 [0.512]	13
H	10 [0.394]	11 [0.433]	13 [0.512]	16 [0.63]	20 [0.787]	20 [0.787]	24 [0.945]
h	7 [0.276]	7	9 [0.354]	9	11 [0.433]	13 [0.512]	13
k	1 [0.039]	1	1.6 [0.063]	1.6	1.6	2 [0.787]	2
Size of screws (as per IS 2269)	M6 × 14	M6 × 14	M8 × 20	M8 × 20	M10 × 25	M12 × 25	M12 × 30

Tolerances for important dimensions of face plates in **[Fig. 5.7d]**.

$D_{in.}$	Tolerance
d_1	+ 0.003 [0.00012] for spindle − 0.005 [0.0002] sizes 3, 4 + 0.004 [0.00016] for sizes 5, 6 − 0.006 [0.00024], + 0.004 for sizes 6, 8, 11 − 0.008 [0.00031]
d_5	+ 2 + 2.5 [0.079, 0.098]
d_4	+ 0.4 + 0.5 [0.016, 0.020]
l_1	+ 0.025 − 0 [0.0017]
$C, D, [H - l_3 + 1.5]$	± 0.2 [0.008]

isn't enough space for providing inner tapped holes. **Figure 5.7c** gives the details of the drive buttons. Note that the face plate/fixture/chuck back plate will be located on the 7° 7.5′ taper of the location spigot of the spindle.

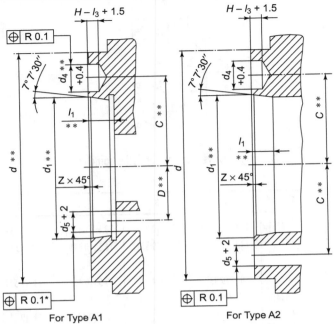

For Type A1   For Type A2

H = Button height [Fig. 5.7c]; l_3 = C' bore in spindle nose [Fig. 5.7a, b]
Other dimensions same as the spindle nose [Fig. 5.7a, b]

Fig. 5.7(d): Face plates for type A_1 and A_2 spindle noses

(1) **Dimensions d, d_1, d_4, C, D, H, l_1, l_3 same as in Figs. 5.7a, b, c, but with different tolerances (2) Tolerances on important dimensions given on page 365 (3) On other dimensions tolerance = ±0.4

Figures 5.8a, b, and c show cam lock type spindle, and cams and studs, and [**Fig. 5.8d**] shows the back plate of the chuck for the same. Note the arched (d_3) cut-out in the stud [**Fig. 5.8c**] and the tapped holes in the back plate for the locking Aln. Scr. The Aln. Scr. is assembled after screwing in the stud into the chuck backplate. The head of the locking Aln. screw prevents the rotation of the stud into the tapped hole in the backplate. After assembling onto the backplate, the studs are inserted into the spindle flange (nose).

The curved cut-out in the cam provides a passage to the long collar of the stud. The cam is provided with a groove. It engages with the dog point of the special Aln. screw. The groove and the screw position the cam axially. Note that the central portion of the cam is eccentric with respect to the ends. This eccentric part acts like a cam.

The ungrooved end of the cam is provided with a square socket, to facilitate rotating the cam with a key. Rotation of the cam clamps the stud into the spindle. The spring at the grooved end of the cam pushes the cam out of the spindle after the retaining screw is withdrawn from the groove. This facilitates the replacement of the cam.

Thus, for mounting/dismantling of the chuck, we only have to rotate the cams by quarter turns to clamp/unclamp the studs. This is much faster than screwing/unscrewing nuts/screws, which call for rotating the wrench by at least 8–10 turns for each nut/screw. The cam lock clamping is very convenient for applications needing frequent changeovers of workholders.

Size No.	5	6	8	11	15	20
d	146	181	225	298	403	546
d_1	82.563 [3.25]	106.375 [4.187]	139.719 [5.5]	196.869 [7.75]	285.775 [11.243]	412.775 [16.25]
@	$\begin{bmatrix}+0.01\\0\end{bmatrix}$ [0.0004]	$\begin{bmatrix}+0.01\\0\end{bmatrix}$	$\begin{bmatrix}+0.012\\0\end{bmatrix}$ [0.0005]	$\begin{bmatrix}+0.014\\0\end{bmatrix}$ [0.00055]	$\begin{bmatrix}+0.016\\0\end{bmatrix}$ [0.00063]	$\begin{bmatrix}+0.02\\0\end{bmatrix}$ [0.00079]
d_2 Max.	45 or Morse 5	65 or Morse 6	85 or Metric 80	135 or Metric 120	210 [8.268]	330 [12.992]
d_3 H8	22 [0.866]	26 [1.0236]	29 [1.142]	32 [1.26]	35 [1.378]	42 [1.653]
d_4 $\begin{bmatrix}+0.05\\0\end{bmatrix}$ [0.002]	19.8 [0.779]	23 [0.905]	26.2 [1.031]	31 [1.22]	35.7 [1.405]	42.1 [1.657]
d_5	10.5 [0.413]	13.5 [0.531]	13.5	13.5	16.5 [0.65]	16.5
d_6	82 [3.228]	106 [4.173]	139 [5.472]	196 [7.716]	285 [11.22]	412 [16.22]
d_7	M6	M8	M8	M8	M10	M10
C	52.4 [2.063]	66.7 [2.62]	85.7 [3.375]	117.5 [4.625]	166.1 [6.539]	231.8 [9.126]
D	32.5 [1.279]	41 [1.614]	57 [2.244]	86 [3.386]	129 [5.079]	190 [7.48]
α	14° 55′	13° 46′	12° 18′	10° 30′	8° 35′	7° 5′
x_1	0.075 [0.003]	0.075	0.075	0.075	0.075	0.075
x_2	0.1 [0.0039]	0.1	0.1	0.1	0.1	0.1
a	1 [0.039]	1	1	1	1	1
b	6 [0.236]	8 [0.315]	8	8	8	9 [0.354]
R	1 [0.039]	1	1	1	1	1
Z	1	1.6 [0.063]	1.6	1.6	2 [0.0787]	2
l_1	13 [0.512]	14 [0.551]	16 [0.63]	18 [0.709]	19 [0.748]	21 [0.827]
l_2	20.6 [0.811]	23.8 [0.937]	27 [1.063]	31.8 [1.252]	36.5 [1.437]	42.9 [1.689]
h_3 $\begin{bmatrix}+0.2\\0\end{bmatrix}$ [0.008]	46 [1.811]	57 [1.244]	64 [2.52]	75 [2.953]	84 [3.307]	94 [3.701]
h_4 [±0.2] [0.008]	—	—	—	—	—	—
h_5	7 [0.276]	9 [0.354]	9	9	11 [0.433]	11
l_6 [±0.1] [0.004]	13.5 [0.531]	15.9 [0.626]	18.25 [0.718]	21.45 [0.844]	24.6 [0.968]	28.6 [1.126]
l Min.	38 [1.496]	45 [1.772]	50 [1.968]	60 [2.362]	70 [2.756]	82 [3.228]

Notes: Refer [Fig. 5.8a]

1. Tolerance on C, D and angular dimensions is controlled by this position tolerance which is the permissible deviation of the hole centers with respect to their theoretical positions.
2. @ Dimension d_1 is taken at the theoretical point of intersection between the generating line of the cone and the face of the flange.
3. Tolerance on untoleranced dimensions: ±0.4.

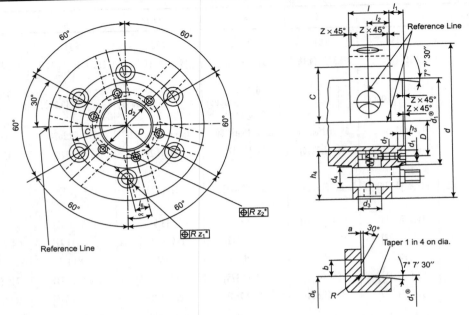

Fig. 5.8(a): Lathe spindles with camlock type nose [Dimensions on page 367]

Size No.	5	6	8	11	15	20
d_1 e 8	22 [0.866]	26 [1.0236]	29 [1.142]	32 [1.26]	35 [1.378]	42 [1.653]
d_2	14 [0.551]	17 [0.669]	21 [0.827]	24 [0.945]	27 [1.063]	33 [1.3]
d_3	7 [0.276]	10 [0.394]	10	10	10	10
$d_4 \pm 0.05$ [0.002]	4.5 [0.177]	6 [0.236]	6	6	8 [0.315]	8
$l_1 \begin{bmatrix} 0 \\ -0.1 \end{bmatrix}$ [0.039]	45 [1.772]	56 [2.205]	63 [2.48]	73 [2.874]	82 [3.228]	92 [3.622]
l_2	35 [1.378]	43 [1.693]	49 [1.929]	59 [2.323]	62 [2.441]	69 [2.716]
l_3 Min.	22 [0.866]	25 [0.984]	28 [1.102]	32 [1.26]	37 [1.457]	43 [1.693]
$l_4 \pm 0.1$ [0.039]	3 [0.118]	4.2 [0.165]	5.3 [0.209]	8.7 [0.342]	6 [0.236]	6
$l_5 \pm 0.1$	5 [0.197]	6.5 [0.256]	6.5	6.5	8.5 [0.335]	8.5
$l_6 \pm 0.1$	2 [0.079]	2.85 [0.112]	3.95 [0.155]	7.35 [0.289]	5.2 [0.205]	5.2
$l_7 \pm 0.2$ [0.008]	22.4 [0.882]	30.2 [1.19]	33.2 [1.307]	39.5 [1.555]	43.6 [1.716]	48.4 [1.905]
$l_8 \begin{bmatrix} 0 \\ -0.2 \end{bmatrix}$	14.2 [0.559]	16.7 [0.657]	18.9 [0.744]	21.2 [0.835]	23.5 [0.925]	27.8 [1.094]
R	11.1 [0.437]	12.7 [0.5]	14.2 [0.559]	16.7 [0.657]	19 [0.748]	22.2 [0.874]
S D_{12}	11 [0.433]	12 [0.472]	14 [0.551]	17 [0.669]	17	22 [0.866]
h_1	13 [0.512]	15 [0.590]	15	15	15	15
h_2	11 [0.433]	12 [0.472]	14 [0.551]	16 [0.63]	16	20 [0.787]

Refer [Fig 5.8 (b)]

	[0.012]	1.45 [0.057]	2.56 [0.101]	2.46 [0.097]	2.44 [0.096]	2.35 [0.092]	3.1 [0.0122]
	[0.004]	0	0.45 [0.018]	0.36 [0.014]	0.28 [0.011]	0.2 [0.008]	0.5 [0.020]
α_1		15°	20°	20°	20°	20°	20°
α_2		10°	10°	10°	15°	15°	15°
Slope on α^*		1.9	2.64	2.64	2.64	2.64	3.18

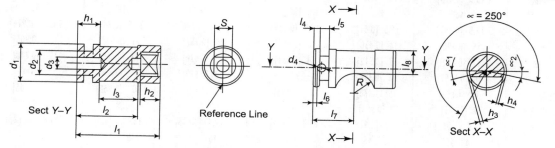

Fig. 5.8(b): Cams for camlock system

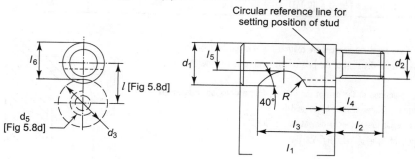

Fig. 5.8(c): Special stud for camlock system

Size No.	5	6	8	11	15	20
$d_1 \begin{bmatrix} 0 \\ -0.1 \end{bmatrix}$ [0.004]	19 [0.748]	22.2 [0.874]	25.4 [1.0]	30.2 [1.189]	34.9 [1.374]	41.3 [1.626]
d_2	M12 × 1	M16 × 1.5	M20 × 1.5	M22 × 1.5	M24 × 1.5	M27 × 2
d_3	11 [0.433]	14 [0.551]	14	14	14	14
l_1	43 [1.693]	49 [1.929]	55.5 [2.185]	67 [2.638]	76 [2.992]	89 [3.504]
l_2	22 [0.866]	27 [1.063]	30.5 [1.201]	35 [1.378]	40 [1.575]	44 [1.732]
$l_3 \pm 0.2$	35.7 [1.405]	40.5 [1.594]	44.5 [1.752]	53.2 [2.094]	58.7 [2.311]	69 [2.716]
$l_4 \pm 0.2$	4.8 [0.189]	4.8	4.8	6.4 [0.252]	6.4	6.4
$l_5 \pm 0.1$	11.9 [0.468]	14.3 [0.563]	16.7 [0.657]	20.6 [0.811]	24.6 [0.964]	28.6 [1.126]
$l_6 \pm 0.1$	16.5 [0.65]	19.6 [0.772]	23.2 [0.913]	26.8 [1.055]	32 [1.26]	38.5 [1.516]
R	11.25 [0.443]	12.7 [0.5]	14.3 [0.563]	15.9 [0.626]	17.5 [0.689]	20.6 [0.817]

Size No.	5	6	8	11	15	20
d	146 [5.748]	181 [7.126]	225 [8.858]	298 [11.732]	403 [15.866]	546 [21.496]
d_1	82.563 [3.25]	106.375 [4.188]	139.719 [5.5]	196.869 [7.75]	285.775 [11.25]	412.775 [16.25]
	$\begin{bmatrix}+0.004\\-0.006\end{bmatrix}\begin{bmatrix}.00016\\.00024\end{bmatrix}$	$\begin{bmatrix}+0.004\\-0.006\end{bmatrix}$	$\begin{bmatrix}+0.004\\-0.008\end{bmatrix}\begin{bmatrix}.00016\\.0003\end{bmatrix}$	$\begin{bmatrix}+0.004\\-0.01\end{bmatrix}\begin{bmatrix}.00016\\.0004\end{bmatrix}$	$\begin{bmatrix}+0.004\\-0.012\end{bmatrix}$	$\begin{bmatrix}+0.005\\-0.015\end{bmatrix}\begin{bmatrix}.0002\\.0006\end{bmatrix}$
d_2	19.4 [0.764]	22.6 [0.89]	25.8 [1.016]	30.6 [1.205]	35.4 [1.394]	41.6 [1.638]
d_3	M12 × 1	M16 × 1.5	M20 × 1.5	M22 × 1.5	M24 × 1.5	M27 × 2
d_4	10.5 [0.413]	13.5 [0.531]	13.5	13.5	13.5	13.5
d_5**	M6	M8	M8	M8	M8	M8
l**	12.5 [0.492]	15.5 [0.61]	17.5 [0.689]	18.7 [0.736]	21.5 [0.846]	24.8 [0.976]
h Min.	15 [0.590]	16 [0.630]	18 [0.709]	20 [0.787]	21 [0.827]	23 [0.905]
h_1	12 [0.472]	13 [0.512]	14 [0.551]	16 [0.630]	17 [0.669]	19 [0.748]
h_2	8 [0.315]	9.5 [0.374]	9.5	13 [0.512]	13	13
h_3	30 [1.181]	35 [1.378]	38 [1.496]	45 [1.772]	50 [1.968]	55 [2.165]
h_4	7 [0.275]	9 [0.354]	9	9	9	9
C	52.4 [2.063]	66.7 [2.626]	85.7 [3.374]	117.5 [4.626]	165.1 [6.5]	231.8 [9.126]
α	14° 55′	13° 46′	12° 18′	10° 30′	8° 35′	7° 5′
x_1	0.1 [0.004]	0.1	0.1	0.1	0.1	0.1
Z	1 [0.039]	1.6 [0.063]	1.6	1.6	2 [0.078]	2

Notes:

1. *Tolerance on C and angular dimensions are controlled by this position tolerance, which is the permissible radial deviation of the hole centers, with respect to their theoretical positions.
2. @ Tolerance of position of the axis of the hole d_5 (radial deviation with respect to the theoretical position defined by α and l).
3. General tolerance for un-toleranced dimensions: ± 0.4.
4. ** For position of hole d_5 ref. [Fig. 5.8c].

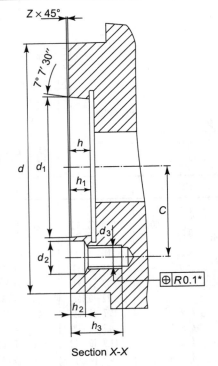

Section X-X

Fig. 5.8(d): Face plates for camlock spindles

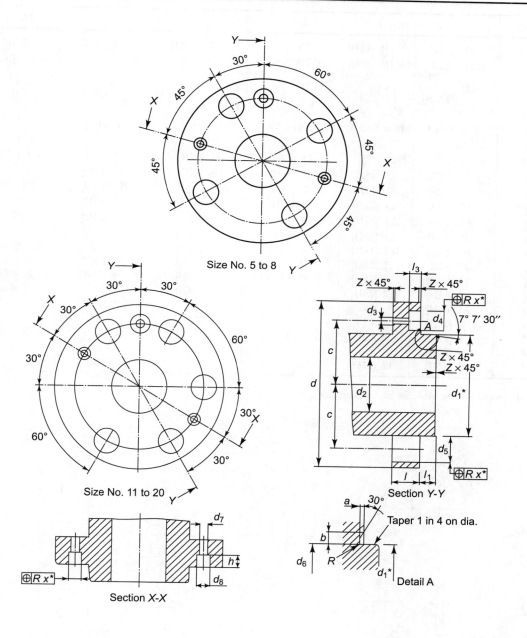

Fig. 5.9(a): Bayonet type lathe spindle noses (See table on p. 372 for dimensions)

Size No.	5	6	8	11	15	20
d	135 [5.315]	170 [6.693]	220 [8.661]	290 [11.417]	400 [15.748]	540 [21.260]
d_1*	82.563 [3.25]	106.375 [4.188]	139.719 [5.5]	196.869 [7.75]	285.775 [11.25]	412.775 [16.25]
	$\begin{bmatrix}+0.01\\0\end{bmatrix}$ [0.0004]	$\begin{bmatrix}+0.01\\0\end{bmatrix}$	$\begin{bmatrix}+0.012\\0\end{bmatrix}$ [0.0005]	$\begin{bmatrix}+0.014\\0\end{bmatrix}$ [0.00055]	$\begin{bmatrix}+0.016\\0\end{bmatrix}$ [0.00063]	$\begin{bmatrix}+0.02\\0\end{bmatrix}$ [0.0008]
d_2 Max.	50 or Morse 5	70 or Morse 6	100 or Metric 100	150 or Metric 140	220 [8.661]	300 [11.811]
d_3	M6	M8	M8	M10	M12	M12
d_4 H8	15.9 [0.626]	19.05 [0.75]	23.8 [0.937]	28.6 [1.126]	34.9 [1.374]	41.3 [1.626]
d_5	21 [0.827]	23 [0.905]	29 [1.142]	36 [1.417]	43 [1.693]	43
d_6	82 [3.228]	106 [4.173]	139 [5.472]	196 [7.716]	285 [11.220]	412 [16.22]
d_7	6.4 [0.252]	8.4 [0.331]	10.5 [0.413]	10.5 [0.413]	13 [0.512]	13
d_8	10.4 [0.409]	13.5 [0.531]	16.5 [0.650]	16.5	19 [0.748]	19
h	10 [0.394]	11 [0.433]	12 [0.472]	13 [0.512]	15 [0.590]	15
C	52.4 [2.063]	66.7 [2.626]	85.7 [3.374]	117.5 [4.626]	165.1 [6.5]	231.8 [9.126]
l	22 [0.866]	25 [0.984]	28 [1.102]	35 [1.378]	42 [1.653]	48 [1.89]
l_1	13 [0.512]	14 [0.551]	16 [0.630]	18 [0.709]	19 [0.748]	21 [0.827]
l_3	6 [0.236]	8 [0.315]	10 [0.394]	12 [0.472]	12	16 [0.63]
x	0.1 [0.004]	0.1	0.1	0.1	0.15 [0.006]	0.15
a	1 [0.039]	1	1	1	1	1
b	6 [0.236]	8 [0.315]	8	8	8	9 [0.354]
R	1 [0.039]	1	1	1	1	1
Z	1 [0.039]	1.6 [0.063]	1.6	1.6	2 [0.079]	2

* At intersection of cone generation line and flange face

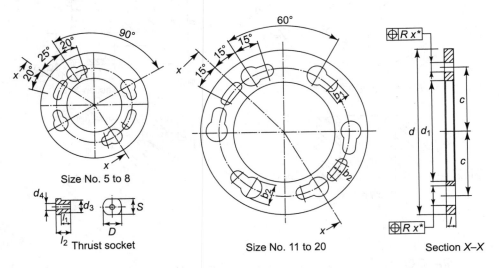

Fig. 5.9(b): Bayonet plates

Machine Tool Elements

Bayonet Plates [Fig 5.9b]

Size No.	5	6	8	11	15	20
d	145 [5.709]	180 [7.087]	230 [9.055]	300 [11.811]	410 [16.142]	550 [21.653]
d_1 H8*	80 [3.150]	100 [3.937]	130 [5.118]	185 [7.283]	270 [10.63]	400 [15.748]
d_2	21 [0.827]	23 [0.905]	29 [1.142]	36 [1.417]	43 [1.693]	43
C	52.4 [2.063]	66.7 [2.626]	85.7 [3.374]	117.5 [4.626]	165.1 [6.5]	231.8 [9.126]
$l \begin{bmatrix} 0 \\ -0.1 \end{bmatrix}$ [0.004]	8 [0.315]	10 [0.394]	12 [0.472]	16 [0.63]	18 [0.709]	22 [0.866]
b_1	11.5 [0.413]	14 [0.551]	18 [0.709]	23 [0.905]	27 [1.063]	27
b_2	11.5 [0.413]	14 [0.551]	18 [0.709]	18	23 [0.905]	23
x	0.1 [0.004]	0.1	0.1	0.1	0.15 [0.006]	0.15
d_3	11 [0.433]	13 [0.512]	17 [0.669]	17	22 [0.866]	22
d_4	M6	M8	M10	M10	M12	M12
$l_1 \begin{bmatrix} +0.2 \\ 0 \end{bmatrix}$	8.2 [0.323]	10.2 [0.401]	12.2 [0.48]	16.2 [0.378]	18.3 [0.72]	22.3 [0.878]
l_2	12 [0.472]	15 [0.590]	18 [0.709]	22 [0.866]	26 [1.024]	30 [1.181]
D	16 [0.63]	19 [0.748]	25 [0.984]	25	32 [1.26]	32
S	14 [0.551]	17 [0.669]	22 [0.866]	22	27 [1.063]	27
HEX SOC.HD. SCR.	M6 × 25	M8 × 30	M10 × 35	M10 × 45	M10 × 55	M12 × 65

Face Plates for Bayonet type lathe spindlenoses [Fig. 5.9c] page 374

Size No.	5	6	8	11	15	20
d	135 [5.315]	170 [6.693]	220 [8.661]	290 [11.417]	400 [15.748]	540 [21.260]
d_1	82.563 [3.25]	106.375 [4.188]	139.719 [5.5]	196.869 [7.75]	285.775 [11.25]	412.775 [16.25]
	$\begin{bmatrix} +0.004 \\ -0.006 \end{bmatrix} \begin{bmatrix} .00016 \\ .00024 \end{bmatrix}$	$\begin{bmatrix} +0.004 \\ -0.006 \end{bmatrix}$	$\begin{bmatrix} +0.004 \\ -0.008 \end{bmatrix} \begin{bmatrix} .00016 \\ .00031 \end{bmatrix}$	$\begin{bmatrix} +0.004 \\ -0.01 \end{bmatrix} \begin{bmatrix} .00016 \\ .00039 \end{bmatrix}$	$\begin{bmatrix} +0.004 \\ -0.012 \end{bmatrix} \begin{bmatrix} .00016 \\ .00047 \end{bmatrix}$	$\begin{bmatrix} +0.005 \\ -0.015 \end{bmatrix} \begin{bmatrix} .0002 \\ .0006 \end{bmatrix}$
$d_2 \begin{bmatrix} +0.1 \\ 0 \end{bmatrix}$ [0.004]	16.3 [0.642]	19.45 [0.767]	24.25 [0.955]	29.4 [1.157]	35.7 [1.405]	42.1 [1.657]
d_3	M10	M12	M16	M20	M24	M24
d_4 Max.	79.6	103.2	136.2	192.9	281.5	408
C	52.4 [2.063]	66.7 [2.626]	85.7 [0.374]	117.5 [4.626]	165.1 [6.5]	231.8 [9.126]
x	0.1 [0.004]	0.1	0.1	0.1	0.15 [0.006]	0.15
l	6.5 [0.256]	6.5	8 [0.315]	10 [0.394]	10	10
l_1	12 [0.472]	13 [0.512]	14 [0.551]	16 [0.63]	17 [0.669]	19 [0.748]
h_1	15 [0.59]	18 [0.709]	24 [0.945]	30 [1.181]	36 [1.417]	36
h_2	18 [0.709]	22 [0.866]	28 [1.102]	34 [1.339]	40 [1.575]	40
Size of stud	M10 × 43	M12 × 50	M16 × 60	M20 × 75	M24 × 90	M24 × 100
z	1 [0.039]	1.6 [0.063]	1.6	1.6	2 [0.079]	2

Notes: 1. Tolerance on C and angular dimensions are controlled by this position tolerance, which is the permissible radial deviation of the hole centers, with respect to their theoretical positions.

2. Tolerance on un-toleranced dimensions: ± 0.4.

3. Dimension d_1 is taken at the theoretical point of intersection between the generating line of the cone and the face of the flange.

Dimensions of face plates for bayonet spindles. [**Fig. 5.9 (c)**]

Bayonet-type clamping [Fig. 5.9] also facilitates easy changeover of workholders. In this system, a bayonet disk is used. The disk is provided with holes, slightly bigger than the heads of the fixing screws. The clear holes are connected with slots, slightly (0.5) bigger than the threaded diameter. The bayonet disk acts like a special washer. Rotating the disk to align the bigger holes with screw centers, allows removal (or insertion) of the plate, without removing the entire screws (or nuts). Rotation of the disk in the opposite direction brings the smaller slots to the screw centers. Tightening the screws in this position, secures the workholder to the spindle nose. Thus, we only have to loosen screws/nuts by a quarter turn, to remove the bayonet disk and the workholder. Similarly, securing workholder requires the rotation of the bayonet disk through a small angle (15°), and turning the clamping screws/nuts through a small angle, to tighten them.

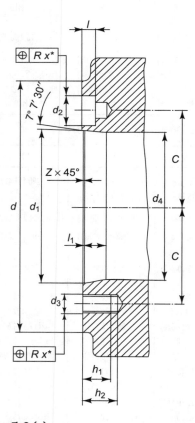

Fig. 5.9(c): Face plates for bayonet spindles

Machine Tool Elements

Drilling machines

Figure 5.10a shows the spindle of a drilling machine. Drills above 12 ϕ, generally have tapered shanks. As drilling machines usually have vertical spindles, the drills are subjected to a gravitational fall. This problem is overcome by using self-holding tapers called **Morse tapers**. They approximately have a 1.5° taper. It differs marginally according to the number (size) of the shank. During drilling, the drilling thrust presses the shank against the taper in the spindle. Although the small taper prevents the gravitational fall, it cannot prevent slip under the heavy drilling torque. This problem is overcome by providing a slot in the spindle. The slot engages with the flats provided on the shank of the drill. This helps prevent the slip due to the torque.

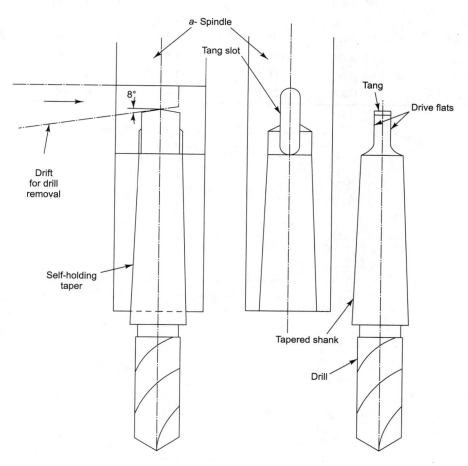

Fig. 5.10(a): Drilling machine spindle

Designation of taper	D	d_5 H11	d_6	g A13	h
Metric 4	4 [0.157]	3 [0.118]	—	2.2 [0.866]	8 [0.315]
Metric 6	6 [0.236]	4.6 [0.181]	—	3.2 [0.126]	12 [0.472]
Morse 0	9.045 [0.356]	6.7 [0.264]	—	3.9 [0.153]	15 [0.59]
Morse 1	12.065 [0.475]	9.7 [0.319]	7 [0.276]	5.2 [0.205]	19 [0.748]
Morse 2	17.78 [0.7]	14.9 [0.587]	11.5 [0.413]	6.3 [0.248]	22 [0.866]
Morse 3	23.825 [0.937]	20.2 [0.795]	14 [0.551]	7.9 [0.311]	27 [1.063]
Morse 4	31.267 [1.231]	26.5 [1.043]	18 [0.709]	11.9 [0.468]	32 [1.26]
Morse 5	44.399 [1.748]	38.2 [1.504]	23 [0.905]	15.9 [0.626]	38 [1.496]
Morse 6	63.348 [2.494]	54.6 [2.15]	27 [1.063]	19 [0.748]	47 [1.85]
Metric 80	80 [3.15]	71.5 [2.815]	33 [1.3]	26 [0.024]	52 [2.047]
Metric 100	100 [3.937]	90 [3.543]	39 [1.533]	32 [1.26]	60 [2.362]
Metric 120	120 [4.724]	108.5 [4.272]	39 [1.533]	38 [1.496]	70 [2.756]
Metric 160	160 [6.992]	145.5 [5.728]	52 [2.047]	50 [1.968]	90 [3.543]
Metric 200	200 [7.874]	182.5 [7.185]	52 [2.047]	62 [2.44]	110 [4.331]

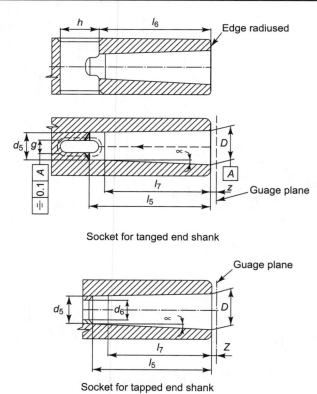

Fig. 5.10(b): Female sockets for self-holding tapers. Refer page 377 for other lengths & taper details

Machine Tool Elements

Refer [Fig. 510(b)] on page 376

Designation of taper	l_5 Min.	l_6	l_7 Approx.	Z^*	α	Taper on dia.
Metric 4	25 [0.984]	21 [0.827]	20 [0.787]	0.5 [0.020]	1°25′56″	1:20
Metric 6	34 [1.339]	29 [1.142]	28 [1.102]	0.5	1°25′56″	1:20
Morse 0	52 [2.047]	49 [1.929]	45 [1.772]	1 [0.039]	1°29′27″	1:19.212
Morse 1	56 [2.205]	52 [2.047]	47 [1.85]	1	1°25′43″	1:20.047
Morse 2	67 [2.638]	62 [2.441]	58 [2.283]	1	1°25′50″	1:20.02
Morse 3	84 [3.307]	78 [3.071]	72 [2.835]	1	1°26′16″	1:19.922
Morse 4	107 [4.213]	98 [3.858]	92 [3.622]	1.5 [0.059]	1°29′15″	1:19.254
Morse 5	135 [5.315]	125 [4.921]	118 [4.645]	1.5	1°30′26″	1:19.002
Morse 6	188 [7.402]	177 [6.968]	164 [6.457]	2 [0.079]	1°29′36″	1:19.18
Metric 80	202 [7.953]	186 [7.323]	170 [6.693]	2	1°25′56″	1:20
Metric 100	240 [9.45]	220 [8.661]	200 [7.874]	2	1°25′56″	1:20
Metric 120	276 [10.866]	254 [10.0]	230 [9.055]	2	1°25′56″	1:20
Metric 160	350 [13.779]	321 [12.638]	290 [11.417]	3 [0.118]	1°25′56″	1:20
Metric 200	424 [16.693]	388 [15.276]	350 [13.779]	3	1°25′56″	1:20

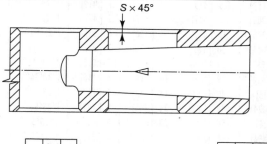

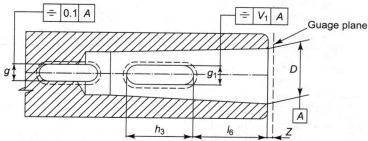

Fig. 5.10(c): Additional slot to prevent tool fall during tapping and back facing

Designation of taper	D	$gA13$	$g_1 A13$	h_3	l_6	S	Z^*	V_1
Morse 3	23.825 [0.938]	7.9 [0.311]	8.2 [0.323]	31 [1.22]	34 [1.339]	1.6	1 [0.039]	0.2 [0.008]
Morse 4	31.267 [1.231]	11.9 [0.468]	8.2	36 [1.417]	30 [1.181]	1.6	1.5	0.2
Morse 5	44.399 [1.748]	15.9 [0.626]	12.2 [0.48]	41 [1.614]	30	2 [.079]	1.5	0.2
Morse 6	63.348 [2.494]	19 [0.748]	16.2 [0.638]	46 [1.811]	30	2.5 [0.98]	2 [0.079]	0.2
Metric 80	80 [3.15]	26 [1.024]	19.5 [0.768]	44 [1.732]	30	3 [.118]	2	0.2
Metric 100	100 [3.937]	32 [1.26]	26.5 [1.039]	52 [2.047]	30	4 [.157]	2	0.3 [0.012]
Metric 120	120 [4.724]	38 [1.496]	32.6 [1.283]	60 [2.362]	30	4	2	0.3
Metric 160	160 [6.299]	50 [1.968]	44.8 [1.764]	76 [2.992]	40 [1.575]	4	3 [.118]	0.4 [0.016]
Metric 200	200 [7.874]	62 [2.441]	56.8 [2.236]	92 [3.622]	60 [2.362]	4	3	0.4

Note: * Z is the maximum permissible deviation of diameter D from the end face.

Designation of taper	D	D_1^*	d_2^*	d_3 Max.	l_3	l_4 Max.
Morse 0	9.045 [0.356]	9.2 [.362]	6.1 [.247]	6 [0.236]	56.5 [2.224]	59.5 [2.342]
Morse 1	12.065 [0.475]	12.2 [.48]	9 [0.354]	8.7 [.342]	62 [2.441]	65.5 [2.579]
Morse 2	17.78 [0.7]	18 [0.709]	14 [0.551]	13.5 [0.531]	75 [2.953]	80 [3.15]
Morse 3	23.825 [0.938]	24.1 [0.949]	19.1 [.752]	18.5 [.728]	94 [2.701]	99 [3.898]
Morse 4	31.267 [1.231]	31.6 [1.244]	25.2 [.992]	24.5 [.965]	117.5 [4.626]	124 [4.882]
Morse 5	44.399 [1.748]	44.7 [1.76]	36.5 [1.437]	35.7 [1.405]	149.5 [5.886]	156 [6.141]
Morse 6	63.348 [2.494]	63.8 [2.512]	52.4 [2.063]	51 [2.008]	210 [8.268]	218 [8.583]
Metric 80	80 [3.15]	80.4 [3.165]	69 [2.716]	67 [2.638]	220 [8.661]	228 [8.976]
Metric 100	100 [3.937]	100.5 [3.957]	87 [3.425]	85 [3.346]	260 [10.236]	270 [10.63]
Metric 120	120 [4.724]	120.6 [4.748]	105 [4.134]	102 [4.016]	300 [11.811]	312 [12.283]
Metric 160	160 [6.299]	160.8 [6.331]	141 [5.551]	138 [5.433]	380 [14.961]	396 [15.591]
Metric 200	200 [7.874]	201 [7.913]	177 [6.968]	174 [6.85]	460 [18.11]	480 [18.898]

Designation of taper	α	$b h_{13}$	e Max.	R Max.	R_1	α	Taper on dia	% Taper
Morse 0	3	3.9 [.153]	10.5 [.413]	4 [.157]	1 [.039]	1°29′27″	1:19.212	5.205
Morse 1	3.5	5.2 [.205]	13.5 [.531]	5 [.197]	1.2 [.47]	1°25′43″	1:20.047	4.988
Morse 2	5	6.3 [.248]	16 [.63]	6 [.236]	1.6 [.63]	1°25′50″	1:20.02	4.995
Morse 3	5	7.9 [3.11]	20 [.787]	7 [.276]	2 [.078]	1°26′16″	1:19.922	5.02
Morse 4	6.5	11.9 [.468]	24 [.945]	8 [.315]	2.5 [.098]	1°29′15″	1:19.254	5.194
Morse 5	6.5	15.9 [.626]	29 [1.102]	10 [.394]	3 [.118]	1°30′26″	1:19.022	5.263
Morse 6	8	19 [.748]	40 [1.575]	13 [.512]	4 [.157]	1°29′36″	1:19.18	5.214
Metric 80	8	26 [1.024]	48 [1.89]	24 [.945]	5 [.197]	1°25′56″	1:20	5
Metric 100	10	32 [1.26]	58 [2.283]	30 [1.181]	5	1°25′56″	1:20	5
Metric 120	12	38 [1.496]	68 [2.677]	36 [1.417]	6 [.236]	1°25′56′	1:20	5
Metric 160	16	50 [1.968]	88 [3.464]	48 [1.89]	8 [.315]	1°25′56″	1:20	5
Metric 200	20	62 [2.441]	108 [4.252]	60 [2.362]	10 [.394]	1°25′56″	1:20	5

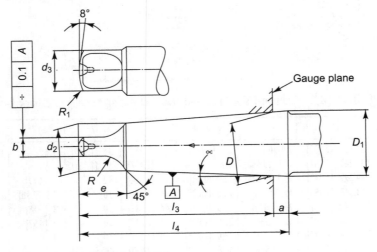

Fig. 5.10(d): Tapered shanks for drills

The drill shank is also provided an approximate 8° angle on the shank end-face. This end angle is used for removing the drill, jammed in the spindle. A wedge called the **drift** is used to extract the jammed drill from the spindle. **Figure 5.10b** shows female morse tapers for drilling m/c spindles. Note that the sizes of the diameter exceeding 64, use metric tapers which have a fixed 1° 25′ 56″ angle, corresponding to 1 : 20 taper.

Drilling machines are also used for tapping holes. Tapping, subjects the tap holder to a downward force, during tap withdrawal. This is prevented by providing additional slots in the spindle and the tool shank [**Fig. 5.10c**]. A wedge inserted into the drill shank and the spindle, prevents the fall of the drill under a downward force. This arrangement is also used for back-facing: spot-facing in an upward direction.

Figure 5.10d shows standard (male) tapered shanks used for drills. Sizes above 65 ϕ have metric tapers.

Self-holding tapers are also used for end mills, slot drills, etc., used on vertical milling machines. Their helix angle generates a downward force. This tends to pull the cutter downwards. A fall of the cutter during operation can damage the costly cutter. This is prevented by replacing the tang by a tapped hole [**Fig. 5.10e**]. Milling m/cs are provided with a drawbolt. Its threads engage with the tapped hole in the cutter shank, to prevent the extraction of the cutter during operation.

Morse tapers are also used in drill-holders, used on capstans and turret lathes.

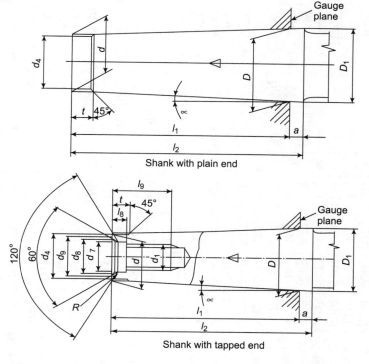

Fig. 5.10(e): Tapered shanks with tapped holes for milling cutters

Designation of taper	D	D_1^*	d^*	d_4 Max.	d_1	d_7	d_8 Max.	d_9 Max.	l_1 Max.	l_2 Max.	l_9 Min.	t Max.	l_8	a	R	α	Taper	% Taper on dia.
Metric 4	4 [0.157]	4.1 [0.161]	2.9 [0.114]	2.5 [0.098]	—	—	—	—	23 [0.905]	25 [0.984]	—	2 [0.079]	—	2	0.2	1°25′56″	1:20	5
Metric 6	6 [0.236]	6.2 [0.244]	4.4 [0.173]	4 [0.157]	—	—	—	—	32 [1.26]	35 [1.378]	—	3 [0.118]	—	3	0.2	1°25′56″	1:20	5
Morse 0	9.045 [0.356]	9.2 [0.362]	6.4 [0.252]	6 [0.236]	—	—	—	—	50 [1.968]	53 [2.087]	—	4 [0.157]	—	3	0.2	1°29′27″	1:19.212	5.205
Morse 1	12.065 [0.475]	12.2 [0.48]	9.4 [0.37]	9 [0.354]	M6	6.4 [0.252]	8 [0.315]	8.5 [0.335]	53.5 [2.106]	57 [2.244]	16 [0.63]	5 [0.197]	4 [0.157]	3.5	0.2	1°25′43″	1:20.047	4.988
Morse 2	17.78 [0.7]	18 [0.709]	14.6 [0.575]	14 [0.552]	M10	10.5 [0.413]	12.5 [0.492]	13.2 [0.52]	64 [2.52]	69 [2.716]	24 [0.945]	5 [0.197]	5 [0.197]	5	0.2	1°25′50″	1:20.02	4.995
Morse 3	23.825 [0.938]	24.1 [0.949]	19.8 [0.779]	19 [0.748]	M12	13 [0.512]	15 [0.59]	17 [0.669]	81 [3.189]	86 [3.386]	28 [1.102]	7 [0.276]	6 [0.236]	5	0.6	1°26′16″	1:19.922	5.02
Morse 4	31.267 [1.231]	31.6 [1.244]	25.9 [1.020]	25 [0.984]	M16	17 [0.669]	20 [0.787]	22 [0.866]	102.5 [4.035]	109 [4.291]	32 [1.26]	9 [0.354]	8 [0.315]	6.5	1	1°29′15″	1:19.254	5.194
Morse 5	44.399 [1.748]	44.7 [1.76]	37.6 [1.480]	35.7 [1.405]	—	21 [0.826]	26 [1.024]	30 [1.181]	129.5 [5.098]	136 [5.354]	40 [1.575]	10 [0.394]	11 [0.433]	6.5	2.5	1°30′26″	1:19.002	5.263
Morse 6	63.348 [2.494]	63.8 [2.512]	53.9 [2.122]	51 [2.008]	M20	25 [0.984]	31 [1.22]	36 [1.417]	182 [7.165]	190 [7.48]	50 [1.968]	16 [0.63]	12 [0.472]	8	4	1°29′36″	1:19.18	5.214
Metric 80	80 [3.15]	80.4 [3.165]	70.2 [2.764]	67 [2.638]	M24	31 [1.22]	38 [1.496]	45 [1.772]	196 [7.716]	204 [8.031]	65 [2.559]	24 [0.945]	14 [0.551]	8	5	1°25′56″	1:20	5
Metric 100	100 [3.937]	100.5 [3.957]	88.4 [3.48]	85 [3.346]	M30	37 [1.457]	45 [1.772]	52 [2.047]	232 [9.134]	242 [9.528]	80 [3.15]	30 [1.181]	16 [0.63]	10	5	1°25′56″	1:20	5
Metric 120	120 [4.724]	120.6 [4.748]	106.6 [4.197]	102 [4.016]	M36	37 [1.457]	45 [1.772]	52 [2.047]	268 [10.551]	280 [11.024]	80 [3.15]	36 [1.417]	16 [0.63]	12	6	1°25′56″	1:20	5
Metric 160	160 [6.299]	160.8 [6.331]	143 [5.63]	138 [5.433]	M36	50 [1.968]	60 [2.362]	68 [2.677]	340 [13.386]	356 [14.016]	100 [3.937]	48 [1.89]	20 [0.787]	16	8	1°25′56″	1:20	5
Metric 200	200 [7.874]	201 [7.913]	179.4 [7.063]	174 [6.85]	M48	50 [1.968]	60 [2.362]	68 [2.677]	412 [16.22]	432 [17.008]	100 [3.937]	60 [2.362]	20 [0.787]	20	10	1°25′56″	1:20	5

Dimensions of tapered shanks with tapped holes for milling cutters for [Fig. 5.10(e)] on p. 379

Milling machine spindles

Milling machine spindles are provided with **Quick release tapers [Fig. 5.11]**. They have a much bigger 8°17′50″ taper angle, corresponding to a 7/24 taper. The taper is used to locate (centralize) the arbor. For transmitting the torque, the machine spindle is provided with two facial slots. These are fitted with tenons, which engage with the slots provided in the arbor flange. They act as keys, to transmit the heavy milling torque.

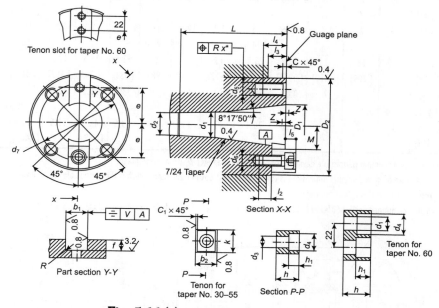

Fig. 5.11(a): Milling machine spindle noses

Self-release	Internal details					Current head insertion details					Tenon slot			
7/24 taper No.	D_1 basic	d_1 H12	d_2 Min.	L Min.	‡ Z	D_2 h5	l_3 Min.	d_7	x^*	d_5	l_4 Min.	C	b_1 M6	f Min.
30	31.75 [1.25]	17.4 [.685]	17 [.669]	73 [2.874]	0.4 [.016]	69.832 [2.75]	12.5 [.492]	54 [2.126]	0.075 [.003]	M10	16 [.629]	2 [.079]	15.9 [.626]	8 [.315]
40	44.45 [1.75]	25.3 [.996]	17	100 [3.937]	0.4	88.882 [3.5]	16 [.629]	66.7 [2.626]	0.075	M12	20 [.787]	2	15.9	8
45	57.15 [2.25]	32.4 [1.275]	21 [.827]	120 [4.724]	0.4	101.6 [4]	18 [.709]	80 [3.15]	0.075	M12	25 [.984]	2	19 [.748]	9.5 [.374]
50	69.85 [2.75]	39.6 [1.559]	27 [1.063]	140 [5.512]	0.4	128.57 [5.062]	19 [.748]	101.6 [4.0]	0.1 [.004]	M16	25	3 [.118]	25.4 [1.0]	12.5 [.492]
55	88.9 [3.5]	50.4 [1.984]	27	178 [7.708]	0.4	152.4 [6.0]	25 [.984]	120.6 [4.75]	0.1	M20	30 [1.181]	3	25.4	12.5
60	107.95 [4.25]	60.2 [2.37]	35 [1.378]	220 [8.661]	0.4	221.44 [8.718]	38 [1.496]	177.8 [7.0]	0.1	M20	30	3	25.4	12.5

Self-release 7/24 taper No.	Tenon slot					Tenon									HEX. SOC. HD. SCR
	e^{**} ±0.2	d_6	l_2	R	V	b_2 h_5	d_3	d_4	h	h_1	k Max.	l_5 Max.	M Min.	C_1	
30	24.75 [.974]	M6	9 [0.354]	1.6 [.063]	0.06 [.0024]	15.9 [.626]	6.4 [.252]	10.4 [.409]	16 [.63]	7 [.275]	16.5 [.65]	8 [.315]	16.5 [.65]	1.6 [.063]	M6 × 16
40	32.75 [1.289]	M6	9	1.6	0.06	15.9	6.4	10.4	16	7	19.5 [.768]	8	23 [.905]	1.6	M6 × 16
45	39.75 [1.565]	M6	9	1.6	0.06	19 [.748]	6.4	10.4	19 [.748]	7	19.5	9.5 [.374]	30 [1.181]	1.6	M6 × 20
50	49.25 [1.939]	M12	18 [.709]	2 [0.079]	0.08 [.0031]	25.4 [1.0]	13 [.512]	19 [.748]	25 [.984]	13.5 [.531]	26.5 [1.043]	12.5 [.492]	36 [1.417]	2	M12 × 25
55	61.25 [2.411]	M12	18	2	0.08	25.4	13	19	25	13.5	26.5	12.5	48 [1.89]	2 [.079]	M12 × 25
60	72.75 [2.864]	M12	18	2	0.08	25.4	13	19	25	13.5	45.5 [1.791]	12.5	61 [2.401]	2	M12 × 25

Notes:
1. ‡ Z is the maximum permissible deviation of position of diameter D_1 from guage plane.
2. *Tolerance of d_7 and angular dimensions is controlled by this position tolerance which is the permissible radial deviation of hole centers with respect to their theoretical positions.
3. **Not mentioned in IS 6681-1972. Calculated from the values of M and k. [Refer **Fig 5.11(a)**]

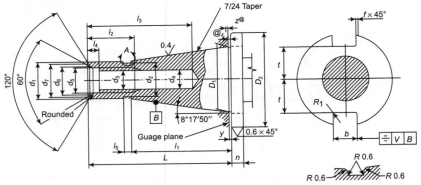

Fig. 5.11(b): Milling machine arbors

The helix angle of some milling cutters, generates an axial force which tends to extract the milling arbor from the spindle. This is prevented by providing a tapped hole in the rear-end of the arbor. The arbor is secured to the spindle by a draw-bolt which engages with the female threads at the rear-end of the arbor.

Grinding machines spindles

These too are provided with a 1:5 taper, on the wheel seating portion. **Figure 5.12** shows standard grinding m/c spindles. Care should be taken to ensure that rotation of the spindle does not loosen the securing nut/screw. This often calls for the usage of left-hand threads. Pedestal grinders with grinding wheels at both ends, require the threads of the opposite hand at the ends. Looking from the threaded end, if the direction of the rotation is anti-clockwise,

Machine Tool Elements

Self release 7/24 taper No.	External taper				Collar and driving slot					Z @	Internal Details														
	D_1 basic	d_1	Tolerance on d_1, a_0		d_2	L Max	l_1	l_5	y		D_2	n	b H/12	t	v Max	R_1	l	d_3	d_4	d_5	d_6	d_7	l_2	l_3 Min	l_4
30	31.75 [1.25]	17.4 [.685]	−0.29 [.011] −0.36 [.014]		16.5 [.65]	70 [2.756]	50 [1.968]	3 [.112]	1.6 [.063]	0.4 [.016]	50 [1.968]	8 [.315]	16.1 [.634]	16.2 [.638]	0.12 [.005]		1.6	1.6 [.063]	M12	10.25 [.403]	12.5 [.492]	15.5 [.61]	16 [.63]	24 [.945]	50 [1.968]	6 [.236]
40	44.45 [1.75]	25.3 [.996]	−0.3 [.012] −0.38 [.015]		24 [.945]	95 [3.74]	67 [2.638]	5 [.197]	1.6	0.4	63 [2.48]	10 [.394]	16.1	22.5 [.886]	0.12		1.6	1.6	M16	14 [.551]	17 [.669]	19.5 [.768]	23 [.905]	30 [1.181]	70 [2.756]	8 [.315]
45	57.15 [2.25]	32.4 [1.275]	−0.31 [.0122] −0.41 [.016]		30 [1.181]	110 [4.33]	86 [3.386]	6 [.236]	3.2 [.126]	0.4	80 [3.15]	10 [.394]	19.3 [.76]	29 [1.142]	0.12		1.6	1.6	M20	17.5 [.689]	21 [.827]	24	30 [1.181]	38 [1.496]	70	10 [.394]
50	69.85 [2.75]	39.6 [1.559]	−0.31 [.0122] −0.41 [.016]		38 [1.496]	130 [5.118]	105 [4.134]	8 [.315]	3.2	0.4	100 [3.937]	12 [.472]	25.7 [1.012]	35.3 [1.39]	0.2 [.008]		2	2 [.079]	M24	21 [.827]	25 [.984]	28	36 [1.417]	45 [1.772]	90 [3.543]	11 [.433]
55	88.9 [3.5]	50.8 [2.0]	−0.34 [.0134] −0.46 [.018]		48 [1.89]	168 [6.614]	130 [5.118]	9 [.354]	3.2	0.4	130 [5.118]	14 [.55]	25.7	45 [1.772]	0.2		2	2	M24	21	25 [.984]	28 [1.102]	36	45 [1.772]	90	11
60	107.95 [4.25]	60.2 [2.37]	−0.34 [.0134] −0.46 [.018]		58 [2.283]	210 [8.268]	165 [6.496]	10 [.394]	3.2	0.4	160 [6.299]	16 [.63]	25.7	60 [2.362]	0.2		2	2	M30	26.5 [1.043]	31 [1.22]	35 [1.378]	45 [1.772]	56 [2.205]	110 [4.33]	14 [.551]

Dimensions of Milling Machine Arbors with Quick Release 7 : 24 Taper [Fig. 5.11b].

the thread should be right-handed. If the spindle rotates clockwise, the threads should be left-handed. Smaller spindles, up to 63 ϕ, can have male or female threads.

Nominal diameter of grinding spindle $d \pm 0.1$ [.004″]	l	Type A			Type B
		d_1	d_2	l_2	d_3 **
32 [1.26″]	40 [1.575″] 50 [1.968″]	M16	13 [.512″]	1.5 [.059″]	M12 × 1.5
40 [1.575″]	40 50	M24	19.8 [.779″]	1.5	M16 × 1.5
50 [1.968″]	50 63 [2.48″]	M24	19.8	3 [.118″]	M16 × 1.5
63 [2.48″]	63 80 [3.15″]	M36 × 3	31.8	3	M20 × 1.5
80 [3.15″]	80 100 [3.973″]	M36 × 3	31.8 [1.256″]	4 [.157″]	—
100 [3.937″]	100 125 [4.921″]	M48 × 3	43.8 [1.724″]	4	—

Notes:
1. * Dimensions l_1 and l_3 are according to the design of clamping.
2. ** As a protection against the self-releasing of grinding wheel flanges, the threads should be either left-hand or right-hand, depending upon the direction of rotation. In such cases where the direction of rotation is changing, protection against self-releasing may be done by keying, or similar means, according to the designer's choice. All dimensions in mm.

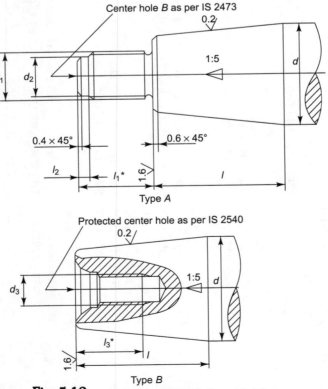

Fig. 5.12: Grinding machine spindles with 1:5 taper

5.1.3.3 Spindle supports: Bearing

Spindles are supported by bearings. These can be broadly classified as:
1. Sliding friction sleeve bearings
2. Rolling friction (Anti-friction) bearings.

The following table compares sliding and rolling friction bearings:

Table 5.4: Comparison of sliding and rolling friction bearings

1	Criterion	Sliding friction sleeve bearing	Rolling friction ball/roller/needle bearing
1a	Maximum rotary speed	Up to 100,000 R.P.M.	Depends upon the size and type of bearing, lubrication (grease/oil) method, and type of load (radial, axial, combined). More details in the next section.
b	Linear speed	Up to 6 meters/min. only [19.7 ft/min] for simple sleeve/bearing.	
2	Space required	More axial space Less Radial Space	More radial space Less axial space
3	Efficiency	Less than 95%	Up to 99.5 %
4	Temperature rise	High	Low
5	Cost	High due to costly materials	Low due to mass production
6	Load capacity	High	Limited
7	Impact and vibration-resistance	Good	Low
8	Serviceability	Replacement costly and time consuming	Easy replacement
9	Noise	Less	More

Clearance Ratio $[\psi]$

$= \dfrac{\text{Diametrical clearance}}{\text{Diameter [nominal]}}$

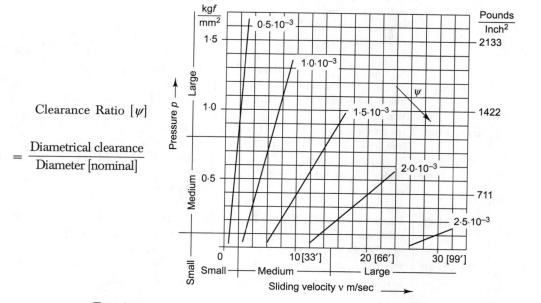

Fig. 5.13: Clearance ratio for various BRG pressures and sliding speeds

Sliding friction bearings can be further divided into:

1) **Simple sleeve bearings** with partial metal to metal contact, due to inadequate lubrication, i.e. boundary lubrication:
2) **Hydro-dynamic bearings** have a full fluid film between the shaft and the bearing, and no metal to metal contact. When the m/c is started, initially even the hydro-dynamic bearing operates under boundary lubrication. This changes slowly, to full film.

Table 5.5: Materials for sliding friction bearings

S. No.	Material	Applications	Max. mean specific pressure kgf/mm²	Modulus of elasticity E kgf/mm²	Allowed contact Stress f_c kgf/mm²	Hardness HB
1	Aluminium bronze	Resistant to wear, used in heavy-duty bearing, requiring high strength and good impact-resistance.	4 [5690 psi]	11,250 [15.93 × 10⁶ psi]	6–10 [8530 – 14220]	202
2	Phosphor bronze	Excellent corrosion-resistance, with moderate strength. Very good bearing material when working under reasonable to good lubrication conditions. Used for plain bearing bushes, running against steel shafts where lubrication is adequate and shaft misalignment is minimum.	6.1 [8680 psi]	11,200 [15.86 × 10⁶]	1.8–3 [2560 – 4270]	60
3	Cast bronze		8 [11380 psi]	12,000 [17.07 × 10⁶]	3–6 [4270 – 8530]	100–120
4	Tin bronze	Very good running properties. Used for heavily loaded bearings.	10 [14220 psi]	10,000 [14.22 × 10⁶]	1.2–5.5 [1700 – 7820]	40–110
5	Leaded bronze	Less sensitive to impact loads, less wear.	7.5 [10670 psi]	8000 [11.38 × 10⁶]	0.6–3 [853 – 4270]	19–60
6	White metals	Good anti-seizure properties. Sensitive to impact loads, used for overlays for lightly loaded bearings.	3 [4270 psi]	3100–6300 [4.41 × 10⁶ – 8.96 × 10⁶]	0.2–1.5 [280 – 2130]	6–30
7	Low tin lead alloy	Better resistance to impact loads compared to white metals, higher wear-resistance.	2 [2840 psi]	1600–3100 [2.276 × 10⁶ – 4.41 × 10⁶]	0.5–1.8 [710 – 2560]	18–36
8	Aluminium alloys	Higher wear-resistance. Sensitive to edge loading.	4 [5690 psi]	6900 – 7500 [9.81 × 10⁶ – 10.67 × 10⁶]	1.2–9.0 [1700 – 12800]	40–180

Note: Hardness of journal should be at least 100 Brinel more than the bearing hardness.

Machine Tool Elements

Simple sleeve bearings can do with intermittent lubrication.

Contact stress $f_c = \sqrt{\dfrac{\psi WE}{2.88\, l\, d}}$ (Eqn. 5.14)

ψ = Clearance ratio (From [**Fig. 5.13**])
W = Radial load on bearing (kg) = 0.4536 × Lbs
E = Elasticity modulus of bearing material–kg/mm^2
(From Table 5.5) = 7.0307 × 10^{-4} × Lbs/Inch^2
l = Length of bearing (mm) = 25.4 × inches
d = Diameter of journal (spindle portion, mating with bearing) = 25.4 × Inches

Specific pressure $(p) = \dfrac{W}{l\,d} \leq \dfrac{1}{10}$ the value in Table 5.5 (Eqn. 5.15)

Table 5.5 gives the recommended specific pressures for various bearing materials.

$\dfrac{l}{d} = 0.8 - - 1.2$ (2 Max in exceptional cases)

Journal bending stress $(f_b) = \dfrac{5Wl}{d^3}$ (Eqn. 5.16)

f_b = Bending stress for spindle material (Table 5.1). = 7.0307 × 10^{-4} × Lbs × Inch^2

Example 5.11: Design a sleeve bearing for a 40 ϕ shaft, subjected to 1,500 kg radial load. R.P.M. = 500.

Solution: Assuming $l = d$

Linear sliding speed = $\dfrac{\pi d N}{60{,}000} = \dfrac{\pi \times 40 \times 500}{60{,}000}$

= 1.02 m/sec

Specific pressure $(p) = \dfrac{1500}{40 \times 40} = 0.938$ kg/mm^2

From [**Fig. 5.13**], for the above values, ψ is less than 0.0005.
From Eqn. 5.13,

$f_c = \sqrt{\dfrac{\psi WE}{2.88\, l\, d}} = \sqrt{\dfrac{0.0005 \times 1500 \times E}{2.88 \times 40 \times 40}} = \sqrt{\dfrac{E}{6144}}$.

As specific pressure $[p]$ should be less than or equal to $\dfrac{1}{10}$ the value stated in Table 5.5
Max Mean Specific Pressure $\geq 0.938 \times 10$ i.e. ≥ 9.38 kg/mm^2

Referring the table only Tin Bronze can meet the requirement. For this bearing metal E = 10,000 kg/mm^2

$$f_c = \sqrt{\frac{10,000}{6144}} = 1.276 \text{ kg/mm}^2 \text{ i.e.} < 5.5 \text{ kg/mm}^2$$

Hydro-dynamic Bearings or **fluid film bearings** provide for high precision and a long life span due to their inherent good damping and heat dissipation properties, and the absence of any metal to metal contact. The journal (spindle) acts like a pump, delivering oil into the clearance between the shaft and the bearing. However, the narrowing clearance in the direction of rotation, causes pressure rise till the hydro-dynamic force counter-balances the load on the bearing [**Fig. 5.14**].

The load capacity of the hydro-dynamic bearing depends upon:

1. $\frac{1}{d}$ ratio (preferably 0.4–1.0)
2. Surface finish which determines the minimum film thickness (h_o), necessary to prevent rupture of the film due to surface unevenness. The minimum film thickness should be 20 to 50% higher than the critical film thickness (h_c).

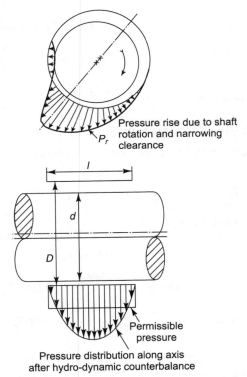

Fig. 5.14: Equilibrium between Brg. load and hydro-dynamic pressure

$$\frac{h_o}{h_c} = 1.2 - 1.5 \quad \text{(Eqn. 5.17)}$$

$$h_c = R_s + R_b + \delta \quad \text{(Eqn. 5.18)}$$

R_s = Spindle (journal) roughness [Peak to valley]

$= 4\, R_{as} = 4$ Roughness of spindle [Average: mean]

$R_b = 4 R_{ab} = 4$ Roughness of bearing bore [Average]

Roughness can be found from Table 5.6.

$\delta =$ Shaft deflection at bearing [**Fig. 5.16**]

$$= \frac{W(L-A)l}{12\,E\,I\,L}(-0.25\,L^2 - A^2 + 2AL) \qquad \text{(Eqn. 5.19)}$$

Table 5.6: Roughness (R_a) for various surface finish grades (*N*)

Finish grade	Roughness R_a-microns	Machining processes
N2	0.05 [2 Microinches]	Lapping, cloth-buffing
N3	0.10 [3.9]	Honing, super-finishing
N4	0.20 [7.9]	Lapping, honing, super-finishing
N5	0.4 [15.7]	Lapping, honing, super-grinding
N6	0.8 [31.5]	Die casting, rolling, broaching reaming, shaving, tumbling.
N7	1.6 [63]	Moulding, turning, boring, shaping, planing, milling, filing
N8	3.2 [126]	Milling, hacksaw cutting forging
N9	6.3 [248]	Sand casting, drilling

Grades rougher than N9, are not used for bearings and have therefore, been excluded from the table. The journal should have a better finish than the bearing bore

Lubrication	Journal finish [μ]	Bearing finish μ
Boundary [partly metal to metal contact]	0.2–0.8 [7.9 – 31.5]	0.8–1.6 [31.5 – 63]
Full film [No metal to metal contact]	0.15–0.3 [5.9 – 11.8]	0.3–0.5 [11.8 – 19.7]

Lubrication

Lubricants can be supplied through a circumferential groove [**Fig. 5.15a**] which divides the bearing into two shorter bearings. There is more eccentricity between the journal and the bearing. It is widely used in reciprocating loads, like connecting rod bearing.

For surface speeds above 50 m/sec, cylindrical overshot bearings [**Fig. 5.15b**] are used. In these, there are two axial lubricant grooves connected by a wide groove in the upper half of the circumference. This help eliminate oil shearing action in the upper half and provides for cooler operation.

Pressure bearing [**Fig. 5.15c**] has an oil groove in only half (45°) of the top bearing. Suitable for low-load, high-speed operations, these bearings are vulnerable to the presence of dirt in the oil.

The oil flow can be increased by providing more grooves in the multiple-groove bearings [**Fig. 5.15d**]. Elliptical bearings [**Fig. 5.15e**] too run cooler, due to an increase in the oil flow. Three-lobe bearings [**Fig. 5.15f**] centralize the spindle through hydro-dynamic pressure, developed in oil wedges at 3 places.

Hydro-static bearings have many oil pockets **[Fig. 5.15g]**. Lubricating oil pumped in at considerable pressure, changes metal-to-metal friction into fluid friction, even at low speeds. In air bearings lubricating oil is replaced by aerostatic pressure. Although friction is very low, rigidity is also very low, as viscosity of air is only $\frac{1}{2000}$ of the oil viscosity.

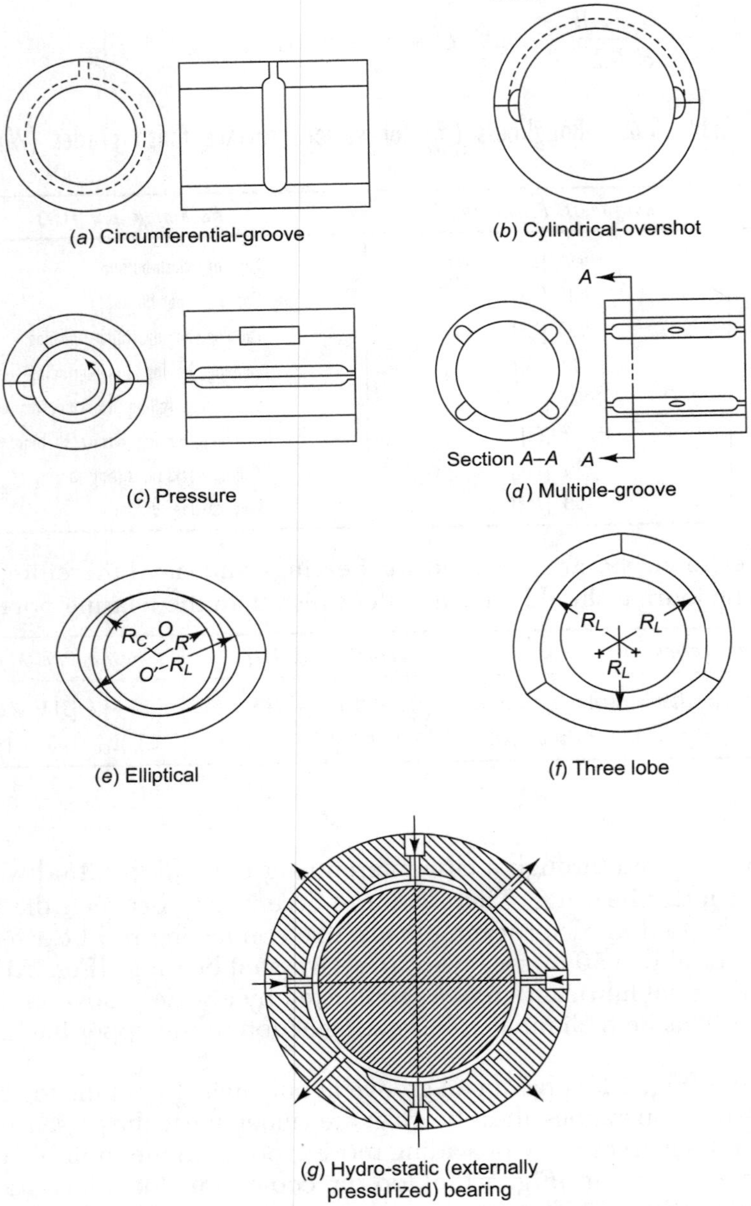
(a) Circumferential-groove
(b) Cylindrical-overshot
A
A
Section A–A
(c) Pressure
(d) Multiple-groove
R_C
O
R
O′
R_L
(e) Elliptical
R_L
R_L
R_L
(f) Three lobe
(g) Hydro-static (externally pressurized) bearing

Fig. 5.15: Lubricating sleeve bearings. (Reprinted with permission from Machinery's Handbook, 27th edition.)

Clearance adjustment

Bearings can be split in 2, 3, or more parts. Some parts remain stationary while others are provided with a radial adjustment to vary the clearance. Ease in assembly and disassembly of spindle, makes these bearings convenient for supporting spindles of heavy machine tools.

Load capacity (W)

$$\text{Load capacity } (W) = \frac{1000\, \eta\, Vl}{\psi^2}\, \phi \qquad \textbf{(Eqn. 5.20)}$$

ψ = Clearance ratio (From [**Fig. 5.13**])

η = Lubricant dynamic viscosity (From [**Fig. 5.17**])

Table 5.7: Sommerfeld No. [ϕ] for given $\frac{l}{d}$ and X [Eccentricity ratio from Eqn 5.23]

l/d	\multicolumn{14}{c}{X}
	0.33	0.4	0.5	0.6	0.65	0.7	0.75	0.8	0.83	0.9	0.925	0.95	0.975	0.99
	\multicolumn{14}{c}{Value of ϕ}
1.5	1.37	1.76	2.47	3.5	4.16	5.1	6.42	8.32	10.3	17.9	23.8	37.3	76.7	196.4
1.3	1.2	1.55	2.2	3.17	3.8	4.68	5.96	7.79	9.68	17.2	22.9	36.2	75.2	193.9
1.2	1.11	1.43	2.05	2.97	3.58	4.44	5.68	7.47	9.3	16.66	22.34	35.4	74.1	192
1.1	1	1.3	1.89	2.75	3.33	4.15	5.34	7.08	8.86	16	21.6	34.4	72.7	190
1	0.896	1.17	1.7	2.51	3.05	3.83	4.96	6.61	8.35	15.31	20.7	33.4	71.1	187
0.9	0.781	1.03	1.51	2.24	2.75	3.47	4.54	6.1	7.75	14.43	19.74	32	68.95	183.4
0.8	0.662	0.874	1.3	1.95	2.4	3.06	4.03	5.48	7.02	13.35	18.4	30.2	66.1	178.7
0.7	0.54	0.716	1.07	1.64	2.03	2.61	3.47	4.78	6.18	12	16.8	27.85	66.2	171.8
0.6	0.421	0.562	0.85	1.31	1.64	2.13	2.87	3.98	5.21	10.42	14.75	24.95	57.3	162.5
0.5	0.31	0.416	0.636	0.99	1.25	1.64	2.22	3.15	4.16	8.52	12.33	21.28	50.6	149
0.4	0.208	0.281	0.433	0.705	0.867	1.14	1.57	2.26	3.02	6.43	9.55	16.72	41.65	129.4

$V = \dfrac{\pi\, dN}{60{,}000}$, where d = spindle [journal] ϕ. N = R.P.M.

l = Length of bearing (mm)

ϕ = Sommerfeld No.

$$\phi = \frac{19.11\, p\, \psi^2}{\eta\, N} \qquad \textbf{(Eqn. 5.21)}$$

$p = \dfrac{W}{ld}$ = specific pressure (kg/mm^2)

It should be remembered that temperature rise decreases the dynamic viscosity. This also decreases the load capacity. The capacity can be increased by using lubricants with a higher viscosity, or by decreasing the clearance ratio.

$$\psi = \frac{\text{Brg. Bore} - \text{Journal } \phi}{\text{Journal } \phi} = \frac{D-d}{d}$$

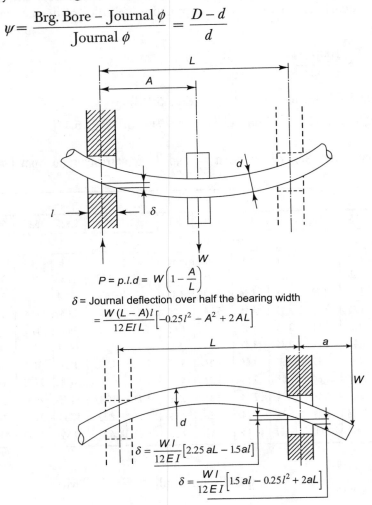

Fig. 5.16: Deflections at Brg. edge

As both, the bearing bore as well as the journal dia have manufacturing tolerances, there will be two values of ψ.

$$\psi_{max} = \frac{D_{max} - d_{min}}{d_{nominal}} \qquad \text{(Eqn. 5.22a)}$$

$$\psi_{min} = \frac{D_{min} - d_{max}}{d_{nom}} \qquad \text{(Eqn. 5.22b)}$$

$$\text{The eccentricity ratio } (X) = 1 - \frac{2h_o}{d\psi} \quad \text{(Eqn. 5.23)}$$

Table 5.7 gives the values of Sommerfeld No. (ϕ) for various values of $\frac{1}{d}$ and X.

Example 5.12: Find the load capacity of a $\phi\, 35 \times 20$ lg bearing, if R.P.M. = 2200 Oil dynamic viscosity (η) = 7×10^{-9} kgf. sec/mm^2

Brg. Bore is finished to $R_a = 0.4$ microns, while the journal diameter finish is $R_a = 0.2$ microns

$\psi = 0.001$; Deflection at Brg edge (δ) = 0.005

Solution:

$$V = \frac{\pi\, dn}{60{,}000} = \frac{\pi \times 35 \times 2200}{60{,}000} = 4.03 \text{ m/sec.}$$

From Eqn. 5.18,

$$h_c = R_b + R_s + \delta$$
$$= 4(0.0004 + 0.0002) + 0.005$$
$$= 0.0024 + 0.005 = 0.0074$$

From Eqn. 5.17, $h_o = 1.2 - 1.5\,[h_c] = 1.2 - 1.5\,(0.0074)$
$$= 0.0089 - 0.00111$$

From Eqn. 5.23, $X = 1 - \dfrac{2\,h_o}{d\,\psi}$

Taking the mean value, 0.0095 of h_o,

$$X = 1 - \frac{2 \times 0.0095}{35 \times 0.001} = 0.457$$

$$\frac{1}{d} = \frac{20}{35} = 0.57 \approx 0.6$$

From Table 5.7 for $\dfrac{1}{d} = 0.6$ and $X = 0.46$,

$$\phi = 0.706 \text{ (by interpolation)}$$

From Eqn. 5.20, Load $(W) = \dfrac{1000\,\eta v l}{\psi^2}\,\phi$

$$= \frac{1000 \times 7 \times 10^{-9} \times 4.03 \times 20}{0.001^2} \times 0.706$$

$$= 398.3 \text{ kg}$$

Temperature rise

Under stable conditions, the heat generated by friction between the lubricant oil layers in the clearance, is equal to the heat withdrawn from the bearings.

$$H = \frac{WV\mu}{427} = H_1 + H_2$$

$$= \text{Sp. heat of oil} \times \text{Density of oil} \times \text{Oil flow} \times (t_m - t_i) + \frac{\alpha \pi \, dl(t_m - t_i)}{10^6}$$

t_m = Mean Brg temp (°C); t_i = oil inlet temp (°C)
α = Heat transfer coefficient (kcal/m² − sec. °C)
μ = Coefficient of friction

For most of the lubricating oils,

Sp. heat × density = 405 kcal/m³ °C, and,

$$(t_m - t_i) = 5.78 \frac{p \cdot \frac{\mu}{\psi}}{\frac{Q \times 10^6}{\psi V l d} + \frac{B}{\psi V}}, \qquad \textbf{(Eqn. 5.24)}$$

where Q = Oil flow thru Brg (m³/sec)

$$B = \frac{\alpha \pi}{405} \text{ (from Table 5.7)}$$

Table 5.8: Values of α and B for lubricants

Brg Type	α kcal/m² sec°C	$B = \alpha\pi/405$
Light Brg with high ambient temp.	0.013 [0.0008112 BTU/Ft^3 sec °F]	1 × 10⁻⁴
Std. Brg	0.018 [0.00112]	1.4 × 10⁻⁴
Heavy Brg with intensive heat dissipation (water/oil cooling)	0.033 [0.00206]	2.5 × 10⁻⁴

$$\frac{\mu}{\psi} = 0.150 + 1.92 \, (1.119 - X) \left[\left(1 + 2.31 \frac{d^2}{l^2}\right) - (1.052 + X) \right] \qquad \textbf{(Eqn. 5.25)}$$

$$\frac{Q \times 10^6}{\psi V l d} = \left(0.285(0.2035 + X) \frac{0.072 \frac{l^2}{d^2} - 1.05 + X}{0.433 \frac{l^2}{d^2} + 1.05 - X} \right) \quad \text{(Eqn. 5.26)}$$

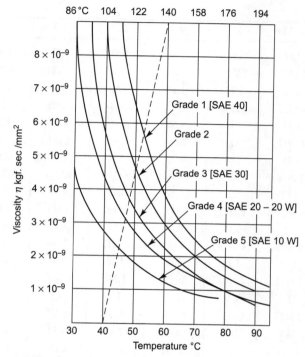

Fig. 5.17: Lub. oil viscosity at various temperatures

If the bearing housing is made of non-metallic material, the low conductivity of the material cannot dissipate heat effectively. This results in a higher ambient temperature.

$$(t_m - t_i) = 5.78 \left(\frac{p \cdot \frac{\mu}{\psi}}{Q \, 10^6 / \psi V l d} \right) \quad \text{(Eqn. 5.27)}$$

Table 5.9 gives the values of $\frac{\mu}{\psi}$ and $\frac{Q}{\psi V l d}$, for various values of $\frac{1}{d}$ and the eccentricity ratio X.

Example 5.13: Find the temperature rise and oil flow for a hydro-dynamic bearing for ϕ 25 journal, with $R_a = 0.2$ and 20 long bearing finished to $R_a = 0.2$. The spindle is running at 1000 R.P.M. and the lubricant has 5×10^{-9} kg.sec/mm² viscosity. The deflection at the bearing edge = 0.0006. Bearing load = 200 kg.

Table 5.9: $\dfrac{\mu}{\psi}$ and $\dfrac{Q \times 10^6}{\psi V l/d}$, for given $\dfrac{l}{d}$ and X

$\dfrac{l}{d}$	Ratio	X							
		0.33	0.5	0.7	0.8	0.9	0.95	0.975	0.99
1.5	μ/ψ	2.87	1.92	1.24	0.97	0.67	0.48	0.35	0.22
	$Ql/\psi V l d$	0.0697	0.0896	0.1038	0.1009	0.0888	0.077	0.067	0.576
1.3	μ/ψ	3.26	2.13	1.33	1.02	0.7	0.49	0.35	0.22
	$Ql/\psi V l d$	0.0775	0.1006	0.1171	0.1134	0.0986	0.093	0.0718	0.0605
1	μ/ψ	4.34	2.71	1.58	1.16	0.76	0.52	0.36	0.23
	$Ql/\psi V l d$	0.0917	0.1215	0.144	0.1415	0.122	0.1017	0.0843	0.0686
0.9	μ/ψ	4.96	3.04	1.72	1.24	0.79	0.54	0.37	0.23
	$Ql/\psi V l d$	0.0968	0.1295	0.1563	0.1537	0.1334	0.1108	0.0909	0.0729
0.8	μ/ψ	4.83	3.5	1.92	1.35	0.84	0.56	0.38	0.24
	$Ql/\psi V l d$	0.1023	0.1382	0.1695	0.1685	0.147	0.122	0.0997	0.0784
0.7	μ/ψ	7.1	4.18	2.21	1.52	0.91	0.6	0.4	0.24
	$Ql/\psi V l d$	0.1078	0.1474	0.184	0.1852	0.164	0.1367	0.1115	0.0864
0.6	μ/ψ	9.09	5.22	2.66	1.78	1.02	0.65	0.42	0.26
	$Ql/\psi V l d$	0.1132	0.1566	0.1995	0.204	0.1842	0.1552	0.1267	0.0974
0.5	μ/ψ	12.3	6.94	3.39	2.2	1.2	0.73	0.46	0.27
	$Ql/\psi V l d$	0.1183	0.1656	0.2152	0.224	0.208	0.1785	0.1473	0.1133
0.4	μ/ψ	18.3	10.1	4.47	2.98	1.54	0.89	0.53	0.13
	$Ql/\psi V l d$	0.123	0.174	0.232	0.245	0.235	0.207	0.175	0.136

Note: 'Q' in the table is in $m^3 \times 10^6$ i.e. in cubic cms [e.c]. [1 cc = .061 Inch^3 = 2.642 × 10^{-4} U.S. gallon]

Solution:

$$V = \frac{\pi \times 25 \times 1000}{60,000} = 1.31 \text{ m/sec}$$

$$h_c = 4\,(0.0002 + 0.0002) + 0.0006$$
$$= 0.0022$$

$$h_o = 0.2 - 1.5\,h_c$$

Taking the higher value,

$$h_o = 1.5 \times 0.0022$$
$$= 0.0033$$

$$\frac{l}{d} = \frac{20}{25} = 0.8;\quad p = \frac{200}{25 \times 20} = 0.4 \text{ kg/mm}^2$$

From **[Fig. 5.13]**, for $p = 0.4$ and $V = 1.31$ m/s

$$\psi \approx 0.5 \times 10^{-3} = 0.0005,$$

and, from Eqn. 5.23

$$X = 1 - \frac{2 h_o}{d \psi} = 1 - \frac{2 \times 0.0033}{25 \times 0.0005} = 0.528 \approx 0.53$$

From Table 5.9, for $\frac{1}{d} = 0.8$ and $X = 0.53$,

$$\frac{\mu}{\psi} = 3.5, \text{ and } \frac{Q}{\psi V \, 1d} = 0.1382$$

From Table 5.8, for a standard medium duty bearing $B = 1.4 \times 10^{-4}$,
From Eqn. 5.24,

$$t_m - t_1 = 5.78 \frac{P \cdot \frac{\mu}{\psi}}{\frac{Q}{\psi V \, ld} + \frac{B}{\psi V}} = 5.78 \frac{0.4 \times 3.5}{0.1382 + \frac{1.4 \times 10^{-4}}{5 \times 10^{-4} \times 1.31}}$$

$$= 5.78 \frac{1.4}{0.1382 + 0.2137} = 23°C$$

Oil flow required can be found as follows:

$$\frac{Q \times 10^6}{\psi V \, 1d} = \frac{Q \times 10^6}{5 \times 10^{-4} \times 1.31 \times 25 \times 20} = 0.1382$$

$$\therefore \quad Q = 0.04526 \text{ cm}^3/\text{sec}$$

For a non-metallic housing, from Eqn. 5.27,

$$t_m - t_1 = 5.78 \frac{P \cdot \frac{\mu}{\psi}}{\frac{Q \times 10^6}{\psi V 1d}} = 5.78 \frac{1.4}{0.1382}$$

$$= 58.5°C$$

Sliding friction thrust bearings

These take the axial thrust along the face. Parallel or flat plate thrust bearings **[Fig. 5.18a]** can be used for occasional or very light loads (0.05 – 0.1 kg/mm²). Provision of a step **[Fig. 5.18b]** increases the load capacity two to three times. However, it is not suitable for bigger sizes, as it is vulnerable to misalignment. Tapered land thrust **[Fig. 5.18c]** bearing, though costlier, is more suitable for bigger sizes and heavier loads, and must be properly aligned. Kingsbury thrust bearing [**Fig. 5.18d**], which is even costlier, uses a tilting pad. It can withstand a considerable degree of misalignment. Stepped, tapered land, and Kingsbury thrust bearings are normally used with 0.14 kg/mm² bearing pressure. The maximum pressure should not exceed 0.34 kg/mm².

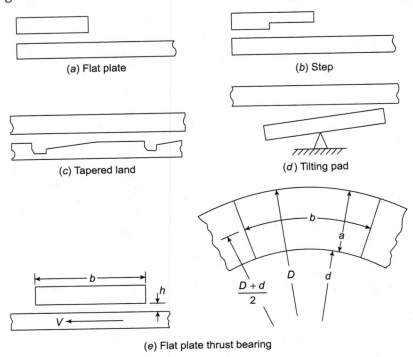

Fig. 5.18: Sliding friction thrust bearings

(Reprinted with permission from Machinery's Handbook, 27th edition)

The inside diameter (d) should be bigger than the spindle size, to provide an adequate radial clearance.

Brg. outside dia $(D) = \sqrt{\dfrac{5W}{\pi p} + d^2}$ (Eqn. 5.28)

p = Pressure (kg/mm²); W = Load (kg)

The velocity (V) is taken as the velocity at mean (pitch) diameter of annular area.

Velocity $(V) = \dfrac{\pi N}{60,000}\left(\dfrac{D+d}{2}\right)$ (Eqn. 5.29)

N = R.P.M.

Bearings are provided with a groove for lubricating oil. If the groove width is 'S', the number of pads (i) can be found as follows:

Machine Tool Elements

$$i = \frac{\pi(D+d)}{(D-d)+2S} \quad \text{(Eqn. 5.30)}$$

The calculated value of i should be changed to the nearest even number. Then the pad width (b) can also be found.

$$b = \frac{\pi(D+d) - 2is}{2i} \quad \text{(Eqn. 5.31)}$$

and Bearing pressure $(p) = \frac{2W}{ib(D-d)} \quad \text{(Eqn. 5.32)}$

The approximate frictional power loss can be found from the following equation:

$$P_f = \frac{ib(D-d)M}{1730} \quad \text{(Eqn. 5.33)}$$

M = Constant for bearing pressure (p) and pitch line velocity (m/sec) from [Fig. 5.19]

The necessary oil flow (Q) depends upon the permissible temperature rise $(t_m - t_i)$.

$$Q = \frac{7.57 \, P_f}{J(t_m - t)} \quad \text{(Eqn. 5.34)}$$

Q = Oil flow required (Lits/sec)
P_f = Frictional power (kW) from Eqn. 5.33
J = Lubricant specific heat kilo joules/kg/°C [1 Joule = 0.239 Calories]

Example 5.14: Design a flat thrust bearing for 400 kg axial load, at 4200 RPM, with lubricant having specific heat of 14.6 kilo Joules/kg/°C. The shaft has a ϕ 70 diameter. Assume oil groove width = 5 and temperature rise = 20°C.

Solution:

Let us take the bearing inside ϕ, i.e. $d = 74$;

From Eqn. 5.28, outside dia $(D) = \sqrt{\dfrac{5w}{\pi p} + d^2} \quad \text{(Eqn. 5.35)}$

Assuming minimum pressure of 0.05 kg/mm^2,

$$D = \sqrt{\frac{5 \times 400}{\pi \times 0.05} + 74^2} = \sqrt{18206.74}$$

$$= 134.9 \approx 135$$

From Eqn. 5.29, Pitch line velocity $(V) = \dfrac{\pi N}{60,000}\left(\dfrac{D+d}{2}\right) \quad \text{(Eqn. 5.36)}$

$$= \frac{\pi \times 4200}{60,000}\left(\frac{74+135}{2}\right) = 23 \text{ m/s}$$

From Eqn. 5.30, No. of Pads $(i) = \dfrac{\pi(74+135)}{(135-74)+(2\times 5)} = 9.25$ **(Eqn. 5.37)**

Rounding up to the nearest even number, $i = 10$

From [**Fig. 5.19**], for 0.05 kg/mm² pressure and 23 m/sec velocity, $M = 0.26$ (approx.)

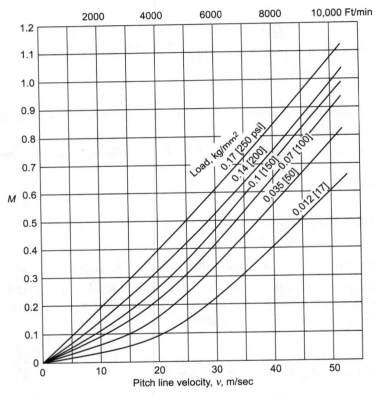

Fig. 5.19: Friction power constant M for various pressures and velocities (Reprinted with permission from Machinery's Handbook, 27th edition)

From Eqn. 5.31,
$$b = \frac{\pi(D+d) - 2is}{2i}$$
$$= \frac{\pi(135+74) - 2(10\times 5)}{2\times 10} = 27.83$$

From Eqn. 5.32,
$$p = \frac{2W}{ib(D-d)} = \frac{2\times 400}{10\times 27.83(135-74)}$$
$$= 0.047$$

From Eqn. 5.33,
$$P_f = \frac{i\,b\,(D-d)\,M}{1730}$$
$$= \frac{10 \times 27.83\,(135-74)\,0.21}{1730}$$
$$= 2.06 \text{ kW}$$

From Eqn. 5.34,
$$Q = \frac{7.57\,P_f}{J\,(t_m - t_i)} \quad \textbf{(Eqn. 5.38)}$$

For 20°C temperature rise,
$$Q = \frac{7.57 \times 2}{14.6 \times 20} = 0.052 \text{ l/s}$$

Note: Lubricating oil pump capacity is generally specified in lits/min.

∴ $Q = 0.052 \times 60 = 3.11 \approx 3$ lits /min.

Anti-friction (Rolling contact) bearings

These use hardened steel balls, rollers, or needles as the rolling elements. **Figure 5.20** shows the commonly used types. **Deep groove ball bearings [Fig. 5.20a]** are mainly used for radial load. But they can also take on axial load up to 25% of the radial load. **Spherical self-aligning ball bearings [Fig. 5.20b]** are convenient where there is likelihood of misalignment of up to 2°–3°. Roller bearings **[Fig. 5.20c]** can take much more radial load than ball bearings, but very little axial load. For heavy radial as well as axial loads, **tapered roller bearings [Fig. 5.20d]** are used. For limited space and heavy radial load, at low angular speed (RPM/w), **needle roller bearings [Fig. 5.20e]** are used. In these, the roller length is more than 4 times the diameter. Needle roller bearings are very vulnerable to axial loads. For such applications, a bearing which combines radial and thrust bearings **[Fig. 5.20f]**, is used.

Angular contact ball bearing [Fig. 5.20g] can take on much radial and axial load, in one direction. For axial loads in two opposite directions, it is necessary to use two angular contact bearings and mount them face to face or back to back **[Fig. 5.20h]**. Angular contact ball bearings can, however, take on lesser load than tapered roller bearings.

Load capacity can, be increased by increasing the ball/roller size which also increases the overall size. Most of the bearings are available in light, medium, and heavy (load) series. Bearings' capacity can also be increased by using double-row ball/roller bearings **[Fig. 5.20i]**.

For pure axial load, thrust bearings are used **[Fig. 5.20j]**. **Figure 5.20k** shows the needle bearing for flat surfaces of slides and guideways

As the rolling elements have a hardened steel ring on the outside (outer race) and the inside (inner race), they don't come in contact with the shaft or housing. So there is no wear of the shaft or housing. Hence, the name anti-friction bearings. After wear of balls/rollers/race, the entire bearing is replaced by a new bearing.

The choice of bearing depends upon the intensity and the type of load, space available, maximum rotary speed (RPM), and cost. It is often convenient to use two tandem bearings: one for the radial load, and the other for the axial thrust. **Deep groove ball bearings** are the cheapest and therefore, should be considered first.

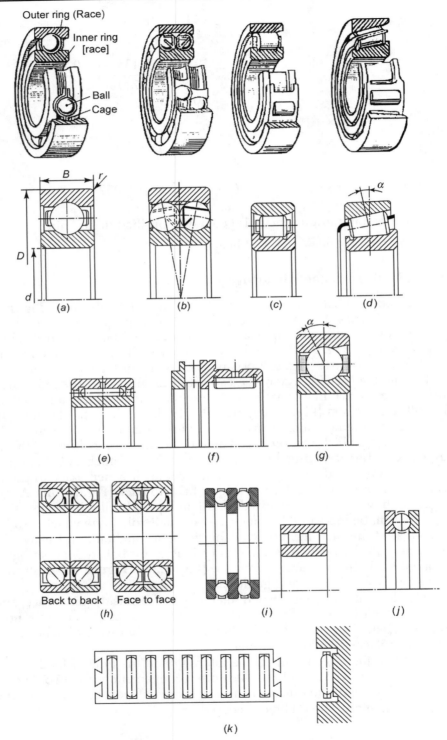

Fig. 5.20: Various types of ball, roller, and needle bearings

Load and life

The rated dynamic load (C), specified in the bearings tables, is the constant stationary load (C) that 90% of the bearings can withstand, for one million (10^6) revolutions. However, it is more convenient to specify the bearing life in hours (L_h).

$$L_h = \frac{10^6}{60N}\left[\frac{C}{P}\right]^\varepsilon \qquad \text{(Eqn. 5.39)}$$

$$= 20,000 - 30,000 \text{ Hrs. for single shift working}$$

$$= 50,000 - 60,000 \text{ Hrs. for three shift working}$$

Equivalent dynamic load (P) also take into account the axial load.

$$P = X W_r + Y W_a \qquad \text{(Eqn. 5.40)}$$

Values of X and Y depend upon the type of bearing (Table 5.12). Needle bearings and roller bearings without a flange (shoulder) on the inner ring (races), cannot take on any axial load. They can only take on radial load. Roller bearings with shoulders on both the rings, can take on a limited axial load. As there is a sliding contact between the shoulder and roller face, the axial thrust capacity depends upon the type of lubrication, and the inside (raceway) diameter (D_1) of the outer ring (race).

For grease lubrication

$$W_a = f_a f_b D_i^2 \left(2 - \frac{ND_i}{10^5}\right) \qquad \text{(Eqn. 5.41)}$$

In oil lubrication W_a depends upon the product ND_i.

$$W_a = f_a f_b D i^2 \left(2 - \frac{ND_i}{10^5}\right), \text{ when } ND_i \leq 1,20,000 \qquad \text{(Eqn. 5.42a)}$$

$$W_a = f_a f_b D i^2 \left(1 - \frac{ND_i}{6 \times 10^5}\right), \text{ when } ND_i \geq 1,20,000 \qquad \text{(Eqn. 5.42b)}$$

W_a = Maximum axial load (kg)

N = R.P.M.; D_i = I.D. of outer race (mm) = $\frac{D+d}{2} - d_b$ [or d_r]

d_b = Ball ϕ = 0.3 [$D - d$]; d_r = Roller ϕ = 0.25 [$D - d$]

f_a, f_b = Factors dependent on the types of load and bearing (from Tables 5.10 and 5.11)

Table 5.10: 'f_a' for various types of axial loads

Axial load factor	Constant continuous $W_a \leq 0.4\ W_r$	Fluctuating intermittent	Sudden shock
f_a	0.6	0.4	0.2

Table 5.11: f_b for various types of roller bearings

	Roller Brg Type Series Nos.	f_b
1	Roller anchored in outer race. Shoulder on one side of inner race.	
	Series NJ/RJ – 2 –22	0.24
	NJ/RJ – 3 – 23	0.30
	NJ/RJ – 4	0.33
2	Roller anchored in outer race. Inner race split in 2 pieces to provide shoulders on both sides of rollers.	
	Series NUP/RT – 2 – 22	0.24
	– 3 – 23	0.30
	4	0.33

For satisfactory functioning of roller bearings under axial load,
$W_a \leq 0.13 \, W_r$ for 02 and 03 series
$W_a \leq 0.2 \, W_r$ for 22 and 23 series
Under no circumstances axial load should exceed 40% of the radial load, i.e.

$$W_{a\,max} = 0.4 \, W_r$$

In thrust ball bearings, Equivalent load $(P) = W_a$

Spherical roller thrust bearings can take radial load up to 55% of the axial load, i.e.
$W_{r\,max} = 0.55 \, W_a$

$$\text{Equivalent load } (P) = W_a + 1.2 \, W_r$$

Varying loads at varying speeds are common in general purpose machine tools, using workpieces with wide variations in size and material hardness. Past experience shows that the possible equivalent loads ($P_1, P_2, P_3 \cdots$), rotary speeds ($N_1, N_2, N_3 \cdots$), and duration of each load in percentage ($q_1, q_2, q_3, q_4 \cdots$) of the total time, and the life expectancy for the various loads ($L_1, L_2, L_3 \cdots$) can be computed from the following equations:

$$L_h = \frac{10^6}{60N}\left(\frac{C}{P}\right)^3 \quad \text{for ball bearings} \quad \textbf{(Eqn. 5.43a)}$$

$$L_h = \frac{10^6}{60N}\left(\frac{C}{P}\right)^{3.333} \quad \text{for roller bearings} \quad \textbf{(Eqn. 5.43b)}$$

Then estimated life $(L_h) = \dfrac{q_1}{L_1} + \dfrac{q_2}{L_2} \cdots \dfrac{q_n}{L_n}$

Static load capacity

In bearings used for oscillatory motions or very low rotary speed ($N < 10$ RPM), the load can be considered static. Static load capacity, stated in the bearing tables, is the uniformly distributed load (C_o) which produces contact stress (f_c) of 400 kg/mm² [0.0001[0.000004″] over all deformation]

Machine Tool Elements

Equivalent static load $(P_o) = X_o W_r + Y_o W_a$ (Eqn. 5.44a)

The value of X_0 and Y_0 depends upon the type of bearing. The required static rating (C_o) of the bearings depends upon the type of the load.

$$C_o = S_o P_o$$ (Eqn. 5.44b)

| Load S_o | Steady. No vibrations 0.5 | Normal, quiet running 1 | Shock loads 2 | Spherical roller thrust bearing 2 |

Table 5.12: Constants X, Y, X_o, and Y_o for various types of bearings

No.	Brg. type	Static load		Dynamic load			
		X_o	Y_o	$\dfrac{W_a}{C_o}$	e	For $\dfrac{W_a}{W_r} \geq e^*$	
						X	Y
1	Deep groove ball brg.	0.6	0.5	0.025	0.22	0.56	2
				0.04	0.24		1.8
				0.07	0.24		1.6
				0.13	0.31		1.4
				0.25	0.37		1.2
				0.5	0.44		1
2	Angular contact Ball brg series 70C 72C 70CG 72CG	0.5	0.46	0.015	0.38	0.44	1.47
				0.029	0.40		1.4
				0.058	0.43		1.3
				0.087	0.46		1.23
				0.12	0.47		1.19
				0.17	0.5		1.12
				0.29	0.55		1.02
				0.44	0.56		1.0
	Series 72 B 72 BG	0.5	0.26	—	1.14	0.35	0.57
3	Tapered roller brg	0.5		—		0.4	
	32206–8		0.9		0.37		1.6
	32209		0.8		0.4	0.4	1.5
	32210		0.8		0.43		1.4
	32211–13				0.4		1.5
	32214–28				0.43		1.4

* When $\dfrac{W_a}{W_r} < e$, $X = 1.0$ and $Y = 0$

Notes: 1. For cylindrical roller and needle brgs, $P = P_o = W_r$
2. For thrust ball brgs, $P = P_o = W_a$

Example 5.16: Select bearing for $\phi 25$ shaft for single shift (8 Hrs) running at 700 RPM, if radial load = 300 kg and axial thrust = 100 kg.

Solution:
Assuming that dynamic load (P) is equal to the radial load, from Eqns. 5.39, and 5.43a

$$L_h = 25000 = \frac{10^6}{60\,N}\left(\frac{C}{P}\right)^3 = \frac{10^6}{60 \times 700}\left(\frac{C}{300}\right)^3$$

$$\frac{C}{300} = \sqrt[3]{\frac{25000 \times 60 \times 700}{10^6}} = 10.16$$

$$C = 3049.18$$

Refer to the bearings tables for Heavy-duty deep groove ball brg [p 430] 6405 ($\phi 25 \times \phi 80 \times 21$) has static capacity ($C_o$) of 2000 kg and dynamic capacity (C) of 2825 kg.

$$\frac{W_a}{C_o} = \frac{100}{2000} = 0.05$$

From Table 5.12, for $\frac{W_a}{C_o} = 0.05$, $e = 0.24$

Now, $\quad \frac{W_a}{W_r} = \frac{100}{300} = 0.33$, i.e. more than 0.24

So, $\quad X = 0.56$, and $Y = 1.7$ (approx.)

Equivalent dynamic load (P) = $XW_r + YW_a$
$$= (0.56 \times 300) + (1.7 \times 100)$$
$$= 338 \text{ kg}$$

Brg life (L_h) = $\frac{10^6}{60 \times 700}\left(\frac{2825}{338}\right)^3 = 13901.3$ Hrs

This is only half of the desired life. If we use 2 bearings,

$$\frac{W_a}{C_o} = \frac{100}{2 \times 2000} = 0.025, \text{ and } e = 0.22$$

As $\quad \frac{W_a}{W_r} = 0.33$, i.e. > 0.22

$X = 0.56$ and $Y = 2$
$P = XW_r + YW_a = (0.56 \times 300) + (2 \times 100)$
$\quad = 368$

$$L_h = \frac{10^6}{60 \times 700} \left(\frac{2 \times 2825}{368}\right)^3 = 86169.3, \text{ i.e. much higher than 25000 Hrs.}$$

If we use 2 medium-duty deep groove ball bearings, [p 429] i.e. No. 6305 ($\phi 25 \times \phi 62 \times 17$) $C = 1660$ and $C_o = 1040$

$$\frac{W_a}{C_o} = \frac{100}{2 \times 1400} = 0.036, \text{ and from Table 5.12,}$$

$$e < 0.24, \text{ i.e. } < 0.33, \text{ i.e. } \frac{W_a}{W_r}$$

So, $\quad X = 0.56$ and $Y = 1.85$ (approx.)

$$P = (0.56 \times 300) + (1.85 \times 100) = 353 \text{ kg}$$

$$L_h = \frac{10^6}{42000} \left(\frac{2 \times 1660}{353}\right)^3 = 19808 \text{ Hrs.} \approx 20,000$$

Therefore, 2 Brgs of No. 6305 can be used. Let us check the static equivalent load. From Table 5.12

$$P_o = X_o W_r + Y_o W_a = (0.6 \times 300) + (0.5 \times 100) = 230$$

From Eqn. 5.44b, Reqd. $C_o = S_o P_o$. For shock loads, $S_o = 2$

Reqd. $C_o = 2 \times 230 = 460$

Reqd. C_o is much lesser than the static rating (C_o) of 1040 kg, of even a single 6305 bearing. We can also use 2 angular contact bearings No. 7305 B [p 435] ($\phi 25 \times \phi 62 \times 17$). With $C = 1930$ and $\quad C_o = 1250$,

$$\frac{W_a}{C_o} = \frac{100}{2 \times 1250} = 0.04, \text{ and } (e) = 0.41,$$

as $\quad \frac{W_a}{W_r} = 0.33$ is less than 0.44 (e)

$$X = 1.0 \text{ and } Y = 0; \text{ so } P = W_r = 300$$

$$L_h = \frac{10^6}{42,000} \left(\frac{2 \times 1930}{300}\right) = 50716 \text{ Hrs., i.e. very high.}$$

If we use light-duty angular contact bearings, i.e. No. 7205 ($\phi 25 \times \phi 52 \times 15$) $C = 1160$, and $C_o = 780$.

$$\frac{W_a}{C_o} = \frac{100}{2 \times 780} = 0.064, \text{ and } (e) = 0.44$$

As $\quad \frac{W_a}{W_r} = 0.33 < 0.44$

$X = 1.0$, $Y = 0$, and $P = W_r = 300$

$$L_h = \frac{10^6}{42,000}\left(\frac{2 \times 1160}{300}\right)^3 = 11012 \text{ Hrs., i.e.} < 20,000$$

Let us consider roller bearings.

Roller $\phi [d_r] = 0.25 [D - d]$

For N J 205 bearing, overall sizes are $\phi 25 \times \phi 52 \times 15$.

So, $d_r = 0.25 [52 - 25] = 6.75 \approx 7$

Outer ring raceway Dia $[D_i] = \dfrac{25 + 52}{2} + 7 = 45.5$

From Tables 5.10 and 5.11, $f_a = 0.2$ and $f_b = 0.24$

From Eqn. 5.41, for grease lubrication,

$$W_a = f_a f_b D_i^2 \left[2 - \left(\frac{ND_i}{10^5}\right)\right]$$

$$= 0.2 \times 0.24 \times 45.5^2 \left[2 - \left(\frac{700 \times 45.5}{10^5}\right)\right]$$

$$= 99.37 \left[2 - \frac{31850}{10^5}\right] = 167.1 \text{ kg} > 100$$

As the ND_i value 31850 is less than 1,20,000, the value of W_a will be same for oil lubrication. [Eqn. 5.42a]. Therefore, a single roller bearing can be used.

Elastic deformation (Deflection)

Bearing deformation affects the spindle rigidity and overall rigidity of the machine tool. The deformation depends upon the size of the rolling element which in turn depends upon the bearing size. In radial ball bearings,

Ball Dia $(d_b) = 0.3 (D - d)$

Roller Dia $(d_r) = 0.25 (D - d)$

D = Brg outside dia (mm), d = Brg inside dia (mm).

The calculated size must be rounded up to the nearest standard ball/roller size.

Standard ball sizes: 0.3, 0.4, 0.5, 0.6, 0.7, 0.8, 1.0, 1.2, 1.5, 2 to 9, with difference of 0.5, 9, 10, 11, 11.5, 12 to 26, with difference of 1.0, 28, 30, 32, 34, 35, 36, 38, 40, 45, 50, 55, 60, 65.

Roller or needle size should be rounded to the nearest integer. The length of the roller is 1–1.5 times the diameter. In tapered rollers the diameter is taken as the mean diameter at mid-length. The length of the tapered roller is 1.6 times the mean diameter. The cone angle at the outer race is usually 16°.

Distortion of bearing also depends upon the contact angle (α) which depends upon the radial clearance (Δ_r) in microns (0.001 mm). The radial clearance depends upon the bearing bore (d), the grade (group) of the bearing, and the type of the bearing. Normal grade is 'CO' grade. Grade (C2) has lesser radial clearance, while grade (C1) has the least. Grade (C3) has more clearance than the normal (CO), while grade C4) has even more radial clearance.

Table 5.16: Formulae for calculating radial [δ_r] and axial [δ_a] displacements

Type of bearing	Loading conditions	
	$W_a \leq 1.25\ W_r \tan \alpha$	$W_r = 0$
Self-aligning ball bearings	$\delta_r = \dfrac{3.2}{\cos \alpha} \sqrt[3]{\dfrac{Q^2}{d_b/d_r}}$	—
Deep groove ball bearings	$\delta_r = 2 \sqrt[3]{\dfrac{Q^2}{d_b/d_r}}$	—
Angular contact ball bearings	$\delta_r = \dfrac{2}{\cos \alpha} \sqrt[3]{\dfrac{Q^2}{d_b/d_r}}$	$\delta_a = \dfrac{2}{\sin \alpha} \sqrt[3]{\dfrac{Q^2}{d_b/d_r}}$
Roller bearings with line contact of both raceways	$\delta_r = \dfrac{0.6}{\cos \alpha} \dfrac{Q^{0.9}}{l_a^{0.8}}$	$\delta_a = \dfrac{0.6}{\sin \alpha} \dfrac{Q^{0.9}}{l_a^{0.8}}$
Roller bearing with line contact for one raceway and point contact for the other	$\delta_r = \dfrac{1.2}{\cos \alpha} \dfrac{Q^{0.75}}{l_a^{0.5}}$	$\delta_a = \dfrac{1.2}{\sin \alpha} \dfrac{Q^{0.75}}{l_a^{0.5}}$
Thrust ball bearings	—	$\delta_a = \dfrac{2.4}{\sin \alpha} \sqrt[3]{\dfrac{Q^2}{d_b/d_r}}$

Tables 5.13 and 5.14 [p 411 – 12] give radial clearances for ball, and cylindrical roller bearings of various grades. **Figure 5.21** shows a graph for finding the contact angle (α) from the radial clearance, and the rolling element (ball, roller) diameter. Table 5.16 gives the formulae for computing bearing distortion from contact angle (α), roller/ball dia (d_r/d_b), and the load per rolling element (Q). The number of rolling elements (Z) in the bearing depends upon the mean $\phi \left(\dfrac{D+d}{2} \right)$, the element size ($d_b/d_r$), and the space taken by the cage, positioning (separating) the rolling elements. Radial load/element = Q_r

$$Q_r = \frac{5 W_r}{iZ \cos \alpha} \quad \text{(Eqn. 5.45a)}$$

Axial load/element $(Q_a) = \dfrac{W_a}{Z \sin \alpha}$ \quad **(Eqn 5.45b)**

i = No. of rows of rolling elements

W_r, W_a = Radial and axial load (kg)

$$Z = \frac{1.57 (D + d)}{1.25 \, d_b \, (\text{or } d_r)}$$

d_b = Ball ϕ; d_r = roller ϕ; D = Brg. outside;

d = Brg. inside

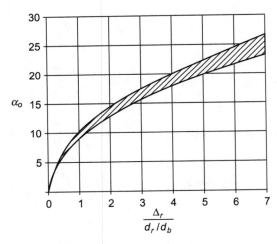

Fig. 5.21: Radial clearance and contact angle

Example 5.17: Find the elastic deformation for 1 kg load for $\phi\, 40 \times \phi\, 68 \times 15$ deep groove ball bearing of normal (CO) grade. Assume the cage clearance between the balls to be 25% of the ball size.

Solution:

From Eqn. 5.44a, $d_b = 0.3\,(68 - 40) = 8.4$

So, \hspace{2em} $d_b = 8$ (nearest standard ball size).

$$Z = \frac{1.57\,(68 + 40)}{8 \times 1.25} = 16.96 \approx 17$$

From Table 5.13, [p 411] for $d = 40$ and grade $(C\,0)$, maximum $\Delta_r = 20$

Table 5.13: Radial internal clearances in deep groove ball bearings [microns]

Bearing bore diameter d mm		Group 2 ($C2$)				Group 0 ($C0$) (Normal group)				Group 3 ($C3$)				Group 4 ($C4$)			
		Manufacturing limits		Acceptance limits		Manufacturing limits		Acceptance limits		Manufacturing limits		Acceptance limits		Manufacturing limits		Acceptance limits	
Over	Upto	Min.	Max.	Min.	Max.	Min.	Max.	Min.	Max.	Min.	Max.	Min.	Max.	Min.	Max.	Min.	Max.
10	18	0	8	—	9	5	15	3	18	13	23	11	25	20	30	18	33
18	24	0	9	—	10	7	17	5	20	15	25	13	28	23	33	20	36
24	30	0	10	—	11	8	18	5	20	15	25	13	28	25	38	23	41
30	40	0	10	—	11	8	18	6	20	18	30	15	33	30	43	28	46
40	50	0	10	—	11	8	20	6	23	20	33	18	36	33	48	30	51
50	65	3	13	—	15	10	25	8	28	25	41	25	43	41	58	38	61
65	80	3	13	—	15	13	28	10	30	28	48	25	51	48	69	46	71
80	100	3	15	—	18	15	33	12	36	33	56	30	58	56	81	53	84
100	120	3	18	—	20	18	38	15	41	38	63	36	66	63	94	61	97
120	140	3	20	—	23	20	46	18	48	46	76	41	81	76	109	71	114
140	160	3	20	—	23	20	51	18	53	51	86	46	91	86	124	81	130
160	180	3	23	—	25	23	58	20	61	58	97	53	102	97	140	91	147
180	200	3	28	—	30	28	69	25	71	69	112	63	117	112	157	107	163
200	225	4	32	—	—	32	82	—	—	82	132	—	—	132	187	—	—
225	250	4	36	—	—	36	92	—	—	92	152	—	—	152	217	—	—

Table 5.14: Radial internal clearances in cyl. roller bearings

Deviations in microns

Bearing bore diameter d mm		Group 1 (C1) Limits		Group 2 (C2)				Group 0 (C0)(Normal group)				Group 3 (C3)				Group 4 (C4)			
				Inter-changeable		Matched		Inter-changeable		Matched		Inter-changeable		Matched		Inter-changeable		Matched	
Over	Up to	Min.	Max.	Min.	Max.	Min.	Max.	Min.	Max.	Min.	Max.	Min.	Max.	Min.	Max.	Min.	Max.	Min.	Max.
10	18	—	—	0	30	10	20	10	40	20	30	25	55	35	45	35	65	45	55
18	24	5	15	0	30	10	20	10	40	20	30	25	55	35	45	35	65	45	55
24	30	5	15	0	30	10	25	10	45	25	35	30	65	40	50	40	70	50	60
30	40	5	15	0	35	12	25	15	50	25	40	35	70	45	55	45	80	55	70
40	50	5	18	5	40	15	30	20	55	30	45	40	75	50	65	55	90	65	80
50	65	5	20	5	45	15	35	20	65	35	50	45	90	55	75	65	105	75	90
65	80	10	25	5	55	20	40	25	75	40	60	55	105	70	90	75	125	90	110
80	100	10	30	10	60	25	45	30	80	45	70	65	115	80	105	90	140	105	125
100	120	10	30	10	65	25	50	35	90	50	80	80	135	95	120	105	160	120	145
120	140	10	35	10	75	30	60	40	105	60	90	90	155	105	135	115	180	135	160
140	160	10	35	15	80	35	65	50	115	65	100	100	165	115	150	130	195	150	180
160	180	10	40	20	85	35	75	60	125	75	110	110	175	125	165	150	215	165	200
180	200	15	45	25	95	40	80	65	135	80	120	125	195	140	180	165	235	180	220
200	225	15	50	30	105	45	90	75	150	90	135	140	215	155	200	180	255	200	240
225	250	15	50	40	115	50	100	90	165	100	150	155	230	170	215	205	280	215	265

1 micron = 39.4 micro-inches

1 micro-inch = 1×10^{-6} Inch = 0.000001"

Machine Tool Elements

$$\frac{\Delta_r}{d_b} = \frac{20}{8} = 2.5$$

From [Fig. 5.21], $\alpha = 15°$

From Eqn. 5.45a, $$Q = \frac{5 \times 1}{1 \times 17 \times \cos 15°} = 0.304 \text{ kg}$$

From Table 5.16, $$\delta_r = 2 \sqrt[3]{\frac{0.304^2}{8}} = 0.45 \text{ microns}$$

$$\delta_r = 0.00045 \text{ mm/kg radial load}$$

Example 5.18: Find the deflection/kg for roller bearings (a) $\phi 45 \times \phi 100 \times 36$, and (b) $\phi 40 \times \phi 68 \times 15$ for $C2$ grade. Roller brgs will have line contact with one race, and point contact with the other.

Solution:

From page 408,

(a)
$$d_r = 0.25 (D - d)$$
$$d_r = 0.25 (100 - 45) = 13.75 \approx 14$$
$$\text{Length} = 1.5 \times 14 = 21$$

From page 410 No. of rollers
$$(Z) = \frac{1.57 (d + D)}{1.25 \, d_r}$$

$$= \frac{1.57 (45 + 100)}{1.25 \times 14}$$

$$= 13$$

From Table 5.14, for $d = 45$ and Grade $C2$, $\Delta_r = 40$

$$\frac{\Delta_r}{d_r} = \frac{40}{14} = 2.86 \approx 3$$

From [Fig. 5.21], $\alpha = 16 - 17.5 \approx 16.5°$

From Eqn. 5.45a, $$Q_r = \frac{5 \, W_r}{i \, z \, \cos \alpha} = \frac{5 \times 1}{1 \times 13 \times \cos 16.5}$$

$$= 0.40$$

From Table 5.16, [p 409] for line contact with one race, and point contact with the other,

$$\delta_r = \frac{1.2}{\cos \alpha} \times \frac{Q^{0.75}}{1_a^{0.5}} = \frac{1.2}{\cos 16.5°} \times \frac{0.4^{0.75}}{21^{0.5}}$$

$$= 0.137 \text{ microns}$$
$$= 0.000137 \text{ mm}$$
$$d_r = 0.25 \, (68 - 40) = 7$$
$$l_a = 7 \times 1.5 = 10.5$$

(b)

$$Z = \frac{1.57 \, (68 + 40)}{1.25 \times 7}$$
$$= 19.37$$
$$\approx 19$$

From Table 5.14, for $d = 40$ and Grade $C2$, $\Delta_r = 35$

$$\frac{\Delta_r}{d_r} = \frac{35}{7} = 5$$

From [**Fig. 5.21**], $\quad \alpha = 21°$

From Eqn. 5.45a,
$$Q_r = \frac{5 \, W_r}{i \, z \, \cos\alpha} = \frac{5 \times 1}{1 \times 19 \times \cos 21°}$$
$$= 0.282 \text{ kg}$$

From Table 5.21,
$$\delta_r = \frac{1.2}{\cos 21°} \cdot \frac{0.28^{0.75}}{10.5^{0.5}}$$
$$= 0.154 \text{ microns}$$
$$= 0.000154 \text{ mm}$$

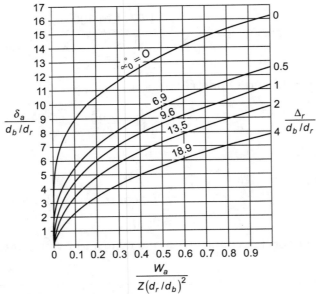

Fig. 5.22: Axial displacement [δ_a], radial clearance [Δ_r]

Figure 5.22 shows the relationship between $\dfrac{\delta_a}{d_b/d_r}$, $\dfrac{W_a}{z\{d_b/d_r\}^2}$, and contact angle α. It is sometimes necessary to also calculate the axial displacement (δ_a) for angular contact bearings, roller bearings, and thrust bearings. Table 5.16 [p 409] gives the formulae for calculating radial and axial deflections [displacements] for various types of bearings.

Pre-loading thrust bearing

Thrust bearings call for axial pre-loading to prevent sliding friction between the rolling elements and races.

$$\text{Preload } (W_{ap}) \geq K \left[\dfrac{NC_o}{10^6}\right]^2 \quad \textbf{(Eqn. 5.46)}$$

C_o = Static capacity (kg); K = Factor for brg type.
N = R.P.M.

Table 5.17: Preload factor (K) for various types of brgs.

Brg. type	Thrust ball brg.	Angular contact ball brg.	Roller thrust brg.
K	0.08	0.03	0.004

Speed limits for rolling bearings

The maximum limiting speed for oil lubrication is higher than that of grease lubrication, as is evident from the maximum permissible RPM, stated in the bearings tables. For speeds higher than the permissible limit, it is necessary to use oil jet/mist lubrication, or provide for a circulation and cooling arrangement (radiator/heat exchanger) for the lubricating oil.

Some high-speed applications use special bearings with higher precision (class 5, 6), more radial clearance, and special ball/roller cages. The highest speeds possible in such bearings can be computed, by multiplying the maximum permissible speed for oil lubrication, by the factors stated in the following table.

Brg. type	Ball brgs				Roller brgs	
	Deep groove	Self-aligning	Angular contact	Thrust	Cylindrical	Spherical thrust
Multiplying factor	3	1.5	1.5	1.4	2.2	2

Correction for combined load

Speed limits, given in the bearings tables are applicable for 10% of the rated dynamic load (C), stated in the tables. Also, these loads should only be radial for radial bearings, and axial for thrust bearings. Speed limits for combined radial and axial loads, will be lesser. These depend upon the load angle β.

$$\beta = \tan^{-1} \frac{W_a}{W_r} \quad \text{(Eqn. 5.47)}$$

Figure 5.23 gives the correction factors for various load angles and different types of rolling bearings.

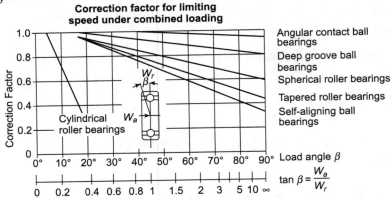

Fig. 5.23: Correction factor for combined load

Mounting rolling friction bearing

A mounting arrangement should:

1. **Position** the shaft correctly, axially [**Fig. 5.24a**]. The working (nose) end bearing should be axially secured. For a rotating shaft, the outer race is axially clamped by brg. covers.

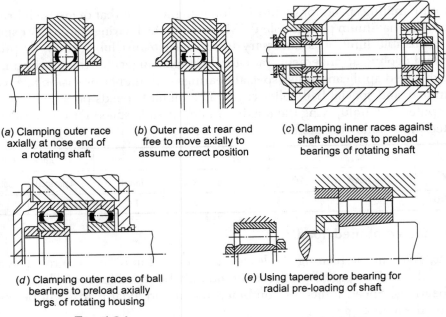

(a) Clamping outer race axially at nose end of a rotating shaft

(b) Outer race at rear end free to move axially to assume correct position

(c) Clamping inner races against shaft shoulders to preload bearings of rotating shaft

(d) Clamping outer races of ball bearings to preload axially brgs. of rotating housing

(e) Using tapered bore bearing for radial pre-loading of shaft

Fig. 5.24: Mounting and pre-loading ball and roller bearings

Machine Tool Elements

2. **Axial adjustment** is necessary to allow for axial expansion. The outer race of bearing at the rear end (opposite to the nose end) should be able to move axially, to assume a suitable position [**Fig 5.24b**].
3. **Proper fits** should ensure that there is no sliding friction between the bearing and shaft/housing. However, where necessary (usually the rear end), the bearing should be able to move axially.
4. **Lubrication arrangement** should be suitable for both, load and the duration of usage. Grease is suitable for slow speeds and low loads. For higher speeds and loads, oil lubrication is necessary. Simple oil immersion (as in gearboxes and engine crack cases) is satisfactory in medium-duty applications. For very heavy loads and very high speeds, an oil circulation arrangement and a pump are necessary. The eventuality of a high temperature-rise, calls for some kind of an oil cooling arrangement.

Dimensions for oil seals [Fig 5.25 (c)]
(All dimensions in millimeters) [1mm = .03937″]

Shaft Diameter d_1	Nominal Bore Diameter of Housing D	$b \pm 0.2$ [.008″] Types A and B	Chamfer c Min	Shaft Diameter d_1	Nominal Bore Diameter of Housing D	$b \pm 0.2$ [.008″] Types A and B	Chamfer c Min
6 [.2362″]	16 [.63″] 22 [.8661″]	7 [.275″]	0.3 [.012″]	15 [.5905″]	24 [.9449″] 26 [1.0236″] 30 [1.811] 32 [1.26] 35 [1.3779]	7 [1.276″]	0.3 [.012″]
7 [.2756″]	16 22	7	0.3	16 [.63″]	28 [1.1024] 30 32 35	7	0.3
8 [.315″]	16 22 24 [.9449″]	7	0.3	17 [.6693″]	28 30 32 35 40 [1.5748]	7	0.3
9 [.3543″]	22 24 26 [1.0236″]	7	0.3	18 [.7087″]	30 32 35 40	7	0.3
10 [.3937″]	19 [.748″] 22 24 26	7	0.3	20 [.7874″]	30 32 35 40 47 [1.8504]	7	0.3
11 [.4331″]	22 26	7	0.3				
12 [.4724″]	22 24 28 [1.1024″] 30 [1.1811″]	7	0.3				
14 [.5512″]	24 28 30 35 [1.3779″]	7	0.3				

Shaft Diameter d_1	Nominal* Bore Diameter of Housing	b ± 0.2 Types A and B	b ± 0.2 Type C	‡ C Min	Shaft Diameter d_1	Nominal* Bore Diameter of Housing	b ± 0.2 Types A and B	b ± 0.2 Type C	‡ C Min
22 [.8661"]	32 [1.26"] 35 [1.3779] 40 [1.5748] 47 [1.8504]	7 [.276"]	— 9 [.354]	0.3 [.012"]	32 [1.26"]	45 [1.7716"] 47 [1.8504] 52 [2.0472]	7 [.276"]	— 9 [.354"]	0.4 [.016"]
24 [.9449"]	35 37 [1.4567] 40 47 [1.8504]	7	— 9	0.3	35 [1.3779"]	47 50 [1.9685] 52 62 [2.4409]	7	— 9	0.4
25 [.9842"]	35 40 42 [1.6535] 47 52 [2.0472]	7	— 9	0.4 [.016"]	36 [1.4173"]	47 50 52 62	7	— 9	0.4
26 [1.0236"]	37 42 47	7	— 9	0.4	38 [1.4961"]	52 55 [2.1653] 62	7	— 9	0.4
28 [1.1024"]	40 47 52	7	— 9	0.4	40 [1.5748"]	52 55 62 72 [2.8346]	7	— 9	0.4
30 [1.1811"]	40 42 47 52 62 [2.4409]	7	— 9	0.4	42 [1.6535"]	55 62 72	8 [.315]	— 10 [.394"]	0.4

IS : 5129 — 1969
For limits of housing bore, see [Fig. 5.25a]

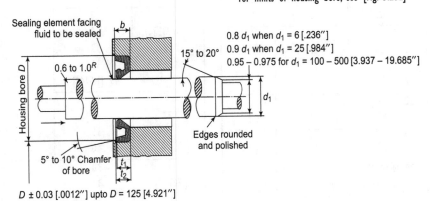

Fig. 5.25(a): Using oil seal to prevent lubricant leakage

Dimension for oil seals (Contd.)

Shaft Diameter d_1	Nominal Bore Diameter of Housing	b ± 0.2 Types A and B	b ± 0.2 Types C	c Min
45 [1.7716"]	60 [2.5984"]	8 [.315"]	—	0.4 [.016"]
	62 [2.4409]		[.394"]	
	65 [2.5991]		10	
	72 [2.8346]			
48 [1.8898"]	62	8	—	0.4
	72		10	
50 [1.9685"]	65	8	—	0.4
	68 [2.6771]		10	
	72			
	80 [3.1496]			
52 [2.0472"]	68	8	—	0.4
	72		10	
55 [2.1653"]	70	8	—	0.4
	72		10	
	80			
	85 [3.3465]			
56 [2.2047"]	70 [2.7569]	8	—	0.4
	72		10	
	80			
	85			
58 [2.2835"]	72	8	—	0.4
	80		10	
60 [2.5984"]	75 [2.9528]	8	—	0.4
	80		10	
	85			
	90 [3.5433]			
62 [2.4409"]	85	10	12	0.5
	90			
63 [2.4803"]	85	10	[.472]	0.5
	90		12 [.020"]	
65 [2.5591"]	85	10	12	0.5
	90			
	100 [3.937]			
[2.6771] 68	90	10	12	0.5
	100			
70 [2.7559]	90	10	12	0.5
	100			
75 [2.9528]	95 [3.7402]	10	12	0.5
	100			
78 [3.0708]	100	10	12	0.6 [.024]
80 [3.1496]	100	10	12	0.5
	110 [4.3307]			
85 [3.3465]	110	12 [.472]	15 [.59]	0.8 [.031]
	120 [4.7244]			
90 [3.5433]	110	12	15	0.8
	120			
95 [3.7402"]	120 [4.7244"]	12 [.472"]	15 [.59]	0.8 [.03]
	125 [4.9213]			
100 [3.937]	120	12	15	0.8
	125			
	130 [5.1181]			
105 [4.1339]	130	12	15	0.8
	140 [5.5118]			
110 [4.3307]	130	12	15	0.8
	140			
115 [4.5276]	140	12	15	0.8
	150 [5.9055]			
120 [4.7244]	150	12	15	0.8
	160 [6.2992]			
125 [4.9213]	150	12	15	0.8
	160			
130 [5.1181]	160	12	15	0.8
	170 [6.6929]			
135	170	12	15	0.8
140 [5.5118]	170	15	15	1 [.039"]
145	175 [5.7087]	15	15	1
150 [5.9055]	180 [7.0866]	15	15	1
160 [6.2992]	190 [7.4803]	15	15	1
170 [6.6929]	200 [7.874]	15	15	1
180 [7.0866]	210 [8.2677]	15	15	1
190 [7.4803]	220 [8.6614]	15	15	1
200 [7.874]	230 [9.0551]	15	15	1
210 [8.2677]	240 [9.4488]	15	15	1
220 [8.6614]	250 [9.8425]	15	15	1
230 [9.0551]	260 [10.2362]	15	15	1
240 [9.4488]	270 [10.63]	15	15	1
250 [9.8425]	280 [11.0236]	15	15	1
260 [10.2362]	300 [11.811]	20 [.787]	20	1
280 [11.0236]	320 [12.5984]	20	20	1
300 [11.811]	340 [13.3858]	20	20	1
320 [12.5984]	360 [14.1732]	20	20	1
340 [13.3858]	380 [14.9606]	20	20	1
360 [14.1732]	400 [15.748]	20	20	1
380 [14.9606]	420 [16.5354]	20	20	1
400 [15.748]	440 [17.3228]	20	20	1
420 [16.5354]	460 [18.1102]	20	20	1
440 [17.3228]	480 [18.8976]	20	20	1
460 [18.1102]	500 [19.685]	20	20	1
480 [18.8976]	520 [20.4724]	20	20	1
500 [19.685]	540 [21.2598]	20	20	1

For limits of housing bore see [Fig. 5.25a]

5. **Pre-loading** of bearings is necessary to completely prevent sliding friction.
   Axial pre-loading is accomplished by clamping the inner races of the bearings [**Fig. 5.24c**] onto the shaft shoulder faces. In angular contact bearings, clamping two bearings against each other is enough to preload them.

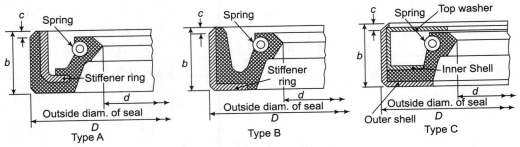

Fig. 5.25(c): Standard oil seals

Radial pre-loading can be effected by using interference fits, with the shaft and housing. NNK-type, double-row roller bearings, often used for machine tools, have 1:12 tapered bores. For pre-loading, the tapered bearings are forced onto the tapered seat on the shafts [**Fig. 5.24e**].

6. **Seals** should be used where necessary [**Fig. 5.25**], to prevent leakage of lubricant.
7. **Rotating housing [Fig 5.24d]** should be axially secured by clamping the outer races at both ends of shafts. Then the inner race of the rear bearing should be made free, to move axially, to assume the correct position.

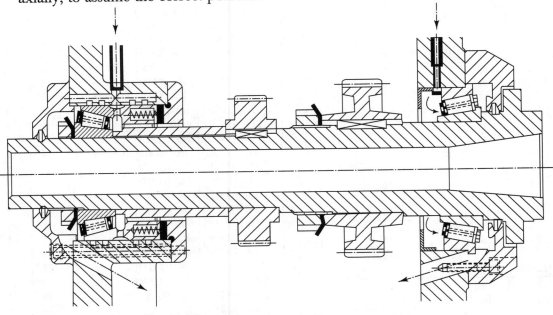

Fig. 5.26: Lathe spindle with tapered roller bearings

Figure 5.26 shows a lathe spindle with tapered roller bearings. Note the collar on the outer race of front bearing, and the oil passage provided in housing and the bearing covers. Tightening the nut at the rear-end, clamps and preloads the bearings axially. The lubricant flow direction is highlighted by the arrows. The springs at the rear-end allow the rear bearing to assume a suitable position.

Bearings' fits

Table 5.18a gives the recommended fits for the housings to be assembled with normal class (CO) radial bearings. These fits are suitable for steel or cast iron housings. For non-ferrous housings with lesser strength, a tighter fit is required. It should be noted that when outer ring displacement is necessary, H7 or J6 grade bore is used. For preventing movement of the outer ring, fits with more interference (K6, M6, N6) are used. Also, roller bearings with an outside diameter above 125, call for more interference (N6), than bearings with a lesser diameter (M6). When the outer ring (and housing) is stationary, H7 grade is satisfactory for all types of loads. Generally, split housing should not be used for applications in which the outer ring has to rotate.

Table 5.18a: Housing fits for radial ball/roller bearings

Housing	Outer Ring		Load	Tolerance	Remarks
Solid or split	Stationary	Axially displaceable	All	H7	
	"	"	Shock	J6	
	"	"	Heating	G6	High temperature.
Solid	Rotating	Not displaceable	Light loads	M6	Roller Brg ≤ 125
	Rotating	Not displaceable	Heavy loads	N6	Roller Brg > 125
	Rotating	Not displaceable	Heavy with shock	P6	Also for thin walled Hsg
			Not determinable	K6	For accurate silent run

Table 5.18b: Shaft fits for radial ball/roller bearings

Inner ring	Bearing Bore [Divide by 25.4 for Inch equivalents]			Brg. axial displacement	Load example	Tolerance
	Ball brg.	Roller brg. All sizes All sizes	Spherical roller			
Stationary	All sizes	"		Necessary not necessary	Wheels Pulleys	g6 n6
Rotating	≤ 18 18–100 100–200 –	≤ 40 40–100 140–200	≤ 40 40–100 100–200	light load 6–7% of dynamic capacity	Machine tools, Pumps Electrical Machines	h5 j6 k6 m6

Inner ring	Bearing Bore [Divide by 25.4 for Inch equivalents]			Examples	Load	Tolerance
	Ball brg.	Roller brg. All sizes All sizes	Spherical roller			
Rotating	≤ 18 [.71"] [.71]18–100[3.94"] [3.94]100–140 [5.51]140–200 [7.87]200–280 – [11.02]	– ≤ 40 [1.57"] 40–100 [3.94] 100–140 [5.51] 140–200 [7.87] 200–400 [15.75]	– ≤ 40 [1.57"] 40–65 [2.56] 65–100 [3.94] 100–140 [5.51] 140–280 [11.02] > 280	General motors Pumps Compressors I.C. Engines	Normal Heavy More than 10% of (C)	j5 k5 m5 m6 n6 p6 r6
Rotating	–	[1.97"]50–140[5.51"] [5.51]140–200[7.87]	[1.97"]50–100[3.94"] 100–140[5.51] > 140	Railway axles, Traction Motors	Heavy with shock	n6 p6 r6

Table 5.18c: Housing fits for thrust bearings

Brg. Type	Load	Condition	Tolerance
Thrust ball brg. Spherical Roller Thrust Brg.	Only axial Combined - Axial and Radial Heavy— Radial Load	Hsg. Stationary -Hsg. Rotating -Hsg. Rotating	H8 J7 K7 M7

Table 5.18d: Shaft fits for thrust bearings

Load	Brg. Bore	Tolerance
Only axial Combined axial and radial stationary load	All sizes All sizes	j6 j6
Rotating load	≤ 200 [7.87"] 200–400 [7.87"–15.75"] > 400	k6 m6 n6

Table 5.18e gives shaft and housing fits for higher precision special bearings, with special ball/roller cages. Generally, ball and roller bearings' bores range between J5 – J6 class transition fits. The outside diameter of the outer race diameter has less than 5th grade tolerance, only 10% of it is interference, and the rest clearance. P6 grade bearings have 12 to 20% lesser tolerance, about 10–15 % of it is interference. Grade P5 and P4 have even lesser, i.e. 60% and 50% of the normal grade tolerance, and zero interference.

Table 5.18e: Tolerances for mounting special precision bearings

	Bearing type	Shafts *				Requirement	Bearing housing		
		Diameter range mm	Bearing tolerance class				Bearing tolerance class		
			P6	P5, SP	UP		P6	P5, SP	UP
Radial bearings	Radial ball bearing	Divide by 25.4 for Inch equivalents				Free bearings, normal and light loads	P6	P5, SP	UP
		≤ 18 [.71"]	h5	h4	h3		j6	js5	js4
		(18) – 100 [3.94"]	j5	js4	js3				
		100	k5	k4	–	Rotating outer ring load	M6	M5	M4
	Cylindrical roller bearings **	≤ 40 [1.57"]	j5	js4	–	Normal and light load	K6	K5	K4
		(40) – 140 [5.51"]	k5	k4	–	Heavily loaded and rotating	M6	M5	M4
		(140) – 200 [7.87"]	m5	m5	–	Outer ring			
	Taper roller bearings	≤ 40 [1.57"]	j6	j5	–	Adjustable	J6	Js5	–
		(40) – 140 [5.51"]	k6	k6	–	Located	K6	K5	–
		(140) – 200 [7.87"]	m6	m5	–	Rotating outer ring load	M6	M5	–
Thrust bearings	Angular contact thrust ball bearings	All diameters	–	h5	h4	Bearings have common housing seating with cylindrical roller bearings NN 30K or NU 49K	K6	K5	K4
	Thrust ball bearings	All diameter	h6	h5	–		H8	H7	–

* Only applicable to cylindrical bore bearings ** Tapered bearings seating checked with ring gauge j6 or j5, fit to be selected for adjustable, lightly-loaded bearings. Tighter fit than that obtained with given tolerances should not be selected, although this may be recommended for cylindrical roller bearings.

Precision of spindle components

Form, position, and alignment errors of all the parts mounted on the machine tool spindle must be within certain limits. The necessary precision depends upon the tolerance on the spindle and housing, for the mounted bearings.

	Precision grade			
Spindle/Brg. housing	6	5	4	3
Mounted parts	IT3	IT2	IT1	IT0

Tolerances for IT grades can be found from the following table.

Tolerances for IT grades in microns [1 micron = 0.001mm = 0.00004″ = 39.4 micro-inches]

Dim	IT0	IT1	IT2	IT3
up to 3 [.118″]	0.5 [19.7 μ in]	0.8 [31.5 μ in]	1.2 [47.2 μ in]	2 [78.7 μ in]
3–6 [.236″]	0.6 [23.6 ″]	1 [39.4 ″]	1.5 [59.1 ″]	2.5 [98.4 ″]
6–10 [.394″]	0.6 [23.6 ″]	1 ″	1.5 [59.1 ″]	2.5 [98.4 ″]
10–18 [.709″]	0.8 [31.5 ″]	1.2 [47.2 ″]	2 [78.7 ″]	3 [118.1 ″]
18–50 [1.968″]	1 [39.4 ″]	1.5 [59.1 ″]	2.5 [98.4 ″]	4 [157.5 ″]
50–80 [3.15″]	1.2 [47.2 ″]	2 [78.7 ″]	3 [118.1 ″]	5 [196.9 ″]
80–120 [4.724″]	1.5 [59.1 ″]	2.5 [98.4 ″]	4 [157.5 ″]	6 [236 ″]
120–180 [7.087″]	2.0 [78.7 ″]	3.5 [137.8 ″]	5 [196.9 ″]	8 [315 ″]
180–250 [9.842″]	3 [118.1 ″]	4.5 [177.2 ″]	7 [276 ″]	10 [394 ″]
250–315 [12.402″]	4 [157.5 ″]	6 [236 ″]	8 [315 ″]	12 [472 ″]
315–400 [15.748″]	5 [196.9 ″]	7 [276 ″]	9 [354 ″]	13 [512 ″]
400–500 [19.685″]	6 [236 ″]	8 [315 ″]	10 [394 ″]	15 [591 ″]

Figure 5.29 gives recommendations for the maximum permissible deviations in form (roundness/ovality), taper (cylindricity/parallelism), alignment (centrality of both ends), and face squareness (runout), for accurate and very accurate bearing arrangements.

B Width of bearing ring, mm

d Bearing nominal bore diameter, mm

Δd Deviation from diameter d, microns (μm)

d_1 Nominal diameter at large end of bore $= d + 0.083333B$, mm

Δd_1 Deviation from diameter d_1, microns (μm)

α Nominal taper angle 2° 23′ 9.4″

Δx Angle deviation $= 1.716 \, (\Delta d_1 - \Delta d)/B$ [minutes]

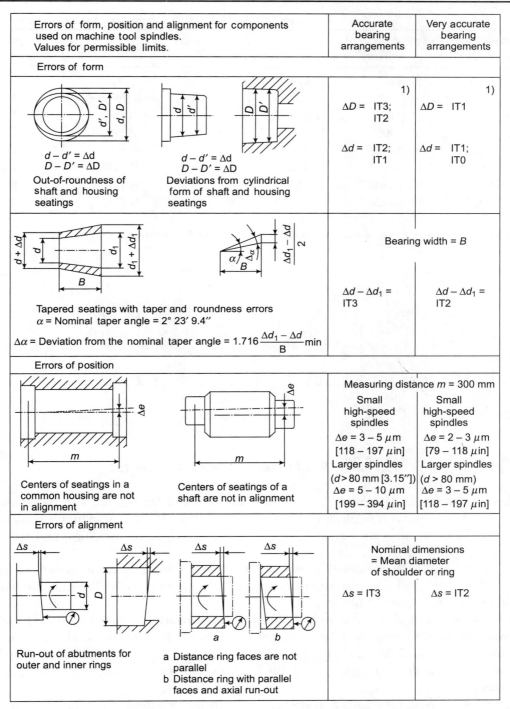

Fig. 5.29: Permissible deviations in parts mounted on m/c tool spindles

1 Micron = 0.001mm = 0.0000394" = 39.4 Micro-inch *(Deviations in microns (μm))*

Nominal bore diameter d mm Divide by 25.4 for inch equivalents		Bore diameter d deviations (Tolerance H_8)		($\Delta d_1 - \Delta d$) (tolerance grade IT1)	
Over	Up to	High	Low	High	Low
10 [.3937"]	18 [.7087"]	27 [1063 μ in]	0	18 [709 μ in]	0
18 [.7087]	30 [1.1811]	33 [1299 "]	0	21 [827 "]	0
30 [1.1811]	50 [1.9685]	39 [1535 "]	0	25 [984 "]	0
50 [1.9685]	80 [3.1496]	46 [1811 "]	0	30 [1181 "]	0
80 [3.1496]	120 [4.7244]	54 [2128 "]	0	35 [1378 "]	0
120 [4.7244]	180 [7.0866]	63 [2482 "]	0	40 [1575 "]	0
180 [7.0866]	250 [9.8425]	72 [2837 "]	0	46 [1811 "]	0

Figure 5.30 gives tolerances for diameters, and taper angle of bearings, with 1:12 (2° 32¢ 9.4²) tapered bores.

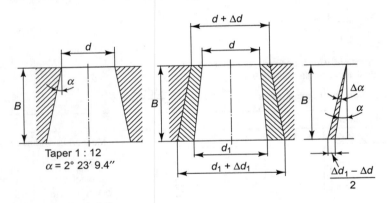

Fig. 5.30: Normal tolerances for tapered bore, taper 1:12

Tables 5.19, 5.20, 5.21, and 5.22 give the dimensions of 60, 62, 63, and 64 series deep groove ball bearings, with minimum spindle shoulder size (D_1), and maximum housing step size (D_2). The tables also state static (Co) and dynamic (C) load capacities, and maximum permissible speeds (R.P.M.).

Tables 5.23, and 5.24 give details of self-aligning ball bearings, with cylindrical and tapered bores. Tables 5.25 and 5.26 list angular contact ball bearings, while Table 5.27 details the double row angular contact bearings.

Tables 5.28 to 5.31 give the sizes of cylindrical roller bearings while Tables 5.32 and 5.33 detail tapered roller bearings. Table 5.34 gives the dimensions of spherical roller bearings, and Tables 5.35, 5.36, 5.37, and 5.38 cover needle roller bearings. Table 5.39 details combined radial and axial needle bearings, while Table 5.40 covers flat needle cages, used for slides and guideways.

Tables 5.41 to 5.43 give sizes and capacities for thrust bearings.

Machine Tool Elements

Table 5.19: Specifications of 60 series deep groove ball bearings

Bearing of basic design No. (SKF)	d mm	D_1 mm	D mm	D_2 max	B mm	r ≈ mm	r_1 mm	Basic capacity, kgf Static $C0$	Basic capacity, kgf Dynamic C	Max. Speed for Lubrication by Oil	Max. Speed for Lubrication by Grease
6000	10	12	26	24	8	0.5	0.3	190	360	20000	17000
01	12	14	28	26	8	0.5	"	220	400	20000	17000
02	15	17	32	30	9	0.5	"	255	440	20000	17000
6003	17	19	35	33	10	0.5	"	285	465	20000	17000
04	20	23	42	36	12	1.0	0.6	450	735	16000	13000
05	25	28	47	44	12	"	"	520	780	16000	13000
6006	30	35	55	50	13	1.5	1.0	710	1040	13000	10,000
07	35	40	62	57	14	1.5	"	880	1250	13000	10,000
08	40	45	68	63	15	1.5	"	980	1320	10000	8500
6009	45	50	75	70	16	1.5	"	1270	1630	10000	8500
10	50	55	80	75	16	1.5	"	1370	1700	8000	6700
11	55	61	90	84	18	2	"	1800	2200	8000	6700
6012	60	66	95	89	18	2	"	1930	2280	8000	6700
13	65	71	100	94	18	2	"	2120	2400	8000	6700
14	70	76	110	104	20	2	"	2550	3000	6000	5000
6015	75	81	115	109	20	2	"	2800	3100	6000	5000
16	80	86	125	119	22	2	"	3350	3750	6000	5000
17	85	91	130	124	22	2	"	3600	3900	5000	4300
6018	90	97	140	133	24	2.5	1.5	4150	4550	5000	4300
19	95	102	145	138	24	2.5	"	4500	4750	5000	4300
20	100	107	150	143	24	2.5	"	4500	4750	4000	3400
6021	105	114	160	151	26	3	2	5400	5700	4000	3400
22	110	119	170	161	28	3	"	6100	6400	4000	3400
24	120	129	180	171	28	3	"	6550	6700	3000	2400
6026	130	139	200	191	33	3	"	8300	8300	3000	2400
28	140	149	210	201	33	3	"	9000	8650	3000	2400
30	150	160	225	215	35	3.5	"	10400	9800	2500	1750
6032	160	170	240	230	38	3.5	"	11800	11200	2500	1750
34	170	180	260	250	42	3.5	"	14300	13200	2500	1750
36	180	190	280	270	46	3.5	"	16600	15000	2000	1500
6038	190	200	290	280	46	3.5	"	18000	15300	2000	1500
40	200	210	310	300	51	3.5	"	20000	17000	2000	1500

r = Corner radius on bearing
[Divide by 25.4 for Inch equivalents]

r_1 = corner radius on shaft/housing
1 kg = 2.2046 Lbs.

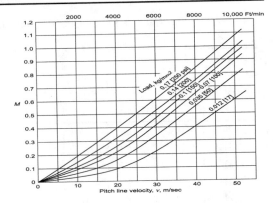

Table 5.20: Specification of 62 series deep groove ball bearings

[Divide by 25.4 for Inch equivalents]   1 kg = 2.2046 Lbs

ISI No.	Bearing of basic design No. (SKF)	d	D_1	D	D_2	B	r	r_1 '	Basic capacity, kgf		Max R.P.M	
							≈		Static	Dynamic	Lubricant	
		mm	min	mm	max	mm	mm	mm	C_0	C	Oil	Grease
10BC02	6200	10	14	30	26	9	1	0.6	224	400	20000	17000
12BC02	01	12	16	32	28	10	1	0.6	300	540	20000	17000
15BC02	02	15	19	35	31	11	1	0.6	355	610	16000	13000
17BC02	6203	17	21	40	36	12	1	0.6	440	750	16000	13000
20BC02	04	20	26	47	41	14	1.5	1.0	655	1000	16000	13000
25BC02	05	25	31	52	46	15	1.5	1.0	710	1100	13000	10000
30BC02	6206	30	36	62	56	16	1.5	1.0	1000	1530	13000	10000
35BC02	07	35	42	72	65	17	2	1.0	1370	2000	10000	8500
40BC02	08	40	47	80	73	18	2	1.0	1600	2280	10000	8500
45BC02	6209	45	52	85	78	19	2	1.0	1830	2550	8000	6700
50BC02	10	50	57	90	83	20	2	1.0	2120	2750	8000	6700
55BC02	11	55	64	100	91	21	2.5	1.5	2600	3400	8000	6700
60BC02	6212	60	69	110	101	22	2.5	1.5	3200	4050	6000	5000
65BC02	13	65	74	120	111	23	2.5	1.5	3550	4400	6000	5000
70BC02	14	70	79	125	116	24	2.5	1.5	3900	8400	5000	4300
75BC02	6215	75	84	130	121	25	2.5	1.5	4250	5200	5000	4300
80BC02	16	80	91	140	129	26	3	2.0	4550	5700	5000	4300
85BC02	17	85	96	150	139	28	3	2.0	5500	6550	4000	3400
90BC02	6218	90	101	160	149	30	3	2.0	6300	7500	4000	3400
95BC02	19	95	107	170	158	32	3.5	2.0	7200	8500	4000	3400
100BC02	20	100	112	180	168	34	3.5	2.0	8150	9650	3000	2400
105BC02	6221	105	117	190	178	36	3.5	2.0	9300	10400	3000	2400
110BC02	22	110	122	200	188	38	3.5	2.0	10400	11200	3000	2400
120BC02	24	120	132	215	203	40	3.5	2.0	10400	11400	3000	2400
130BC02	6226	130	144	230	216	40	4	2.5	11600	12200	2500	1750
140BC02	28	140	154	250	236	42	4	2.5	12900	12900	2500	1750

Machine Tool Elements

ISI No.	Bearing of basic design No. (SKF)	d mm	D_1 min mm	D mm	D_2 max mm	B Mm	r ≈ mm	r1' mm	Basic capacity, kgf Static C_0	Basic capacity, kgf Dynamic C	Max R.P.M Lubricant Oil	Max R.P.M Lubricant Grease
150BC02	6230	150	164	270	256	45	4	2.5	14300	13700	2500	1750
160BC02	32	160	174	290	276	48	4	2.5	15600	14300	2000	1500
170BC02	6234	170	187	310	293	52	5	3.0	19000	16600	2000	1500
180BC02	36	180	197	320	303	52	5	3.0	20400	17600	1600	1100
190BC02	38	190	207	340	323	55	5	3.0	24000	20000	1600	1100
200BC02	40	200	217	360	343	58	5	3.0	26500	21200	1600	1100

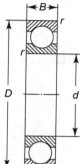

D_1, abutment diam. on shaft [Refer page 428]
D_2 abutment diam. on housing
r_1 corner radii on shaft and housing.

Table 5.21: Specifications of 63 series deep groove ball bearings

[Divide by 2.2046 for Inch equivalents] 1 kg = 2.2046 Lbs

ISI No.	Bearing of basic design No. (SKF)	d mm	D_1 min mm	D mm	D_2 max mm	B mm	r ≈ mm	r_1 mm	Basic capacity, kgf Static C_0	Basic capacity, kgf Dynamic C	Max R.P.M Lubricant Oil	Max R.P.M Lubricant Grease
10BC03	6300	10	14	35	31	11	1.5	0.6	360	630	16000	13000
12BC03	01	12	18	37	31	12	1.5	1	430	765	16000	13000
15BC03	02	15	21	42	36	13		1	520	880	16000	13000
17BC03	6303	17	23	47	41	14	1.5	1	630	1060	13000	10,000
20BC03	04	20	27	52	45	15	2	1	765	1250	13000	10,000
25BC03	05	25	32	62	55	17	2	1	1040	1660	10000	8500
30BC03	6306	30	37	72	65	19	2	1	1460	2200	10000	8500
35BC03	07	35	44	80	71	21	2.5	1.5	1760	2600	8000	6700
40BC03	08	40	49	90	81	23	2.5	1.5	2200	3200	8000	6700
45BC03	6309	45	54	100	91	25	2.5	1.5	3000	4150	8000	6700
50BC03	10	50	61	110	99	27	3	2.0	3550	4800	6000	5000
55BC03	11	55	66	120	109	29	3	2.0	4250	5600	6000	5000
60BC03	6312	60	72	130	118	31	35	2.0	4800	6400	5000	4300
65BC03	13	65	77	140	128	33	3.5	2.0	5500	7200	5000	4300
70BC03	14	70	82	150	138	35	3.5	2.0	6300	8150	5000	4300
75BC03	6315	75	87	160	148	37	3.5	2.0	7200	9000	4000	3400

ISI No.	Bearing of basic design No. (SKF)	d mm	D_1 min mm	D mm	D_2 max mm	B Mm	r ≈ mm	r1' mm	Basic capacity, kgf		Max R.P.M Lubricant	
									Static C0	Dynamic C	Oil	Grease
80BC03	16	80	92	170	158	39	3.5	2.0	8000	9650	4000	3400
85BC03	17	85	99	180	166	41	4	2.5	8800	10400	4000	3400
90BC03	6318	90	104	190	176	43	4	2.5	9800	11200	3000	2400
95BC03	19	95	109	200	186	45	4	2.5	11200	12000	3000	2400
100BC03	20	100	114	215	201	47	4	2.5	13200	13700	3000	2400
105BC03	6321	105	119	225	211	49	4	2.5	14300	14300	2500	1750
110BC03	22	110	124	240	226	50	4	2.5	16600	16000	2500	1750
120BC03	24	120	134	260	246	55	4	2.5	17000	16300	2500	1750
130BC03	6326	130	147	280	263	58	5	3	19600	18000	2500	1750
140BC03	28	140	157	300	283	62	5	3	22400	20000	2000	1500
150BC03	30	150	167	320	303	65	5	3	25500	21600	2000	1500

D_1, abutment diam. on shaft
D_2, abutment diam. on shaft and housing
r_1, corner radii on shaft and housing

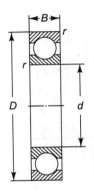

Table 5.22: Specifications for 64 series deep groove ball bearings

Divide by 25.4 for Inch equivalents

1 kg = 2.2046 Lbs

No.	d	D_1 min	D	D_2 max	B	r Brg	r_1 Shaft/Hsg	Basic capacity, kgf		Max RPM Lubricant	
								Static C0	Dynamic C	Oil	Grease
6403	17	26	62	53	17	2	1	1280	1800	10000	8500
6404	20	29	72	63	19	2	1	1660	2400	8000	6700
6405	25	36	80	69	21	2.5	1.5	2000	2825	7100	6100
6406	30	41	90	79	23	2.5	1.5	2400	3350	6300	5300
6407	35	46	100	89	25	2.5	1.5	3250	4300	5600	4800
6408	40	53	110	97	27	3	2	3800	5000	5000	4300
6409	45	58	120	107	29	3	2	4650	5850	4500	3800
6410	50	64	130	116	31	3.5	2	5300	7000	4000	3400
6411	55	69	140	126	33	3.5	2	6400	7850	4000	3400
6412	60	74	150	136	35	3.5	2	7100	8450	3600	3000

Machine Tool Elements

No.	d	D_1 min	D	D_2 max	B	r Brg	r_1 Shaft/Hsg	Basic capacity, kgf		Max RPM Lubricant	
								Static C_0	Dynamic C	Oil	Grease
6413	65	79	160	146	37	3.5	2	8000	9150	3200	2600
6414	70	86	180	164	42	4	2.5	9100	10000	2800	2200
6415	75	91	190	174	45	4	2.5	10160	12000	2800	2200
6416	80	96	200	184	48	4	2.5	12800	13000	2500	1750
6417	85	105	210	190	52	5	3	13800	13800	2500	1750
6418	90	110	225	205	54	5	3	16600	15200	2200	1800

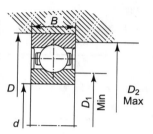

Table 5.23: Specifications of 22 series self-aligning (Ball brgs with straight and tapered bore)

Divide by 25.4 for Inch equivalents   1 kg = 2.2046 Lbs

Bearing with cylindrical bore no. (SKF)	Bearing with taper* bore no. (SKF)	d mm	D mm	B mm	r ≈ mm	Basic capacity, kgf		Max R.P.M Lubricant	
						Static C_0	Dynamic C	Oil	Grease
2200		10	30	14	1	180	570	20000	17000
01		12	32	14	1	200	585	20000	17000
02		15	35	14	1	216	600	16000	13000
2203	2204 K	17	40	16	1	280	765	16000	13000
04	05 K	20	47	18	1.5	390	980	16000	13000
05		25	52	18	1.5	425	980	13000	10000
2206	2206 K	30	62	20	1.5	560	1200	13000	10000
07	07 K	35	72	23	2	800	1700	10000	8500
08	08 K	40	80	23	2	915	1760	10000	8500
2209	2209 K	45	85	23	2	1020	1800	8000	6700
10	10 K	50	90	23	2	1080	1800	8000	6700
11	11 K	55	100	25	2.5	1270	2080	8000	6700
2212	2212 K	60	110	28	2.5	1600	2650	6000	5000
13	13 K	65	120	31	2.5	2040	3400	6000	5000
14	—	70	125	31	2.5	2160	3450	5000	4300

Bearing with cylindrical bore no. (SKF)	Bearing with taper* bore no. (SKF)	d mm	D mm	B mm	r ≈ mm	Basic capacity, kgf		Max R.P.M Lubricant	
						Static C0	Dynamic C	Oil	Grease
2215	2215 K	75	130	31	2.5	2240	3450	5000	4300
16	16 K	80	140	33	3	2500	3800	5000	4300
17	17 K	85	150	36	3	3000	4550	4000	3400
2218	2218 K	90	160	40	3	3600	5500	4000	3400
19	19 K	95	170	43	3.5	4300	6550	4000	3400
20	20 K	100	180	46	3.5	5100	7650	3000	2400
2221	–	105	190	50	3.5	5600	8500	3000	2400
22	2222 K	110	200	53	3.5	6400	9800	3000	2400

* Taper 1: 12 on diameter.
Note: Abutment diameters D_1, D_2 and corner radii r, are the same as for the 62 series.

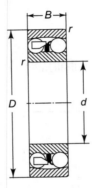

Table 5.24: Specifications of self-aligning 23 series ball brgs with straight and tapered bore

Divide by 25.4 for Inch equivalents

1 kg = 2.2046 Lbs

Bearing with cylindrical bore no. (SKF)	Bearing with taper* bore no. (SKF)	d mm	D mm	B mm	r ≈ mm	Basic capacity, kgf		Max R.P.M Lubricant	
						Static C0	Dynamic C	Oil	Grease
2301		12	37	17	1.5	300	915	16000	13000
02		15	42	17	1.5	335	930	13000	10000
03		17	47	19	1.5	415	1120	13000	10000
2304	2304 K	20	52	21	2	550	1400	10000	8500
05	05 K	25	62	24	2	765	1900	10000	8500
06	06 K	30	72	27	2	1020	2450	8000	6700

Bearing with cylindrical bore no. (SKF)	Bearing with taper* bore no. (SKF)	d mm	D mm	B mm	r ≈ mm	Basic capacity, kgf		Max R.P.M Lubricant	
						Static C0	Dynamic C	Oil	Grease
2307	2307 K	35	80	31	2.5	1320	3050	8000	6700
08	08 K	40	90	33	2.5	1600	3550	6000	5000
09	09 K	45	100	36	2.5	1960	4250	6000	5000
2310	2310 K	50	110	40	3	2400	5000	5000	4300
11	11 K	55	120	43	3	2850	5850	5000	4300
12	12 K	60	130	46	3.5	3350	6800	4000	3400
2313	2313 K	65	140	48	3.5	3900	7500	4000	3400
14	—	70	150	51	3.5	4500	8500	4000	3400
15	15 K	75	160	55	3.5	5200	9500	3000	2400
16	16 K	80	170	58	3.5	5850	10600	3000	2400
2317	2317 K	85	180	60	4	6200	11000	2500	1750
18	18 K	90	190	64	4	6950	11800	2500	1750

* Taper 1:12 on diameter.
Note: Abutment diameters D_1, D_2 and corner radii r, are the same as for 63 series

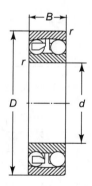

Table 5.25: Series 02 [skf 72] angular contact bearings specifications

Divide by 25.4 for Inch equivalents   1 kg = 2.2046 Lbs

ISI No.	Bearing No. (SKF)	d mm	D mm	B mm	r ≈ mm	r_1 ≈ mm	a ≈ mm	Basic capacity, kgf		Max. RPM Lubricant	
								Static C0	Dynamic C	Oil	Grease
15BA02	7202 B	15	35	11	1	0.5	16	375	630	13000	10000
17BA02	03 B	17	40	12	1	0.8	18	475	780	13000	10000

ISI No.	Bearing No. (SKF)	d mm	D mm	B mm	r ≈ mm	r_1 ≈ mm	a ≈ mm	Basic capacity, kgf		Max. RPM Lubricant	
								Static C_0	Dynamic C	Oil	Grease
20BA02	7204 B	20	47	14	1.5	0.8	21	655	1040	10000	8500
25BA02	7205 B	25	52	15	1.5	0.8	24	780	1160	10000	8500
30BA02	06 B	30	62	16	1.5	0.8	27	1120	1600	10000	8500
35BA02	07 B	35	72	17	2	1	31	1530	2120	8000	6700
40BA02	7208 B	40	80	18	2	1	34	1900	2500	8000	6700
45BA02	09 B	45	85	19	2	1	37	2160	2800	6000	5000
50BA02	10 B	50	90	20	2	1	39	2360	2900	6000	5000
55BA02	7211 B	55	100	21	2.5	1.2	43	3000	3650	6000	5000
60BA02	12 B	60	110	22	2.5	1.2	47	3650	4400	5000	4300
65BA02	13 B	65	120	23	2.5	1.2	50	4300	5000	5000	4300
70BA02	7214 B	70	125	24	2.5	1.2	53	4750	5400	5000	4300
75BA02	15 B	75	130	25	2.5	1.2	56	5000	5600	4000	3400
80BA02	16 B	80	140	26	3	1.5	59	5700	6300	4000	3400
85BA02	7217 B	85	150	28	3	1.5	64	6550	7100	4000	3400
90BA02	18 B	90	160	30	3	1.5	67	7650	8300	4000	3400
95BA02	19 B	95	170	32	3.5	2	71	8800	9500	3000	2400
100BA02	7220 B	100	180	34	3.5	2	76	9300	10200	3000	2400
105BA02	21 B	105	190	36	3.5	2	80	10400	11000	2500	1750
110BA02	22 B	110	200	38	3.5	2	84	11600	12000	2500	1750

Note: Abutment diameters D_1, D_2 and corner radii r, are the same as for the 62 series.

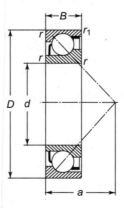

Table 5.26: Series 03 [SKF 73] angular contact bearing specifications

Divide by 25.4 for Inch equivalents   1 kg = 2.2046 Lbs

ISI No.	Bearing No. (SKF)	d mm	D mm	B mm	r ≈ mm	r_1 ≈ mm	a ≈ mm	Basic capacity, kgf		Max. RPM Lubricant	
								Static C0	Dynamic C	Oil	Grease
17BA03	7303 B	17	47	14	1.5	0.8	21	720	1160	10000	8500
20BA03	04 B	20	52	15	2	1	23	830	1370	10000	8500
25BA03	05 B	25	62	17	2	1	27	1250	1930	10000	8500
30BA03	7306 B	30	72	19	2	1	31	1700	2450	8000	6700
35BA03	07 B	35	80	21	2.5	1.2	35	2040	2850	8000	6700
40BA03	08 B	40	90	23	2.5	1.2	39	2550	3550	6000	5000
45BA03	7309 B	45	100	25	2.5	1.2	43	3400	4550	6000	5000
50BA03	10 B	50	110	27	3	1.5	47	4050	5300	6000	5000
55BA03	11 B	55	120	29	3	1.5	52	4750	6200	5000	4300
60BA03	7312 B	60	130	31	3.5	2	55	5500	7100	5000	4300
65BA03	13 B	65	140	33	3.5	2	60	6300	8000	5000	4300
70BA03	14 B	70	150	35	3.5	2	64	7350	9000	4000	3400
75BA03	7315 B	75	160	37	3.5	2	68	8150	9800	4000	3400
80BA03	16 B	80	170	39	3.5	2	72	9150	10600	4000	3400
85BA03	17 B	85	180	41	4	2	76	10200	11400	3000	2400
90BA03	7318 B	90	190	43	4	2	80	11400	12200	3000	2400
95BA03	19 B	95	200	45	4	2	84	12500	13200	2500	1750
100BA03	20 B	100	215	47	4	2	90	15300	15000	2500	1750
105BA03	7321 B	105	225	49	4	2	94	16600	16000	2500	1750
110BA03	22 B	110	240	50	4	2	99	19300	17600	2500	1750

Note: Abutment diameters D_1, D_2 and corner radii r, are the same as for 63 series.

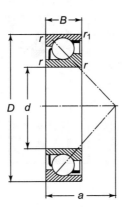

Table 5.27: Double row angular contact ball brgs 33 series: specifications

Divide by 25.4 for Inch equivalents

1 kg = 2.2046 Lbs

Bearing No (SKF)	d mm	D mm	B mm	$r \approx$ mm	Basic capacity, kgf		Max. R.P.M Lubricant	
					Static C_0	Dynamic C	Oil	Grease
3302	15	42	19	1.5	930	1400	10000	8500
03	17	47	22.2	1.5	1290	1930	8000	6700
04	20	52	22.2	2	1400	1930	8000	6700
3305	25	62	25.4	2	2000	2650	6000	5000
06	30	72	30.2	2	2750	3550	6000	5000
07	35	80	34.9	2.5	3600	4500	5000	4300
3308	40	90	36.5	2.5	4550	5500	5000	4300
09	45	100	39.7	2.5	5600	6700	4000	3400
10	50	110	44.4	3	7350	8150	4000	3400
3311	55	120	49.2	3	8000	8800	4000	3400
12	60	130	54	3.5	9650	10200	3000	2400
13	65	140	58.7	3.5	11200	11800	3000	2400
3314	70	150	63.5	3.5	12900	13700	3000	2400
15	75	160	68.3	3.5	14000	14300	2500	1750
16	80	170	68.3	3.5	16000	16300	2500	1750
3317	85	180	73	4	18000	18000	2500	1750
18	90	190	73	4	21200	20400	2500	1750

Note: Abutment diameters D_1, D_2 and corner radii r, are the same as for the 63 series.

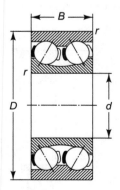

Table 5.28: Dimension of series 02 [SKF 200] single roller bearings

Divide by 25.4 for Inch equivalents   1 kg = 2.2046 Lbs   (All dimensions in mm)

IS designation				d	D	B	r	r₁	E	Basic load rating kgf		Limiting speed rpm		Equivalent SKF designation			
Type RN	Type RJ	Type RU	Type RT							Dynamic C	Static C0	Grease	Oil	Type RN (N)	Type RJ (NJ)	Type RU (NU)	Type RT (NUP)
15 RN 02	15 RJ 02	15 RU 02	—	15	35	11	1	0.5	29.3	830	430	19000	24000	N 202	NJ 202	NU 202	—
17 RN 02	17 RJ 02	17 RU 02	—	17	40	12	1	0.5	33.9	965	520	17000	20000	N 203	NJ 203	NU 203	—
20 RN 02	20 RJ 02	20 RU 02	20 RT 02	20	47	14	1.5	1	40	1040	695	15000	18000	N 204	NJ 204	NU 204	NUP 204
25 RN 02	25 RJ 02	25 RU 02	25 RT 02	25	52	15	1.5	1	45	1200	850	12000	15000	N 205	NJ 205	NU 205	NUP 205
30 RN 02	30 RJ 02	30 RU 02	30 RT 02	30	62	16	1.5	1	53.5	1560	1160	10000	13000	N 206	NJ 206	NU 206	NUP 206
35 RN 02	35 RJ 02	35 RU 02	35 RT 02	35	72	17	2	1	61.8	2360	1700	9000	11000	N 207	NJ 207	NU 207	NUP 207
40 RN 02	40 RJ 02	40 RU 02	40 RT 02	40	80	18	2	2	70	3100	2300	8500	10000	N 208	NJ 208	NU 208	NUP 208
45 RN 02	45 RJ 02	45 RU 02	45 RT 02	45	85	19	2	2	75	3250	2500	7500	9000	N 209	NJ 209	NU 209	NUP 209
50 RN 02	50 RJ 02	50 RU 02	50 RT 02	50	90	20	2	2	80.4	3400	2700	7000	8500	N 210	NJ 210	NU 210	NUP 210
55 RN 02	55 RJ 02	55 RU 02	55 RT 02	55	100	21	2.5	2	88.5	4150	3250	6300	7500	N 211	NJ 211	NU 211	NUP 211
60 RN 02	60 RJ 02	60 RU 02	60 RT 02	60	110	22	2.5	2.5	97.5	4800	4000	5600	6700	N 212	NJ 212	NU 212	NUP 212
65 RN 02	65 RJ 02	65 RU 02	65 RT 02	65	120	23	2.5	2.5	105.6	5700	4750	5300	6300	N 213	NJ 213	NU 213	NUP 213
70 RN 02	70 RJ 02	70 RU 02	70 RT 02	70	125	24	2.5	2.5	110.5	6000	5000	5000	6000	N 214	NJ 214	NU 214	NUP 214
75 RN 02	75 RJ 02	75 RU 02	75 RT 02	75	130	25	2.5	2.5	116.5	6950	5850	4800	5600	N 215	NJ 215	NU 215	NUP 215
80 RN 02	80 RJ 02	80 RU 02	80 RT 02	80	140	26	3	3	125.3	7800	6800	4500	5300	N 216	NJ 216	NU 216	NUP 216
85 RN 02	85 RJ 02	85 RU 02	85 RT 02	85	150	28	3	3	133.8	9000	7800	4300	5000	N 217	NJ 217	NU 217	NUP 217
90 RN 02	90 RJ 02	90 RU 02	90 RT 02	90	160	30	3	3	143	11000	9300	3800	4500	N 218	NJ 218	NU 218	NUP 218
95 RN 02	95 RJ 02	95 RU 02	95 RT 02	95	170	32	3.5	3	151.5	12500	11000	3600	4300	N 219	NJ 219	NU 219	NUP 219
100 RN 02	100 RJ 02	100 RU 02	100 RT 02	100	180	34	3.5	3.5	160	14000	12200	3400	4000	N 220	NJ 220	NU 220	NUP 220
105 RN 02	105 RJ 02	105 RU 02	105 RT 02	105	190	36	3.5	3.5	168.8	18371	14062	3200	3800	N 221	NJ 221	NU 221	NUP 221
110 RN 02	110 RJ02	110 RU 02	110 RT 02	110	200	38	3.5	3.5	178.5	22226	16556	3000	3600	N 222	NJ 222	NU 222	NUP 222
120 RN 02	120 RJ 02	120 RU 02	120 RT 02	120	215	40	3.5	3.5	191.5	24041	18371	2800	3400	N 224	NJ 224	NU 224	NUP 224
130 RN 02	130 RJ 02	130 RU 02	130 RT 02	130	230	40	4	4	204	25855	20639	2600	3200	N 226	NJ 226	NU 226	NUP 226
—	140 RJ 02	140 RU 02	140 RT 02	140	250	42	4	4	221	29711	24041	2400	3000	—	NJ 228	NU 228	NUP 228
—	150 RJ 02	150 RU 02	150 RT 02	150	270	45	4	4	238	34700	28123	2000	2600	—	NJ 230	NU 230	NUP 230

Note: Dimensions D1 and D2 same as 62 series ball bearings [Table 5.19]

Table 5.29: NU23 cylindrical roller bearings: Specifications

Bearing No. (SKF)	d mm	D mm	B mm	r ≈ mm	r_1 ≈ mm	F mm	Basic capacity, kfg		Max RPM lubricant	
							Static C_0	Dynamic C	Oil	Grease
NU 2305	25	62	24	2	2	35	2280	3200	10000	8500
2306	30	72	27	2	2	42	2800	3750	10000	8500
2307	35	80	31	2.5	2	46.2	3250	4500	8000	6700
NU 2308	40	90	33	2.5	2.5	53.5	4800	6100	8000	6700
2309	45	100	36	2.5	2.5	58.5	5700	7500	8000	6700
2310	50	110	40	3	3	65	7500	9300	6000	5000
NU 2311	55	120	43	3	3	70.5	8500	11000	6000	5000
2312	60	130	46	3.5	3.5	77	10600	13200	5000	4300
2313	65	140	48	3.5	3.5	83.5	12200	14600	5000	4300
NU 2314	70	150	51	3.5	3.5	90	14000	16300	5000	4300
2315	75	160	55	3.5	3.5	95.5	17300	20400	4000	3400
2316	80	170	58	3.5	3.5	103	19000	22000	4000	3400
NU 2317	85	180	60	4	4	108	20000	23200	4000	3400
2318	90	190	64	4	4	115	22800	26000	3000	2400
2319	95	200	67	4	4	121.5	26500	30000	3000	2400
NU 2320	100	215	73	4	4	129.5	31500	34500	3000	2400
2322	110	240	80	4	4	143	43000	46500	2500	1750
2324	120	260	86	4	4	154	54000	57000	2500	1750
NU 2326	130	280	93	5	5	167	65500	68000	2500	1750
2328	140	300	102	5	5	180	73500	75000	2000	1500
2330	150	320	108	5	5	193	83000	83000	2000	1500

Note: Abutment Diameters D_1, D_2 and Corner Radii r, are the same as for 63 Series

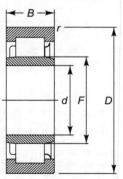

Divide by 25.4 for Inch equivalents   1 kg = 2.2046 Lbs

Table 5.30: Dimension of series 30 double roller bearing

IS designation	Boundary dimensions				Abutment dimensions			Basic load rating kfg		Limiting speed rpm		Equivalent DIN and CSN designation
	d	D	B	r Nom.	D1 Min.	D2 Max.	r1	Dynamic C	Static CO	Lubrication		
										Grease	Oil	
30 RD 30 K	30	55	19	1.5	35	50	1	2540	1814	11000	14000	NN 3006 K
35 RD 30 K	35	62	20	1.5	40	57	1	3266	2495	9500	12000	NN 3007 K
40 RD 40 K	40	68	21	1.5	45	63	1	3765	2858	9000	11000	NN 3008 K
45 RD 30 K	45	75	23	1.5	50	70	1	4218	3402	8000	9500	NN 3009 K
50 RD 30 K	50	80	23	1.5	55	75	1	4536	3765	7500	9000	NN 3010 K
55 RD 30 K	55	90	26	2	61.5	83.5	1	5988	4990	6700	8000	NN 3011 K
60 RD 30 K	60	95	26	2	66.5	88.5	1	6350	5443	6300	7500	NN 3012 K
65 RD 30 K	65	100	26	2	71.5	93.5	1	6486	5670	6000	7000	NN 3013 K
70 RD 30 K	70	110	30	2	76.5	103.5	1	8437	7530	5300	6300	NN 3014 K
75 RD 30 K	75	115	30	2	81.5	108.5	1	8437	7530	5000	6000	NN 3015 K
80 RD 30 K	80	125	34	2	86.5	118.5	1	10342	9435	4800	5600	NN 3016 K
85 RD 30 K	85	130	34	2	91.5	123.5	1	10704	10161	4500	5300	NN 3017 K
90 RD 30 K	90	140	37	2.5	98	132	1.5	12474	11794	4300	5000	NN 3018 K
95 RD 30 K	95	145	37	2.5	103	137	1.5	12928	12247	4000	4800	NN 3019 K
100 RD 30 K	100	150	37	2.5	108	142	1.5	13608	13154	3800	4500	NN 3020 K
105 RD 30 K	105	160	41	3	114	151	2	16556	16103	3600	4300	NN 3021 K
110 RD 30 K	110	170	45	3	119	161	2	19958	19278	3400	4000	NN 3022 K
120 RD 30 K	120	180	46	3	129	171	2	20639	20639	3200	3800	NN 3024 K
130 RD 30 K	130	200	52	3	139	191	2	25855	25402	2800	3400	NN 3026 K
140 RD 30 K	140	210	53	3	149	201	2	27216	27670	2600	3200	NN 3028 K
150 RD 30 K	150	225	56	3.5	161	214	2	29711	30391	2400	3000	NN 3030 K
160 RD 30 K	160	240	60	3.5	171	225	2	33340	34700	2200	2800	NN 3032 K
170 RD 30 K	170	260	67	3.5	181	249	2	40824	43090	2000	2600	NN 3034 K
180 RD 30 K	180	280	74	3.5	191	269	2	50803	53525	1900	2400	NN 3036 K
190 RD 30 K	190	290	75	3.5	201	275	2	53525	57607	1900	2400	NN 3038 K
200 RD 30 K	200	310	82	3.5	211	299	2	58514	62143	1800	2200	NN 3040 K
220 RD 30 K	220	340	90	4	233	327	2.5	73937	79834	1700	2000	NN 3044 K
240 RD 30 K	240	360	92	4	253	347	2.5	78473	86184	1500	1800	NN 3048 K
260 RD 30 K	260	400	104	5	276	384	3	94349	103420	1300	1600	NN 3052 K
280 RD 30 K	280	430	106	5	296	404	3	97978	111132	1200	1500	NN 3056 K
300 RD 30 K	300	460	118	5	316	444	3	117936	136080	1100	1400	NN 3060 K
320 RD 30 K	320	480	121	5	336	464	3	122472	140616	1000	1300	NN 3064 K
340 RD 30 K	340	520	133	6	360	500	4	147420	172368	950	1200	NN 3068 K
360 RD 30 K	360	540	134	6	380	520	4	154224	188244	950	1200	NN 3072 K

1. Divide by 25.4 for Inch equivalents

2. 1 kg = 2.2046 Lbs

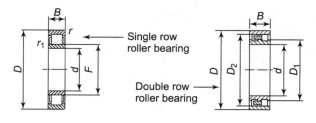
Single row roller bearing

Double row roller bearing

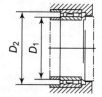

Table 5.31: Ring gauges for RD30K bearings

Divide by 25.4 for Inch equivalents   1 kg = 2.2046 Lbs

Ring gauge No.	Bearing No. SP & UP classes	Ring gauge dimensions			Dimensioning of bearing seating				Tol. on	
		d	D	B	d_1	a^*	b	c	c	d_2
GRA 3006	30 RD 30 K	30	52	19	30.1	4	24	6.2	0.1	32
GRA 3007	35 RD 30 K	35	57	20	35.1	6	25	6.2	0.1	37
GRA 3008	40 RD 30 K	40	62	21	40.1	6	28	8.2	0.1	42
GRA 3009	45 RD 30 K	45	67	23	45.1	8	30	8.2	0.1	47
GRA 3010	50 RD 30 K	50	72	23	50.1	8	30	8.2	0.1	52
GRA 3011	55 RD 30 K	55	77	26	55.15	8	32.5	8.3	0.12	57
GRA 3012	60 RD 30 K	60	82	26	60.15	8	34.5	10.3	0.12	62
GRA 3013	65 RD 30 K	65	88	26	65.15	10	34.5	10.3	0.12	67
GRA 3014	70 RD 30 K	70	95	30	70.15	10	38.5	10.3	0.12	73
GRA 3015	75 RD 30 K	75	100	30	75.15	10	38.5	10.3	0.12	78
GRA 3016	80 RD 30 K	80	105	34	80.15	10	44.5	12.3	0.12	83
GRA 3017	85 RD 30 K	85	112	34	85.2	12	44	12.4	0.15	88
GRA 3018	90 RD 30 K	90	120	37	90.2	12	47	12.4	0.15	93
GRA 3019	95 RD 30 K	95	128	37	95.2	12	47	12.4	0.15	98
GRA 3020	100 RD 30 K	100	135	37	100.2	12	47	12.4	0.15	103

Notes: To allow for final adjustment when mounting, initial width of distance ring should be a + 0.2 mm. Final width should be determined during mounting when the desired internal clearance in the bearing is obtained. All measurement must be made from the polished surface at the large end of the taper bore on the ring gauge.

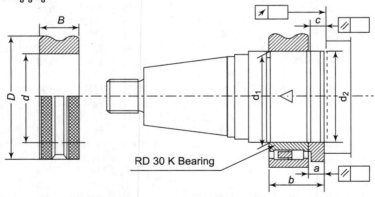

RD 30 K Bearing

The permissible axial runout of shoulders, and errors in parallelism of distance rings and between gauge reference face and shoulder are as follows:

Class of RD 30 K bearings	Values in μm Diameter range in mm (OD)			
	18–50	50–120	120–180	> 180
SP	3	4	5	7
UP	2	3	4	5

Table 5.32: Series 322 taper roller bearings

	Brg No.	d	D_1 max	D_2 min	D	D_3 min	D_4 min	B	C	T	r brg	r_1 brg	r abut	a_1	a_2	a ≈	Basic capacity kgf Static C_0	Basic capacity kgf Dynamic C	Max R.P.M. Lubricant Grease	Max R.P.M. Lubricant Oil
322	06 A	30	37	36	62	52	57	20	17	21.25	1.5	0.5	1	2	4	15	3850	4380	6300	8500
	07 A	35	43	42	72	61	67	23	19	24.25	2	0.5	1	3	5	18	5080	5760	5000	7000
	08 A	40	48	47	80	68	78.5	23	19	24.75	2	0.5	1	3	5.5	19	5440	6350	4500	6000
	09 A	45	53	52	85	73	80	23	19	24.75	2	0.8	1	3	5.5	20	5760	6480	4500	6000
322	10 A	50	58	57	90	78	85	23	19	24.75	2	0.8	1	3	5.5	21	6480	7260	4000	5300
	11 A	55	63	64	100	87	94	25	21	26.75	2.5	0.8	1.5	3	5.5	22	7980	8900	3600	4700
	12 A	60	69	69	110	95	102	28	24	29.75	2.5	0.8	1.5	4	5.5	24	10000	10700	3200	4300
	13 A	65	75	74	120	105	112	31	27	32.75	2.5	0.8	1.5	4	5.5	26	11800	12700	2800	3800
	14 A	70	80	79	125	108	117	31	27	33.25	2.5	0.8	1.5	4	5.5	28	12700	13600	2800	3800
	15 A	75	85	84	130	113	123	31	27	33.25	2.5	0.8	1.5	4	6	29	13600	14000	2500	3500
	16 A	80	90	90	140	122	132	33	28	33.25	3	1	2	4	6	30	15200	16100	2500	3500
	17 A	85	96	95	150	130	140	36	30	38.5	3	1	2	5	7	33	17700	18350	2200	3200
	18 A	90	102	100	160	138	150	40	34	42.5	3	1	2	5	8	36	21100	21500	2000	3000
	19 A	95	108	107	170	146	158	43	37	45.5	3.5	1.2	2	5	8	38	24900	25000	2000	3000
322	20 A	100	114	112	180	155	168	46	39	49	3.5	1.2	2	5	10	41	28600	28100	1800	2600
	21 A	105	120	117	190	163	178	50	43	53	3.5	1.2	2	5	10	44	32650	32200	1800	2600
	22 A	110	125	122	200	171	188	53	46	56	3.5	1.2	2	8	10	46	35400	34000	1600	2200
322	24	120	135	132	215	184	203	58	50	61.5	3.5	1.2	2	9	10	52	34700	36290	1600	2200
	26	130	146	144	230	193	219	64	53	67.75	4	1.5	2.5	9	14	44	41504	43092	1500	2000
	28	140	158	154	250	210	238	68	55	71.75	4	1.5	2.5	9	14	47	48989	49989	1400	1900

Divide by 25.4 for Inch equivalents

1 kg = 2.2046 Lbs

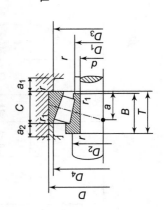

Taper roller bearing

Table 5.33: Series 323 taper roller bearings [Fig. same as for Table 5.32]

Brg No.	d	D_1 max	D_2 min	D	D_3 min	D_4 min	B	C	T	r brg	r_1 brg	r abut	a_1	a_2	a	Basic Capacity kgf Static C_0	Basic Capacity kgf dynamic C	Max R.P.M. lubricant Grease	Max R.P.M. lubricant Oil
323 06 A	30	38	37	72	61	66	27	23	28.75	2	0.8	1	2	5.5	17	5670	6600	5000	6700
07 A	35	43	44	80	68	74	31	25	32.75	2.5	0.8	1.5	3	7.5	20	7255	8300	4500	6000
08 A	40	50	49	90	76	82	33	27	35.25	2.5	0.8	1.5	3	8	23	8750	9800	4000	5300
09 A	45	56	54	100	85	93	36	30	38.25	2.5	0.8	1.5	3	8	25	12000	12700	3600	4700
323 10 A	50	62	60	110	94	102	40	33	42.25	3	1	2	3	9	28	14500	15100	3200	4300
11 A	55	67	65	120	103	111	43	35	45.5	3	1	2	4	10	29	16300	17200	2800	3800
12 A	60	73	72	130	112	120	46	37	48.5	3.5	1.2	2	4	11	31	18800	20000	2500	3500
323 13 A	65	80	77	140	121	130	48	39	51	3.5	1.2	2	4	11.5	33	21500	22200	2500	3500
14 A	70	85	82	150	129	140	51	42	54	3.5	1.2	2	4	11.5	36	25400	25800	2200	3200
15 A	75	91	87	160	138	149	55	45	58	3.5	1.2	2	4	12.5	38	29000	29700	2000	3000
16	80	97	92	170	147	159	58	48	61.5	3.5	1.2	2	4	13	40	24500	26500	2000	3000
17	85	102	99	180	155	167	60	49	63.5	4	1.5	2.5	9	14	42	28600	30400	1800	2600
18	90	108	104	190	163	177	64	53	67.5	4	1.5	2.5	9	14	44	29700	30850	1800	2600
19	95	113	109	200	171	186	67	55	71.5	4	1.5	2.5	10	14	47	36300	37600	1600	2200
323 20	100	121	114	215	183	200	73	60	77.5	4	1.5	2.5	11	16	51	41500	41500	1600	2200
21	105	127	119	225	193	209	77	63	81.5	4	1.5	2.5	11	17	54	46300	46250	1400	1900
22	110	135	124	240	205	222	80	65	84.5	4	1.5	2.5	12	17	56	49900	48000	1400	1900
24	120	145	134	260	219	239	86	69	90.5	4	1.5	2.5	12	18	60	56700	54500	1200	1600

Note: Suffix A, denotes high load carrying capacity. Divide by 25.4 for Inch equivalents. 1 kg = 2.2046 Lbs

Machine Tool Elements

Table 5.34: Series 222 Spherical roller bearings

Bearing with cylindrical bore	d mm	D mm	B mm	E ≈ mm	r ≈ mm	Basic capacity, kgf		Max R.P.M lubricant	
						Static C0	Dynamic C	Grease	Oil
*22205 C	25	52	18	32	1.5	2200	3000	8000	9500
*06 C	30	62	20	38	1.5	3100	4500	6000	7500
07 C	35	72	23	44	2	4050	5200	6000	7500
22208 C	40	80	23	50	2	4800	6200	5000	6000
09 C	45	85	23	55	2	5200	6400	5000	6000
10 C	50	90	23	60	2	5500	6700	4000	4800
22211 C	55	100	25	66	2.5	6800	8300	4000	4800
12 C	60	110	28	73	2.5	8500	10000	4000	4800
13 C	65	120	31	79	2.5	10200	11800	3000	3700
22214 C	70	125	31	84	2.5	10800	12200	3000	3700
15 C	75	130	31	90	2.5	11200	12700	3000	3700
16 C	80	140	33	95	3	14000	15300	2500	3100
22217 C	85	150	36	100	3	16300	18000	2500	3100
18 C	90	160	40	107	3	19600	20800	2500	3100
19 C	95	170	43	113	3.5	22800	24500	2000	2700
22220 C	100	180	46	120	3.5	25500	27000	2000	2700
22 C	110	200	53	132	3.5	34500	34500	2000	2700
24 C	120	215	58	143	3.5	40000	40000	1600	2200
22226 C	130	230	64	153	4	48000	46500	1600	2200
28 C	140	250	68	167	4	54000	53000	1600	2200
30 C	150	270	73	179	4	65500	64000	1300	1800
22232 C	160	290	80	191	4	78000	73500	1300	1800
34 C	170	310	86	204	5	90000	83000	1000	1500
36 C	180	320	86	213	5	95000	88000	1000	1500
22238 C	190	340	92	226	5	102000	95000	1000	1500
40 C	200	360	98	238	5	116000	108000	800	1200
44 C	220	400	108	264	5	143000	129000	800	1200

Note: For bearings with taper bore add K-Example 22208 CK. Taper 1:12 on diameter.
* Not available with taper bore.
Abutment diameters D_1, D_2 and corner radii r, are the same as for the 62 series [Table 5.19].
Divide by 25.4 for Inch equivalents. 1 kg = 2.2046 Lbs

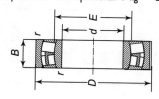

Table 5.35: Light series needle bearings (All dimensions in mm)

Needle bearing with inner ring	Needle bearing without inner ring	d Nom.	di Nom.	D Nom.	B Nom.	r	Basic load rating Dynamic C kgf	Basic load rating Static C_0 kgf	Limiting speed rpm	Equivalent NRB designation Needle bearing with inner ring	Equivalent NRB designation Needle bearing without inner ring
NEA 1012	NES 1012	12	17.6	28	15	1	1150	960	21600	Na 1012	Na 1012 S/Bi
NEA 1015	NES 1015	15	20.8	32	15	1	1285	1110	18300	Na 1015	Na 1015 S/Bi
NEA 1017	NES 1017	17	23.9	35	15	1	1420	1250	15900	Na 1017	Na 1017 S/Bi
NEA 1020	NES 1020	20	28.7	42	18	1	2000	1870	13200	Na 1020	Na 1020 S/Bi
NEA 1025	NES 1025	25	33.5	47	18	1	2200	2160	11100	Na 1025	Na 1025 S/Bi
NEA 1030	NES 1030	30	38.2	52	18	1	2440	2440	10000	Na 1030	Na 1030 S/Bi
NEA 1035	NES 1035	35	44	58	18	1	2700	2790	8600	Na 1035	Na 1035 S/Bi
NEA 1040	NES 1040	40	49.7	65	18	1.5	2930	3130	7600	Na 1040	Na 1040 S/Bi
NEA 1045	NES 1045	45	55.4	72	18	1.5	3160	3470	6900	Na 1045	Na 1045 S/Bi
NEA 1050	NES 1050	50	62.1	80	20	2	3420	3860	6100	Na 1050	Na 1050 S/Bi
NEA 1055	NES 1055	55	68.8	85	20	2	3690	4250	5500	Na 1055	Na 1055 S/Bi
NEA 1060	NES 1060	60	72.6	90	20	2	3820	4460	5200	Na 1060	Na 1060 S/Bi
NEA 1065	NES 1065	65	78.3	95	20	2	4240	5050	4900	Na 1065	Na 1065 S/Bi
NEA 1070	NES 1070	70	83.1	100	20	2	4420	5390	4500	Na 1070	Na 1070 S/Bi
NEA 1075	NES 1075	75	88	110	24	2	6600	8200	4300	Na 1075	Na 1075 S/Bi
NEA 1080	NES 1080	80	96	115	24	2	7000	8850	4000	Na 1080	Na 1080 S/Bi

Divide by 25.4 for Inch equivalents of mm dimensions   1 kg = 2.2046 Lbs

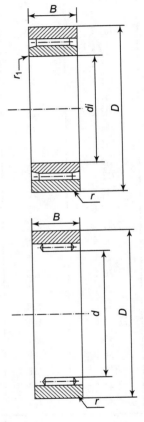

Machine Tool Elements

Table 5.36: Medium series needle bearings [Fig. same as Table 5.35]

Divide mm dimensions by 25.4 for Inch equivalents. 1 kg = 2.2046 Lbs

(All dimensions in mm)

Needle bearing with inner ring	Needle bearing without inner ring	d Nom.	di Nom.	D Nom.	B Nom.	r	Basic load rating Dynamic C kgf	Basic load rating Static C_0 kgf	Limiting speed rpm	Equivalent NRB designation Needle bearing with inner ring	Equivalent NRB designation Needle bearing without inner ring
NEA 2015	NES 2015	15	22.1	35	22	1	2480	2200	17200	Na 2015	Na 2015 S/Bi
NEA 2020	NES 2020	20	28.7	42	22	1	2980	2680	13200	Na 2020	Na 2020 S/Bi
NEA 2025	NES 2025	25	33.5	47	22	1	3220	3220	11100	Na 2025	Na 2025 S/Bi
NEA 2030	NES 2030	30	38.2	52	22	1	3640	3640	10000	Na 2030	Na 2030 S/Bi
NEA 2035	NES 2035	35	44	58	22	1	4000	4150	8600	Na 2035	Na 2035 S/Bi
NEA 2040	NES 2040	40	49.7	62	22	1	4380	4650	9600	Na 2040	Na 2040 S/Bi
NEA 2045	NES 2045	45	55.4	72	22	1.5	4750	5150	6900	Na 2045	Na 2045 S/Bi
NEA 2050	NES 2050	50	62.1	80	22	1.5	6500	7600	6100	Na 2050	Na 2050 S/Bi
NEA 2055	NES 2055	55	68.8	85	28	2	7250	8400	5500	Na 2055	Na 2055 S/Bi
NEA 2060	NES 2060	60	72.6	90	28	2	7550	8600	5200	Na 2060	Na 2060 S/Bi
NEA 2065	NES 2065	65	78.3	95	28	2	8200	9750	4900	Na 2065	Na 2065 S/Bi
NEA 2070	NES 2070	70	83.1	100	28	2	8500	10300	4500	Na 2070	Na 2070 S/Bi
NEA 2075	NES 2075	75	88	110	32	2	10800	13400	4300	Na 2075	Na 2075 S/Bi
NEA 2080	NES 2080	80	96	115	32	2	11500	14600	4000	Na 2080	Na 2080 S/Bi
NEA 2085	NES 2085	85	99.5	120	32	2	11800	15100	3800	Na 2085	Na 2085 S/Bi
NEA 2090	NES 2090	90	104.7	125	32	2	12500	15950	3600	Na 2090	Na 2090 S/Bi
NEA 2095	NES 2095	95	109.1	130	32	2	12100	16600	3500	Na 2095	Na 2095 S/Bi
NEA 2100	NES 2100	100	114.7	135	32	2	12800	17400	3300	Na 2100	Na 2100 S/Bi
NEA 2105	NES 2105	105	119.2	140	32	2	13400	18100	3200	Na 2105	Na 2105 S/Bi
NEA 2110	NES 2110	110	124.7	145	34	2	13900	18900	3000	Na 2110	Na 2110 S/Bi
NEA 2115	NES 2115	115	132.5	155	34	2	14400	20000	2900	Na 2115	Na 2115 S/Bi
NEA 2120	NES 2120	120	137	160	34	2	14800	20600	2800	Na 2120	Na 2120 S/Bi
NEA 2125	NES 2125	125	143.5	165	34	2	15200	21600	2700	Na 2125	Na 2125 S/Bi
NEA 2130	NES 2130	130	148	170	34	2	15500	22100	2600	Na 2130	Na 2130 S/Bi
NEA 2140	NES 2140	140	158	180	36	2	16300	23800	2400	Na 2140	Na 2140 S/Bi
NEA 2150	NES 2150	150	170.5	195	36	2	17200	25300	2200	Na 2150	Na 2150 S/Bi
NEA 2160	NES 2160	160	179.3	205	36	2	17800	26800	2100	Na 2160	Na 2160 S/Bi
NEA 2170	NES 2170	170	193.8	220	42	3	24300	37500	2000	Na 2170	Na 2170 S/Bi
NEA 2180	NES 2180	180	202.6	230	42	3	25100	39200	1900	Na 2180	Na 2180 S/Bi
NEA 2190	NES 2190	190	216	240	42	3	26200	41700	1800	Na 2190	Na 2190 S/Bi
NEA 2200	NES 2200	200	224.1	255	42	3	26800	43200	1700	Na 2200	Na 2200 S/Bi
NEA 2210	NES 2210	210	236	265	42	3	28800	45500	1600	Na 2210	Na 2210 S/Bi
NEA 2220	NES 2220	220	248.4	280	49	3	34400	56800	1500	Na 2220	Na 2220 S/Bi
NEA 2230	NES 2230	230	258.4	290	49	3	35300	59000	1500	Na 2230	Na 2230 S/Bi
NEA 2240	NES 2240	240	269.6	300	49	3	36400	61500	1400	Na 2240	Na 2240 S/Bi
NEA 2250	NES 2250	250	281.9	315	49	3	37500	64500	1300	Na 2250	Na 2250 S/Bi

Table 5.37: Double needle bearings [Fig. on page 447]

Divide mm dimensions by 25.4 for Inch equivalents. 1 kg = 2.2046 Lbs

Brg. No.	d	D_1	D	B	r	Basic Static $C0$	Capacity, kgf dynamic C	Max. speed RPM
NA 69/32	32	40	52	36	1	4550	4100	10000
07	35	42	55	36	1	4750	4200	9000
08	40	48	62	40	1	6400	5800	8000
09	45	52	68	40	1	6800	6000	8000
10	50	58	72	40	1	7400	6200	7000
12	60	68	85	45	1.5	10100	8100	6000
14	70	80	100	54	1.5	14500	11300	5000
16	80	90	110	54	1.5	16000	11900	4000
18	90	105	125	63	2	22800	15100	4000

Table 5.38a: Type-B needle cages [Fig. on page 448]

(All dimension in mm)

NRB designation	C_i	C_e	L	Basic load rating Dynamic C kgf	Basic load rating Static $C0$ kgf	Limiting speed RPM
B 698	6	9	8	288	175	65000
B 61013	6	10	13	473	305	65000
B 81110	8	11	10	418	287	50000
B 91210	9	12	10	440	310	44000
B 10139	10	13	9	336	222	40000
B 101313	10	13	13	530	403	40000
B 121513	12	15	13	580	461	33000
B 50386	12	15	13	580	461	33000
B 50255	12	15	13.7	575	456	33000
B 50114	13	16	14	575	461	31000
B 141813	14	18	13	790	605	28000
B 50160	14.4	20.4	19.9	1400	1030	28000
B 50320	14.8	19.8	10	585	368	27000
B 151913	15	19	13	820	645	27000
B 5019015 × 19 × 20	15	19	20	970	785	27000
B 50113	15.2	22.21	12	1010	645	26000
B 162013	16	20	13	850	685	25000
B 5079216 × 20 × 20	16	20	20	1155	1010	25000
B 50139	16	21	10	680	453	25000
B 16221216 × 22 × 12	16	22	12	990	650	25000
B 162217.2	16	22	17.2	1160	815	25000
B 172115	17	21	15	1030	885	23000
B 172717	17	27	17	1780	1365	23000
B 182213	18	22	13	910	760	22000
B 5078718 × 22 × 21.6	18	22	21.6	1400	1290	22000
B 182420	18	24	20	1650	1310	22000
B 182421	18	24	21	1650	1310	22000

NRB designation	Ci	Ce	L	Basic load rating		Limiting speed RPM
				Dynamic C kgf	Static C_0 kgf	
B 182616	18	26	16	1660	1150	22000
B 182816	18	28	16	1950	1310	22000
B 192313	19	23	13	940	795	21000
B 202413	20	24	13	965	835	20000
B 202417	20	24	17	1290	1210	20000
B 202613.6	20	26	13.6	1080	770	20000
B 222613	22	26	13	985	875	18000
B 50650	22	26	14	985	875	18000
B 222617	22	26	17	1320	1270	18000
B 5074322 × 29 × 15.6	22	29	15.6	1730	1310	18000

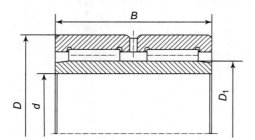

Double needle bearings. Dimensions on p 446.

Table 5.38b: Type-B needle cages [Contd.]

Divide mm dimensions by 25.4 for Inch values. 1 kg = 2.2046 kg Lbs (All dimensions in mm)

NRB designation	Ci	Ce	L	Basic load rating		Limiting speed RPM
				Dynamic C kgf	Static C_0 kgf	
B 5079322.9 × 28.9 × 13.7	22.9	28.9	13.7	1375	1040	17000
B 50010	23.95	27.92	19.2	1200	1150	17000
B 5079424 × 28 × 10	24	28	10	780	660	16000
B 50468	24.8	33.8	25.2	3180	2580	16000
B 252913	25	29	13	1060	985	16000
B 50173	25	29	16	1100	1020	16000
B 252917	25	29	17	1420	1440	16000
B 50008	25	29	18.4	1100	1020	16000
B 253020	25	30	20	1840	1750	16000
B 253124	25	31	24	2350	2160	16000
B 283317	28	33	17	1770	1710	14000
B 283327	28	33	27	2780	3060	14000
B 283825.2	28	38	25.2	3890	3230	14000
B 284421	28	44	21	4110	2850	14000

NRB designation	C_i	C_e	L	Basic load rating		Limiting speed RPM
				Dynamic C kgf	Static C_0 kgf	
B 50105	29.5	33.5	18.4	1140	1120	13500
B 303517	30	35	17	1650	1820	13000
B 303527	30	35	27	2830	3180	13000
B 304030	30	40	30	4755	4150	13300
B 323717	32	37	17	1890	1920	12500
B 323727	32	37	27	2960	3430	12500
B 50174	35	40	13	1240	1140	11500
B 354017	35	40	17	1960	2050	11500
B 354027	35	40	27	3070	3670	11500
B 50329	35	42	16	2240	1990	11500
B 374227	37	42	27	3190	3920	11000
B 404517	40	45	17	2100	2330	10000
B 404527	40	45	27	3290	4160	10000
B 424717	42	47	17	2170	2460	9500
B 424727	42	47	27	3400	4400	9500
B 425430.7	42	54	30.7	5500	4800	9500
B 454919.2	45	49	19.2	1840	2320	8900
B 455017	45	50	17	2230	2600	9000
B 455021	45	50	21	2745	3430	9000
B 455027	45	50	27	3500	4650	9000
B 455218	45	52	18	2630	2600	9000
B 455221	45	52	21	4450	4750	9000
B 455320	45	53	20	4450	4750	9000
B 475217	47	52	17	2290	2740	8500
B 475227	47	52	27	3600	4890	8500
B 485317	48	53	17	2280	2740	8500
B 505520	50	55	20	2490	3090	8000
B 505530	50	55	30	3650	5050	8000
B 505623	50	56	23	3110	3640	8000

Divide mm dimensions by 25.4 for Inch equivalents. 1 kg = [2.2046 kg] Lbs

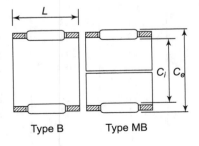

Type B   Type MB

Table 5.38c: Types-B, MB needle cages [Fig. on p448]

Divide mm dimensions by 25.4 for Inch equivalents (All dimensions in mm)

NRB designation	C_i	C_e	L	Basic load rating		Limiting speed RPM
				Dynamic C kgf	Static C_0 kgf	
B 505825	50	58	25	4390	4730	8000
B 556020	55	60	20	2580	3340	7500
B 556030	55	60	30	3550	5000	7500
B 556324.9	55	63	24.9	4280	4680	7500
B 606520	60	65	20	2720	3660	6700
B 606530	60	65	30	3980	6000	6700
B 606825	60	68	25	4820	5600	6700
B 657020	65	70	20	2800	3900	6000
B 657023	65	70	23	3480	6100	6000
B 657030	65	70	30	4100	6400	6000
B 657323	65	73	23	4600	5475	6000
B 657325	65	73	25	5050	6100	6000
B 707620	70	76	20	3270	4310	5700
B 707630	70	76	30	4920	7300	5700
B 707825	70	78	25	5200	6400	5700
B 737920	73	79	20	3350	4500	5500
B 758330	75	83	30	6100	8050	5300
B 808620	80	86	20	3530	4970	5000
B 808830	80	88	30	6400	8700	5000
B 859220	85	92	20	3650	4700	4700
B 859330	85	93	30	6500	9100	4700
B 909830	90	98	30	6700	9500	4500
MB 50265	19	23	17.2	900	745	21000
MB 283317	28	33	17	1710	1650	14000
MB 323714.2	32	37	14.2	1250	1130	12500
MB 50629	32	37	20	1880	1905	12500
MB 50133	39	44	24	2450	2830	10200

1 kg = 2.2046 Lbs

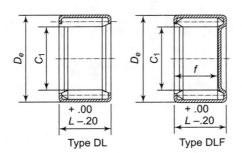

Type DL   Type DLF   Type DLH

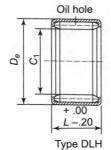

Oil hole

For dimensions refer page 450

Table 5.38d: Types DL, DLH, DLF, DLFH needle bushes (See figures on p. 449)

Divide mm dimensions by 25.4 for Inch equivalents.   1 kg = 2.2046 Lbs

NRB designation								Basic load rating*		
DL DLH DLF DLFH	DL/. . . . DLH/. . . . DLF/. . . . DLFH//. .	Di	Ci	De	L	F	Dynamic C kgf	Static C_0 kgf	Limiting speed RPM
610	–	–	6	12	10	7.8	309	274	50000
810	–	–	8	14	10	7.8	386	347	37500
DLC910	–	–	9	13	10	–	570	480	33000
914 12	–	–	9	14	12	9.8	630	562	33000
1012	–	–	10	16	12	9.8	640	593	30000
1210	–	–	12	18	10	–	525	496	25000
1212	1212/08	8	12	18	12	9.8	735	699	25000
1212	1212/09	9	12	18	12	–	–	–	–
1312	1312/09	9	13	19	12	9.8	766	737	23000
1412	1412/10	10	14	20	12	9.8	825	797	21500
1412	1412/11	11	14	20	12	–	–	–	–
1512	1512/11	11	15	21	12	9.8	855	840	20000

* For shaft raceway hardness of 60 *HRC*.

Note: Needle bushes types DL, DLF, and DLH do not have inner rings. Type DL is otherwise similar to type DLH, while type DLF is similar to type DLFH.

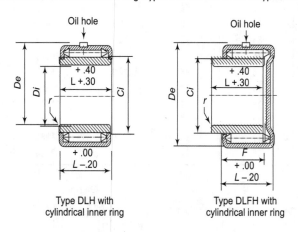

Type DLH with cylindrical inner ring   Type DLFH with cylindrical inner ring

Table 5.38e: Types DL, DLH, DLF, DLFH needle bushes [Contd.]

DL DLH DLF DLFH	DL/... DLH/... DLF/... DLFH//..	D_i	C_i	D_e	L	F	Basic load rating		Limiting speed RPM
							Dynamic C kgf	Static C_0 kgf	
1512	1512/12	12	15	21	12	–	855	840	20000
DLC 1516	–	–	15	21	16	–	1200	1180	20000
1612	1612/12	12	16	22	12	9.8	912	902	18500
1712	1712/13	13	17	23	12	9.8	945	940	17500
1812	1812/13	13	18	24	12	9.8	992	1000	16500
1816	1816/13	13	18	24	16	13.8	1555	1575	16500
2012	2012/15	15	20	26	12	9.8	1070	1100	15000
2016	2016/15	15	20	26	16	13.8	1680	1730	15000
2216	2216/17	17	22	28	16	13.8	1795	1890	13500
2516	2516/20	20	25	33	16	13.8	1680	1890	12000
2520	2520/20	20	25	33	20	17.8	2390	2690	12000
2820	2820/23	23	28	36	20	17.8	2590	2970	11000
3016	3016/25	25	30	38	16	13.8	1900	2210	10000
3020	3020/25	25	30	38	20	17.8	2720	3160	10000
3025	–	–	30	38	25	–	3720	4340	10000
3542 16P	–	–	35	42	16	–	2240	2690	8500
3516	3516/30	30	35	43	16	13.8	2120	2560	8500
3520	3520/30	30	35	43	20	17.8	3020	3640	8500
3520	3520/28	28	35	43	20	–	”	”	”
4016*	4016/35	35	40	48	16	13.8	2330	2900	7500
4020	4020/35	35	40	48	20	17.8	3210	4120	7500
4416	4416/40	40	44	52	16	13.8	2500	3160	6800
4516(P)	4516/40	40	45	52	16	–	2750	3450	6500
4716	–	–	47	55	16	–	2620	3370	6400
5020	5020/45	45	50	58	20	–	3900	5030	6000
5520	5520/50	50	55	63	20	17.8	4025	5550	5500

Table 5.39: Combined radial and thrust needle bearings

Brg designation		d_r	D	D_3	H	L	a	l	r	Basic capacity kgf				Max. speed RPM
										Static		Dynamic		
										radial C_0	axial C_{0a}	radial C	axial C_a	
NKXR	15	15	24	28.2	9	23	25.2	6.5	0.5	750	1210	950	1050	6500
	17	17	26	30.2	9	25	27.2	8	0.5	870	1320	1050	1100	6000
	20	20	30	35	10	30	30.7	10.5	0.5	1350	2600	1440	1910	5000
	25	25	37	44	11	30	39.7	9.5	1	1680	4050	1660	2550	4000
	30	30	42	49	11	30	44.7	9.5	1	1980	4500	1980	2700	3000
	35	35	47	55	12	30	50.7	9	1	2270	4950	2140	3150	2500
	40	40	52	67	13	32	61.7	10	1	2550	9900	2290	5300	2000
	45	45	58	71.4	14	32	67.2	9	1	2850	9100	2420	5300	1800
	50	50	62	76.4	14	35	72.2	10	1	4000	10100	3350	5600	1600

* The inner race can be ordered separately.
† The above bearings are also available with dust shield, when designated as NKXR Z.
Divide mm dimensions by 25.4 for Inch equivalents.

1 kg = 2.2046 Lbs

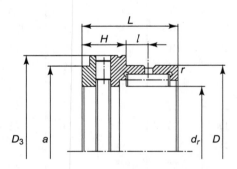

Combined radial and thrust needle bearings dimensions in Table 5.39

Table 5.40: Flat cages (Linear needle bearings for slideways) series FF AND FF···ZW

(All dimensions in mm)

Brg designation		b	B	L	a	No. of needles	d × l	E +0.1	H −0.2	Basic capacity, kgf	
										Co	C
FF	2010	10	—	32	2	7	2 × 6.8	10.3	1.7	870	540
	2515	15	—	45	2.4	8	2.5 × 9.8	15.3	2.2	1810	1190
	3020	20	—	60	3	9	3 × 13.8	20.4	2.7	3490	2430
	3525	25	—	75	3.2	10	3.5 × 17.8	25.4	3.2	5900	4300
FF	2025 ZW	10	25	32	2	14	2 × 6.8	25.3	1.7	1740	930
	2535 ZW	15	35	45	2.4	16	2.5 × 9.8	35.3	2.2	3620	2000
	3045 ZW	20	45	60	3	18	3 × 13.8	45.4	2.7	6980	4100
	3555 ZW	25	55	75	3.2	20	3.5 × 17.8	55.4	3.2	11800	7360

Divide mm dimensions by 25.4 for Inch equivalents.   1 kg = 2.2046 Lbs

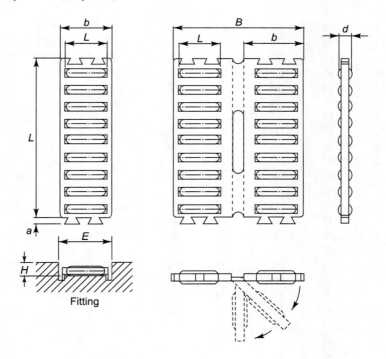

Fitting

Table 5.41a: 512 series single thrust bearings [See figure on p. 455]

Divide mm dimensions by 25.4 for Inch equivalents.

1 kg = =2.2046 Lbs

ISI No.	Bearing No. (SKF)	d mm	D mm	H mm	d_2 min. mm	r ≈ mm	Basic capacity, kgf Static C_0	Basic capacity, kgf Dynamic C	Max. permissible speed rpm
10 TA 12	51200	10	26	11	10.2	1	1400	1000	10000
12 TA 12	01	12	28	11	12.2	1	1560	1040	8000
15 TA 12	02	15	32	12	15.2	1	2040	1220	8000
17 TA 12	51203	17	35	12	17.2	1	2200	1270	8000
20 TA 12	04	20	40	14	20.2	1	3100	1730	6000
25 TA 12	05	25	47	15	25.2	1	4150	2160	6000
30 TA 12	51206	30	52	16	30.2	1	4800	2280	6000
35 TA 12	07	35	62	18	35.2	1.5	6400	3050	5000
40 TA 12	08	40	68	19	40.2	1.5	7650	3450	5000
45 TA 12	51209	45	73	20	45.2	1.5	8650	3650	4000
50 TA 12	10	50	78	22	50.2	1.5	9150	3750	4000
55 TA 12	11	55	90	25	55.2	1.5	13200	5500	3000
60 TA 12	51212	60	95	26	60.2	1.5	14600	5700	3000
65 TA 12	13	65	100	27	65.2	1.5	15600	5850	2500
70 TA 12	14	70	105	27	70.2	1.5	16300	6000	2500
75 TA 12	51215	75	110	27	75.2	1.5	17300	6100	2500
80 TA 12	16	80	115	28	80.2	1.5	18000	6200	2000
85 TA 12	17	85	125	31	85.2	1.5	22000	7500	2000
90 TA 12	51218	90	135	35	90.2	2	27000	9150	2000
100 TA 12	20	100	150	38	100.2	2	34000	11400	1600
110 TA 12	22	110	160	38	110.2	2	37500	12000	1600
120 TA 12	51224	120	170	39	120.2	2	39000	12000	1300
130 TA 12	26	130	190	45	130.3	2.5	51000	16000	1300
140 TA 12	28	140	200	46	140.3	2.5	54000	16300	1000
150 TA 12	51230	150	215	50	150.3	2.5	60000	17600	1000
160 TA 12	32	160	225	51	160.3	2.5	62000	18000	1000
170 TA 12	34	170	240	55	170.3	2.5	73500	21200	800
180 TA 12	51236	180	250	56	180.3	2.5	76500	21600	800
190 TA 12	38	190	270	62	190.3	3	91500	25000	800
200 TA 12	40	200	280	62	200.3	3	96500	25500	600
	51244	220	300	63	220.3	3	104000	26000	600
	48	240	340	78	240.3	3.5	143000	33500	500
	51252	260	360	79	260.3	3.5	156000	34500	500
	56	280	380	80	280.3	3.5	163000	35500	400
	60	300	420	95	300.3	4	224000	45500	400
	51264	320	440	95	320.4	4	236000	46500	400
	68	340	460	96	340.4	4	245000	47500	300
	72	360	500	110	360.4	5	315000	58500	300

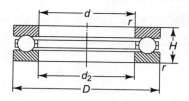

Table 5.41b: Single thrust bearing: Series 513

All dimensions in mm; For nomenclature refer Table 5.41a

Brg. No.		d	D	H	d_2 B	r ≈	Basic capacity, kgf		Max. speed RPM
							Static C_0	dynamic C	
513	05	25	52	18	27	1.5	5000	2900	3600
	06	30	60	21	32	1.5	6350	3330	2800
	07	35	68	24	37	1.5	8600	4370	2500
	08	40	78	26	42	1.5	10900	5350	2000
	09	45	85	28	47	1.5	13800	6620	2000
513	10	50	95	31	52	2	16300	8000	1600
	11	55	105	35	57	2	19900	9250	1400
	12	60	110	35	62	2	21500	9660	1400
	13	65	115	36	67	2	23100	10000	1400
	14	70	125	40	72	2	27700	12000	1200
	15	75	135	44	77	2.5	31500	13600	1100
	16	80	140	44	82	2.5	34000	14000	1100
	17	85	150	49	88	2.5	40800	16000	1000
	18	90	155	50	93	2.5	40800	16000	1000
513	20	100	170	55	103	2.5	49000	18800	900
	22	110	187	63	113	3	57600	21500	800
	24	120	205	70	123	3.5	74000	25400	710
	26	130	220	75	134	3.5	84300	28100	630
	28	140	235	80	144	3.5	98000	30400	630
	30	150	245	80	154	3.5	103000	31300	560
	32	160	265	87	164	4	122500	35400	560

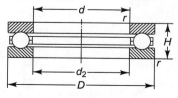

Divide mm dimensions by 25.4 for Inch equivalents. 1 kg = 2.2046 Lbs

Table 5.42a: Series 522 double thrust ball bearings

Brg. No. (SKF)	d mm	d_1 mm	D mm	H mm	d_2 min. mm	a mm	r ≈ mm	r_1 ≈ mm	Basic Capacity kfg		Max speed RPM
									Static C_o	Dynamic C	
52202	15	10	32	22	15.2	5	1	0.5	2040	1220	8000
04	20	15	40	26	20.2	6	1	0.5	3100	1730	6000
05	25	20	47	28	25.2	7	1	0.5	4150	2160	6000
52206	30	25	52	29	30.2	7	1	0.5	4800	2280	6000
07	35	30	62	34	35.2	8	1.5	0.5	6400	3050	5000
08	40	30	68	36	40.2	9	1.5	1	7650	3450	5000
09	45	35	73	37	45.2	9	1.5	1	8650	3650	4000
10	50	40	78	39	50.2	9	1.5	1	9150	3750	4000
52211	55	45	90	45	55.2	10	1.5	1	13200	5500	3000
12	60	50	95	46	60.2	10	1.5	1	14600	5700	3000
13	65	55	100	47	65.2	10	1.5	1	15600	5850	2500
14	70	55	105	47	70.2	10	1.5	1.5	16300	6000	2500
15	75	60	110	47	75.2	10	1.5	1.5	17300	6100	2500
52216	80	65	115	48	80.2	10	1.5	1.5	18000	6200	2000
17	85	70	125	55	85.2	12	1.5	1.5	22000	7500	2000
18	90	75	135	62	90.2	14	2	1.5	27000	9150	2000
20	100	85	150	67	100.2	15	2	1.5	34000	11400	1600

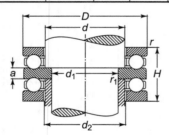

Divide mm dimensions by 25.4 for Inch equivalents. 1 kg = 2.2046 Lbs

Table 5.42b: Series 523 double thrust ball bearings

Brg. No.	d	d_1	D	H	d_2 min.	a	r ≈	r_1 ≈	Basic Capacity kfg		Max speed RPM
									Static C_o	Dynamic C	
523 05	25	20	52	34	27	8	1.5	0.5	5000	2900	3600
06	30	25	60	38	32	9	1.5	0.5	6350	3330	2800
07	35	30	68	44	37	10	1.5	0.5	8600	4370	2500

Machine Tool Elements

Brg. No.		d	d_1	D	H	d_2 min.	a	r ≈	r_1 ≈	Basic Capacity kfg		Max speeed RPM
										Static C_o	Dynamic C	
	08	40	30	78	49	42	12	1.5	1	10900	5350	2000
	09	45	35	85	52	47	12	1.5	1	13600	6620	2000
523	10	50	40	95	58	52	14	2	1	16300	7980	1600
	11	55	45	105	64	57	15	2	1	20000	9250	1400
	12	60	50	110	64	62	15	2	1	21300	9600	1400
	13	65	55	115	65	67	15	2	1	23100	9980	1200
	14	70	55	125	72	72	16	2	1.5	27700	12000	1200
	15	75	60	135	79	77	18	2.5	1.5	31500	13600	1100
	16	80	65	140	79	82	18	2.5	1.5	34000	14000	1100
	17	85	70	150	87	88	19	2.5	1.5	40800	16100	1000
	18	90	75	155	88	93	19	2.5	1.5	40800	16100	1000
523	20	100	85	170	97	103	21	2.5	1.5	49000	18800	900
	22	110	95	190	110	113	24	3	1.5	57600	21500	800
	24	120	100	210	123	123	27	3.5	2	73900	25400	710
	26	130	110	225	130	134	30	3.5	2	84300	28100	630
	28	140	120	240	140	144	31	3.5	2	98000	30400	630

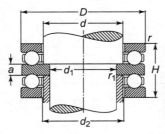

Divide mm dimensions by 25.4 for Inch equivalents. 1 kg = 2.2046 Lbs

Table 5.43: Series 532 thrust ball brgs with spherical hsg. washer and seating ring [See figure on p. 459]

Brg. No.		d	d_1	D_w	h	r	d_2	D_u	a	H	D_2 min	r_1 max abut	R	A	Basic capacity kgf		Max. speed RPM
															Static C_0	dynamic C	
532	05U	25	27	47	16.7	1	36	50	5.5	19	38	0.6	40	19	4080	2170	4000
	06U	30	32	52	17.8	1	42	55	5.5	20	43	0.6	45	22	4700	2270	3600
	07U	35	37	62	19.9	1.5	48	65	7	22	51	1.0	50	24	6350	3040	2800
	08U	40	42	68	20.3	1.5	55	72	7	23	57	1.0	56	28.5	8000	3700	2500
	09U	45	47	73	21.3	1.5	60	78	7.5	24	62	1.0	56	26	8750	3920	2200
532	10U	50	52	78	23.5	1.5	62	82	7.5	26	67	1.0	64	32.5	9800	4220	2000
	11U	55	57	90	27.3	1.5	72	95	9	30	76	1.0	72	35	13600	5670	1800
	12U	60	62	95	28	1.5	78	100	9	31	81	1.0	72	32.5	14520	6000	1600
	13U	65	67	100	28.7	1.5	82	105	9	32	86	1.0	80	40	15420	6080	1600
	14U	70	72	105	28.8	1.5	88	110	9	32	91	1.0	80	38	16100	6210	1400
	15U	75	77	110	28.4	1.5	92	115	9.5	32	96	1.0	90	49	17000	6350	1400
	16U	80	82	115	29.5	1.5	98	120	10	33	101	1.0	90	46	18100	6490	1400
	17U	85	88	125	33.1	1.5	105	130	11	37	109	1.0	100	52	22700	8300	1200
	18U	90	93	130	38.5	2	110	140	13.5	42	117	1.0	100	45	27700	10160	1100
532	20U	100	103	150	40.9	2	125	155	14	45	130	1.0	112	52	34000	12250	1000
	22U	110	113	160	40.2	2	135	165	14	45	140	1.0	125	65	38500	12930	900
	24U	120	123	170	40.8	2	145	175	15	46	150	1.0	125	61	40000	12930	900
	26U	130	133	187	47.9	2.5	160	195	17	53	166	1.5	140	67	50800	16330	800
	28U	140	143	197	48.6	2.5	170	210	17	55	176	1.5	160	87	53500	16550	710
	30U	150	153	212	53.3	2.5	180	225	20.5	60	189	1.5	160	79	58500	18370	710
	32U	160	163	222	54.7	2.5	190	235	21	61	199	1.5	160	74	62140	18820	630

Divide mm dimensions by 25.4 for Inch equivalents. 1 kg = 2.2046 Lbs

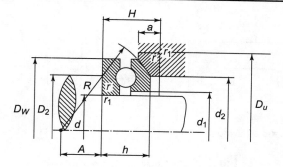

5.2 GUIDEWAYS FOR MACHINE TOOLS

Guideways ensure that work or tool moves in a straight line or a circular path. Usually, they also support tool, work and limit the distortion that is caused by cutting, feeding, and other forces. Since guideways are subjected to sliding friction, they should be wear-resistant. They should be provided with wear-compensation devices like adjustable gibs.

5.2.1 Requisites of good guideways

1. **Precision:** High degree of straightness, flatness, roundness, and consistent surface finish.
2. **Wear resistance:** Durability, ingenious design (shape) to free precision from being affected by wear, low frictional losses through good surface finish and wear-resistant material.
3. **Rigidity:** Minimum structural distortion under various forces, to limit deviations in processed work.

5.2.2 Guideway profiles

Closed guideways [**Fig. 5.31a**] constrain the motion of the saddle (carriage) completely. Open guideways [**Fig. 5.31b**] rely on the weight of the carriage and the pressure on it, to keep the saddle in contact with the guideway, in one (vertical) direction. Guideway shapes can be broadly classified as flat [**Fig. 5.32a**], vee [**Fig. 5.32b**], dovetail [**Fig. 5.32c**], or cylindrical [**Fig. 5.32d**]. Although flat ways are easiest to manufacture, they tend to accumulate dirt, and also have poor lubricant-retention, particularly on vertical surfaces. Vee ways do not lose centrality even after wear. When the Vee apex is in an upward position, neither the saddle nor the guideway accumulate any dirt, nor does it retain the lubricant. When the Vee apex points downward, lubricant retention is better, but dirt accumulation increases. Vees are usually made un-symmetrical, to compensate for unequal pressure

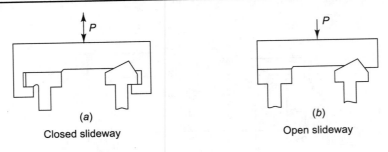

Fig. 5.31: Closed and open slideways

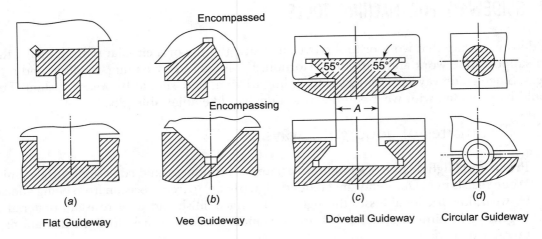

Fig. 5.32: Guideway profiles

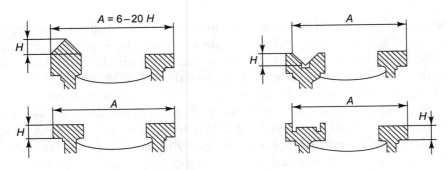

Fig. 5.33: Combination guideways

Machine Tool Elements

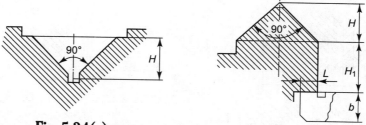

Fig. 5.34(a): Prismatic symmetrical guideways (triangular)

Bracketed dimensions give inch equivalents   *(All dimensions in mm)

H	10 [0.394]	12 [0.472]	16 [0.63]	20 [0.787]	25 [0.984]	32 [1.26]	40 [1.475]	50 [1.968]	63 [2.48]	80 [3.15]	100 [3.937]
	10	14 [5.551]	18 [0.709]	22 [0.866]	28 [1.102]	36 [1.417]	45 [1.772]	56 [2.205]	71 [2.795]	90 [3.543]	110 [4.331]
H1	12	16	20	25	32	40	50	63	80	100	125 [4.921]
	14	18	22	28	36	45	56	71	90	110	140 [5.512]
Min.*	5 [0.197]	6 [0.236]	8 [0.315]	10	12	16	20	25	32	40	50
L	8	10	12	16	20	25	32	40	50	63	80
b	6	8	10	12	16	20	25	30 [1.181]	40	50	60 [2.362]

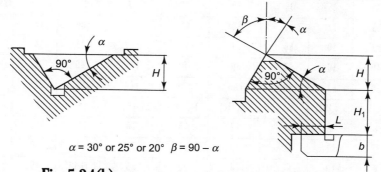

$\alpha = 30°$ or $25°$ or $20°$ $\beta = 90 - \alpha$

Fig. 5.34(b): Prismatic unsymmetric guideways (triangular)

*(All dimensions in mm)

H	20	25	32	40	50	63	80	100	125	160 [6.299]	200 [7.874]
	22	28	36	45	56	71	90	110	–	–	–
H1	25	32	40	50	63	80	100	125	–	–	–
	28	36	45	56	71	90	110	140	–	–	–
Min.*	10	12	16	20	25	32	40	50	60	80	100
L	16	20	25	32	40	50	63	80	–	–	–
b	12	16	20	25	30	40	50	60	–	–	–

* Bracketed dimensions shows Inch equivalents.

Inch values appearing in table below [**Fig. 5.34 (a)**] are not repeated in the above table.

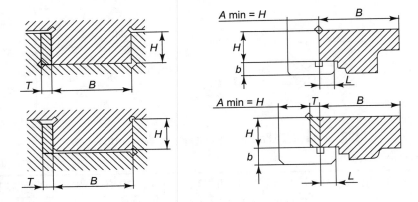

Fig. 5.34(c): Flat guideways

(All dimensions in mm)

H	10	12	16	20	25	32	40	50	63	80	100
	[0.394]	[.472]	[0.63]	[0.787]	[0.984]	[1.26]	[1.575]	[1.968]	[2.48]	[3.15]	[3.937]
B	16	20	25	32	40	50	63	80	100	125 [4.921]	160 [6.299]
	20	25	32	40	50	63	80	100	125	160	200 [7.874]
	25	32	40	50	63	80	100	125	160	200	250 [9.842]
	32	40	50	63	80	100	125	160	200	250	320 [12.598]
	40	50	63	80	100	125	160	200	250	320	400 [15.748]
T	3 [0.118]	4 [.157]	5 [.197]	5	6 [.236]	8 [.815]	10	12	16	20	25
Min.*	5	6	8	10	12	16	20	25	32	40	50
L	8	10	12	16	20	25	32	40	50	63	80
b	6	8	10	12	16	20	25	30 [1.181]	40	50	60 [2.362]

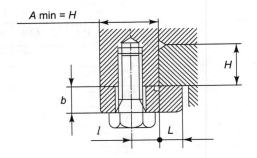 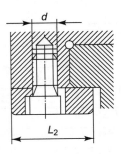

Fig. 5.34(d): Guide strips for flat guideways (without gibs)

Machine Tool Elements

Guide strips for flat guideways [Fig. 5.34d] (All dimensions in mm)

H	10 [.394]	12 [.472]	16 [0.63]	20 [.787]	25 [.984]	32 [1.26]	40 [1.575]	50 [1.968]	63 [2.48]	80 [3.15]	100 [3.937]
d	M5	M6	M6	M8	M10	M12	M12	M16	M16	M20	M24
Min.*	5 [.197]	6 [.236]	8 [.315]	10	12	16	20	25	32	40	50
L	8	10	12	16	20	25	32	40	50	63	80
b	6	8	10	12	16	20	25	30 [1.181]	40	50	60 [2.362]
Min.*	15 [0.59]	18 [0.709]	24 [.945]	30	37 [1.457]	48 [1.89]	60 [2.362]	75 [2.953]	95 [3.74]	120 [4.724]	150 [5.905]
L_2	18	22	28 [1.102]	36 [1.417]	45 [1.772]	57 [2.244]	72 [2.835]	90 [3.543]	113 [4.449]	143 [5.63]	180 [7.087]
l	4	5	5	6	8	10	10	12	12	15 [0.59]	18

distribution. Combination of flat and vee surfaces, i.e. dovetail guideways, are more compact. Cylindrical ways are used for very low loads, due to their low rigidity. Also, they can only be secured at the ends and their wear compensation is quite complicated.

Combination guideways Usually, machine tools use two guideways to pilot the saddle: a vee way and a flat way [**Fig. 5.33**]. The distance over the guideways (A) ranges from 6 to 20 times the guideway height (H). It should be at least 3.5 times the height of center above the saddle top face

Std. guideway heights (H): 10, [.394]   12, [.472]   16, [.63]   20, [.787]   25, [.984]   32, [1.26]   40, [1.575]
50, [1.968]   63, [2.48]   80, [3.15]   100 [3.937]

Std. guideway widths (A): 63, [2.48]   71, [2.795]   80, [3.15]   90, [3.543]   100,   110, [4.33]   125, [4.921]
140, [5.512]   160, [6.23]   180, [7.087]   200, [7.874]   220, [.866]   250, [9.842]
280, [11.024]   320, [12.598]   360, [14.173]   400, [15.748]   450, [17.716]   500, [19.685]
560, [22.047]   630, [24.083]   710, [27.559]   800, [31.496]   900, [35.433]   1000, [39.37]
1120, [44.094]   1250, [49.213]   1400, [55.118]   1600, [66.992]   1800, [70.866]   2000. [78.740]

While [**Figs. 5.34a and 5.34b**] give the dimensions of standard symmetrical and unsymmetrical vee guideways, [**Fig. 5.34c**] gives the specifications of standard flatways. **Figure 5.35(a)** gives the dimensions of standard straight gibs and guide strips for flatways, while [**Fig. 5.35 (b)**] gives the dimensions of headed tapered gibs. Enclosed tapered gibs are shown in [**Fig. 35c**]. The locking insert is usually made of softer brass, to prevent damage by the adjusting screw. **Figure 5.36a** gives the dimensions of straight gibs for dovetail ways and [**Fig 5.36**] states the dimensions of tapered gibs.

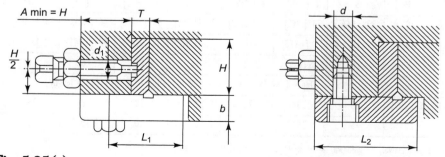

Fig. 5.35(a): Guide strips for flat guideways (without straight gibs) [Dimensions on p 464]

(All dimensions in mm)

H	10 [.3937]	12 [.472]	16 [.63]	20 [.787]	25 [.984]	32 [1.26]	40 [1.575]	50 [1.968]	63 [2.48]	80 [3.15]	100 [3.937]
b	6 [.236]	8 [.315]	10	12	16	20	25	30 [1.181]	40	50	60 [2.362]
d	M6	M6	M6	M8	M10	M12	M12	M16	M16	M20	M24
d_1	M4	M5	M6	M8	M10	M12	M12	M16	M16	M20	M20
T	3 [.118]	4 [.157]	5 [.197]	5	6	8	10	12	16	20	25
Min.*	12	15 [.59]	18 [.709]	21 [.827]	26 [1.024]	34 [1.338]	40	49 [1.929]	60	75 [2.953]	93 [3.661]
L_1	15	19 [.748]	22 [0.866]	27 [1.063]	34	43 [1.693]	52 [2.047]	64 [2.52]	78 [3.071]	98 [3.858]	123 [4.842]
Min.*	18 [.709]	22	29 [1.142]	35 [1.378]	43	56 [2.205]	70 [2.756]	87 [3.425]	111 [4.37]	140 [5.512]	175 [6.89]
L_2	21	26	33 [1.299]	41 [1.614]	51 [2.008]	65 [2.559]	82 [3.228]	102 [4.016]	129 [5.079]	163 [6.417]	205 [8.071]

* Min. values are used where unit pressure on surface L is very small.
Bracketed dimensions give inch values

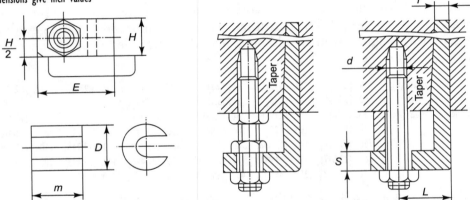

Fig. 5.35(b): Headed type taper gibs for flat guideways

(All dimensions in mm)

H	16 [.63]	20 [.787]	25 [.984]	32 [1.26]	40 [1.575]	50 [1.968]	63 [2.48]	80 [3.15]	100 [3.937]
d	M8	M10	M10	M12	M12	M16	M16	M20	M20
T	5 [.197]	5	6 [.236]	8 [.315]	10 [.394]	12 [.472]	16 [.63]	20 [.787]	25
l	20	22 [.866]	24 [.945]	28 [1.102]	30 [1.181]	38 [1.496]	46 [1.811]	58 [2.283]	65 [2.559]
E	30	32	35 [1.378]	40	42 [1.811]	55 [2.165]	62 [1.441]	78 [3.071]	85 [3.346]
S	8	10	10	12	12	16	16	20	20
Taper 1: 50 for L/H <10 1: 100 for L/H >10; L = S > ide length									
D	18 [.709]	20	20	22	22	28	28	35	35
m*	20	25	25	32	32	40	40	50	50

* 'm' will be reduced progressively as guideway wears and gib is adjusted to compensate the wear.
Bracketed dimensions give Inch equivalents.

Machine Tool Elements

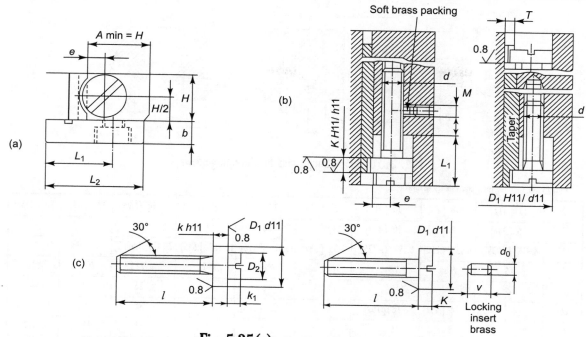

Fig. 5.35(c): Enclosed tapered gibs

Dimensions in mm. Inch equivalents in brackets.

H	16 [.63]	20 [.787]	25 [.984]	32 [1.26]	40 [1.575]	50 [1.968]	63 [2.48]	80 [3.15]	100 [3.937]
d	M6	M8	M8	M10	M12	M16	M16	M20	M24
D_1	14 [.551]	18 [.709]	18	25	32	40	40	50	63
l_1	25	25	25	30 [1.181]	35 [1.378]	40	40	45 [1.772]	50 [2.362]
K	5 [.197]	6 [.236]	6	8 [.315]	8	10 [.394]	10	12 [.472]	12
L_1	24 [.945]	30	34 [1.338]	42 [1.811]	56 [2.205]	68 [2.677]	80	100	125 [4.921]
L_2	32	42 [1.811]	50	62 [2.441]	80	100	120 [4.724]	150 [5.905]	190 [7.48]
b	10	12	16	20	25	30	40	50	60 [2.362]
e	5	6	6	8	10	12	12	14	16
M	M5	M6	M6	M8	M10	M12	M12	M16	M16
T	5	5	6	8	10	12	16	20	25
	Taper 1 : 50 for $L/H < 10$, 1 : 100 for $L/H \geq 10$ where L is length of slide								
l	30	35	35	40	45 [1.772]	55 [2.165]	55	65 [2.559]	70 [2.756]
D_2	9 [.354]	10	10	14	18	22	22	26 [1.024]	30
K_1	4	5	5	6	6	6	6	8	8
D_0	3.5 [.138]	4.2 [.181]	4.2	5.8 [.228]	7.5 [.295]	9	9	12	12
v	Length of locking insert to suit dimension A, and length of screw								

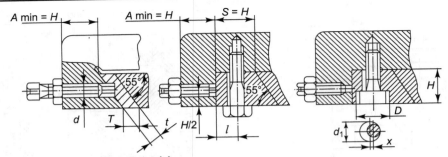

Fig. 5.36(a): Straight gibs for dovetail guideways

Dimensions in mm. Inch equivalents in brackets

H	12 [.472]	16 [.63]	20 [.787]	25 [.984]	32 [1.26]	40 [1.575]	50 [1.968]	63 [2.48]	80 [3.15]
T	6 [.236]	8 [.315]	10 [.394]	12	16	20	25	—	—
t*	4.915 [.1935]	6.553 [.258]	8.192 [.3225]	9.83 [.387]	13.106 [.516]	16.383 [.645]	20.479 [.8062]	—	—
d	M5	M6	M8	M10	M12	M12	M16	M16	M20
l	—	9	12	14 [.551]	16	20	25	32	40
D	—	14 [.551]	18 [.709]	20	22 [.866]	22	30	30 [1.181]	35 [1.378]
d_1	—	8	10	12	14	14	18	18	23 [.905]
x	—	1	1	1	1	1	1	1	1.5 [.059]

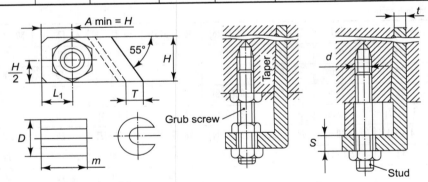

Fig. 5.36(b): Headed type taper gibs for dovetail guideways**

Dimensions in mm. Inch equivalents in brackets

H	16 [.63]	20 [.787]	25 [.984]	32 [1.26]	40 [1.575]	50 [1.968]	63 [2.48]	80 [3.15]
T	5	5 [.157]	6 [.236]	8 [.315]	10 [.394]	12 [.472]	16	20 [.787]
t*	4.096 [.161]	4.096	4.915 [.1935]	6.553 [.258]	8.192 [.3225]	9.83 [.387]	13.106 [0516]	16.385 [.645]
S	8	10	10	12	12	16	16	20
d	M8	M10	M10	M12	M12	M16	M16	M20
l_1	10	12	16	20	20	25	25	32
Taper 1: 50 for $L/H < 10$, 1 : 100 for $L/H \geq 10$, where L = Length of slide								

Machine Tool Elements

D	18 [.709]	20 [.787]	20	22 [.866]	22	28 [1.102]	28	35 [1.378]
m*	20	25	25	32 [1.26]	32	40 [1.575]	40	50 [1.968]

* Theoretical values of gib thickness without the dovetail guideway tolerance and grinding allowance.
** Taper normal to gib thickness. 1:61.039 for 1:50, and 1:122.078 for 1:100.

Width (A) of the dovetail way is measured at the small end of the dovetail **[Fig 5.32c]**. The width (A) ranges from 5 to 10 times the dovetail height (H). The depth of a female dovetail way is 0.5–1 more than the height (H) of the male dovetail.

Std. Dovetail Heights (H): 6, 8, 10, 12, 16, 20, 25, 32, 40, 50, 63, 80.
[.236] [.315] [.394] [.472] [.63] [.787] [.984] [1.26] [1.575] [1.968] [2.48] [3.15]

Std. Dovetail Widths (A): 32, 36, 40, 45, 50, 56, 63, 70,
[1.26] [1.417] [1.575] [1.968] [1.968] [1.205] [2.48] [2.756]
80, 90, 100, 110, 125, 140, 160,
[3.15] [3.543] [3.937] [4.331] [4.921] [5.112] [6.299]
180, 200, 220, 250, 280, 315,
[7.087] [7.874] [8.661] [9.242] [11.024] [12.402]
355, 400, 450, 500, 560, 630, 710, 800
[13.976] [15.748] [17.716] [19.685] [22.047] [24.803] [27.953] [31.496]

5.2.3 Guideway materials

To minimize the wear of a longer, and costlier fixed guideway (bed), its hardness should be more than the shorter, moving, saddle. Grey cast iron can be flame/induction-hardened to RC 40–52. Nodular cast iron can be hardened more, up to RC 55. Iron can be cast into suitable shapes to minimize machining. It is thus clear why most slideway beds are made of cast iron.

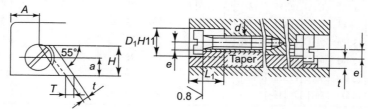

Fig. 5.36(c): Setup of taper gibs for dovetail guideways—enclosed type

Tapered gibs for dovetail guideways [Fig. 5.36c]   (All dimensions in mm)

H	6 [.236]	8 [.315]	10 [.394]	12 [.472]	16 [.63]	20 [.787]	25 [.866]	32 [1.26]	40 [1.575]	50 [1.968]
a	8	8	8	10	10	12	15 [0.59]	20	25	30 [1.181]
A	12	12	16	16	16	20	25	32	40	50
d	M5	M5	M5	M6	M6	M8	M8	M10	M12	M16
D_1	12	12	12	14 [.551]	14	18 [.709]	18	25	32	40
e	4 [.157]	4	4	5 [.197]	5	6	6	8	10	12
L_1	15	15	15	25 [.59]	25	25	25	30 [1.181]	35 [1.378]	40

Taper 1 : 50 for L/H < 10, 1 : 100 L/H ≥ 10 where L = Length of slide										
T	3	3	4	4	5	5	6	8	10	12
t*	[.118] 2.457 [.0967]	2.457	3.277 [.129]	3.277	4.096 [.161]	4.096	4.915 [.1935]	6.553 [.258]	8.192 [.3225]	9.83 [.387]

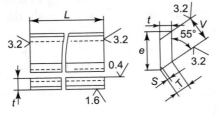

All dimensions in mm

Fig. 5.36(d): Straight gibs for dovetail guideways   **Fig. 5.36(e):** Tapered gibs for dovetail guideways

V	T	t*	e	S
6 [.236]	3 [.118]	2.457 [.0967]	9.045 [.356]	1 [.039]
8 [.315]	4 [.157]	3.277 [.129]	12.061 [.475]	1
10 [.394]	5 [.197]	4.096 [.161]	15.076 [.593]	1
12 [.472]	6	4.915 [.1935]	18.091 [.712]	1
16 [.63]	8	6.553 [.258]	24.121 [.95]	1
20 [.787]	10	8.192 [.3225]	30.151 [1.187]	1.5 [.059]
25 [.945]	12	9.83 [.387]	37.402 [1.472]	2 [.079]
32 [1.26]	16	13.106 [.516]	48.242 [1.899]	2
40 [1.575]	20	16.383 [.643]	60.303 [2.374]	3 [.118]
50 [1.968]	25	20.479 [.806]	75.378 [2.968]	3
63 [2.48]	30 [1.181]	24.574 [.967]	94.116 [3.705]	3

V	T	t*	S
6	3	2.457	1
8	3	2.457	1
10	4	3.277	1
12	4	3.277	1
16	5	4.096	1
20	5	4.096	1.5
25	6	4.915	2
32	8	6.553	2
40	10	8.192	3
50	12	9.83	3
63	16	13.106	3

Theoretical values of gib thickness without the dovetail guideway tolerance and grinding allowance.

It is also possible to use harder (RC 60) steel liners, by screwing or welding them to unhardened beds [**Fig. 5.37**]. The latest trend is to coat beds with a thin layer of polymer by spraying, or gluing a thin film. The hardness along the guideway length, should not vary more than 25 BHN on a single bed, and 45 BHN for sectional beds.

5.2.3.1 Surface finish

Machining, grinding, lapping, and scraping are used to obtain the requisite surface finish and precision. An average number of bearing spots are measured over the entire slideway. It is customary to measure the contact spots on the bed by using a marking compound. The following table gives the number of spots for 25 sq. area, for various applications.

Machine Tool Elements

Table 5.44: No. of recommended contact (bearing) spots and surface finish for various applications

No.	Application	No. of spots in 25 sq. Area	Surface finish Ra microns
1	Guideways of heavy machinery, rotary tables, gearbox covers	4 ± 1	3.2
2	Guideways of heavy machine tools, scraped tables	7 ± 1	1.6
3	Guideways of heavy machine tools, bearing bushes, jigs and fixtures, when width $A \leq 250$	10 + 2 − 1	0.8
4	Precision machines	16 + 7 − 3	0.4
5	High-precision machines	25 + 7 − 1	0.2

Note: Average of spots over 100 cm² is used for measurements.

5.2.3.2 Clearance adjustment

Saddles (carriages) sliding over guideways, are provided with parallel or tapered gibs which can be adjusted by screws, to allow for a sliding motion without affecting the precision of the travel. Straight gibs are packings, adjusted by the setting screws square with the guideway. **Figure 5.35a** gives the dimensions of straight gibs used for flatways. The gibs are locked in a set position, by locking the set screw with the help of a check nut.

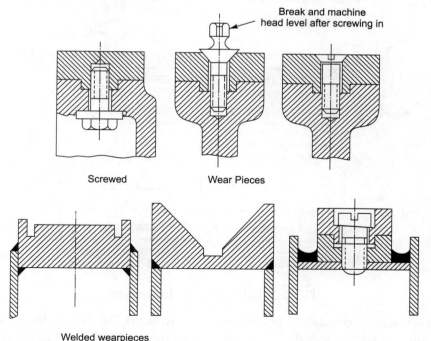

Fig. 5.37: Fixing hardened wearpieces to guideways

When a tapered gib is used, it is also necessary to taper the gib-side of the saddle. As both the saddle and the gib have matching tapers, the mating (guiding) side of the gib remains parallel to the guideway. The clearance is adjusted by a screw with an axis, parallel to the saddle travel [**Fig. 5.35b**]. The L-shaped gib has a head, to facilitate engagement with the faces of the nuts used, to lock the gib in a set position. For enclosed gibs, it is necessary to provide a slot, to facilitate an engagement with the collar of the special adjusting screw [**Fig. 5.35c**]. Gibs for dovetail slides, are shaped to match the dovetail angle [**Figs. 5.36 a and b**].

5.2.3.3 Lubrication

Good lubrication minimizes friction. Manual lubrication by the operator is widely used. The working (mating) faces of a guideway are provided with grooves, which retain the lubricant. The grooves are shaped in such a way that the motion of the saddle spreads the lubricant all over the contacted faces [**Fig. 5.38a**]. For slow feeding speeds, the lubricant requires special additives like oleic and stearic acid, which help the oil adhere to surfaces. The oils for slideways should have 30–150 C St. viscosity, at 40°C. For very slow and intermittent operations, grease can be used, as an exception.

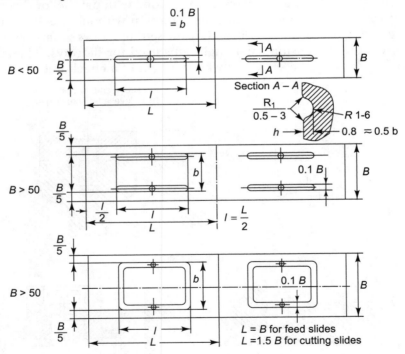

Fig. 5.38(a): Oil grooves for guideways

The grooves, size, and spacing depend upon the width (B) of the guideway, the stroke of the saddle, and the sliding speed. The groove width ranges from 3 to 16 mm and the depth from 0.8 to 5 mm [**Fig. 5.38b**]. Width-wise margin from the longitudinal edge is one-fifth the guideway width, i.e. $\frac{B}{5}$. For low-sliding speeds, the pitch is equal to the guideway width 'A'. The pitch can be 50% more, for higher sliding speeds. The pitch, however, should not exceed the often-used stroke length. The lengthwise margin from the nearest edge should be half of the longitudinal pitch.

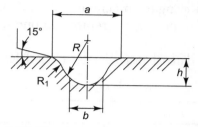

Fig. 5.38(b): Grooves for plane sliding surfaces

Nominal size of groove h	R	R_1	a	b	Recommended for width of guideway B	
					Over	Up to
0.8 [.031"]	1 [.039"]	0.5 [.02]	3	1.6	25 [.984"]	40
1.2 [.047"]	1.6 [.063"]	0.8	4	2.5	40 [1.575]	60
2 [.079"]	2.5 [.098"]	1.2	7 [.275]	4	60 [2.362]	100
3 [.118"]	4 [.157"]	2	10 [.394]	6	100 [3.937]	160
5 [.197"]	6 [.236]	3	16 [.63]	10	160 [6.299]	250 [9.842"]

In manually lubricated guideways, the situation is similar to that of boundary lubrication in sliding bearing sleeves. In this kind of semi-liquid friction, there is only a partial separation caused by the lubricant film, and much metal-to-metal contact.

The design of a guideway mainly entails determining its size, to contain the pressure on the guiding surfaces, to the limits stated in the following table.

Table 5.45: Permissible pressure (kg/mm²) on guideways

Sr. No.	Machines	Speed	Recommended pressure (kg/mm^2)		Max. pressure (kg/mm^2)
			C.I. guideway C.I./Steel saddle	Steel guideway Steel saddle	
1.	Lathes, Milling M/cs	Slow (feedings)	0.125 [178 psi]	0.15 [213 psi]	0.30 [427 psi]
2.	Planers, Shapers, Slotters	High (cutting)	0.04 [57]	0.05 [71]	0.08 [144]
3.	Grinding	Medium	0.0035 [5]	0.005 [7]	0.007 [10]
4.	Heavy	Slow	0.05 [71]	0.06 [85]	0.1 [142]
5.	Heavy	High	0.02 [28]	0.025 [35]	0.04 [57]

Bracketed values give pressure in Lbs/inch² [psi]

Slideway design

For determining the pressure on a guideway, it is necessary to compute cutting, gravitational, and other forces, causing pressure on the guideway. This involves computing the maximum forces on the guideway. This can be done by assuming a maximum cut width/depth and a maximum feed rate, i.e. the metal removal rate. The rate depends upon the power capacity: the kW rating of the motor, powering the machine tool. The following table gives the amount of material (in cc) one kilowatt of power can remove in one second.

Table 5.46: Material removal rates (cc/sec) for 1 kW power

[1 HP = .746 kw] [1 kw = 1.34 H.P.]

Hardness BHN	Steels	Cast iron	Non-ferrous materials
100	0.58 [.0354 Inch³]		
120	0.66 [.0403]	1.31 [.08 Inch³]	
140	0.62 [.0378]	1.04 [.0635]	
160	0.60 [.0366]	1.07 [.0653]	0.45 [.0275 Inch³]
180	0.58 [.0354]	0.90 [.0549]	0.48 [.0293]
200	0.55 [.0336]	0.59 [.036]	0.4 [.0244]
240	0.51 [.0311]	0.40 [.0244]	0.37 [.0226]
260	0.46 [.0281]	0.40	0.39 [.0238]
300	0.41 [.025]		
320	0.36 [.022]		
340	0.36		

The cutting force (F) can be found by dividing the power rating by the cutting speed.

$$\text{Cutting force } (F) \text{ kg} = \frac{kW \times 100}{V} = \frac{100\,P}{V} \qquad \text{(Eqn. 5.48)}$$

P = Motor rating (kW); V = Cutting speed (meters/sec.).

[1 meter = 3.2802 feet]

Forces on flat guideways [Fig. 5.39a]

The cutting force (F_z) acts in the direction of cutting. In lathes, this is tangential to the workpiece diameter at the cutting point, i.e. vertical. Due to heavy chip friction in metal-cutting, the horizontal force (F_y) can be as high as 76% of the cutting force (F_z). As these forces are in equilibrium, forces on the slideway faces and edges, can be found by solving equations $\Sigma Y = 0$, $\Sigma X = 0$, $\Sigma Z = 0$, and $\Sigma M_x = 0$. The solutions are:

Force on guideway edge face $D = F_y = D$

Force on guideway face $C = \dfrac{F_z Y_f - F_y h}{b} + \dfrac{G}{2} = C$

Machine Tool Elements

Force of face $A = F_z + \dfrac{G}{2} - \left(\dfrac{F_z Y_f - F_y h}{b}\right) = A$

Feed force $= F_x + \mu (F_y + F_z + G)$

$b =$ Distance between guideways centers [mm]

$F_y =$ Forces in the direction, perpendicular to the guideway.

$F_z =$ Downward forces on slideway.

$F_x =$ Forces parallel to guideway.

$A, C, D =$ Forces on slideway faces A, C, and edge face D [kg]

$G =$ Wt. of carriage (kg)

$F_x, F_y, F_z =$ Cutting forces in directions X, Y, and Z [kg]

$\mu =$ Coefficient of friction.

$Y_F =$ Distance of vertical force F_z from the center line of the slideway face A.

$h =$ Height of spindle center from guideways center line (mm).

The average pressure on the slideways can be computed by dividing force A, C, D by the slideway areas in contact with the saddle (carriage), traversing the guideways. The areas depend on the guideway widths and saddle length.

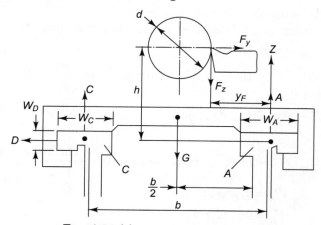

Fig. 5.39(a): Forces on flat guideway

Pressure of face C $(P_C) = \dfrac{C}{W_C L} = \dfrac{\dfrac{F_z Y_f - F_y h}{b} + \dfrac{G}{2}}{W_C L}$

Pressure on face A $(P_A) = \dfrac{A}{W_A L} = \dfrac{F_z + \dfrac{G}{2} - \dfrac{F_z Y_f - F_y h}{b}}{W_A L}$

Pressure on edge $D(P_D) = \dfrac{D}{W_D L} = \dfrac{F_y}{W_D L}$

W_A, W_C, W_D = Widths of faces A, C, D (mm)

L = Length of saddle (mm)

Example 5.19: Find the forces on flat guideways on a lathe, if guideways are 25 thk. and 50 wide. The center distance between the guideways is 400. The machine has a 110 height above the guideway top faces. The machine is powered by a 5.5 kW motor. The machine mostly shapes steel workpieces at a speed of 24 meter/min. The tool frictional force (F_y) is 30% of the cutting force (F_z).

Wt. of saddle = 40 kg; Length of saddle = 250

Solution:

From Eqn 5.48, $\quad F_Z = \dfrac{100P}{V} = \dfrac{100 \times 5.5}{\dfrac{24}{60}}$

$= 1375$ kg

$F_y = 0.30 \times 1375 = 412.5$ kg

Y_f depends upon the workpiece cutting diameter.

Assuming the maximum possible diameter, i.e. twice the center height of 110, workpiece diametrer = 2×110.

$b = 400$; So $\dfrac{b}{2} = 200$ and $\dfrac{d}{2} = \dfrac{220}{2} = 110$

So, $\quad Y_f = \dfrac{b}{2} - \dfrac{d}{2} = 200 - 110 = 90$

h = Height above guideway + $\dfrac{\text{Thickness}}{2} = 110 + \dfrac{25}{2} = 122.5$

$W_A = W_C = 50$

$W_D = 25$; $G = 40$

Pressure on face $C(P_C) = \dfrac{\dfrac{F_z Y_f - F_y h}{b} + \dfrac{G}{2}}{W_C L}$

$= \dfrac{\dfrac{1375 \times 90 - 412.5 \times 122.5}{400} + \dfrac{40}{2}}{50 \times 250}$

$= \dfrac{183.05 + 20}{12500} = 0.016$ kg/mm^2

Pressure on face $A(P_A) = \dfrac{F_z + \dfrac{G}{2} + \dfrac{F_z Y_f - F_y h}{b}}{W_A L}$

$= \dfrac{1375 + 20 + 183.05}{12500} = 0.126 \text{ kg/mm}^2$

Pressure on edge $D(P_D) = \dfrac{F_y}{W_D L} = \dfrac{412.5}{25 \times 250} = 0.066 \text{ kg/mm}^2$

If the right-hand slideway is replaced by a non-symmetrical Vee guideway, with angles α and β, with horizontal face of guideway 'C' [**Fig. 5.39b**], the values of the forces will be as follows:

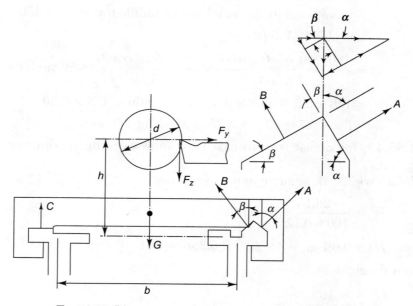

Fig. 5.39(b): Forces on flat and Vee combination guideway

Force on face $A = A = \dfrac{F_z(d+b)\cos\alpha}{2b} + F_y \dfrac{h\cos\alpha}{b} - F_y \sin\alpha + \dfrac{G\cos\alpha}{2}$ \hfill (**Eqn. 5.49**)

Force on face $B = B = \dfrac{F_z(d+b)\cos\beta}{2b} + F_y \dfrac{h\cos\beta}{b} + F_y \sin\beta + \dfrac{G\cos\beta}{2}$ \hfill (**Eqn. 5.50**)

Force on face $C = C = \dfrac{F_z(b-d)}{2b} - F_y \dfrac{h}{b} + \dfrac{G}{2}$ \hfill (**Eqn. 5.51**)

Machine Tools Handbook

Example 5.20: A lathe is subjected to a 160 kg vertical force (F_z), and a 50 kg horizontal force (F_y), at the cutting point on the 200 diameter. Design a V and Flat combination guideway, if the saddle length (L) is 100. Height of the spindle center above the flat guideway = 150, Saddle wt. = 35 kg.

Solution:

While using an unsymmetrical Vee, the face with more width should be placed on the spindle side.

Selecting tentatively $\alpha = 60°$ and $\beta = 30°$; From Eqn. 5.49,

Force on face
$$A = A = \frac{F_z(d+b)\cos\alpha}{2b} + \frac{F_y h \cos\alpha}{b} - F_y \sin\alpha + \frac{G\cos\alpha}{2}$$

Taking
$$b = 3.5 \times \text{height of center above saddle top} = 3.5 \times 150$$
$$= 525 \approx 560 \text{ [std.]}$$

$$A = \frac{160(200+560)\cos 60°}{2 \times 560} + \frac{50 \times 150 \times \cos 60}{560} - 50 \sin 60 + \frac{35\cos 60}{2}$$

$$= 108 \times 0.5 + 13.4 \times 0.5 - 50 \times 0.866 + 17.5 \times 0.50$$
$$= 54 + 6.7 - 43.3 + 8.75 = 26.15 \text{ kg}$$

From Table 5.45, for a cast iron to cast iron slideway, the recommended pressure $(P) = 0.125 \text{ kg/mm}^2$.

So the guideface width (W_a) can be found as follows:

$$W_a = \frac{26.15}{100 \times 0.125} = 2.09$$

V height $(H) = 2.09 \sin 60 = 2.09 \times 0.866 = 1.81$

Selecting standard height 10,

$$W_a = \frac{10}{\sin 60°} = \frac{10}{0.866} = 11.54$$

Actual pressure $(P_A) = \dfrac{26.15}{100 \times 11.54} = 0.0227 \text{ kg/mm}^2$

Pressure on Vee face B $(P_B) = \dfrac{B}{100 \times \dfrac{10}{\sin 30}} = \dfrac{B}{2000}$

From Eqn. 5.50, $B = \dfrac{F_z(d+b)\cos\beta}{2b} + \dfrac{F_y h \cos\beta}{b} + F_y \sin\beta + \dfrac{G\cos\beta}{2}$

So, $B = \dfrac{160(200+560)\cos 30}{2 \times 560} + \dfrac{50 \times 150 \cos 30}{560} + 50 \sin 30 + 17.5 \cos 30$

$$= 108.6 \times 0.866 + 13.4 \times 0.866 + 50 \times 0.5 + 17.5 \times 0.866$$
$$= 94.05 + 11.6 + 25 + 15.15 = 145.8$$

$$W_b = \frac{10}{\sin 30} = 20$$

$$P_B = \frac{145.8}{100 \times 20} = 0.073 \text{ kg/mm}^2.$$

This is less than the recommended pressure of 0.125 for CI/CI or C.I./Stl guideway/saddle combination.

Force on flatway $(C) = \dfrac{F_z(b-d)}{2b} - \dfrac{F_y h}{b} + \dfrac{G}{2}$

$$= \frac{160(560-200)}{2 \times 560} - \frac{50 \times 150}{560} + \frac{35}{2} = 51.43 - 13.4 + 17.5 = 55.5 \text{ kg}$$

The minus sign indicates that the force tends to lift the saddle off the guideway. From **[Fig. 5.35a]**, for a 10 high flatway, engagement width (L) of the guide strip can be 5 or 8. Selecting the latter,

$$\text{Pressure on guide strip} = \frac{55.5}{100 \times 8} = 0.069 \text{ i.e.} < 0.125$$

The saddle can be made of C.I. or steel, while the bed (slideway) can be made of grey cast iron.

Rigidity of guideway

Displacement of guideway surfaces due to pressure, and rotation of the saddle due to unequal displacement of two guideways cause radial displacement of the tool-cutting edge. This leads to a dimensional error: an increase in the workpiece diameter.

For 2 flatways combination **[Fig. 5.40]**,

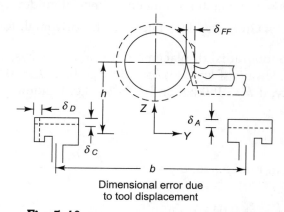

Dimensional error due to tool displacement

Fig. 5.40: Tool displacement on flat guideway

Tool displacement $(\delta_{FF}) = KP_D + K\dfrac{(P_A - P_c)h}{b}$ (Eqn. 5.52a)

For the combination of one vee and one flat guideway [Fig. 5.41],

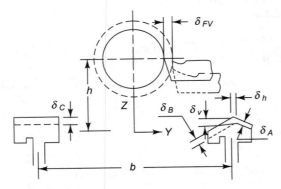

Fig. 5.41: Tool displacement on flat and Vee combination guideway

Tool displacement $(\delta_{FF}) = K(P_B \sin\beta - P_A \sin\alpha) + \dfrac{kh}{b}(P_B \cos\beta + P_A \cos\alpha - P_C)$ (Eqn. 5.52b)

K = Constant for slide and saddle combination
   = 0.1 mm²/kg for computations
P_A = Pressure on guideway face A (kg/mm²)   1 kg/mm² = 1422 psi
P_C = Pressure on guideway C (kg/mm²)
P_D = Pressure on guideway edge D (kg/mm²)
P_B = Pressure on guideway face B (kg/mm²)
h = Spindle center height (mm)   1mm = 0.0337 Inch
b = distance between center lines of guideways (mm)
α, β = Guideway face angles with horizontal, for Vee guideway.

Example 5.21: Find the tool radial displacement for (a) 2 flatways, and (b) one flatway one veeway combination guideways, if $b = 200$; $h = 150$; $P_A = 0.12$; $P_B = 0.15$; $P_C = -0.08$, $\alpha = 60°$, $\beta = 30°$. Find the rigidity, if $F_z = 160$ kg; and $F_y = 110$ kg: Guideway thickness = 25; saddle length = 100.

Solution:

(a) For the flatway, from Eqn. 5.52a, $\delta_{FF} = KP_D + \dfrac{Kh(P_A - P_C)}{b}$

$$P_D = \dfrac{110}{100 \times 25} = 0.044 \text{ kg/mm}^2$$

So, $\delta_{FF} = 0.1 \times 0.044 + \dfrac{0.1 \times 150}{200}(0.12 + 0.08)$

$= 0.0044 + 0.015 = 0.0194$

Rigidity $= \dfrac{160}{0.0194} = 8247$ kg/mm deflection

or $\dfrac{0.0194}{160} = 0.000121$ mm/kg load.

(b) For a flat–vee combination guideway,
From Eqn. 5.52b

$d_{FF} = K(P_B \sin b - P_A \sin a) + \dfrac{Kh}{b}(P_B \cos b + P_A \cos a - P_C)$

$= 0.1(0.15 \sin 30 - 0.12 \sin 60) + \dfrac{0.1 \times 150}{200}(0.15 \cos 30 + 0.12 \cos 60 + 0.08)$

$= 0.1(0.15 \times 0.5 - 0.12 \times 0.866) + 0.075(0.15 \times 0.866 + 0.12 \times 0.5 + 0.08)$

$= -0.0029 + 0.0202 = 0.0173$

Radial displacement for a flat and vee combination guideway, is 89% of the displacement for a two flatway combination.

Rigidity $= \dfrac{0.0173}{160} = 0.000108$ mm/kg load.

Hydraulic support

If we use a pump for pressurizing the lubricating oil, the oil pressure lifts the saddle off the guideway, to create an oil film between the sliding surfaces.

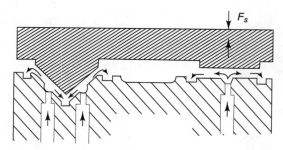

Fig. 5.42: Hydro-statically supported and lubricated guideway

Hydro-static lubrication [Fig. 5.42]

Pressurized oil is pumped into closed oil grooves to generate thrust. This tends to lift the saddle off the guideway. As the oil escapes through the gap created by the oil film, there is a drop in pressure. Actual force (thrust) developed is only 1/3 to 1/2 of the force that a leakage-free system can create.

$$\text{Leakage Factor } (\psi) = \left(1/3 + \frac{l}{6L} + \frac{b}{6B} + \frac{lb}{3LB}\right) = 1/3 - 1/2 \quad \textbf{(Eqn. 5.53a)}$$

Refer to **[Fig. 5.38]** for explanation of symbols l, L, b, B, etc.

Load capacity of saddle $(F_s) = pA\psi$ **(Eqn. 5.53b)**

p = Lubricating oil pressure in oil groove (kg/mm^2) 1 kg/mm^2 = 1422 psi

A = Contact area between guideway and saddle [mm^2] = BL

p_p = Pressure of lubricating oil at pump outlet [kg/mm^2]

ψ = Leakage factor from Eqn. 5.53

F_s = Load capacity [kg]; h = oil film thickness [mm] 1mm = 0.03937 Inch

$$\text{Rigidity} = 3\left(1 - \frac{p}{p_p}\right)\frac{F_s}{h} \quad \textbf{(Eqn. 5.53c)}$$

Generally, rigidity of a hydro-static slideway is more than 100 kg/micron.

Example 5.22a: Find the load capacity and rigidity of a guideway, if slideway width $(B) = 40$. Pitch of grooves $(L) = 40$. Pump pressure = 150 kg/cm^2, Oil groove pressure = 100 kg/cm^2; Oil film thickness $[h] = 0.015$.

Solution: Refer **[Fig. 5.38a]**

Length of oil groove $\quad (l) = 0.5\ L = 0.5 \times 40 = 20$

Groove width $\quad (b) = 0.1 B = 0.1 \times 40 = 4$

Leakage factor

$$(\psi) = \left(1/3 + \frac{1}{6L} + \frac{b}{6B} + \frac{lb}{3LB}\right)$$

$$= \left(1/3 + \frac{20}{6 \times 40} + \frac{4}{6 \times 40} + \frac{20 \times 4}{3 \times 40 \times 40}\right)$$

$$= 0.45$$

Load capacity

$$(F_s) = p A \psi = \frac{100}{100} \times 40 \times 40 \times 0.45$$

$$= 720 \text{ kg (per 40 length)}$$

$$\text{Rigidity} = 3\left(1 - \frac{100}{150}\right)\frac{720}{0.015} = 48,000 \text{ kg/mm}$$

or, 0.000021 mm/kg

Hydro-dynamic lubrication

At high speeds, dynamic action of the lubricant film generates a hydro-dynamic force which lifts the saddle off the guideway. For good hydro-dynamic action, the sliding bodies should be inclined towards each other. They should form a liquid wedge. Most of the hydro-dynamic guideways have wedge-like inclined passages, on both sides of the lubrication grooves [**Fig. 5.43**]. Nearly 20% of the guideway width is left straight, since even hydro-dynamic guideways operate under semi-liquid friction, at the beginning of the machine operation and during braking.

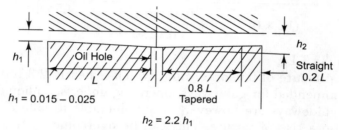

Fig. 5.43: Fluid wedge in hydro-dynamic slideway

Load capacity (F_{hw}) of a liquid-wedged guideway can be found from the following equation:

$$F_{hw} = n \times \frac{0.158\ \mu V L^2 W_W}{h_1^2 \left(1 + \frac{0.64 L^2}{W_W^2}\right)} \qquad \textbf{(Eqn. 5.53d)}$$

n = No. of wedged passages; μ = Coefficient of dynamic viscosity kgs/m^2
V = Sliding speed (meters/sec) \hfill 1 Feet = 0.305 m
L = Overall length of wedge and flat portion (m) \hfill 1 m = 3.281 Ft
W_W = Width of inclined passage (m)
h_1 = Film thickness at flat land (meters)
  = 0.01 – 0.02 mm—Medium machine tools \hfill 1 mm = 0.03937 Inch
  0.06 – 0.10 mm—Heavy machine tools \hfill 1 Inch = 25.4 mm

Example 5.22b: Find the load capacity of two hydraulic wedges, if the overall length of the wedge (L) = 200 mm, sliding velocity (V) = 20 meters/min; wedge width (W_W) = 160 mm, film on flat land (h_1) = 0.015 mm; $\mu = 4 \times 10^{-9}$ kg sec/mm^2.

Solution:

All data must be converted to meters and seconds.

$$L = \frac{200}{1000} = 0.2 \text{ m};\ V = \frac{20}{60} = 0.33 \text{ m/s}$$

$$W_W = \frac{160}{1000} = 0.16 \text{ m};\ h_1 = \frac{0.015}{1000} = 1.5 \times 10^{-5} \text{ m}$$

$n = 2$ (given); $\mu = 4 \times 10^{-9} \times 10^6 = 4 \times 10^{-3}$ kg sec/m^2

From Eqn. 5.53,

$$F_{hw} = n \times \frac{0.158\, \mu V L^2 W_W}{h_1^2 \left(1 + \dfrac{0.64 L^2}{W_W^2}\right)}$$

$$= 2 \times \frac{0.158 \times 4 \times 10^{-3} \times 0.33 \times 0.2^2 \times 0.16}{(1.5 \times 10^{-5})^2 \left(1 + \dfrac{0.64 \times 0.2^2}{0.16^2}\right)} = \frac{2.67 \times 10^{-6}}{4.5 \times 10^{-10}}$$

$$= 5933.3 \text{ kg.}$$

This results in a pressure of 0.093 kg/mm^2, which is nearly 2.3 times the pressure of 0.04 kg/mm^2, recommended for guideways, operating under semi-liquid friction.

Hydro-dynamic guideways are however, very vulnerable to failures of the lubrication system. There is also a risk of seizure. This can be overcome by using seizure-resistant materials' pair, such as cast iron slideway and non-ferrous saddle.

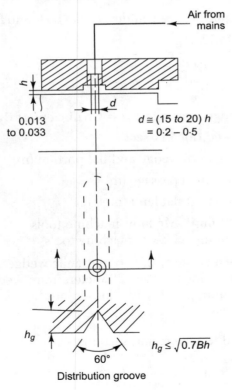

Fig. 5.44: Aerostatic guideway

Aerostatic guideways [Fig. 5.44]

In these, an air cushion replaces the oil film. Filtered air at 3–4 kg/cm² pressure, is delivered through a 0.2–0.5 mm orifice into the air pocket, which discharges it into the atmosphere through a gap of 0.013 to 0.033. The air distribution groove is often made vee-shaped. The depth of the groove depends upon the guideway width and the air gap between the sliding surfaces.

Groove depth $(h_g) \leq \sqrt{0.7Bh}$
B = Guideway width (mm), h = Air gap (mm)

Air-supported guideways have a very low rigidity, of only 10 kg/micron. Compressibility of air, makes matters worse. Aerostatic guideways are hence, not used for cutting or feeding motions. But they are conveniently used in jig drilling and jig boring machines.

Anti-friction guideway [Fig. 5.45]

Usage of hardened balls and rollers between the mating guideway faces, reduces friction drastically. Coefficient of friction (μ) plunges from 0.05–0.06 (sliding friction), to 0.001–0.0025 (rolling friction). Anti-friction bearings enhance the machine life much, and facilitate precise uniform motions even at slow speeds. This is due to the elimination of stiction (friction due to sticking). Even the rigidity can be increased significantly by pre-loading the anti-friction bearings. Hardened, and precisely-machined steel guideways are necessary. Usually, 20 Cr 1 Mn 60 Si 27 Ni 25 steel case-hardened 0.8–1 deep to RC 60–62 is satisfactory. Naturally, anti-friction guideways are costlier. Furthermore, the rolling elements lag behind the saddle [Fig. 5.46a]. For long strokes, it is necessary to provide for a recirculation arrangement for the balls/rollers [Fig. 5.46b].

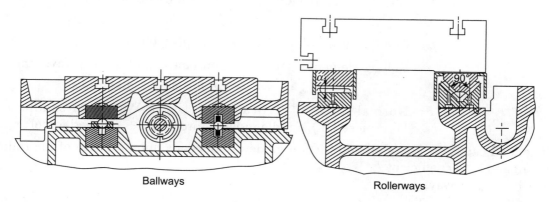

Ballways   Rollerways

Fig. 5.45(a): Open anti-friction ways

Like sliding friction guideways, anti-friction guideways can be open [Fig. 5.45a] or enclosed [Fig. 5.45b]. For heavy-duty machines calling for high rigidity, pre-loadable closed guideways are used.

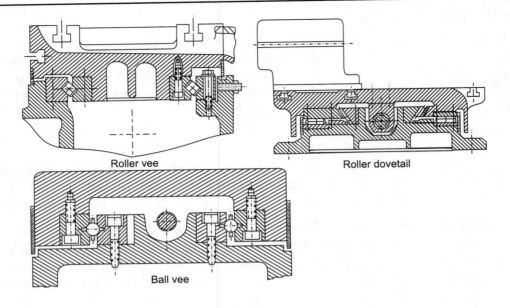

Fig. 5.45(b): Closed anti-friction ways

Friction force $\quad (F_{fa}) = nF_o + \dfrac{\mu K\, F}{r}$ **(Eqn. 5.54)**

n = No. of guideway faces; F_o = Constant frictional force
$\qquad\qquad\qquad\qquad\qquad\quad$ = 0.4–0.5 kg [.9 – 1.1 Lbs]

μ = Coefficient of rolling friction = 0.001–0.0025

r = Rolling element radius; K = Constant = 1.5 for open ways, and,
$\qquad\qquad\qquad\qquad\qquad\quad$ = 2.8 for closed guideways

F = Normal force = Vertical force (F_y) + Weight of saddle and Workpiece (W); Saddle moving force = $F_{fa} + F_x$ (Feed force)

The load capacity (Fl_g) of an anti-friction guideway depends upon the size of the rolling element and load coefficient L.

For balls guideway,
$\qquad F_{lbg} = K\, d^2$ for a ballway $\qquad\qquad\qquad\qquad$ **(Eqn. 5.55a)**

For rollers guideway,
$\qquad F_{lrg} = Kbd$ $\qquad\qquad\qquad\qquad\qquad\qquad\qquad$ **(Eqn. 5.55b)**

d = Dia. of ball/roller (mm); b = Roller length (mm)
K = Load coefficient (kg/mm^2)

For balls, $K = 0.06$ for ground steel way, and 0.002 for C.I. way

For rollers, $K = 1.5-2$ for ground steel way, and 0.13–0.2 for C.I. way

The maximum load on a rolling element (F_{mr}) should be commensurate with the permissible pressure on the guideway (Table 5.45 p 471).

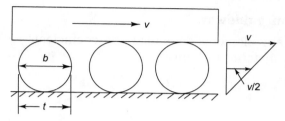

Fig. 5.46(a): Velocity distribution along the height of a rolling element

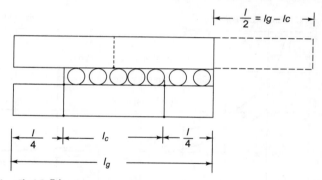

Fig. 5.46(b): Schematic diagram of a limited travel anti-friction way

$$F_{mr} = b.t.\ P_{max} \qquad \text{(Eqn. 5.56)}$$

b = Ball dia/roller length (mm), t = Pitch of balls/rollers (mm)

$\qquad$ = 1.5–2.5 ball/roller ϕ (mm)

P_{max} = Maximum permissible pressure on guideway (kg/mm^2) (From Table 5.45 p 471)

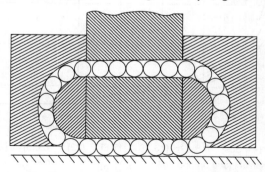

Fig. 5.46(c): Schematic diagram of an unlimited travel way with recirculating rolling elements

No. of rolling elements

Anti-friction guideways call for high precision. Even a 0.015–0.020 error in straightness over the contact length, is considered high. Variation in the diameters of rolling elements aggravate the error. These shortcomings can be compensated for, to satisfactory extent, by keeping the number of rolling elements over 16.

Rigidity of anti-friction guideways

Displacement (δ) of rolling elements depends upon coefficients C_r (for rollers) and C_b (for balls), and the guideway material and finish.

For balls, $\delta_b = C_b F_{lb}$ (Eqn. 5.57)

For rollers, $\delta_r = C_r q$ (Eqn. 5.58)

F_{lb} from Eqn. 5.55a : q = Load for unit length of roller
Value of coefficients C_b and C_r can be found from the graphs in [Fig. 5.47].

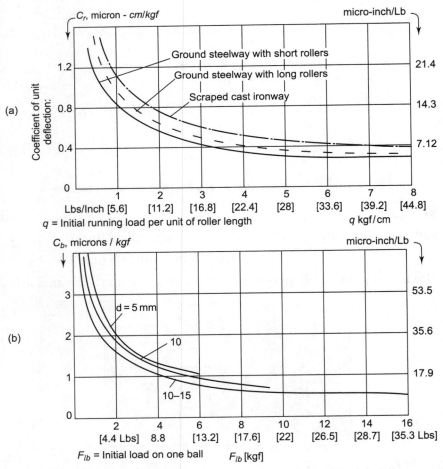

Fig. 5.47: Deflection coefficients for rollers and balls in anti-friction guideways

Machine Tool Elements

Example 5.23: Design an anti-friction roller guideway, with 4 faces, if the load on the guideway $(F_y) = 400$ kg, weight of saddle and work $(W) = 80$ kg, and feed force $(F_x) = 40$ kg. Hardened and ground guideways are to be used. Saddle length = 350. The saddle operates at a cutting speed of 20 m/min. Find the frictional force for (a) open, and (b) closed guideways. Permissible pressure $[p] = 0.08$ kg/mm².

Solution:

From Eqn. 5.55b, $\quad F_{lrg} = K\,bd$

For steel guideways, $\quad K = 1.5 - 2.0 \approx 1.75$

$$F_{lrg} = 1.75\,bd$$

Load on saddle $= F_y + W + F_x = 400 + 80 + 40 = 520$ kg. This will be shared by 4 guideways

$$F_{lrg} = 1.75\,bd = 520; \text{ and } bd = \frac{130}{1.75} = 74.3$$

Taking tentatively, roller length $(b) = 1.6\,d$

$$1.6\,d \times d = 74.3 \quad \text{or} \quad ; d^2 = \frac{74.3}{1.6} = 46.43, \text{ and } d = 6.8 \approx 7$$

$$b = 1.6\,d;\ b = 1.6 \times 7 = 11.2 \approx 11;$$

$$\text{No of rollers} = \frac{350}{1.5 \times 7} = 33.3 \text{ max.} \approx 33$$

there should be at most 33 rollers under the 350 long saddle.

$$\text{Min number of rollers} = \frac{350}{2.5 \times 7} = 20$$

So, $\quad t = \dfrac{350}{33} = 10.6 \approx 1.515$ roller ϕ

From Eqn. 5.56, Max load/roller $[F_{mr}] = b.t.p_{max}$.

$$= 11 \times 10.6 \times 0.08 = 9.33 \text{ kg}$$

Load/cm of roller length $[q] = \dfrac{9.33}{1.1} = 8.48$ kg ≈ 8.5 kg

From **[Fig. 5.47a]**, for $q = 8.5$ and long roller guideway of ground steel,

$$C_r = 0.3,$$

So, roller deflection $[\delta_r] = C_{rq} = 0.3 \times 8.5 = 2.55$ microns

$$\text{Rigidity} = \frac{520}{0.002.55} = 203921 \text{ kg/mm}$$

$$= 204 \text{ kg/micron}$$

Stiffness $= 4.9 \times 10^{-6}$ mm/kg

From Eqn. 5.54, $K = 1.5$ for open guideway and 2.8 for closed guideways

a) $$F_{fa} = nF_o + \frac{\mu KF}{r}$$

$$= 4 \times 0.5 + \frac{0.0025 \times 1.5 \times 520}{7/2} = 2.0 + 0.56 \approx 2.56 \text{ kg} \quad [\text{open way}]$$

b) $$F_{fa} = 2 + \frac{0.0025 \times 2.8 \times 520}{3.5} = 2 + 1.04 = 3.04 \text{ kg} \quad [\text{closed way}]$$

Force needed to effect feed (F_x), should be added to the frictional force.

Total force for moving (traversing) the slide $F_t = F_x + F_{fA} = 40 + 3.04 = 43.04$ kg for closed guideways.

Figure 5.48 gives the formulae for computing traversing (feeding) forces, for commonly used configurations of anti-friction guideways.

	Type of ways	Traversing force Q kgf
1	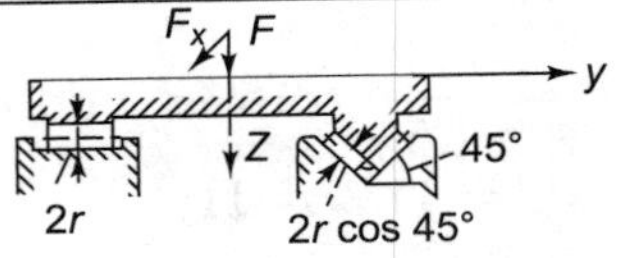 (diagram with F_x, F, z, $2r$, $2r\cos 45°$, $45°$)	$F_a = F_x + 3F_o + \dfrac{1.5\mu}{r} F$ $F = F_z + G_1 + G_2$
2	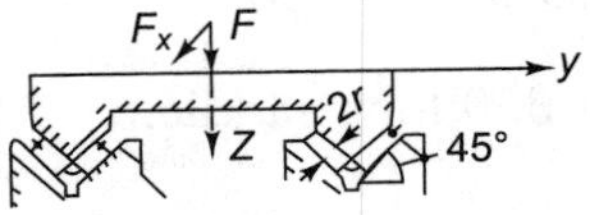 (diagram with F_x, F, $2r$, z, $45°$)	$F_a = F_x + 4F_o + \dfrac{1.4\mu}{r} F$
3	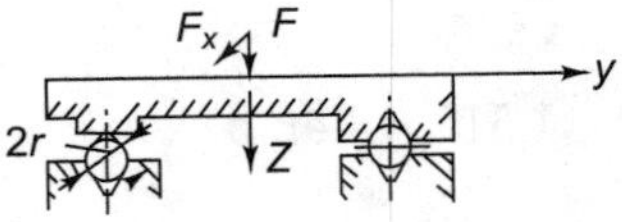 (diagram with F_x, F, $2r$, z)	$F_a = F_x + 2F_o + \dfrac{1.5\mu}{r} F$
4(i)	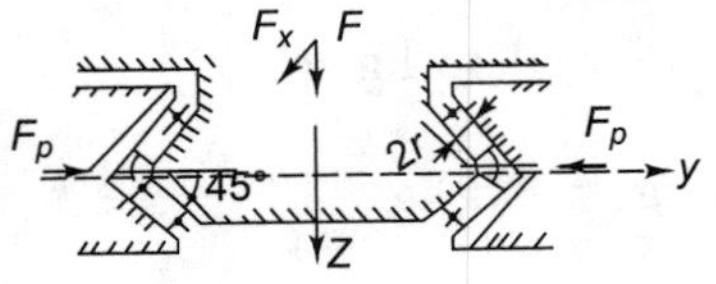 (diagram with F_p, F_x, F, $2r$, $45°$, z, F_p)	$F_a = F_x + 4F_o + \dfrac{2.8\mu}{r} F_p$
4(ii)	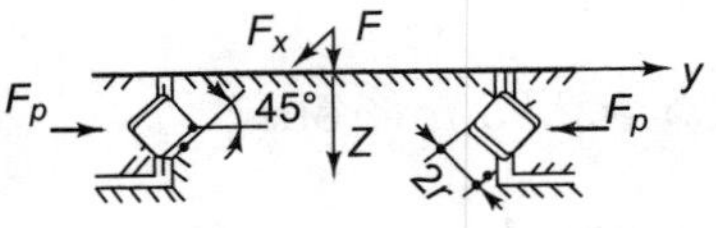 (diagram with F_p, F_x, F, $45°$, z, F_p)	$F_a = F_x + 2F_o + \dfrac{2.8\mu}{r} F_p$
4(iii)	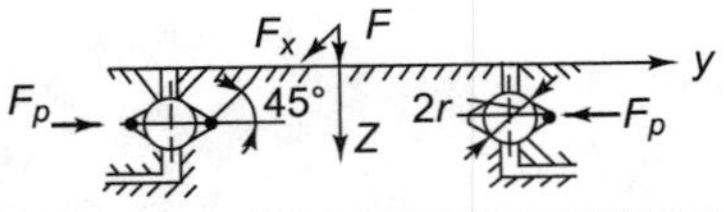 (diagram with F_p, F_x, F, $45°$, $2r$, z, F_p)	

Fig. 5.48: Traversing force $[F_a]$ for anti-friction ways

Notes:
1. The coefficient of rolling friction $\mu = 0.001$ for ground steel ways and $\mu = 0.0025$ for scraped cast iron ways. The initial friction force, referred to one separator, $F_0 = 0.4$ kgf. [.9 Lbs]
2. Because of the low value of the friction forces, a simplified arrangement has been accepted in which the ways are subject only to the feed force F_x, vertical component F_z of the cutting force, table weight G_1 and workpiece weight G_2. The tilting moments, force F_y and the components of the traversing force are not taken into account.
3. In the type 4 ways only the feed force F_x and the preload force F_p are taken into consideration.

Guideways' protection

Entry of dirt, sharp metal chips, and abrasive matter in the gap between the guideway and the carriage, can cause much more harm to the guideway surfaces, and affect precision, than the slow, long-term wear due to friction. Ingress of such harmful matter into the gap can be prevented by seals, covers, and steel strips.

Seals are placed at the ends of the carriage **[Fig. 5.49]**. Felt seals are used in slow-speed applications with a low-frequency usage. Backing felt by more durable rubber seal, can prolong felt seal life. Further, the addition of brass chip-cleaner is even better. Usage of a leaf spring to limit the seal clamping force and provide flexibility, is also quite common.

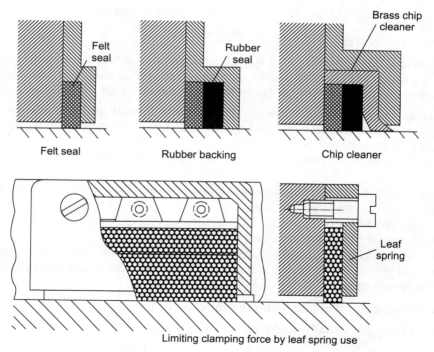

Fig. 5.49: Seals usage for guideway protection

Covers [Fig. 5.50] fixed at the ends of a saddle (carriage) shield the guideways from metal cuttings and dirt. For long strokes, telescopic covers can be used. Heavy covers need roller supports. In machines such as grinding, where the swarf is not sharp, even leather covers can be used.

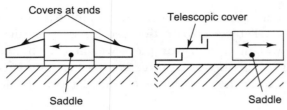

Fig. 5.50: Covers for guideway protection

Steel Strips [Fig. 5.51] of thin gauge are widely used to shield guideways from harmful matter. The strip keeps covered, the exposed part of the guideway, not in contact with the saddle.

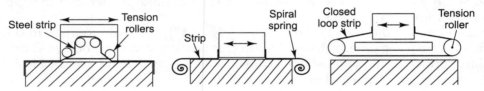

Fig. 5.51: Using steel strip for guideway protection

Machine tool structures

Mounting beds and bases, housings for spindle and gear trains; work/tool holders and movers such as tables, saddles, knees, tailstocks, can be classified as machine tool structures. For satisfactory operation of machine tools, the structures must meet the following requisites:

1. **Precision:** The mounting surfaces should have formal and dimensional accuracy. Dimensions must be precise, and geometrical precision (parallelism, squareness, circularity, flatness) must be ensured.
2. **Stability:** The structure should retain precision till the end of the useful life of the machine tool.
3. **Strength:** Deflections, and stresses due to metal-cutting forces and temperature variations, should be kept within specified limits.
4. **Convenience:** The structure should provide convenient access for inspection, repairs, and replacement of consumable and damaged parts.

    The above requisites call for the considerate selection of materials and a strong, sturdy design to meet the required standards of rigidity: generally, **100 kg/micron 5.6 Lbs/micor-inch or more**.

The choice of material depends upon the function of the part.

1. **Beams:** These are usually designed to limit deflection. The stress for this deflection depends upon the elasticity modulus (E) and $\dfrac{\text{strain}}{\text{stress}}$ $\left(\dfrac{\delta}{f}\right)$ ratio, which depends upon the

span (l) and depth (d) of beam: ratio $\left(\dfrac{l^2}{d}\right)$. For equal span and deflection, the material requirement for a cast iron beam is nearly 13 times the requirement for a steel beam.

2. **Bars** are subjected to direct tension (screws), torsion (shafts), or bending stress.

- **Direct tension:** For the same length and deflection, the material requirements of two materials with elastic moduli E_1 and E_2, and densities 6_1 and 6_2, are in the proportion stated in the following equation:

$$\frac{W_1}{W_2} = \frac{E_2/6_2}{E_1/6_1} = \frac{f_{t2}/6_2}{f_{t1}/6_1}$$

$W_1 = W_t$ of required material 1 (kg)

$W_2 = W_t$ of required material 2 (kg)

$E_1, E_2 =$ Elastic moduli of materials 1 and 2

$6_1, 6_2 =$ Densities of materials 1 and 2

$f_{t1}, f_{t2} =$ Ultimate tensile strengths of materials 1 and 2.

- **Bending:** For bars of equal length, subjected to equal bending moment, and possessing the same factor of safety,

$$\frac{W_1}{W_2} = \frac{f_{t2}^{2/3}/6_2}{f_{t1}^{2/3}/6_1}$$

- **Torsion:** For 2 bars of the same length, subjected to equal torque, and possessing the same factor of safety,

$$\frac{W_1}{W_2} = \frac{f_{s2}^{2/3}/6_2}{f_{s1}^{2/3}/6_2}$$

f_{s1} : Shear strength of material 1
f_{s2} : Shear strength of material 2

Obviously, steel is a better material than cast iron, since the elastic modulus (E) and strength (f_t, f_s) of steel are much more than that of cast iron. However, better vibrations damping, good sliding properties, ease in shaping through casting and the resultant decrease in machining time, makes cast iron a convenient and economical material to use in many low stress applications.

Wall thickness

As stress is inversely proportional to the square of depth $\left(\dfrac{l}{d^2}\right)$, and deflection is inversely proportional to the cube of depth $\left(\dfrac{l}{d^3}\right)$, it is economical to use bigger, overall dimensions and lesser wall thickness (b), to save material. The overall dimensions factor or size factor (S) can be found from the following equation:

$$S = \frac{2l + b + d}{4000}$$

l = length (mm); b = width (mm); d = depth (height) mm

The internal walls of castings require a longer cooling time, and are therefore made 20% thinner than the external walls.

Table 5.47: Wall thickness (t) and size factor (s)

[1 mm = 0.03937″]

Size factor (s)		0.4	0.75	1.0	1.5	1.8	2.0	2.5	3	3.5	4.5
Wall thickness	External (t_e)	6 [.24″]	8 [.31″]	10 [.39″]	12 [.47″]	14 [.55″]	16 [.63″]	18 [.71″]	20 [.79″]	22 [.87″]	25 [.98″]
	Internal (t_i)	5 [.2″]	7 [.27″]	8 [.31″]	10 [.39″]	12 [.47″]	14 [.55″]	16 [.63″]	16 [.63″]	18 [.71″]	20 [.79″]

Fabricated steel structures use thinner walls. Usually, 10–12 thick steel plates are adequate. If walls are only 3–6 mm thick, it is necessary to use stiffening ribs. It is however, easier to use plates of a different thickness in fabrication, by using the correct fillet size. When thickness difference is less than 50% $\left(\frac{t_1}{t_2} < 1.5\right)$, fillet size ($\omega$) can be found from the following equation:

Fillet size (w) = $\left(\frac{1}{6} - \frac{1}{3}\right) t_1$

Economy is the most important criterion for the choice of the material. Additional cost of pattern in castings, and machining cost in steel fabrication, should be compared. It is also advisable to weld simple castings together rather than complicating a pattern. Where quantity is likely to be more, cast iron is preferable. It is also a common practice to use a fabricated construction for the first prototype model, and to change over to casting later, when the shape and size of the structure are found satisfactory during machine manufacture and tryouts.

Static stiffness is the force required for unit deflection (one mm or one micron). When the direction of force is different than the direction of deflection, it is called **cross stiffness**.

All the elements in a machine tool can be considered as a system of springs. In case of a train of gears [**Fig. 5.52a**], the cumulative deflection at the last (output) shaft is maximum. Thus, strengthening the output shaft, increases the cumulative stiffness of the system, appreciably.

If the elements are in parallel [**Fig. 5.52b**], the deflection is the same for all the elements. It is economical to increase the stiffness of the strongest element, to appreciably increase the resultant stiffness.

Dynamic stiffness is the ratio between the dynamic force and dynamic deflection. The rotation of an un-balanced workpiece, or usage of a number of cutters, leads to vibrations. For an un-damped spring-mass system,

Machine Tool Elements

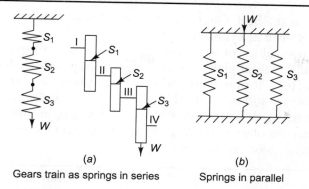

(a) Gears train as springs in series
(b) Springs in parallel

Fig. 5.52: Machine tool elements as a system of springs

$$\text{Natural frequency} = \sqrt{\frac{S}{m}} = \sqrt{\frac{\text{Stiffness}}{\text{Mass}}}$$

Preferably, the lowest natural frequency should be 2.5 times higher than the highest excitation frequency.

In lathes, excitation frequency depends upon R.P.M.

$$\text{Excitation frequency} = \frac{\text{RPM}}{60}$$

Grinding machines run at a very high R.P.M. to provide 30 m/sec linear grinding speed. Bigger (500) grinding wheels with a lower R.P.M., have a lower natural frequency. Such machines use light, but stiff structures. Machines with smaller grinding wheels (internal) have a higher R.P.M. They use heavy constructions with high stiffness, to keep the vibration frequencies low.

In machines with wide variations in frequency, such as milling machines, it is sometimes necessary to mount an additional mass on the arbor, especially if the excitation frequency coincides with the natural frequency of the spindle or arbor, and causes resonance.

Sophisticated equipment is necessary to measure the dynamic stiffness of machines. Measuring the actual static stiffness is comparatively much easier. The ratio between the dynamic and static stiffness (or deflections) is called **amplification factor**.

$$\text{Amplification factor } (A) = \frac{\text{Static stiffness}}{\text{Dynamic stiffness}} = \frac{S_s}{S_d}$$

Also,
$$A = \frac{\text{Dynamic deflection}}{\text{Static deflection}} = \frac{\delta_d}{\delta_s}$$

Usually, $A \leq 2$

Sections for machine tool structures

Machine tool structures are subjected to tensile/compressive, bending, and torsional (shear) stresses. For the same volume of material, the profile which gives a higher sectional modulus and moment of inertia, is the best. Generally, hollow sections are better than solid ones, and rectangular sections better than cylindrical ones.

Figure 5.53 gives the values of the bending moment and torque capacities of various sections, in comparison to a solid plate. Weights of sections per meter length and depth, of all the sections, are equal (100). It may be noted that even a hollow cylinder of 100 diameter is 12% stronger and 15% more stiff in terms of bending, than a solid rectangular bar. It can take on 43 times higher shear stress and 9 times more twist angle, in torsion. An I beam is 80% stronger and stiffer than the plate in terms of bending. It can take on 4.5 times higher shear stress and 1.9 times more twist, in torsion.

Section		Area cm²	Weight kgf/m	Relative value of permissible			
				Bending moment kgf.cm		Torque kgf.cm	
				Stress	Deflection	Stress	Angle of twist
(a) 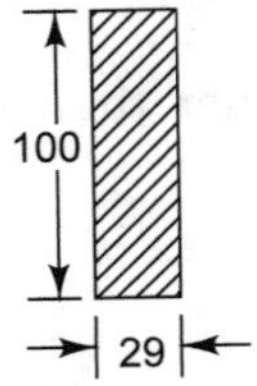	100 × 29 rectangle	29.0	22.8	1	1	1	1
(b) 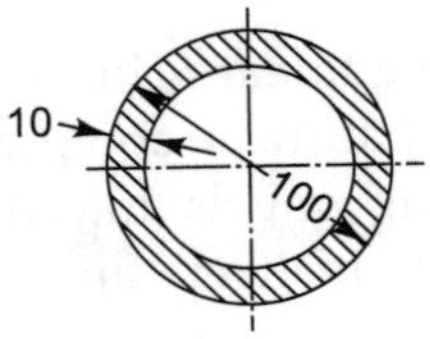	hollow cylinder ⌀100, t=10	28.3	22.2	1.12	1.15	43	8.8
(c) 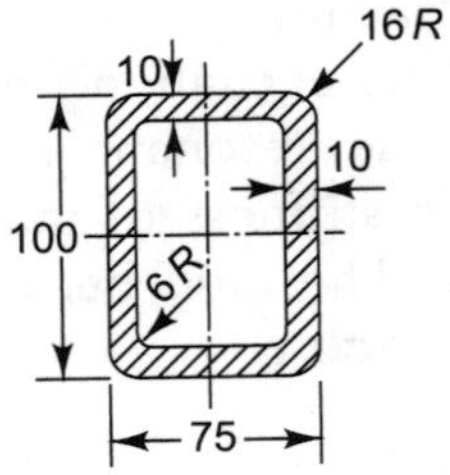	hollow rectangle 100×75, t=10, 16R, 6R	29.07	22.85	1.4	1.6	38.5	31.4
(d) 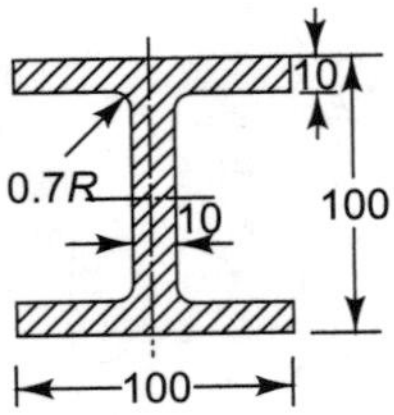	I beam 100×100, t=10, 0.7R	28.4	22.3	1.8	1.8	4.5	1.9

Fig. 5.53: Comparison of sections for stress and deflection/twist angle

Effect of aperture

An aperture in the wall of a section, affects its strength (stress) and stiffness (deflection). Even a circular hole affects the sectional length, equal to double the diameter. The length affected by a slot is equal to the slot length, plus the slot width. The weakening can be corrected to some extent by using a cover. A cover which mates with the cut-out, is better than the cover which does not engage with the cut-out.

Ribs as stiffeners [Figs. 5.54, 5.55]

Well-placed ribs can enhance the stiffness considerably [**Fig. 5.54**]. Ribs that are square/parallel to the sectional walls, increase the bending stiffness up to 17%, and double the torsional stiffness. Diagonal ribs increase the bending stiffness up to 78%, and torsional stiffness up to 3.7 times the stiffness of a ribless box.

Torsional stiffness of even open structures such as lathe bed, can be enhanced substantially, by the provision of ribs [**Fig. 5.55**]. Diagonal ribs are better than parallel or square ones.

Design procedure

Generally, machine tool structures are designed for bending and torsional stiffness (deflection, and twist angle). The design is then checked for strength: tensile/compressive, and torsional (shear) stress.

1. Beds, columns, and other-partially closed profiles are treated as thin-walled bars, and analyzed statically.
2. Closed boxes like gear box housing, are designed for forces, square to the walls.
3. Tables, knees, and other supporting structures are treated as plates.

Permissible distortion

Bending deflection (δ) $\leq 0.003 - 0.005$/meter in direction affecting workpiece precision [$3-5 \times 10^{-6} \times$ length in Inches]

$\leq 0.01 - 0.025$ in cross direction [$1 - 2.5 \times 10^{-5} \times$ length in Inches]

Torsional twist (ϕ) $\leq 0° 30'$/meter   [$0°9'8.65''$/Foot]

The torsional twist angle (ϕ) can be found from the following equation:

Stiffener arrangement	Relative stiffness under		Relative weight	Relative stiffness per weight under	
	Bending	Torsion		Bending	Torsion
1	1.0	1.0	1.0	1.0	1.0
2	1.10	1.63	1.1	1.0	1.48

3 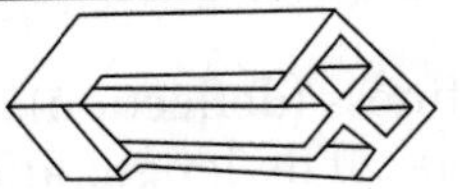		1.17	2.16	1.38	0.85	1.56
4 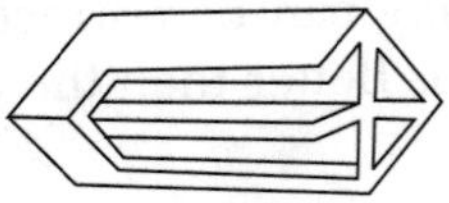		1.78	3.69	1.49	1.20	3.07
5 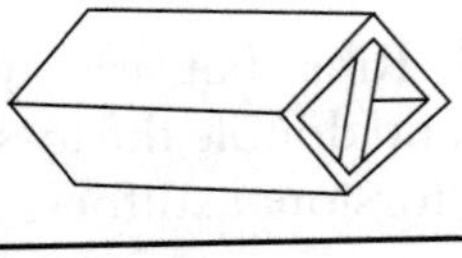		1.55	2.94	1.26	1.23	2.39

Fig. 5.54: Enhancing stiffness by using ribs

Stiffener arrangement		Relative torsional arrangement	Relative weight stiffness	Relative torsional stiffness per unit weight
1 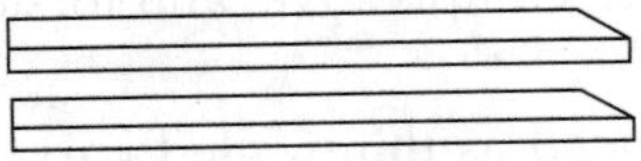		1.0	1.0	1.0
2 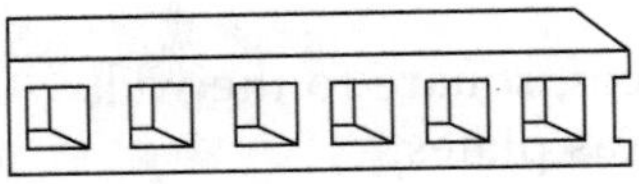		1.34	1.34	1.0
3 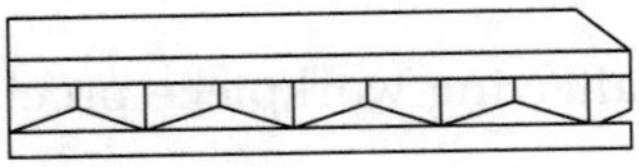		1.43	1.34	1.07
4 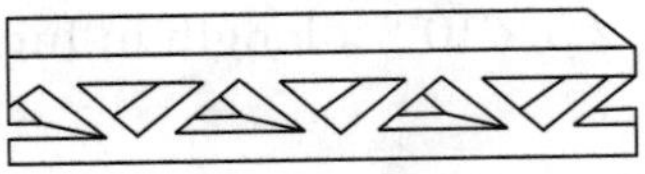		2.48	1.38	1.80
5 		3.73	1.66	2.25

Fig. 5.55: Enhancing torsional stiffness of open sections

$$\frac{\phi}{l} = \frac{T}{GI_P}; \quad [l = \text{length (mm)}; \, T = \text{Torque (kg mm)}] \quad 1 \text{ kg mm} = 0.0866 \text{ Lb. Inch}$$

G = Modulus of rigidity (kg/mm^2) = 0.79×10^4 for M. S. [11.23×10^6 Lbs/ Inch^2]

$$1 \text{ kg/mm}^2 = 1422 \text{ psi}$$

$\quad = 0.89 \times 10^4$ for (0.6% Carbon steel)   [12.65×10^6 Lbs/Inch^2]

$\quad = 0.35 \times 10^4$ for cast iron

I_p = Polar moment of inertia (mm⁴)

$\quad = \dfrac{\pi}{32} D^4$ for solid cylinder of D diameter

$\quad = \dfrac{\pi}{32}(D^4 - d^4)$ for hollow shaft, with inside diameter = d

For the rectangular section,

$$\dfrac{\phi}{l} = \dfrac{KT}{G\,db^3}$$

d = Depth of section (mm), b = Width (mm) for $d > b$

T = Torque (kg mm); K = Constant, dependent on $\dfrac{d}{b}$; G = Rigidity modulus

	1.0	1.5	2.0	2.5	3.0	4	6	8	10	Above 10
K	7.1	5.1	4.36	4.0	3.8	3.56	3.3	3.26	3.2	3.0

Column design

Bending and torsional loads on spindle head and knee of [milling] machines are borne by columns. Vertical lathes and boring machines generate cutting loads, which are unsymmetrical with column shapes. Hollow, box-type columns provide good rigidity. Horizontal stiffeners [**Fig. 5.56**] and vertical ribs are necessary for three-dimensional loads. Although square

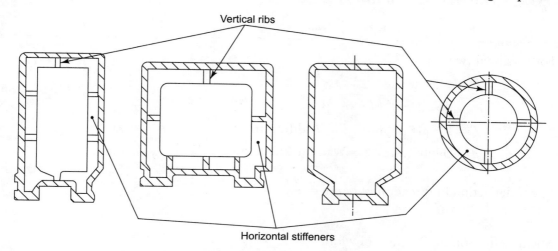

Fig. 5.56: Column sections

columns widely used in boring and milling machines, rectangular columns with a depth 3 times the are width $(d-3b)$, are more economical in vertical lathes and planers, and plano-millers. Columns are usually made tapered, with the biggest section at the base and the smallest section farthest from the base. Sometimes the rectangular section at the base is tapered to a square one at the farthest end, to provide good torsional stiffness, at a more economical cost.

Stresses

Bending and shearing deflections in the two planes, square to the section, should be checked. Bending deflection (δ_b) can be found by treating the column as a cantilever. The shear deflection (δ_s) depends upon the depth : width $\left(\dfrac{d}{b}\right)$ ratio.

$$\delta_{shear} = \lambda \frac{Wl}{GA} \qquad \text{(Eqn. 5.59)}$$

W = Shearing force (kg)

l = Column length (mm)

G = Modulus of rigidity (kg/mm^2) = 0.79×10^4 for M.S.

A = Area of cross section (mm^2)

λ = Constant for $\dfrac{d}{b}$ ratio

$\dfrac{d}{b}$	0.5	0.75	1.0	1.25	1.5	1.75	> 2
λ	4.5	3.0	2.35	2.0	1.75	1.6	1.55

For torsional twist,

$$\frac{\phi}{l} = \frac{2T(b+d-2t)}{4Gt(b-t)^2(d-t)^2} \qquad \text{(Eqn. 5.60)}$$

T = Torque (kgmm); b = Width (mm)

d = Depth (mm); t = Wall thickness (mm) ; $d > b$

Shear deflection in direction of depth = $\dfrac{d\phi}{2}$

Shear deflection in direction of width = $\dfrac{b\phi}{2}$

Cut-outs in columns affect the torsional stiffness. It is advisable to provide covers which engage with the cut-outs, to neutralize the weakening caused due to the cut-outs. Covers can be assembled with the help of suitable screws.

Machine Tool Elements

Example 5.24: Design a column for a drilling machine, if column height is 1.25 meters. maximum torque = 6.6 kgm, feed force = 400 kg. Distance of drill center from column face (throat) = 250 mm.

Solution:

First, we need to find the force at the column top end. Taking bending moments about the column face and column base,

Thrust × Throat = Column length (l) × Force (W)

$$400 \times 250 = 1.25 \times 1000 \times W$$

$$\therefore \quad W = \frac{400 \times 250}{1250} = 80 \text{ kg}$$

Let us consider a square tube. Taking the section tentatively as one-third the height,

$$b = d = \frac{1250}{3} = 416.7 \approx 417$$

Size factor $(s) = \frac{21 + b + d}{4000} = \frac{2500 + 417 + 417}{4000} = 0.83$

From Table 5.47 [p 492], for $S = 0.83$, Wall thickness $(t) = 9$
Ribs and stiffener thickness = 8

$$I = \frac{417^4 - 399^4}{12} = 4.077 \times 10^8$$

Deflection $(\delta) = \frac{W l^3}{3 EI} = \frac{80 \times 1250^3}{3 \times 2.1 \times 10^4 \times 4.077 \times 10^8}$

$$= \frac{1.5625 \times 10^{11}}{25.685 \times 10^{12}} = 0.00608$$

Recommended deflection = 0.005/meter

i.e. $= 1.25 \times 0.005 = 0.00625$, i.e. > 0.00608

If it is necessary to swivel the spindle head around the column, a cylindrical column is necessary. We should consider some standard pipes for the column. Four hundred six (16 inches) diameter pipe has a 381 (15 inches) bore. The next bigger size is 457 outside, and of 429 bore. The outside can be machined to 454 diameter.

$$I \text{[pipe 454]} = \frac{\pi}{64} (454^4 - 429^4) = 4.22 \times 10^8$$

Deflection $(\delta) = \frac{W l^3}{3 EI} = \frac{1.5625 \times 10^{11}}{6.3 \times 10^4 \times 4.22 \times 10^8} = 0.005877$

Let us compute the shear deflection. First, for the square section,

Shear force = W = 80 kg

For a square, $\dfrac{d}{b} = 1$

From table $\lambda = 2.35$

From Eqn. 5.59, shear deflection, for square column

$$\delta_s = 2.35 \, \dfrac{(80 \times 1250)}{0.79 \times 10^4 \times (417^2 - 399^2)}$$

$$= \dfrac{2.35 \times 10^5}{1.1603 \times 10^8} = 0.00202$$

For twist angle of the rectangular section, from p 497

$$\dfrac{\phi}{l} = \dfrac{KT}{Gdb^3}$$

For $\dfrac{d}{b} = 1.0$; $K = 7.1$

From Eqn. 5.60,

$$\dfrac{\phi}{l} = \dfrac{2T(b + d - 2t)}{4Gt(b - 2t)^2 (d - 2t)^2} = \dfrac{2 \times 6.6 \times 1000\,(417 + 417 - 18)}{4 \times 0.79 \times 10^4 \times 9\,(399)^2\,(399)^2}$$

$$= \dfrac{10771200}{7.2081 \times 10^{15}} = 0.149 \times 10^{-8} \text{ radians}$$

$$= 0.149 \times 10^{-8} \times \dfrac{180}{\pi} \text{ degrees} = 8.56 \times 10^{-8} \text{ degrees/mm}$$

For 1 meter length $\phi = 8.56 \times 10^{-8} \times 1000 = 8.56 \times 10^{-5}$ degrees, i.e. $< 0.5°$/meter. For the pipe column of $\phi\,429 \times \phi\,454$,

$$\dfrac{\phi}{l} = \dfrac{T}{GI_P}$$

$$I_p = \dfrac{\pi}{32}\,(454^4 - 429^4) = 8.45 \times 10^8$$

$T = 6.6$ kgm = 6600 kgmm

$$\dfrac{\phi}{l} = \dfrac{6600}{0.79 \times 10^4 \times 8.45 \times 10^8} = 9.88 \times 10^{-10} \text{ radians}$$

For 1 meter length, $\phi = 9.88 \times 10^{-7}$ radians

In degrees $\phi = 9.88 \times 10^{-7} \times \dfrac{180}{\pi}$

$$= 5.66 \times 10^{-5°} \text{ meter, i.e. much less than the limit of } 0.5°/\text{meter}.$$

Figure 5.57 shows the square and round columns.

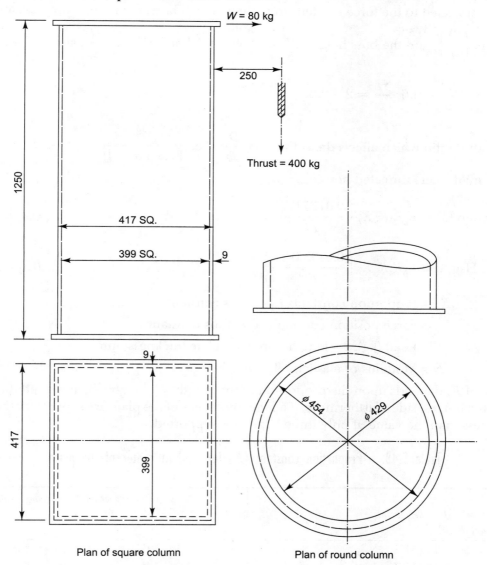

Fig. 5.57: Columns for Example 5.24

Housing design

Casing boxes for gear trains and other mechanisms are called **housing**. Housings which need to be frequently opened for setting mechanisms, are of a split design. Covers for cone pulley drives are usually hinged, to facilitate quick access to the pullies, for belt-shifting. Split casings have only half the stiffness of a solid casing. Displacement (deflection) and stiffness of the housing are measured at a right angle to the plate surface. They depend upon the length (l), height (b), and depth (d) ratio, edges support condition, and position of the force. As the point which is subjected to the force is often strengthened by a boss, the ratio of the boss and hole diameters $\left(\dfrac{D}{d}\right)$, and the boss height ($H$) and plate thickness ($t$), i.e. $\dfrac{H}{t}$, also matter.

Usually $\quad \dfrac{D}{d} = 1.6; \ \dfrac{H}{t} = 2 - 2.2$

The other ratio which affects the stiffness is $\dfrac{D^2}{lb}$, i.e. $\dfrac{D^2}{\text{Plate area } (A)}$

For a mild steel fabricated housing,

$$\text{Deflection } (\delta) = K_p K_b K_w K_r \left(\dfrac{0.0175\, Wl^2}{Et^3} \right) \quad \text{(Eqn. 5.61a)}$$

For C.I. Hsg, $\delta = K_p K_b K_w K_r \left(\dfrac{0.0193\, Wl^2}{Et^3} \right) \quad \text{(Eqn. 5.61b)}$

K_p = Proportion constant; K_b = Boss constant
K_w = Force position constant; K_r = Ribs constant
W = Load (kg); l = length (mm); t = plate thickness (mm)
E = Modulus of elasticity (kg/mm^2)

Value of K_p depends upon the proportion of the length (l), height (b), and depth (d). Note that l is the longest side. Furthermore, if all the four edges of the plate are supported, the value of K_p is less than the value, if only three edges are supported.

Table 5.48: Proportion constant (K_p) for load at center of the plate

Ratio Length (l) : Height (b): Depth (d)	No. of edges supported by adjoining walls	
	4	3
1 : 1 : 1	0.35	0.48
1 : 1 : 0.75	0.44	0.58
1 : 1 : 0.5	0.50	0.65
1 : 0.75 : 0.75	0.30	0.42
1 : 0.75 : 0.5	0.33	0.46

In case of the support of 3 edges if the load is near the corner of an un-supported edge, the value of K_p will be up to 3 times higher.

For central boss with $D = 1.6d$, and boss height 2 to 2.2 times the plate thickness (t), K_b depends upon the ratio,

$$\frac{D^2}{A} = \frac{D^2}{lb}$$

Table 5.49: Boss constant (K_b) for loaded central boss and boss height $H = 2$ to $2.2\ t$

$\dfrac{D^2}{lb}$	No boss	0.02	0.04	0.06	0.08	0.10	0.12	0.14
K_b	1.0	0.85	0.79	0.73	0.69	0.66	0.63	0.61

If the load is offset with the boss, constant K_w must be considered. The following table gives the values of K_w for offset $R = 0.25\ l$ from boss.

Table 5.50: Load (Force) position constant K_w for offset $R = 0.25\ l$ from boss

$\dfrac{D^2}{lb}$	No boss	0.02	0.04	0.06	0.08	0.10
K_w	1.0	0.945	0.92	0.90	0.88	0.86

K_r = Ribs constant

 = 0.85, if ribs are near load (boss)

 = 0.80, if ribs are symmetrically placed on the plate.

Example 5.25: Design a steel-fabricated housing of $400 \times 300 \times 300$ size and find the deflection for a central load of 40 kg, if (i) all the four edges are supported, (ii) top edge is not supported, (iii) also find the deflection if the load is at 100 from the central boss. The ribs are placed symmetrically along the length 400. There is a $\phi\ 20$ hole and a suitable boss.

Solution:

From Table 5.48, for $1 : 0.75 : 0.75$ ratio,

For 4 edges support $\qquad K_{p4} = 0.30$

For 3 edges support $\qquad K_{p3} = 0.42$

Taking $D = 1.6\ d = 1.6 \times 20 = 32$

$$\frac{D^2}{lb} = \frac{32^2}{400 \times 300} = \frac{1024}{120000} = 0.0085$$

From Table 5.49, $K_b = 0.936$ (approx.) by interpolation)

For Central load $K_{wc} = 1.0$

For 100 offset load, from Table 5.50, $K_{wo} = 0.977$ Approx. $\left(\text{for } \dfrac{D^2}{lb} = 0.0085 \right)$

For symmetrical ribs $K_r = 0.8$

(i) For Central load, from Eqn. 5.61(a),

$$\delta = 0.3 \times 0.936 \times 1.0 \times 0.8 \; \dfrac{(0.0175 \times 40 \times 400^2)}{2.1 \times 10^4 \times t^3}$$

From Table 5.47 [p 492], size factor $(s) = \dfrac{2 \times 400 + 300 + 300}{4000} = 0.35$

For $s = 0.4$; $t = 6$, and ribs thickness $= 5$

$$\delta = 0.22464 \; \dfrac{5.333}{6^3} = \dfrac{1.2}{216} = 0.0055$$

(ii) If the top edge is not supported,

$K_p = 0.42$, instead of 0.30

$$\delta = \dfrac{0.42}{0.30} \times 0.0055 = 0.0077$$

(iii) For offset load, from Table 5.50, $K_{wo} = 0.977$ Approx. (by interpolation) instead of $K_{wc} = 1.00$

So, $\delta = 0.977 \times 0.0055 = 0.00537$

If the load is moved near the corner of an un-supported edge

$K_{p3} = 3 \times 0.42 = 1.26$

$\delta = 3 \times 0.0077 = 0.0231$

Table and base design

Table and bases are considered as plates supported at edges. Load of the workpiece is usually considered as uniformly distributed. Machine bases are often also subjected to bending moment.

Rectangular tables

It is convenient to relate thickness (t) to the base/table width (b).

$$t = 0.1 - 0.18 \; b \text{ Preferably, } t = 0.15 \; b$$

$$\text{Stiffness} = \dfrac{W}{\delta} = 7.03 \left[\dfrac{L}{b^3} + \dfrac{Lbs}{\text{Inch}} \right] Et^3 \qquad \textbf{(Eqn. 5.62)}$$

$$\text{Stiffness} \; \dfrac{Lbs}{\text{Inch}} = 56 \times \dfrac{W[\text{kg}]}{\delta[\text{mm}]}$$

Machine Tool Elements

Tables longer than 1000, have a ribbed construction. There are two, top and bottom plates, with connecting ribs at a spacing of 300–400. Stiffness of a ribbed table is lesser than that of a solid table. Area of the ribs (A) is reckoned in the factor ψ.

$$\psi = \frac{15.6\,I}{b^2\,A} \text{ for M.S.} \quad \text{(Eqn. 5.63a)}$$

$$\psi = \frac{14.76\,I}{b^2\,A} \text{ for C.I.} \quad \text{(Eqn. 5.63b)}$$

$$\text{Stiffness} = \frac{EI}{100L\,(1+\psi)} \text{ kgf/mm} \quad \text{(Eqn. 5.64)}$$

Circular tables

For solid plates deflection $(\delta) = \dfrac{0.05525\,WD^2}{Et^3}$ \quad (Eqn. 5.65)

$$\text{Rigidity/stiffness} = \frac{W}{\delta} = \frac{18.1\,Et^3}{D^2} \text{ (kg/mm)}$$

W = Load (kg); δ = deflection (mm)

E = Modulus of elasticity (kg/mm²); t = thickness (mm)

D = Plate diameter (mm)

For ϕ above 1000, two plates of thickness t_1 and t_2, are connected by 10 – 16 radial ribs.

Then,
$$S = \frac{Eh^2}{100(1-p^2)}\left[\frac{t_1\,t_2}{t_1+t_2}\right] \text{ [kg/mm]}$$

$$\psi = \frac{8\,\pi\,S\,(1.2 + p)}{G\,A\,n\,D}$$

$$\text{Stiffness} = \frac{S}{1+\psi}$$

h = distance between (mid-thickness) center lines of plates (mm)
p = Poisson's ratio = 0.3 for M.S. and 0.23 for C.I.
G = Modulus of rigidity (kg/mm²) = 0.79×10^4 (M.S.) [11.23×10^6 Lbs/Inch² for M.S.]
    = 0.35×10^4 (C.I.) [4.98×10^6 Lbs/Inch² for C.I.]

E = Elasticity modulus
   = 2.1×10^4 (M.S.) and 1×10^4 (C.I.) = [30×10^6 Lbs/Inch² for M.S. and 14.22×10^6 Lbs/Inch² for C.I.]

Example 5.26: Design a rectangular M.S. table of 400×700 size, and calculate the stiffness coefficient.

Solution:

Thickness $(t) = 0.15b = 0.15 \times 400 = 60$

$$\text{Stiffness} = \frac{W}{\delta} = 7.03 \left[\frac{L}{b^3} + \frac{2.21}{L^2} \right] Et^3$$

$$= 7.03 \left[\frac{700}{400^3} + \frac{2.21}{700^2} \right] 2.1 \times 10^4 \times 60^3$$

$$= 7.03 \left[1.094 \times 10^{-5} + 4.51 \times 10^{-6} \right] 4.536 \times 10^9$$

$$= 491076.4 \text{ kg/mm} = 491 \text{ kg /micron}$$

Example 5.27: Design a ϕ 700 circular table with 0.0001 mm/kg stiffness.

$$\frac{W}{\delta} = \frac{18.1 \times 2.1 \times 10^4 \, t^3}{700^2} = \frac{1}{0.0001}$$

$$t^3 = \left[\frac{15.6 \times 5795374}{800^2 \times 2304} + \frac{2.21}{1600^2} \right] = 12891.3$$

$$t = 23.44 \approx 23.5$$

Example 5.28: Find stiffness of a fabricated 800×1600 table if top & bottom plates are 12th k and there are 3 Nos 8th to longitudinal ribs.

Solution:

For a solid table $t = 0.15b = 0.15 \times 800 = 120$

$$\text{Stiffness} = 7.03 \left[\frac{1600}{800^3} + \frac{2.21}{1600^2} \right] 2.1 \times 10^4 \times 120^3 = 1017429 \text{ kg/mm from Eqn 5.62}$$

For same height fabricated table,

$$I = \frac{800 \times 120^3 - [800-24][120-24]^3}{12} = 5795374 \text{ mm}^4$$

$$A = [120 - 2 \times 12] \times 8 \times 3 = 2304 \text{ mm}^2$$

From Eqn. 5.63a,

$$\psi = \frac{15.6 \times 5795374}{800^2 \times 2304} = 0.0613$$

$$\text{Stiffness} = \frac{2.1 \times 10^4 \times 5795374}{100 \times 1600 [1 + 0.0613]} = 717587.6 \text{ kg/mm}$$

The stiffness of the fabricated table is 70.5% of the solid table.

CHAPTER 6

MACHINE TOOL DYNAMICS

A machine tool can be compared to a system of springs with stiffness of each spring equal to the rigidity of the corresponding element in the machine tool [**Fig. 5.52**]. The elastic system is affected by the metal-cutting process, frictional forces, and power variations in the source. Also, deflections in the elastic system affect the cutting process, frictional forces, and the power source. These mutual affectations form a closed loop system. The cutting force causes displacement in the elastic system, such as the deflection of the tool post, and carriage. These in turn, reduce the cut depth and cutting force. Thus, there is a mutual dependence, feedback, and ensuing re-adjustment. Frictional forces offset with the lead screw can cause a tilt in the carriage, and change the frictional force. The change leads to a re-adjustment in the elastic system. Similarly, an increase in the cutting force can slow down the motor, which causes power variation, leading to further re-adjustment in the elastic system. **Figure 6.1** shows the closed loop system, with inputs from the cutting process, friction, and power source. Y_1, Y_2, and Y_3 show the feedbacks from the elastic system. The feedbacks alter the cutting force, friction force, and power parameters, which in turn affect the elastic system.

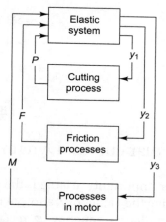

Fig. 6.1: Closed loop system for metal cutting

It is useful to reduce the 3-loop system in [**Fig. 6.1**], to a single-loop system [**Fig. 6.2**], i.e. Equivalent Elastic System (EES).

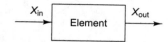

Fig. 6.2: Single-loop equivalent elastic system

6.1 STABILITY OF A SYSTEM

This depends upon the dynamic stability of the EES. In a stable system the variation in the depth of the cut will gradually reduce (damp) the resulting vibrations. In an unstable EES, the vibrations will gradually increase, further aggravating the situation.

The **Nyquist criterion** is useful for studying a system of stable elements. Removing one link, we measure the input and output of various elements, at various frequencies [**Fig. 6.3a**]. By plotting inputs and outputs we can find the amplitudes ratio (A).

$$A = \frac{A_{out}}{A_{in}}$$ (Eqn. 6.1)

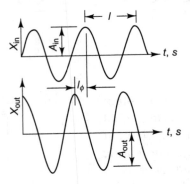

Fig. 6.3(a): Input and output

6.1.1 Polar curves: Dynamic characteristics

The other important variable is the phase angle (ϕ) between the input and output amplitudes. After noting the values of A and ϕ at different frequencies, we can draw the amplitude frequency ($A\ \omega$), and phase frequency ($\phi,\ \omega$) curves [**Fig. 6.3b**]. It is necessary to plot A at various phase angles (ϕ), to obtain the polar phase, amplitude, and the frequency curve [**Fig. 6.3c**] for the element.

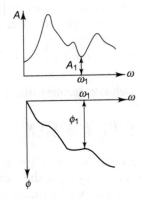

Fig. 6.3(b): Amplitude and phase difference

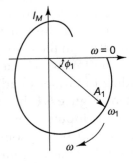

Fig. 6.3(c): Polar curve for amplitude and phasedifference of various frequencies

Polar curves of the two elements in an open loop system can be combined into a resultant curve, with $A = A_1 + A_2$, at angle $\phi = \phi_1 + \phi_2$ [**Fig. 6.3d**].

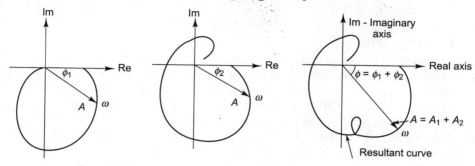

Fig. 6.3(d): Drawing the resultant polar curve for 2 polar curves

If the polar curve intersects the horizontal (real) axis (Re) at a point closer to the origin than point $(-1, 0)$, the closed loop system is stable. If the curve intersects the axis at exactly $(-1, 0)$, it's a borderline case: the threshold of stability. If the curve intersects the negative real axis beyond (farther from the origin) the point, the closed loop system is unstable.

Instability is caused by:
(a) Machining close to critical speed
(b) High slenderness of workpiece or tool (drill, boring bar)

Most of the machine tool elements have inherently stable elastic systems.

Degrees of freedom

Comprising many parts, an elastic system of a machine tool has infinite degrees of freedom. It will be useful to consider only the single-most dominant degree of freedom and disregard the others. The single degree of freedom that can be considered, is the **spring-mass-dashpot** arrangement [**Fig. 6.4**]. The intercept of the polar curve on the positive real axis is a reciprocal of the stiffness (rigidity). It is also called **compliance or unit deflection** (mm/kg). The displacement (deflection) variation lags behind the variation in the cutting force (W). The phase lag (ϕ) increases with an increase in the frequency (ω). The lag is 90° at the natural frequency (ω_n) of the system. The radius in the polar curve is maximum at the resonance frequency and tends to be zero at infinity frequency. So the polar co-ordinate of a single freedom system does not intersect the negative real axis.

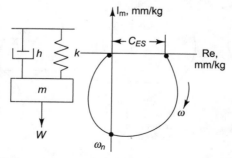

Fig. 6.4: Spring-mass-dashpot arrangement representation

For 'n' degrees of freedom there will be 'n' polar curves. These can be combined into a single resultant curve, as described earlier. Usually, the consideration of two degrees of freedom is enough. The parts which vibrate as a single body are treated as concentrated masses. The stiffness of the joints is considered as the spring constants.

Orthogonal cutting [Fig. 6.5]

In orthogonal cutting in lathes, it is useful to consider two degrees of mode-coupled freedom, in the directions of the maximum and minimum stiffness. [The output parameter (displacement) of EES is the same as the input parameter (undeformed chip thickness) of the cutting process]. By treating the tool as a beam, its deflection (δ) can be found. This deflection will be normal to the tangent, to the rigidity ellipse, at the intersection of the force (W). The deflection (δ) is projected on the axes of the rigidity ellipse. The deflections δ_1 and δ_2 are further projected on the Y axis, to get deflections Y_1 and Y_2 along the Y axis. The resultant deflection (Y) is the difference between Y_1 and Y_2.

$$Y = Y_1 - Y_2$$

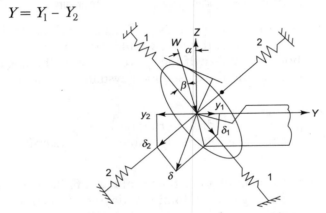

Fig. 6.5: Two degrees of mode coupled freedom in orthogonal cutting

If $Y_2 > Y_1$, the resultant deflection will be negative and the tool will dig into the workpiece instead of deflecting away from it. **Figure 6.6a** shows the polar curve for a positive deflection. **Figure 6.6b** shows the polar curve for a negative resultant deflection. Polar curves for negative and positive deflections can be combined into a single curve [**Fig. 6.6c**].

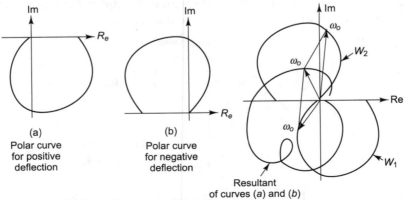

Fig. 6.6: Polar curves for 2 mode-coupled degrees of freedom

If the natural frequencies of the system are close to each other, the tool moves (vibrates) in an elliptical curve. Depending upon the tool position and motion direction, the undeformed chip thickness increases or decreases [**Fig. 6.7**]. The undeformed chip thickness and the cutting force in the direction of the velocity (W_z), increase during the descent of the tool tip. The ascent of the tool leads to a decrease in the undeformed chip thickness and the cutting force (W_z). The undeformed chip thickness is more when the directions of the vibrations and cutting force coincide. The vibrations increase, leading to instability and tool-chatter. When the force of the vibrations is fully compensated by the increasing damping force, the chatter stops. If the tool moves in the opposite direction, the system will be dynamically stable and there will be no chatter.

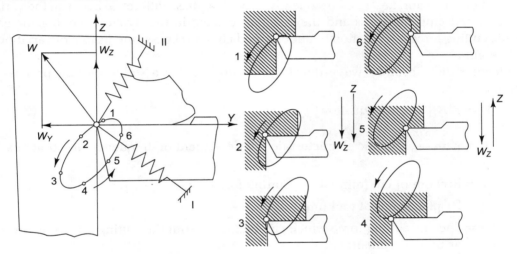

Fig. 6.7: Motion of tool tip in coupled modes system

Dynamic characteristic [Fig. 6.8]

The dynamic charasteristic of a cutting process is the relationship between the dynamic cutting force and change in the undeformed chip thickness, in a direction normal to the cut surface (Y in [**Fig. 6.8**]).

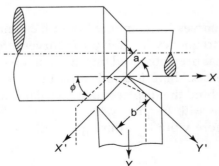

Fig. 6.8: Undeformed chip parameters

$$W = kby \quad \text{(Eqn. 6.2)}$$

W = Dynamic cutting force (kg)

y = Change in undeformed chip thickness (mm)

k = Cutting coefficient = $r + is$ (*is* is an imaginary number)

According to the **harmonious vibrations theory**, the angle between vector K and X axis $(\theta) = \tan^{-1}\dfrac{s}{r}$.

Multiplication of a vector by i is equivalent to its differention. It represents a vector, 90° ahead of the first vector, i.e. 'r'. Consequently, there is a phase difference between the variation in undeformed chip thickness and the resultant change in the dynamic cutting force. The dynamic cutting force is a function of undeformed chip thickness, and its rate of variation: the first differential.

Neglecting the vibratory wave due to previous cut, for a stable cutting process with continuous chip,

$$W = r \, by \, [\text{Tlusty's equation}] \quad \text{(Eqn. 6.3)}$$

$k_s = rb = \dfrac{W}{y}$ = Static characteristic not dependent on frequency of vibrations.

b = Breadth of cut (mm); W = Cutting force (kg)

y = Displacement of tool (mm)

r = Specific cutting force which can be found from the cutting force relationship or by experiment.

= 200 kg/mm² for carbon steel   [1 kg/mm² = 1422 psi]

Change in the undeformed chip thickness does not cause an instantaneous change in the dynamic cutting force. There is always a time lag. Kudinov's analysis considers this phase difference. Furthermore, Kudinov reckons the force on the flank of the tool. A worn tool has a flat land on the flank and causes frictional power loss.

Kudinov's expression

Figure 6.9 shows the polar curves for combined (tool face and flank) force, for 3 different values of flank wear. At zero frequency (ω), the polar curves intercept the real axis R_e, at $x = Ks$, the static characteristic. It has been observed that the flank wear land dampens the vibrations. Also, as the flank wear increases, there is a reduction in the frequency range, in which the dynamic cutting force lags behind the displacement. For flank wear land of 1 mm, frequency (ω) = 1,000. For all vibrations with frequency (ω) ≥ 150 Hz, the cutting process provides a damping effect. There are more vibrations during the idle run of the machine than during actual metal cutting.

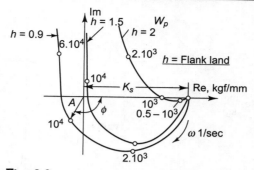

Fig. 6.9: Polar curves reckoning flank wear land

Stability analysis

(a) Static characteristic

Instability manifests in form of self-excited chatter vibrations. The cutting coefficient (K_s) is directly proportional to the undeformed chip width (b). An increase in the chip width beyond a certain limit causes chatter. Refer to **[Fig. 6.4]**, and recall the relationship between the cutting coefficient (K_s) and cut width (b), i.e.

$$K_s = rb = \frac{W}{y} \qquad \text{(Eqn. 6.4)}$$

Polar curve for an open loop system reckoning the cutting process can be obtained by multiplying the amplitude of the EES curve with the cutting coefficient K_s. As the cutting coefficient is equal to the stiffness (rigidity) of the system $\frac{W}{y}$, it is constant. Consequently, for a material with cutting constant (r), $b = \frac{K_s}{r}$.

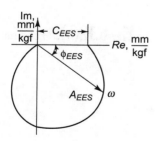

Fig. 6.10(a): Polar curve for single freedom EES

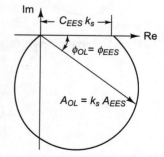

Fig. 6.10(b): Polar curve for open loop

For cut width exceeding the above limit, there will be a resultant chatter.

In general, if cutting conditions can be represented by a static charasteristic of an open loop system, the closed loop system will always be stable.

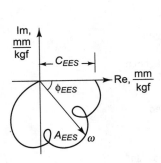 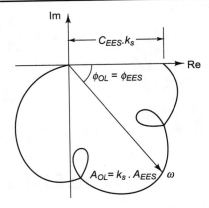

Fig. 6.11(a): Polar curve for multi-freedom EES   **Fig. 6.11(b):** Polar curve for open loop of **[Fig. 6.11a]**

(b) Dynamic characteristic

Dynamic characteristic is a polar curve for resultant of EES, and the cutting process. For plotting the resultant for a given frequency, the amplitudes are multiplied while the phase angles are added **[Fig. 6.12]**. For cut width (b) lesser than the maximum (b_{max}), the system is stable at all cutting speeds. However, if $b > b_{max}$, it is necessary to find two speeds V_1 and V_2, beyond which the system is stable. The system will be unstable between speeds V_1 and V_2.

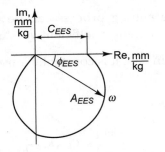 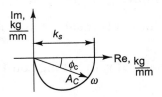

Fig. 6.12(a): Polar curve for single degree of freedom of EES   **Fig. 6.12(b):** Polar curve of cutting

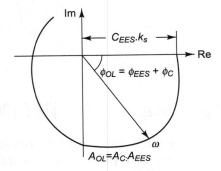

Fig. 6.12(c): Polar curve of open loop system: resultant of **[Figs. 6.12a and 6.12b]**

Figure 6.13 shows the static cutting process characteristic (open loop) for a vibratory system, with 2 degrees of freedom with coupled coordinates. As per **Nyquist's criterion**, if the resultant open loop polar curve intersects on the origin side of the point (–1, 0), the closed loop is stable.

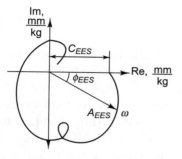

Fig. 6.13(a): Polar curve for 2 degrees freedom of EES **Fig. 6.13(b):** Polar curve of open loop system

Figure 6.14 shows the combined polar curve for EES, and the cutting process, as well as the individual curves for EES, and the cutting process with many degrees of freedom, with coupled coordinates. This curve too is judged by the co-ordinates of the intersecting point with the negative real axis: stable when it is closer to the origin than the point (–1, 0), borderline, when it is on the intersection of the point (–1, 0), and unstable at intercepts farther than the point (–1, 0).

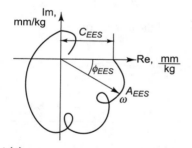 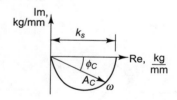

Fig. 6.14(a): Polar curve of multi-degree freedom of EES **Fig. 6.14(b):** Polar curve of cutting

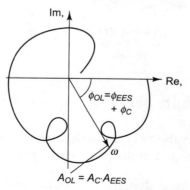

Fig. 6.14(c): Polar curve for open loop: resultant of **[Figs. 6.14a, and 6.14b]**

Regenerative chatter

In multi-teeth tools such as drills and milling cutters, the time between machining by two successive cutting edges (λ), depends upon the RPM (N), and the number of cutting edges (Z).

$$\lambda = \frac{60}{NZ} \text{ seconds} \quad \textbf{(Eqn. 6.5)}$$

Let's suppose that the surface, machined during the previous cut becomes wavy due to vibrations, and the successive cut overlaps the previous cut [**Fig. 6.15**]. This problem is more acute in multi-teeth cutting than in turning. The chatter in multi-teeth cutting is called regenerative chatter. It is generated by a time delay. The delay rotates the radius vector of the transfer function through angle (ϕ) = $\lambda\omega$ in a clock-wise direction [**Fig. 6.16**]. If the polar curve lies within a circle with radius 1.0, rotation by angle ($\lambda\omega$) will not affect the stability. For any value of ($\lambda\omega$), the intercept on the negative axis will be closer to the origin than point ($-1, 0$) on the negative real axis.

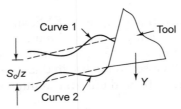

Fig. 6.15: Overlap of wavy cut over the wave of the previous cut

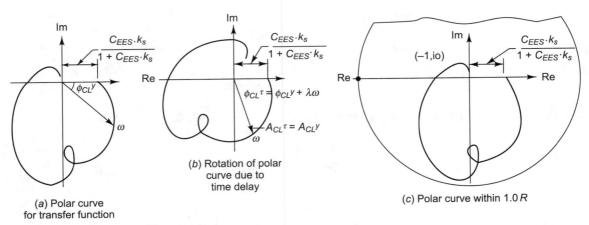

Fig. 6.16: Effect of time delay on polar curve, and stability

If the polar curve intersects the negative axis farther from the origin than point ($-1, 0$), the system will be unstable. If a part of the polar curve projects beyond the circle with radius 1.0 [**Fig. 6.17**], the system will be dynamically stable within the range $\dfrac{\phi_2}{\omega_2}$ to $\dfrac{\phi_1 + 2\pi}{\omega_2}$, and unstable for other values of $\frac{\phi}{\omega}$. It should be noted that ϕ_1 and ϕ_2 are angles of the vector with the negative real axis, at frequencies ω_1 and ω_2. Also, the maximum cut width (b_{max}) for freedom from regenerative chatter, is half of the maximum width for freedom from primary chatter. In unstable zones chatter frequency increases with an increase in the RPM, and the amplitude is maximum in the middle of the unstable zone.

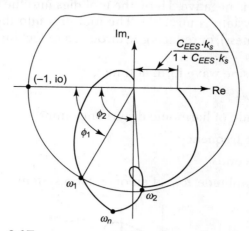

Fig. 6.17: Stability range for polar curve crossing 1.0 RAD

Forced vibrations

Disturbances affecting the *EES* and cutting process cause forced vibrations. Their amplitude (A) and frequency (ω) change the normal cutting speed (V). For circular frequency (ω), the speed will vary between $(V + A\omega)$ to $(V - A\omega)$ in every rotation. This affects the tool life.

Disturbances imposed on the cutting process can originate from any or several of the following sources:
(1) Variation in undeformed chip thickness in:
 (a) Intermittent cutting (milling, broaching)
 (b) Waviness caused in previous cut
 (c) Eccentric turning
 (d) Entry and exit of tool into/from workpieces
(2) Change in cutting speed due to variation in cutting diameter in facing, parting
(3) Change in cutting angles due to change in feed direction

In a parting operation if δ is the deflection for the set undeformed chip thickness [a], then:

Machining accuracy index = $\frac{\delta}{a}$

When frequency (ω) is zero,

$$\frac{\delta}{a} = \frac{C_{EES} \cdot K_s}{1 + C_{EES} \cdot K_s} \quad \textbf{(Eqn. 6.6)}$$

C_{EES} = Compliance (unit deflection) of $EES = \dfrac{\delta}{W}$ (mm/kg)

K_s = Cutting coefficient = rb = Specific cutting force × cut width

If $C_{EES} > 0$, an increase in C_{EES} causes an increase in deflection (δ), causing a decrease in the machining precision.

For $C_{EES} < 0$, deflection is negative. Here, the tool digs into the workpiece.

For $C_{EES}\, Ks = -1$, the system is unstable. The tool digs into the workpiece till it breaks. Increasing the rigidity (stiffness), or reducing K_s through a reduction in the cut width, alleviates the situation.

If the disturbance has a sine wave form at time t,

$$\delta_t = A_y \sin(\omega t)$$

A_y = Amplitude of harmonic disturbance (mm)

ω = Circular frequency

t = time (seconds)

Refer to [**Fig. 6.18**]. For amplitude A_{OL} of an open loop system,

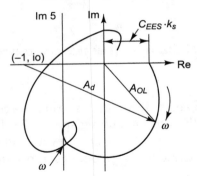

Fig. 6.18: Amplitude [A_d] of forced vibrations due to disturbance in cutting process

If $\left(\dfrac{A_d}{A_{OL}}\right) < 1$, the amplitude of the forced vibrations exceeds the amplitude of the wave from the previous cut. The machining precision decreases with each successive cut.

If $\left(\dfrac{A_d}{A_{OL}}\right) > 1$, the amplitude of the vibratory wave decreases with each pass. An increase in the undeformed chip width (b) increases the ratio $\dfrac{A_d}{A_{OL}}$, and the amplitude of the vibratory wave decreases. This improves the precision of machining.

If $\dfrac{A_d}{A_{OL}} = 1$, amplitude of the vibratory system does not increase or decrease.

Disturbances imposed on equivalent elastic system (*EES*).
These originate from:
(1) **Periodic forces** due to manufacturing errors in transmission elements (gears, belts, splines, bearing races); and lack of balance in rotating parts
(2) **Pulsating forces** due to hydraulic or coolant pumps
(3) **Inertia forces variations** due to reversals of tool/workpiece motions
(4) **External forces** and shocks from other machines/elements transmitted through the machine foundation

If A_f = Amplitude of the resultant wave on the machined surface, and A_i = Amplitude of forced vibrations between the workpiece and cutting tool, during idle run, then,

$$A_f = \frac{A_i}{A_d}$$ (Eqn. 6.7)

Refer to [**Fig. 6.19**].

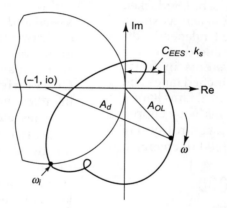

Fig. 6.19: Forced vibrations due to disturbance in *EES*

If $A_d < 1$, an increase in chip width (*b*) reduces the dynamic stability and machining accuracy.
If $A_d > 1$, an increase in (*b*) reduces the stability, but improves machining accuracy.
If $A_d = 1$, amplitude of the vibratory wave is equal to the amplitude of the vibratory wave during machine idle run.

It can be concluded that machining precision can be improved by reducing idle run vibrations, and by minimizing manufacturing errors (eccentricity, unbalance, etc.) of the transmission elements.

CHAPTER 7

MACHINE TOOL OPERATION

Machine tools use mechanical, hydraulic and pneumatic, and electrical and electronic systems for actuation. Some systems must be interlocked with each other, while others such as speed change devices can be operated independently.

Even independently-operable devices are interlocked with other systems in fully-automated machine tools. Punched tapes and cards, magnetic tapes, and optical films [tapes] are gradually replacing drums with copper strips. The basic criteria for a control system design have however, remained unchanged.

7.1 REQUISITES OF A GOOD CONTROL SYSTEM

1. **Safety:** should be ensured, to prevent the **accidental operation** of push buttons and rotary switches, by sinking the switches below the cover surface, or by surrounding them with rings. The emergency stop is of course, an exception.
   Conflicting motions should be interlocked to ensure a **safe sequence**. The wokpiece should be clamped before a cut commences. The spindle should commence rotation before feeding starts. Dead stops should be used to prevent over-travel of traveling elements [saddles], beyond the safety limit.
   Remote controls are necessary to keep the operator at a safe distance from toxic and radioactive workpiece materials.
   Visual and audio signals should be used to draw attention of the operator to undesirable occurrences, warranting attention and corrective action.

2. **Convenience:** Operating devices [handles, wheels, levers, push buttons, switches, etc.] must be in positions, convenient to operate, without the necessity for awkward acrobatic postures. Force required for operating manual devices should not exceed 8 kg [at the most 16 kg]. For frequently operated devices, it should only be 4 kg [at the most 6 kg]. Sizes of knobs, and handles should be such that they can be conveniently operated by average human hand. Also their positions should be suitable for the height and length of the limbs of an average human being. When traveling elements traverse beyond the reach of the operator, actuators for controlling them should be duplicated and placed within easy reach of the operator. This can be easily accomplished by the use of pendant push-button stations. Frequently operated elements should take the minimum time for actuation.

3. **Mnemonic [memory promoting] features:** Coordinate the direction of travel [of the carriage] with the direction of limbs' [hands/feet] motions. Table/carriage motion towards the left of the operator should be accomplished with a leftward motion of the operating lever/handle. It is often convenient to use a single lever/handwheel/push button for a number of direction controls; like the joystick which is used for the turning, ascent, and descent of aircrafts.
4. **Precision:** Prescision of motion should be commensurate with the machine function. The tool head of a planer can be moved within ± 0.5 mm, while the table travel of a jig boring machine table must be precisely within ± 0.005 mm.

We will be discussing mechanical devices first.

7.1.1 Safety devices

Should protect the machine, operator, cutting tool, and workpiece from inadvertent damage.

7.1.1.1 Machine protection

Comprises interlocking devices; means of limiting travel and overload [overstressing] protection.

Interlocks: Prevent simultaneous engagement of two different speeds or feeds, sometimes through a single lever/handwheel control [**Fig. 7.1a**]. Gearboxes with cluster gears, use a single-forked lever to engage the cluster, with only one of the two gears. Similarly, a double-ended clutch [**Fig. 7.1b**] can engage with only one of the 2 gear trains: on the left or right.

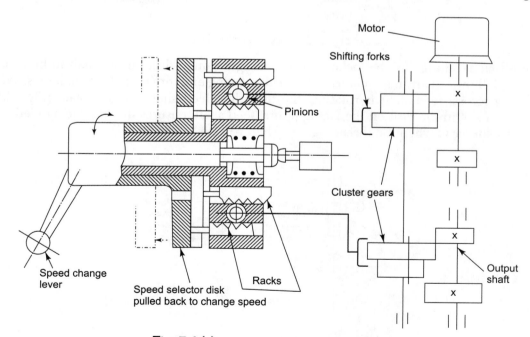

Fig. 7.1(a): Single lever control for gear box

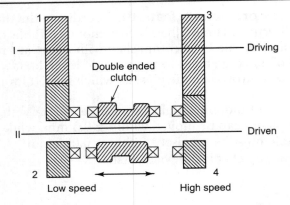

Fig. 7.1(b): Double-ended clutch

Parallel shafts can be interlocked by providing a concave cut-out on the disk to be locked [**Fig. 7.2**]. Engagement of the circular disk [or sector] with the cut out of the other disk, prevents its rotation.

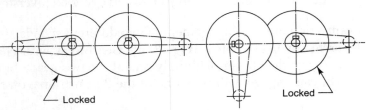

Fig. 7.2: Locking parallel shafts

Perpendicular shafts can be locked by providing a slotted disk on the shaft to be locked. **Figure 7.3** shows an arrangement for interlocking facing and threading feeds in lathes. When the facing [cross] feed shaft is rotating, the threading feed shaft cannot rotate [**Fig. 7.3a**]. Similarly, a slot in the disk on the facing [cross] feed shaft, ensures that the facing feed shaft remains stationary, when the threading feed shaft rotates.

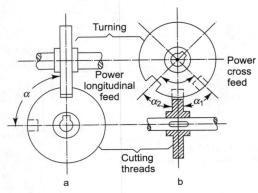

Fig. 7.3: Interlocking perpendicular shafts

The arrangement, shown in **[Fig. 7.4a]**, ensures that only one of the racks can engage with the pinion. It is necessary to shift the pinion axially, to disengage the pinion from one rack, and engage it with the other. The arrangement shown in **[Fig. 7.4b]**, allows the slotted bracket to move axially, only when the pin is engaged in the slot. In the central position shown, the bracket can only move in a direction, square with the shaft.

Hydraulic and electrical interlocks will be described in the later sections on automation.

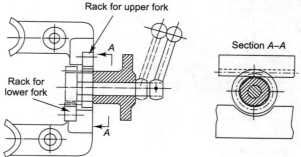

Fig. 7.4(a): Locking components traveling parallel

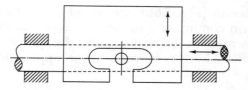

Fig. 7.4(b): Interlocking parts moving square to each other

7.1.1.2 Limiting travel of carriages

The choice of a device depends upon the function and precision of travel.

The combination of a rigid dead stop and a spring-actuated slip clutch **[Fig. 7.5]** is widely used. On touching the dead stop, the spring-backed clutch slips, and prevents further travel. An adjustable rigid stop can be used for setting the travel within ± 0.15 precision.

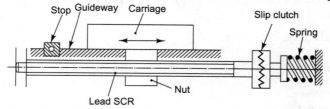

Fig. 7.5: Limiting travel by dead stop and slip clutch

The movable half of a slip clutch can be used for moving a lever, as shown in **[Fig. 7.6a]**. Moving against the spring, the movable half of the clutch swings the swiveling lever. This removes the support to the worm shaft housing, allowing it to fall under gravity. This disengages the worm with the mating wheel teeth.

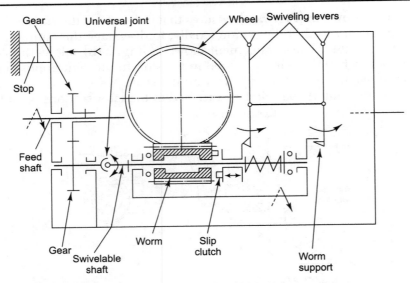

Fig. 7.6(a): Moving lever by using the movable half of the slip clutch

In the arrangement shown in **[Fig. 7.6b]**, the wheel ceases to rotate after a dead stop is reached. However, the worm continues to rotate. The screw action between the worm and the wheel, moves the worm towards the right, against the spring. This moves the bell crank lever in an anti-clockwise direction. As backing to the upper spring has been removed, it disengages the tooth clutch.

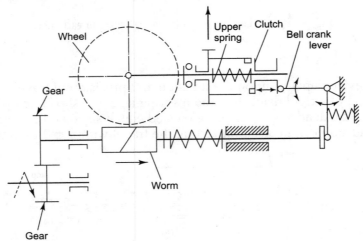

Fig. 7.6(b): Swiveling bell crank lever by the movable half of the slip clutch

Mechanical, travel-limiting devices can give a precision of 0.15–0.2 mm, under load.

Dead stops should be anchored into the tooth of the rack, fitted to a machine bed, to prevent slip. Usually, a micrometric screw is also used on the traveling element, to facilitate precise adjustment. Furthermore, the traveling element can be fitted with indexable stops, which can be set for different strokes of a tool-holding turret.

7.1.1.3 Overload protection

Damage to machine tool elements can be very costly, i.e. if idle time is also taken into account. Motor overload tripping alone is not enough to protect machine elements, for the torque varies according to the speed [RPM] used. Although the slip of the belt under overload provides some protection, additional means are necessary. These can be classified as shear pins and keys, slip clutches [friction, toothed, or ball type], and dropping worms.

Shear pins or keys

Shear pins or keys break off when a machine is overloaded. These are usually designed to fail at 75% of the maximum force or torque, for which the machine element is designed.

Figure 7.7a shows an axial shear pin, while **[Fig. 7.7b]** shows a radial shear pin. The first one is subjected to a single shear, while the latter is subjected to double shear.

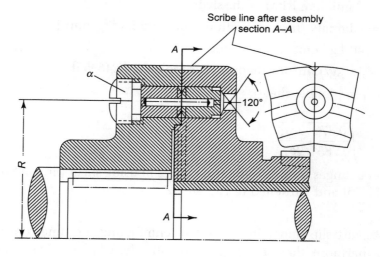

Fig. 7.7(a): Axial shear pin

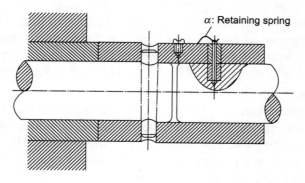

Fig. 7.7(b): Radial shear pin

For an axial shear pin [single shear],

$$\frac{\pi}{4} d^2 = \frac{0.75 \, T_{max}}{R f_s}$$

or

$$d = \sqrt{\frac{0.75}{0.7854} \frac{T_{max}}{R f_s}} \qquad \text{(Eqn. 7.1)}$$

d = Shear pin diameter [mm]; R = Radius of pin center [mm]

T_{max} = Maximum torque [kgmm]

$$= \frac{\text{Motor kW} \times 100}{\text{Minimum RPM of the shaft}}$$

f_s = Ultimate shear strength of pin material [kg/mm^2]

= 40 kg/mm^2 for mild steel

= 52 kg/mm^2 for medium carbon steel [C45/55]

For a radial shear pin [double shear],

$$d_r = \sqrt{\frac{0.375}{0.7854} \frac{T_{max}}{R f_s}} \qquad \text{(Eqn. 7.2)}$$

Usually, shear pin ⌀ ranges from 1.5–10 mm. Bushes housing shear pins, are made of medium carbon steel [C45/55] and hardened to RC 48–52.

Safety clutches just slip under the overload, eliminating the time-consuming shear pin replacement. Furthermore, they re-engage automatically, when the load falls below the slip limit. The construction is similar to that in overtravel clutches. Friction clutches are used more widely than toothed [jaw] clutches. Life of a safety clutch can be enhanced by using the slip motion to switch off the power. **Figure 7.8** shows the various types of safety clutches. The teeth have inclined sides, to facilitate slip. The toothed clutch in [**Fig. 7.8b**] is provided with a threaded sleeve. It engages with the threads of the fixed clutch half. This simplifies the adjustment of the spring [slip] force/torque. The disks backing the spring, act as thrust bearings. The key/splines should be slide fit [H7/g6] with the jaws, otherwise, excessive friction at key/splines is likely to hinder slip.

Friction with key/splines can also be reduced by adding a member, with teeth on both sides [**Fig. 7.8c**]. The intermediate member has no keyway/splines and it engages with the toothed driving half on one side, and with teeth of the driven half, on the other. It is the unkeyed intermediate teethed member, which slides axially against the spring, to disengage with the teethed member on right, to effect the slip.

Balls in the clutch, in [**Fig. 7.8d**], are held in position by riveting.

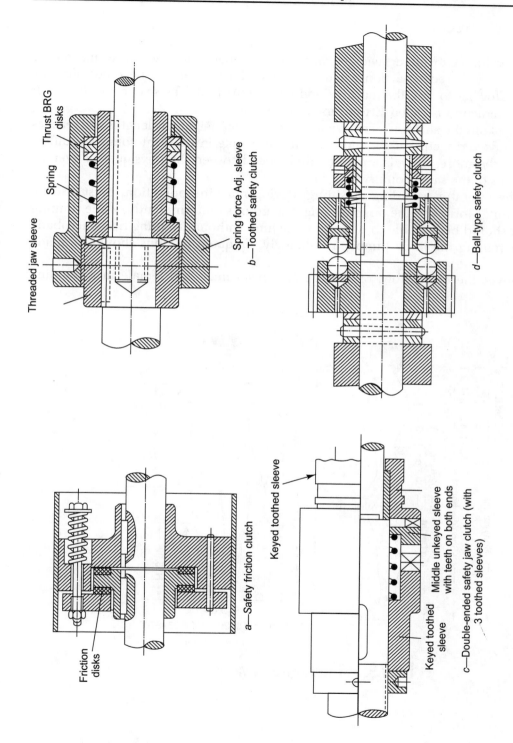

Fig. 7.8: Safety slip clutches

7.1.2 Convenience

The machine operator has to interact with the machine he operates. He observes the machining operations and takes corrective action when necessary. If a cutting tool becomes blunt or breaks, the operator has to stop the machine and replace the tool. The science dealing with man–machine interaction is called **Ergonomics** [work laws].

Ergonomics reckons the anatomy of an average operator: his height, the lengths and reaches of his limbs [hands and feet]. Height and anatomy of people in different countries, differs considerably. Furthermore, there is a marked difference between the physiques of average males and females of the same country.

Posture of the operator [standing or sitting] during machine operation, must also be taken into account. Moreover, the operator might stand or sit intermittently, while running his machine. It should be possible to operate a machine without assuming tiring acrobatic postures, and straining the body [limbs] inordinately. Frequent motions should be made less tiresome.

Figure 7.9 gives the dimensions of average European male and female in a standing position.

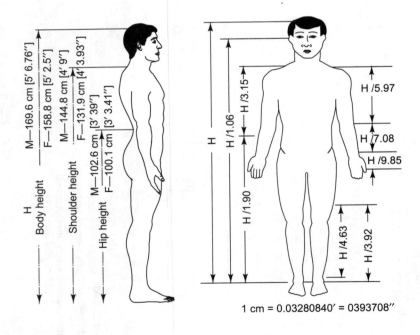

Fig. 7.9: Average European male/female

Machine Tool Operation

Figure 7.10 gives the areas in which he can move his hands, with minimum fatigue and average fatigue.

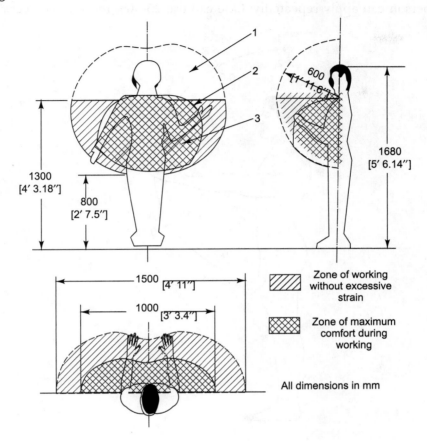

Fig. 7.10: Minimum and average strain areas

Figure 7.11 gives the range of the force a person can apply, without straining limbs excessively. The values should be multiplied by body weight [kg/lbs] to obtain the values of the force a person can apply repeatedly. One can use 25-50% more force occasionally.

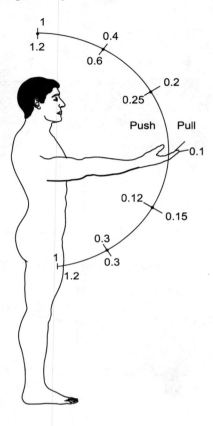

Fig. 7.11: Push/pull force capacities in percentage of body weight [w]

Machine Tool Operation

Table 7.1 states the human force limits for turning, pressing, and lifting; occasionally and frequently.

Table 7.1: Force limits for average human beings

Operation	Force [kg]			
	Male operator		Female operator	
	Frequent, more than 50 times/day	Occasional, up to 10 times/day	Frequent > 50/day	Occasional <10/day
Turning				
Cranks—One hand	8 [17.6 Lbs]	15 [33 Lbs]	5 [11 Lbs]	10 [22 Lbs]
Two hands	12 [26.5]	25 [55]	8 [17.6]	16 [35.3]
Levers, wheels	5 [11]	12 [26.5]	4 [8.8]	8 [17.6]
Knobs	1.5 [3.3]	6 [13.2]	1 [2.2]	4 [8.8]
Pressing				
Foot pedals				
Standing	10 [22]	15 [33]	8 [17.6]	12 [26.5]
Sitting	5 [11]	8 [8.8]	4 [8.8]	6 [13.2]
Toggle switches and push buttons	0.4 [.9 Lbs]	1.2 [2.6]	0.4 [.9]	1 [2.2]
Lifting	15 [33]	20 [44]	10 [22]	15 [33]

Figure 7.12a gives the range of convenient vision angles while working.

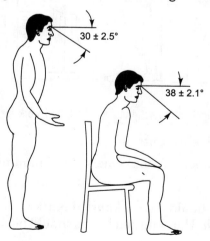

Fig. 7.12(a): Comfortable eyesight angle range

Figure 7.12b gives the recommended size of lettering in displays/dials, at various distances.

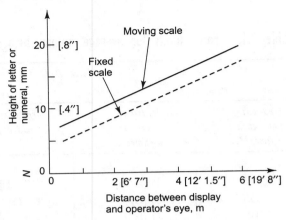

Fig. 7.12(b): Eye distance and letter size

Figure 7.13 states the heights of levers, machine and work, for various working postures.

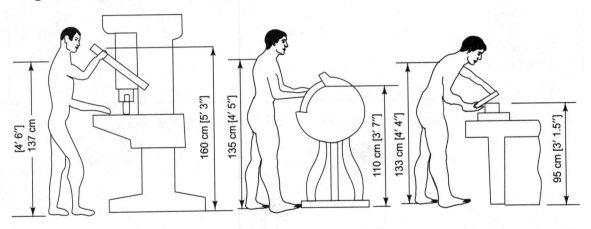

Fig. 7.13: Heights of levers, grinding wheel, and work for various postures

Manually-operated control elements

These also take into account operator convenience and minimization of fatigue.

Knobs

Knobs turned by the fingers, should have a knurled portion, 10–30 [0.4″– 1.6″] in diameter, and 12–25 [0.5″– 1″] in length. Hand-turned knobs should be 35–75 [1.4″– 3″] ϕ, and 20–50 [0.8″– 2″] long.

Cranks

Cranks' lengths should be commensurate with the turning load.

Turning load [kg]	2.5 [5.5 Lbs]	15 [33 Lbs]	25 [5.5 Lbs]
Crank length [mm]	25–75 [1″–3″]	110 [4.33″]	500 [19.7″]

Cranks used for precision-positioning should be 120–200 [4.7″–7.9″] long.

Levers

Levers pulled or pushed towards/away from the body, should have a travel lesser than 175 [7″]. Levers operated by a single hand should be in front of the right shoulder. Levers, which are operated by both hands, should be placed at the center of the operator's body width. Depending upon the operating force, lever length can range from 220–650 [8.7″–21.6″]. However, levers used for changing speeds, etc. are only 150–200 [6″–8″] in length, and have 50–76 [2″–3″] travel. When more than one levers are used, they should be at least 150 [6″] mm apart. All levers should have a round knob at the operating end.

Hand wheels

Hand wheels are used widely for feeding the tool and actuating the feed mechanisms. They are also used for rapid approach. Such wheels are provided with handles. Low speed [<1 RPM], low torque [< 200 kg mm] [1475 Lbs Ft] wheels can be gripped at the rim.

The diameter of a handwheel should be commensurate with the torque it transmits.

Torque [kgmm]	250–500 [1840–3690 Lbs Ft]	500–700 [3690–5160]	750–1110 [5530–8190]	> 1110 [8190 Lbs Ft]
Wheel ϕ [mm]	150 [6″]	250 [10″]	325 [12.8″]	400 [17.75″]

The inside rim of bigger wheels is made wavy, to facilitate good grip.

Star wheels **[Fig. 7.14]** are widely used for effecting strokes of capstan and turret lathes. The handles provide a better grip, and rapid operation; particularly when the stroke can be effected in half a turn or a lesser rotation angle.

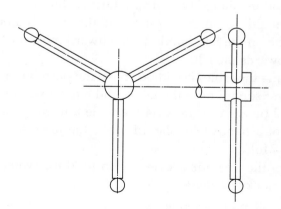

Fig. 7.14: Star wheel

Position of the control elements depends upon the posture of the operator.

Operator posture	Standing	Sitting
Control Element a. Height from floor [mm] b. Maximum horizontal distance from operator [mm]	750–1200 [2′ 5.5″–3′ 11.25″] 1000–1400 [3′–3.4″– 4′ 7.1″]	500–700 [1′–7.7″– 2′ 3.5″] 600–800 [2′–2′ 7.5″]

The height of a star wheel should be the same as the height of an outstretched hand, at a comfortable angle [40°–50°],

i.e. $\dfrac{\text{Operator height}}{1.7} = 1000 [3' \, 3.4'']$, for a 1.7 [5′ 7″] meter tall operator

Moving scales and rotary dials

These should have sub-divisions of 1, 2, or 5. Over-crowding of figures should be avoided, by using figures for only large markings. The figures should be easily readable, when in an operating posture.

$$\text{Letter height} = \frac{\text{Max reading distance (mm)}}{200} = \{0.005 \times \text{Max reading distance [Inches]}\}$$

Displays should be readable without causing any neck strain [**Fig. 7.12a**].

Figure 7.15 shows the symbols, commonly used on control panels. Audio signals such as bleeps and alarm bells are also used to draw the attention of the operator to an undesirable situation, that requires corrective action.

7.1.3 Mnemonic [memory-aiding] considerations

The placement of push buttons should be logical. The push button for moving a table towards the right for instance, should be on the right, while the push button for moving the table towards the left, should be on the left. Similarly, a leftward movement of the operating lever should move the table towards the left.

A push button for raising the table should be above the push button for lowering the table. Upward swing of the lever should raise the table, and a downward swing should lower it.

The same logic should be used for moving the table towards or away from the operator. The clockwise rotation of a handwheel should move the table towards the right/up/away from the operator [**Fig. 7.16**].

A clockwise motion of the operating lever is also used for switching on the machine or rotating a circular table clockwise, and vice versa.

Speed and feed changing mechanisms

Rack and pinion arrangements are widely used to shift the cluster gears in multi-speed gearboxes [**Fig. 7.17**]. A long-lead screw and nut combination, operated by a lever [**Fig. 7.18**],

Sr. No.	Symbol	Function	Sr. No	Symbol	Function
1.	|	On	13.		Manual control
2.	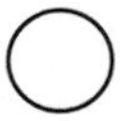 ○	off	14.		Local illumination
3.	 ⊙	On-off	15.		Brakes on
4.	+	Increase (for instance of RPM or cutting speed)	16.		Brakes off
5.	−	Decrease (for instance of RPM or cutting speed)	17.	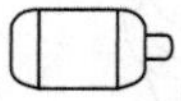	Electric motor
6.	 ∿	Accelerated travel	18.		Cooling fluid
7.		Cross feed	19.		Coolant sump
8.		Vertical feed	20.		Lubricating fluid
9.		Limited rotation	21.		Lubricant pump
10.		Limited reversible rotation	22.		RPM
11.		Stepless regulation of speed or feed	23.		Beware: High voltage (red in colour)
12.	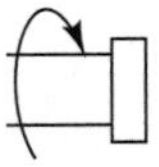	Direction of shaft rotation	24.	!	Attention: Be cautious (Yellow in color)

Fig. 7.15: Symbols for control panels

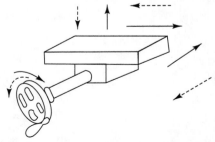

Fig. 7.16: Mnemonic table motion direction for clockwise rotation of wheel

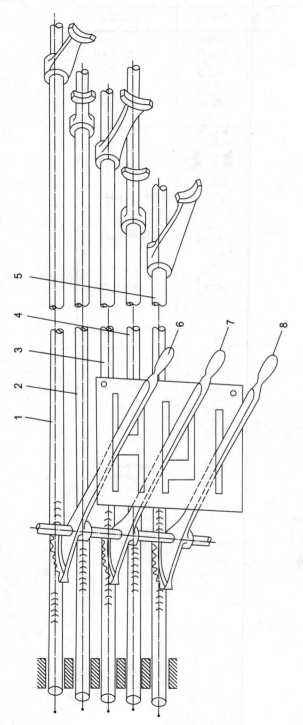

Fig. 7.17: The use of rack transmissions in the control mechanism of a speed gearbox.

is quite common for a 2-position system. **Figure 7.19** shows a single-lever system for obtaining six speeds. An upward or downward shift of the lever moves the upper speed gear [13] towards the left or right. Left/right/center positions of the lever [3] give left/right/center positions of the lower 3 gears-cluster.

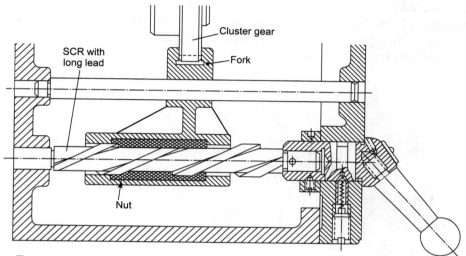

Fig. 7.18: Long-lead screw and nut arrangement for shifting fork-controlling cluster gear

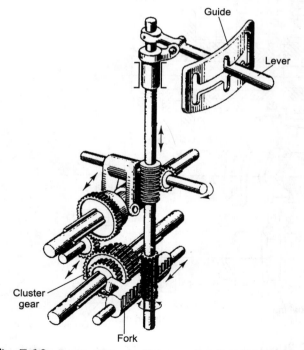

Fig. 7.19: Arrangement for obtainng 6 speeds by a single lever

The sophisticated arrangement shown in [**Fig. 7.20**], uses the Geneva Mechanism and two cylindrical [drum] cams for obtaining 16 speeds. Shaft I has two clusters and 4 positions.

In the position shown in [**Fig. 7.20**], the left-hand gear [A2L] of cluster A2, has meshed with left-hand gear [B2L] on shaft II. Further, left-hand movement of the clusters disengages gears A2L/B2L, and meshes gears A1L and B1L. Right-hand shift of clusters on shaft I, disengages gears A1L and B1L and meshes gears A1R and B1R. The shift of clusters towards the extreme right, engages gears A2R and B2R. Thus, cam I and clusters on shaft I, have 2 right-hand, and 2 left-hand positions.

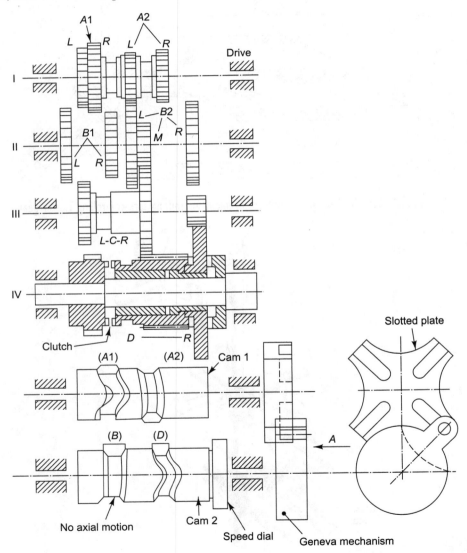

Fig. 7.20: Arrangement for obtaining 16 speeds with a single hand wheel

Cam 2 only has 2 positions: Right-hand and left-hand. In an L.H. position, the single cluster [C] on shaft III, moves towards the left to engage gears CR and B2M, while a rightward motion of the cluster, engages gears CL and B1R. The cluster gear [D] on shaft IV also has 2 positions. In the right-hand position, shown in the figure, gear DR meshes with the extreme right gear on shaft III. Leftward motion of the cluster unmeshes gear DR, and meshes the teeth of the clutch at the left end of shaft IV.

As one rotation of the crank moves the slotted plate of the Geneva mechanism by $1/4^{th}$ turn, we can get 16 positions for the crank. These positions are marked on speed dial, on the shaft for crank and cam 2.

Radial disk [face] cams are more compact and ensure a positive return. Moreover, a single disk can have two face cams on opposite faces **[Fig. 7.21]**.

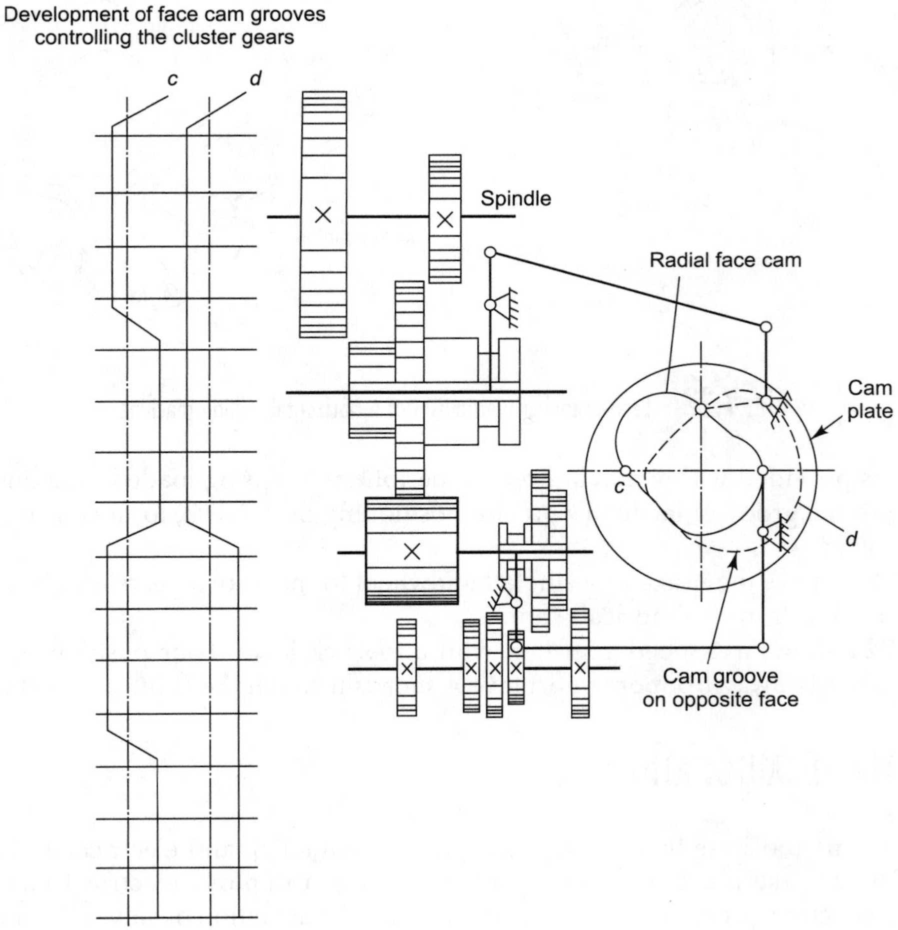

Fig. 7.21: Two radial face cams on opposite faces of a single plate

Disks can have 2 or more face grooves on one side. **Figure 7.22** shows a disk with 2-face cam grooves on the visible side. Pins moving in these grooves, move the gearshift levers suitably, to effect a speed/feed change.

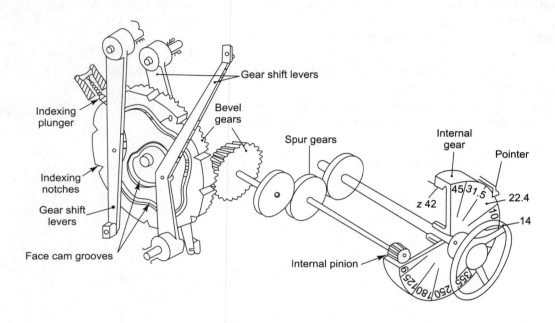

Fig. 7.22: Feed-changing mechanism of a horizontal milling machine

The disk is provided with v notches on the periphery. A spring-loaded indexing plunger, engaging with the grooves, facilitates precise positioning of the disk, for a controlled shift of the gear clusters.

Figure 7.23 shows the usage of a single handwheel to move two gearshift levers, through a double-grooved drum [cylindrical] cam.

Figure 7.24 shows a 4-speed gear box with a joystick lever. Four positions of the lever [left/right, towards/away from operator] give 4 speeds through the shifts of the cluster gears.

7.2 ELECTRICAL AUTOMATION

Modern machine tools use fluid power [pneumatic/hydraulic], and electrical and electronic elements for automation. Even fluid power automation deploys electrical elements like solenoids, contactors, pressure switches, and relays for the actuation of valves. It will therefore be useful to understand the operation of electrical devices, before delving into fluid power automation.

Machine Tool Operation

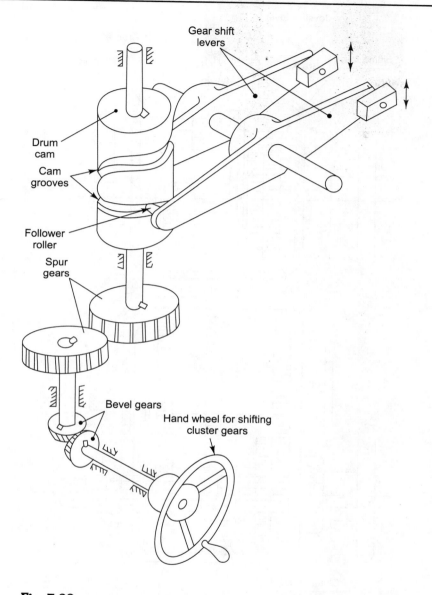

Fig. 7.23: Using single hand wheel to move 2 levers for shifting cluster gears

Electric power is invisible. A look at an electrical cable/wire cannot indicate if it is energized or not. It will be necessary to use a pilot lamp for such an indication. Safety is a very important consideration, as an electric shock is often fatal.

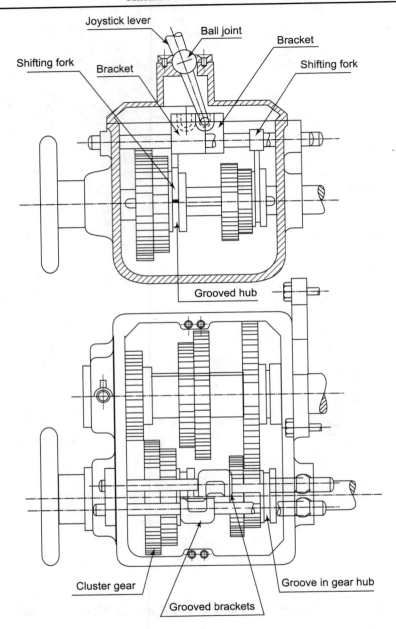

Fig. 7.24: Four-position joy-stick type lever

7.2.1 Safety measures

7.2.1.1 Single power source

Electrical energy should be drawn from a single source so that the entire machine can be **de-energised instantly** during an emergency. A single main switch at the energy-entry point is a must for any machine tool.

If the supply voltage is unsuitable for the operation of some constituents, it is preferable to use a transformer rather than a second source of low-voltage electricity.

Furthermore, a single main switch can be locked, to prevent an unauthorized operation of the machine. In fact, it is a common practice to interlock the electrical control panel with the main switch. When the electrical panel door is opened, the main switch is automatically disconnected to prevent electrical shock to the maintenance personnel. International Standards Organisation [ISO] recommends lesser voltage [80–110 V] for control circuits, to protect electricians from inadvertent shocks.

There are of course, some exceptions. Disconnecting magnetic chucks or brakes by the main switch can be dangerous. Projectile of the released workpiece can cause accidents.

7.2.1.2 Emergency stops

In addition to the main switch, it is necessary to provide additional emergency stop switches at strategic places, for the delay of even a few seconds, can make a difference between life and death. To facilitate instant and easy stoppage, big, red mushroom push buttons are used. However, emergency stops should not de-energize the magnetic chucks/brakes.

7.2.1.3 Overload protection

Accidental short circuit, ensuing from insulation failure or wrong connection, shoots up electrical voltage. This can damage the electrical equipment. This can be prevented by using HRC-type fuses. The choice of the fuse rating should reckon the increase in the current while starting motors. Table 7.2 lists full-load currents, recommended fuse ratings, and the size of copper conductors for 0.06 to 75 kW motors. The other important safety device is the single phasing preventor.

Automatic restarting of motors after the revival of the power supply is prevented by **No Voltage Protection** through contactors, starters, and circuit breakers.

Motor overload is also prevented by bi-metallic overload relays, which trip the motor. The relays can be of an automatically resetting/hand resetting type.

7.2.2 Electrical devices

Electrical devices can be broadly divided into control elements, and operating elements.

7.2.2.1 Control elements

Control elements should be placed in a clean, dry environment. They should be easily accessible, but immune from accidental operation. Supply tubes for air, oil, and gases should not pass through electrical panels or in their vicinity.

Control elements can be broadly classified as below:

a. Switching devices connect/disconnect electrical elements. Limit/micro/pressure switches, push buttons, contactors, starters/circuit breakers are used for energizing/de-energizing electrical devices.

b. Relays are magnetic switches for energizing contactors, electro-magnetic clutches, etc. Time relays can be set to switch/un-switch electric element, after some delay. The time lapse can be varied by setting the relay, suitably.

Auxiliary relays are used for amplification, or for multiplying contacts.

Table 7.2: Capacities of HRC fuses and copper conductors

Fuse rating [Amps] Starting		Recommended copper conductor area mm^2	Motor rating and full load		
Direct Online	Star - Delta		kW	HP	Full load Current [Amps]
2	—	1.5	0.06–0.25	0.08–0.34	0.2–0.8
4	—	1.5	0.37–0.55	0.5–0.75	1.2–1.6
6	—	2.5	0.75–1.1	1–1.5	1.8–2.6
10	—	2.5	1.5	2	3.5
15	—	2.5	2.2–3.7	3–5	5–7.5
25	20	2.5	5.5	7–5	11
30	20	2.5	7.5	10	14
35	25	4	11	15	21
50	35	6	15	20	28
60	50	10	18.5	25	35
80	50	10	22	30	40
100	60	25	30	40	55
125	80	25	37	50	66
160	100	25	45	60	80
200	100–160	35	55–75	75–100	100–135

We will now see how the control elements function.

Machine Tool Operation

a. Switching devices

- **Cam switches:** These comprise cams, mounted on the switch shaft [**Fig. 7.25**]. Made of non-conducting materials [nylon or bakelite], cams actuate movable contact, to separate it from a fixed contact, to disconnect the electrical supply, or allow the spring-loaded movable contact to mate with the fixed contact, to connect the supply.

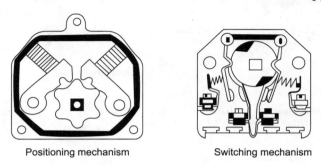

Fig. 7.25: Cam-operated switch

In addition to switching, on and off cam switches are also used for reversals, star/delta starting, changing number of poles of motors [to change speed], distribution, and initiating the following operation:

- **Limit and micro switches [Fig. 7.26]** are actuated by cams, on moving parts of a machine. They are used to control [limit] the stroke [travel] of a moving part. Normal, snap action, overlapping and extended stroke versions, are used for different applications. Limit switches are used for switching on/off electrical elements, and initiating the following operations through contactors and/or relays.

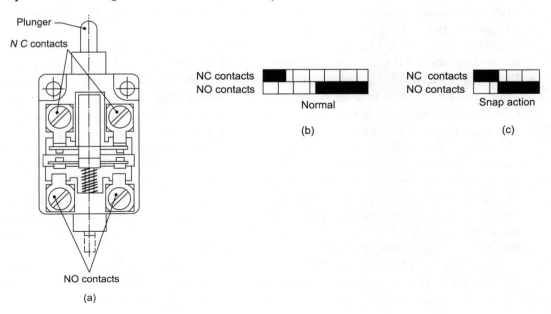

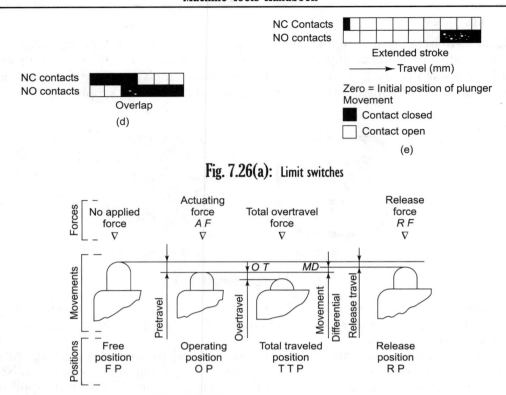

Fig. 7.26(a): Limit switches

Fig. 7.26(b): Types of limit switches

- **Micro switches** are smaller limit switches, with a lower current capacity, shorter stroke, and more precise [0.1] position control.
- **Push buttons,** operated by the finger, are available in many colors, to distinguish their varied functions. They are used for switching on/off, for inching [temporary switching].
- **Pressure switches** are micro switches, operated by a pressure-actuated diaphragm. These can be set to switch on/off, at a required value of pressure.
- **Contactors [Fig. 7.27]** are magnetic switches. On energization of the coil, it acts like an electromagnet, and moves the contactor. This makes or breaks the electrical circuit. Contactors are provided with a number of auxiliary contacts. They are classified as **normally open** [NO], or **normally closed** [NC]. By connecting the electrical elements to these contact points, we can energize some elements and de-energize some other elements, simultaneously. For example, a 2 NO-3NC contactor allows us to close [energize] two circuits and open [de-energize] three circuits, simultaneously. Contactors are actuated by limit/micro/pressure switches. They are widely used for synchronizing and sequencing [through relays] various operations.

Contactors are widely used in motor starters. Pressing the 'start' push button energizes the contactor. This connects the auxiliary [hold-on] contacts. The hold-on contact keeps the motor connected to the power supply, even after the push button is released.

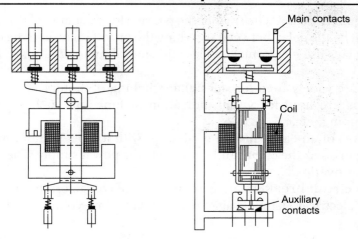

Fig. 7.27: Contactor

Contactor coil functions at 85% to 110% of the normal voltage. When the voltage falls below 85% of the normal, the contactor cannot operate. This disconnects the motor, and protects it against overload.

Combination of a push button and contactor, provides a **no voltage protection**. When the power goes off, the contactor contacts de-engage. The contactor does not switch on automatically. It can be switched on only by pressing the 'start' push button: manual re-setting.

Contactors and relays are most important tools for machine tool automation.

- **Relays [Fig. 7.28]** are also magnetic switches with coils, which become an electromagnet after energization. There is a canti-levered beam, which bends to make a contact under the magnetic force. Relays are used to switch on control circuits for contactors, electromagnetic clutches, etc. Auxiliary relays amplify or multiply contacts.

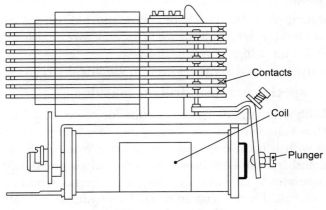

Fig. 7.28: Relay

- **Time relays** switch on/off the controlled circuit after the lapse of the set time. Pneumatic relays are provided with an air cylinder and a flow-control valve. The valve is set to obtain the required 'delay'. The smaller the passage for discharge of the air in the cylinder, the longer will be the delay between switching the on/off relay, and the controlled circuit. Sometimes relays use electronic devices and special motorized units, to vary the delay time.

Time delays are widely used for star-delta switching of motors.

- **Signal lamps** are used for visual indication of functions such as supply on, motor running, clamping, etc.
- **Fuses** are wires of an easily fusible wire/strip, with a cross-section, which melts at the rated current, to break the circuit and disconnect the power supply. They protect electrical elements from overload.
- **Starters and circuit breakers** are combinations of contactors and overload protectors, safeguarding electrical elements against no voltage, low voltage, and overload.

Electrical panels

Electrical panels are installations of electrical elements, mounted in a systematic logical layout in a lockable container. ISO recommends lesser voltage [80–110V] for a control circuit, to safeguard electricians in the event of an inadvertent shock. This calls for the use of a transformer.

Wires

Wires connecting electrical elements should have an adequate cross-section to transmit safely, the full-load current. Table 7.2 gives the recommended cross-sections of wires/cables for 415 Volts, 3 phase, 50 Hz Squirrel cage motors, for operation at 40°C ambient temperature. Fuses recommended in Table 7.2 are designed for starting current, 6 times the full-load current, and 5 seconds starting time for direct online starting. For star-delta starting, the fuses have the capacity of double the full-load current, and a starting time of up to 15 seconds. These values correspond with IS 2208/B588. The full load currents in Table 7.2 are for 1500 RPM motors.

Nameplates on motors, give the details of RPM and kW rating, and full-load current. These should be accounted for, while choosing fuses and wires/cables. Also, high frequency operation of motors, calls for a higher rating of fuses.

Figure 7.29 shows the standard symbols used in electrical circuits.

Electric circuit for a special purpose drilling machine

Figures 7.30a and 7.30b show the electrical circuits for a special-purpose drilling machine. The machine spindle has a single fixed speed and a fixed feed rate. The operator only replaces the drilled workpiece with a fresh one, and presses the 'Start cycle' switch. Clamping the workpiece, drilling it, and spindle-withdrawal are effected automatically.

The control circuit operates at 110 volts. This requires a step-down transformer for supplying power to the control circuit. Feed is commenced by engaging a double-ended clutch. For withdrawal of the spindle, the feed shaft is reversed. The return speed can be kept much higher by using a higher transmission ratio for the return gears. The feed clutch is shifted by a pneumatic cylinder, which moves the shifting fork.

Graphical symbols	Meaning	Graphical symbols	Meaning	Graphical symbols	Meaning
——	Direct current	⏚	Earth	—Z—	Impedance
∿	Alternating current	▨	Frame or chassis connection	—▯— (arrow)	Resistor with moving contact (general symbol)
3 N ∿ 50 Hz 415 V	Alternating current 3 phase with neutral 50 Hz, 415 V.	↗	Variability (general symbol)	—▯— (arrow)	Voltage divider with moving contact
+	Positive polarity	⌐	Variability by steps	—▯— (slash)	Temperature dependent resistor with negative resistance coefficient (thermistor)
−	Negative polarity	∧	Preset adjustment (general symbol)	—▯— (slash)	Variable resistor (general symbol)
△	3 phase winding – delta	⊓	Preset adjustment by steps	—▯—	Resistor with fixed tappings
Y	3 phase winding – star	⌇⌇⌇⌇⌇⌇⌇	Terminal strip	—▯—	Voltage divider with fixed tapping
Y•	3 phase winding – star with neutral broughtout	⊗	Luminous push button	⌒⌒⌒	Inductance or winding
⌇	3 phase winding – zig – zag or interconnected star	—▯—	Resistance, resistor (when not necessary to specify whether it is reactive or not)	─┤├─	Capacitance, capacitor
∿	Flexible conductor	—▭— R	Non-reactive resistor	─┤├─ +	Electrolytic capacitor
○	Terminal				
•─	Junction of conductors				

Fig. 7.29(a): Symbols for electrical circuits

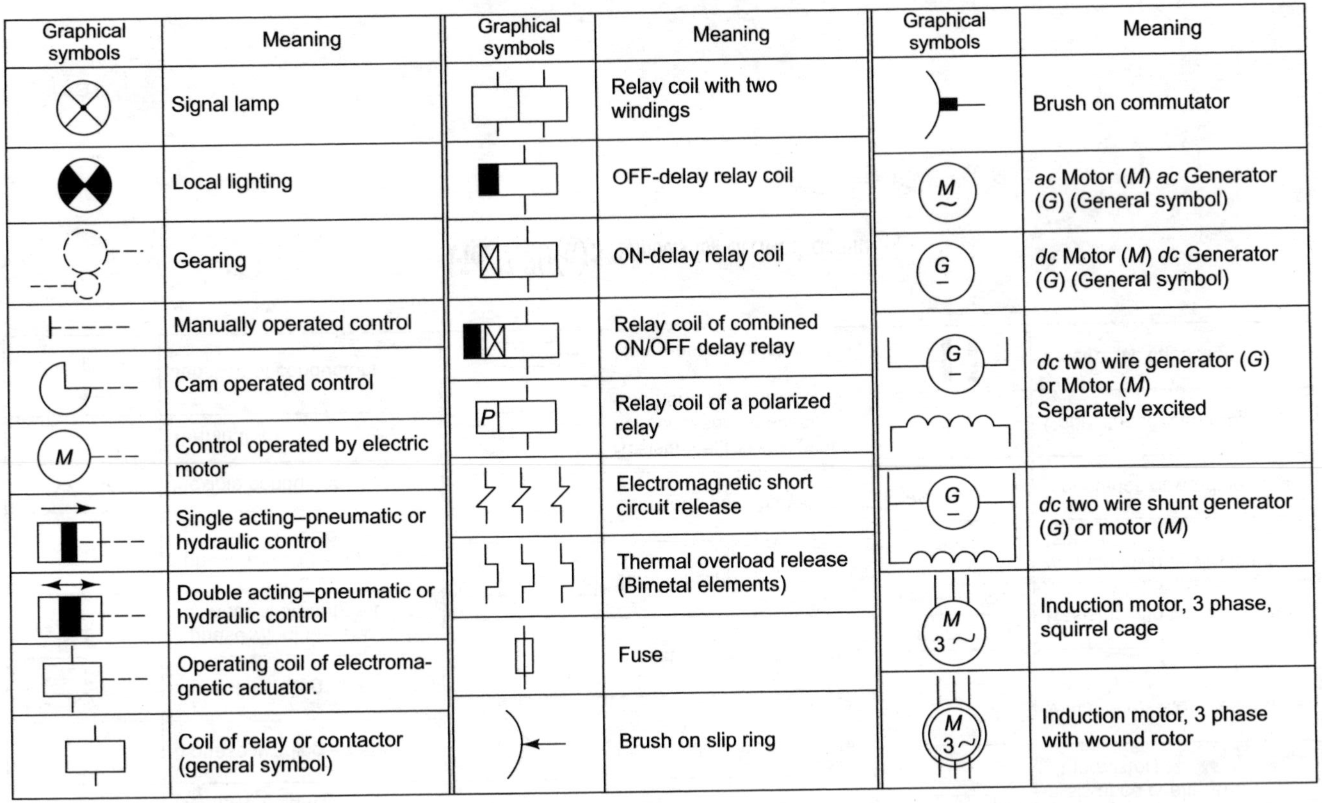

Graphical symbols	Meaning	Graphical symbols	Meaning	Graphical symbols	Meaning
⊗	Signal lamp	▭	Relay coil with two windings	⊥	Brush on commutator
⦻	Local lighting	▭ (filled bottom)	OFF-delay relay coil	(M∼)	ac Motor (M) ac Generator (G) (General symbol)
⊙⊙⊙	Gearing	⊠	ON-delay relay coil	(G−)	dc Motor (M) dc Generator (G) (General symbol)
⊢−	Manually operated control	⊠ (combined)	Relay coil of combined ON/OFF delay relay	(G−)	dc two wire generator (G) or Motor (M) Separately excited
⊂	Cam operated control	[P]	Relay coil of a polarized relay	(G−) with winding	dc two wire shunt generator (G) or motor (M)
(M)	Control operated by electric motor	⟋⟍⟋	Electromagnetic short circuit release	(M 3∼)	Induction motor, 3 phase, squirrel cage
▭ (arrow up)	Single acting–pneumatic or hydraulic control	⊓⊔	Thermal overload release (Bimetal elements)	(M 3∼) with lines	Induction motor, 3 phase with wound rotor
▭ (double arrow)	Double acting–pneumatic or hydraulic control	▭	Fuse		
▭	Operating coil of electromagnetic actuator.	↓	Brush on slip ring		
▭	Coil of relay or contactor (general symbol)				

Fig. 7.29(b): Symbols for electrical circuits

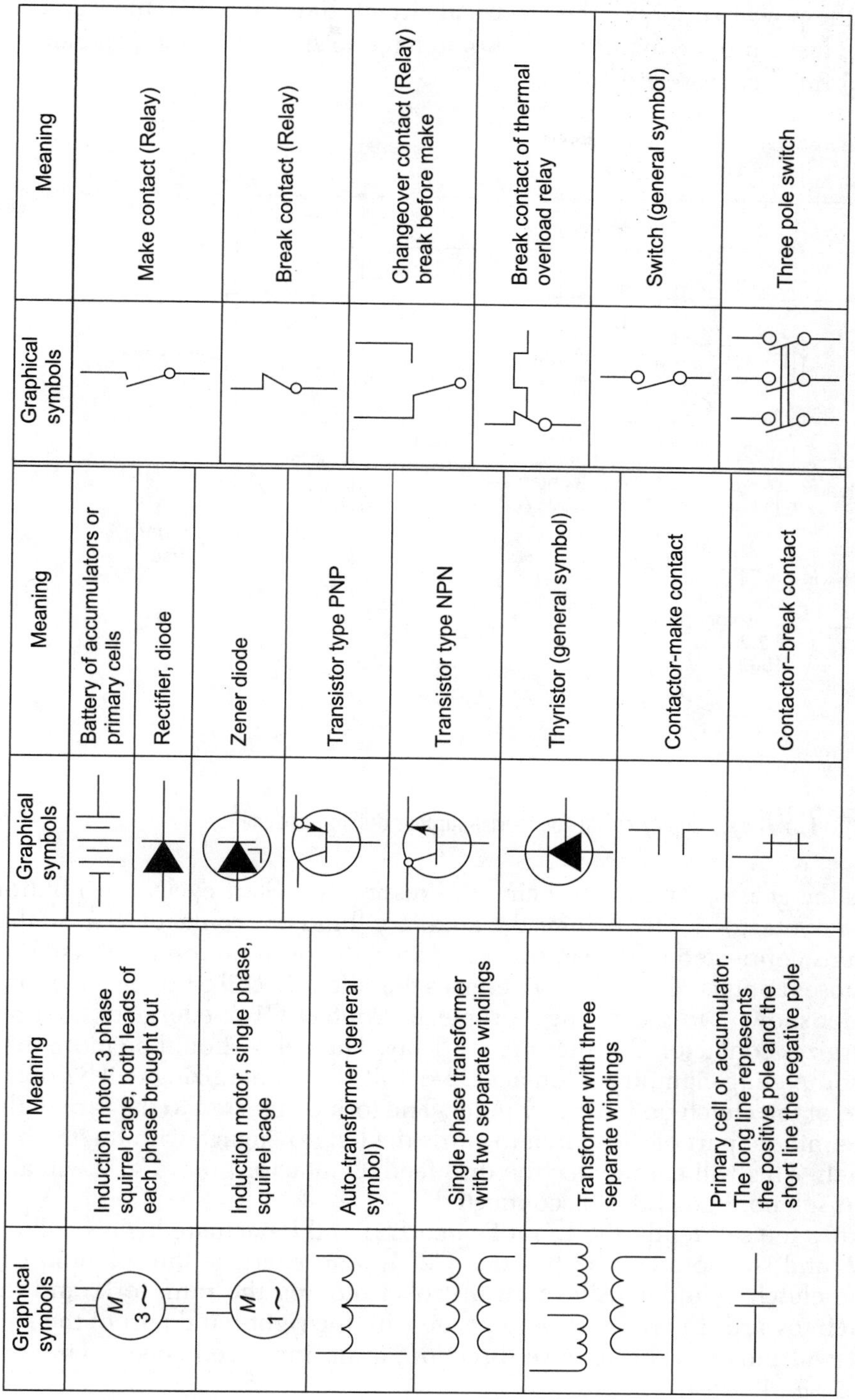

Graphical symbols	Meaning	Graphical symbols	Meaning	Graphical symbols	Meaning
M 3~	Induction motor, 3 phase squirrel cage, both leads of each phase brought out	⊣⊢⊣⊢	Battery of accumulators or primary cells	—	Make contact (Relay)
M 1~	Induction motor, single phase, squirrel cage	▲	Rectifier, diode	—	Break contact (Relay)
(winding)	Auto-transformer (general symbol)	⊕	Zener diode	—	Changeover contact (Relay) break before make
(winding)	Single phase transformer with two separate windings	(PNP)	Transistor type PNP	—	Break contact of thermal overload relay
(winding)	Transformer with three separate windings	(NPN)	Transistor type NPN	—o o—	Switch (general symbol)
⊣⊢	Primary cell or accumulator. The long line represents the positive pole and the short line the negative pole	◁	Thyristor (general symbol)	(three pole)	Three pole switch
		⊐	Contactor–make contact		
		⊣	Contactor–break contact		

Fig. 7.29(c): Symbols for electrical circuits

Figure 7.30a shows the power supply of the electrical circuit. Referring to Table 7.2, for a 2.2 kW motor, the HRC fuses and the main switch should have 15 Amps capacity, and all the wires should have a 2.5 mm² cross-sectional area.

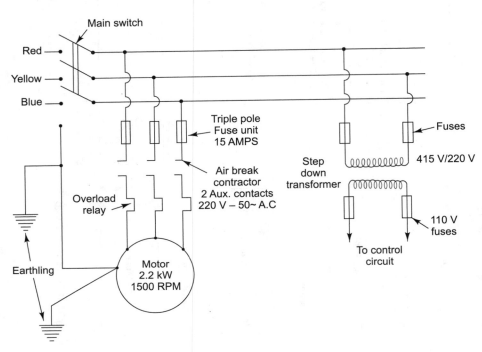

Fig. 7.30(a): Supply circuit for special-purpose drilling machine

Figure 7.30b shows the gearing and control circuit. Pressing the 'Start cycle' push button energizes the coil of the contactor C1 that closes the auxiliary [hold on] contact C1–1, so that the contactor C1 remains connected to the supply, even after the push button is released.

Contactor C1 also closes contacts C1–2. This energizes the solenoid coil S1 of the clamping cylinder D.C. Valve (not shown), to secure the workpiece. Contact C1–2 energizes the time relay coil T1. This energizes contactor C2, after the set delay. The delay should be more than the time required for completing clamping. Contact C2–1 will close, energizing coil S2 of the direction control valve, of the clutch operating cylinder. The fork connected to the ram of the cylinder, will move the middle part of the clutch towards the left, to engage with the teeth of the left part of the clutch. This will commence the drill feed. Contactor point C2–2 will also open up, to ensure that solenoid coil S3 is disconnected.

When the drill reaches the set depth, the Limit Switch LS1 will be actuated. This will de-energize contactor C2, and solenoid coil S2 [contact C2–1]; and energize the solenoid coil S3. This will move the clutch cylinder [A2] in an instroke, moving the ram, fork, and the middle part of the clutch towards the right. It will engage with the right-hand part of the feed clutch. The feed shaft will move in the reverse direction, at an increased speed. The drill spindle will retract in a rapid return.

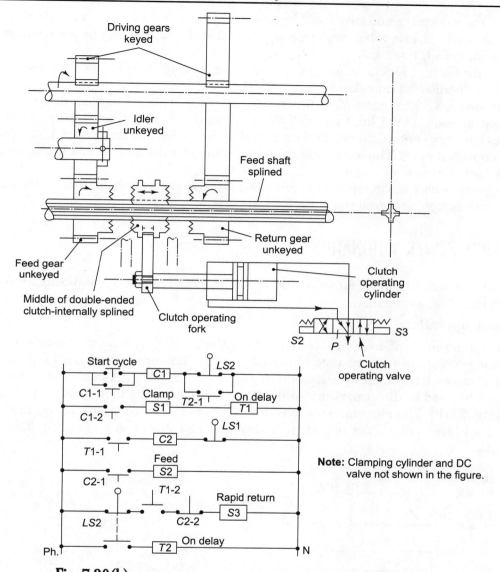

Fig. 7.30(b): Gearing and control circuit for special-purpose drilling machine

It should be noted that the left and the right-end parts of the clutch are not keyed to the splined feed shaft. Furthermore, the left and right-end parts of the clutch are rotating freely, in opposite directions, on the splined feed shaft. Only the middle part of the clutch is keyed [splined] to the feed shaft, so the feed shaft rotates in the same direction as the central part of the double-ended clutch.

At the end of the return stroke, the limit switch LS2 is actuated. This de-energizes the solenoid coil S3, and contactor coil C1. Contact C1–2 opens the de-energizing solenoid S1, to unclamp the workpiece and switch off the timer coil T1, to open contact T1–1. Note that the Timer contact T1–2, prevents energization of solenoid S3 during the delay between energization of contactor coils C1 and C2.

Contactor C2 interlocks energization of solenoids S2 and S3. When one is energized, the other is de-energized automatically.

Limit switch LS2 also energizes timer T2. This timer ensures that contactor C1 is de-energized as soon as the limit switch LS2 is actuated. Also, closing of timer point T2–1 provides auxiliary contact, to energize contactor C1, even when the limit switch LS2 is pressed, and its contact is open. Timer T2 is de-energized soon after the limit switch LS2 opens, after drilling feed commences.

The operator only has to replace the drilled workpiece with a fresh one, and press the 'Start Cycle' push button, to repeat the cycle.

7.3 FLUID POWER AUTOMATION

Usage of pressurized air/oil for actuation of saddle/carriage strokes, indexing, and speed change, facilitates automation.

Solenoid operation

Solenoid operation of hydraulic/pneumatic valves simplifies synchronization and/or sequencing of various motions used in machine tools. **Pilot operation** uses fluid power for operating valves. It is also used for actuating bigger hydraulic valves. A small solenoid operated valve can be used to direct pressurized fluid to either end of the bigger valve, to move the valve [**Fig. 7.31b**]. This eliminates the need for using a bigger solenoid for shifting the heavier spool of a bigger valve. Pilot operation is also very convenient in sequencing fluid power actuators.

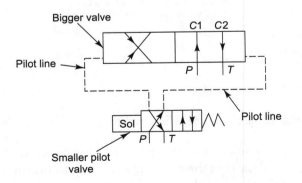

Fig. 7.31(a): Pilot operation of big valve by a small pilot valve

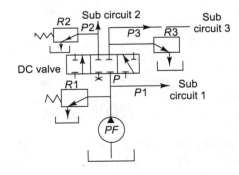

Fig. 7.31(b): Limiting pressure in sub-circuits through additional relief valve and isolating D.C. valve

Pneumatic versus hydraulic systems

Pneumatic actuators take much lesser time than hydraulic actuators. But they are very vulnerable to load variation. As air is easily compressible, speed of a pneumatically-operated ram varies according to load variation. Heavier load compresses the air in the cylinder, and slows down the piston and rod. Some operations such as clamping the workpiece or lifting it, are not affected by fluctuations in speed. Cheap pneumatic power can be used conveniently in such applications. But fluctuations in feeding rate can break milling cutters or grinding wheels. There is no alternative to uniform speed, or uniform force hydraulic power in such applications. We will discuss the important features of hydraulic circuits, before delving into hydraulic automation.

Hydraulic circuits

7.3.1 Force and pressure control

7.3.1.1 Multi-pressure circuits

A pressure relief valve is used to control the maximum pressure and the resultant force. Every hydraulic system must have at least one pressure relief valve to protect the pump from overload. Introduction of more relief valves, in conjunction with direction control valves, facilitates the use of lesser, different pressures in sub-circuits.

Figure 7.31b shows an arrangement for a three-pressures system. Pressure relief valve R1 is set to the maximum system pressure. In addition to protecting the pump, valve R1 also provides maximum pressure in sub-circuit 1, when the Direction Control [D.C] valve is in the central position [as shown in the figure]. As the port P of the D.C. valve is closed in a central position, the maximum pressure in sub-circuit 1 is determined by the relief valve R1. In extreme [R.H. and L.H.] positions, the D.C. valve connects the pump delivery alternatively, to sub-circuits 2 and 3. Relief valve R2 controls the maximum pressure P2 in the subcircuit 2, whereas maximum pressure in subcircuit 3, is determined by the setting of the pressure relief valve R3. The pressure settings of relief valves R2 and R3 must be lesser then the maximum pressure setting on valve R1, for when the sub-circuit 2 is in operation, the two relief valves R1 and R2 are included in sub-circuit 2. Under such circumstances, the pressurized fluid [oil] will take the path of least resistance, and be discharged through the relief valve with a lesser pressure setting, i.e. R2 of sub-circuit 2.

The other method of obtaining a different pressure in a sub-circuit, is to introduce a pressure-reducing valve in the supply line from the pump. As soon as pressure in the sub-circuit branch reaches the pressure setting of the pressure-reducing valve, the valve will close down, to block the flow from the pump, and isolate the sub-circuit to protect it from overpressure. Pressure-reducing valves are widely used to limit the pilot line pressures in the pilot-operated D.C. valves, in high-pressure systems. The system might operate at 300 kg/cm^2, while the pilot pressure is limited to 200 kg/cm^2.

Another method of obtaining different pressures, is to use different pumps for different pressures. Each pump is furnished with an individual pressure relief valve for controlling the maximum pressure, and protection of the pump and the system.

7.3.1.2 Maintenance of minimum pressure

Maintenance of minimum pressure in a hydraulic circuit, can be accomplished by using a preloaded check valve or a counter-balance valve. These valves block the passage till the fluid reaches the pressure intensity, set on the valve. On reaching the set pressure, the valve opens up, to admit pressurized fluid into the system. These valves are convenient for supporting dead weights, and preventing negative loads resulting from dragging the hydraulic actuators.

7.3.2 Speed control

7.3.2.1 Rapid return

Idle return stroke of a hydraulic actuator can be operated at the full-flow speed, even if the speed of the forward-working stroke is reduced. This can be achieved by by-passing the fluid flow through a non-return [check] valve [**Fig. 7.32**]. For a forward stroke pressurized oil has to pass through the flow control valve, as the non-return valve will not allow any fluid to pass in the direction P to C. The flow control valve can be adjusted to obtain the desired speed for the forward stroke. During the return stroke, a large part of the flow is by-passed through the non-return valve, which permits full flow in the direction C to P. Thus, the return stroke speed is maximum for the available flow. Reversal of the direction of the flow of the non-return valve, makes the forward stroke rapid, and the return stroke speed-adjustable.

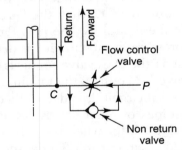

Fig. 7.32: By-passing oil flow through non-return valve for rapid return stroke [instroke]

This method of obtaining two different speeds in opposite directions, is so popular that flow control valves with a built in non-return bypass valve, are readily available.

7.3.2.2 Rapid approach slow working [feed] stroke

A major portion of the working stroke often comprises a long approach to the workpiece, before the actuators commence operation.

An idle approach stroke can be effected rapidly, by **by-passing** the flow through a direction control valve [**Fig. 7.33**]. During rapid approach, the full flow from the pump passes through the DC valve, to the closed end of the cylinder, as shown in the figure. Reaching the preset position in the stroke, the cam attached to the piston rod, operates the limit switch, which energizes the solenoid of the D.C. valve. This closes the valve, to block completely, the free flow through the D.C. valve. The pressurized fluid is compelled to pass through the flow control valve, which can be adjusted, to vary the speed, during the feed part of the stroke. Thus, the operation of the D.C. valve through the cam and limit switch, stops rapid advance and commences the slow part of the stroke. Adjustment of limit switch along the stroke length, facilitates a suitable division of the stroke into rapid and slow parts.

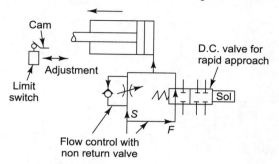

Fig. 7.33: Rapid approach by by-passing oil through a D.C. valve during idle part of outstroke

Alternatively, a cam-operated direction control valve can be used instead of the limit switch. The D.C. valve is operated mechanically, directly by the cam, to effect the speed changeover. Such a D.C. valve is also called a **Deceleration valve**.

Another method of increasing speed is to use re-generating action. In this method, the fluid coming out of the rod-end of the cylinder, is fed to the closed end side, to supplement the flow from the pump. This additional flow, increases the speed of the outstroke of the cylinder [**Fig. 7.34**]. The rod end-side of the cylinder is permanently connected to the supply of the pressurized fluid from the pump. The opposite, closed end-side is connected to the pump for an outstroke. In this position of the DC. Valve, both sides of the piston and pump supply are interconnected. Although both sides of the piston are pressurized, the difference in the areas of closed and rod ends of the piston, pushes the piston in an outstroke. The fluid on the rod end-side is pushed out, and since all other paths are blocked, the emerging fluid flows towards the closed end of the cylinder, supplements the flow from the pump, and increases the speed above the level possible by the flow from the pump only.

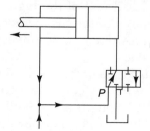

Fig. 7.34: Regenerative circuit

For instroke, the D.C valve is operated to connect the closed end side of the cylinder to the tank. The permanent connection of the pump to the rod end-side, supplies pressurized oil to the rod end side, to effect an instroke, which pushes the oil on the closed end side to the tank, through the DC valve.

A special **composite cylinder** with a small cylinder for low-load, rapid-traverses, and a bigger cylinder for high-load, slow-speed working stroke can be conveniently used for a rapid-slow combination [**Fig. 7.35**].

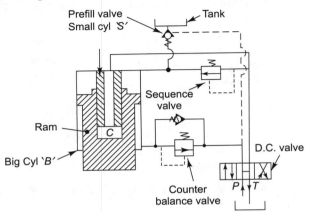

Fig. 7.35: Composite 2-speed cylinder

During outstroke, the small cylinder 'S' is connected to the pressurized oil supply. The oil entering pocket 'C', pushes the ram outwards, at a high speed, determined by the diameter of the small cylinder and the pump delivery. During the outstroke, oil from the tank is sucked into the upper [closed] end of big cylinder 'B' through the prefill valve, which connects the closed end of the big cylinder to the tank. Oil from the rod side of the big cylinder is pushed out, and passes through the counter-balance valve to the D.C. valve, which discharges it into the tank. The counter-balance valve supports the ram against gravitational fall, and ensures a certain back pressure on the rod-end side before the ram begins to move. As soon as the ram meets with resistance from the workpiece, the system pressure rises and closes the prefill valve, blocking the passage between the closed end of the cylinder and the tank.

When the system pressure reaches the intensity set on the sequence valve, the valve connects the pressurized oil supply to the closed end of the big cylinder. The ram moves in a slow working speed, determined by the area of the big cylinder, and the supply from the pump. Thus, the small cylinder provides a low-load, rapid-traverse, and the big cylinder effects the high-load, slow-speed working stroke.

For a return instroke, the D.C. valve is operated to feed the pressurized oil to the rod-end side of the big cylinder. The prefill valve has a pilot connection with the rod-end side of the big cylinder. This opens the prefill valve, when pressurized oil is fed to the rod-end side of the big cylinder. The opening of passage to the tank permits the oil, coming out of the closed end of the big cylinder, to get quickly exhausted through the prefill valve, directly into the tank, without passing through the D.C. valve. The valve is hence, termed **Prefill and exhaust valve**.

Another method of effecting rapid traverses is to use **two pumps**. During rapid approach, both, a high-volume, low-pressure pump, and a low-volume, high-pressure pump, supply pressurized fluid to the system. For a slow-down, the supply from the low-pressure, high-volume pump is cut off from the system, and by-passed directly into the tank. As the high-pressure pump alone can supply lesser fluid, the actuator speed is substantially reduced. At the same time, the high-pressure pump provides the high degree of force, necessary for the working stroke.

7.3.3 Hydraulic automation

This involves the following 4 functions:
1. Sequencing Actuators
2. Synchronising Actuators
3. Interlocking Actuators
4. Hydro-copying

7.3.3.1 Sequencing

When more than one actuator is in use, and these must operate in a certain order, a sequence valve is used. **Figure 7.36** shows the hydraulic circuit for a drilling machine, with hydraulic clamping. The circuit ensures that the workpiece is firmly clamped, before drilling commences.

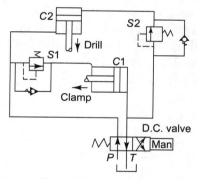

Fig. 7.36: Sequencing

[a] When a D.C. valve is operated manually [or electrically] to an R.H. position, the pressurized oil is fed directly to the closed end of the clamping cylinder C1, that moves in an outstroke, to clamp the workpiece. The sequence valve S2 blocks the oil supply to the closed end of the drilling cylinder C2. When the workpiece is clamped and the system pressure rises to the level set on the sequence valve S2, the valve connects the closed end of the cylinder C2, to the pressurized oil supply, and C2 moves in an outstroke, to commence drilling. Thus, the sequence valve S2 ensures that cylinder C1 operates first, and develops pressure due to work-resistance, before cylinder C2 operates. Similarly, sequence valve S1 ensures that the drilling cylinder C2 withdraws first in an instroke, before the clamping cylinder releases the workpiece. The principle can be extended to any number of actuators with different sequence valves, and pressure settings. Of course, the **setting of the sequence valve must be less than the maximum system pressure**,

set on the pressure relief valve of the pump. Otherwise, the sequence valve will not operate at all. Furthermore, a non-return by-pass valve should be provided with every sequence valve, to ensure a free flow in the opposite direction. The sequence valve with the least pressure setting, will operate first, and one with highest pressure setting last.

[b] Another method of sequencing is using **electrical limit switches** to operate solenoid-operated D.C. valves, controlling the actuators [cylinders]. The cylinder, which should operate first, is attached with a cam, which presses a limit switch during the outstroke. The limit switch operates the solenoid D.C. valve controlling the cylinder, which should operate later.

[c] Another method of sequencing is using **pressure switches** to operate solenoid-controlled valves. Like the sequence valve, a pressure switch can be set to a particular level of pressure. On reaching this pressure level, the pressure switch closes the electrical circuit, to energize the solenoid of the D.C. valve-controlling cylinder. Different pressure switches with different pressure settings can be used for sequencing various cylinders. Naturally, the pressure setting of the pressure switches must be less than the maximum pressure of the pump relief valve. The pressure switch with the least pressure setting, will operate first.

7.3.3.2 Synchronization

[a] **Flow control valves:** Speeds of two or more cylinders can be equalized to a precise level by using flow control valves. Flow to [or from] the cylinders is accurately adjusted to equalize their speeds **[Fig. 7.37]**. This can also be accomplished by a single flow control valve to slow down the faster cylinder.

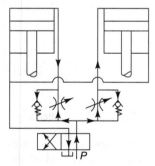

Fig. 7.37: Synchronization with flow controls

[b] Similarly, two cylinders can be synchronized by using **two separate, variable discharge pumps** for their operation. The discharges of both the pumps must be adjusted in such a way that the cylinder speeds are equal. There is lesser heating of oil in this method.

[c] Another method of synchronizing is rigid mechanical coupling of the rams of the two cylinders. The rigid coupling pulls up the ram of the relatively slow cylinder.

[d] Alternatively, the oil to the actuators can be routed through two **coupled hydraulic motors,** with an equal level of displacement. Being coupled, both the motors will deliver

Machine Tool Operation

an equal volume to the two cylinders, and move their rams at an equal speed [**Fig. 7.38**]. Thus, the motors act as flow dividers.

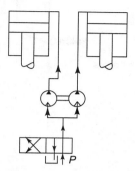

Fig. 7.38: Synchronization with coupled motors

[e] Flow divider valve is a special "T" housing with 3 ports: one inlet, and two outlets. The flow from one of the outlets passes through a built-in flow control valve, which can be adjusted to equalize the flow through the two outlets.

7.3.3.3 Hydraulic interlocks

Hydraulic interlocks use direction control valves to ensure that conflicting motions are prevented.

[a] **Single actuator** can move only in one direction. The cylinder ram can either move in an outstroke, or an instroke. A manually-operated valve automatically ensures this. The operating lever [or pedal] can be moved towards the right or left, moving the ram in outstroke or instroke. Similarly, a pedal can be pressed downward, or allowed to return to a spring-actuated, up position. Manual operation also ensures that a hydraulic motor moves either clockwise or anti-clockwise.

When the Direction Control [D.C.] valve is operated by solenoid [electromagnets], the interlocking is effected by a contactor, which switches off the other coil, when one solenoid is energized. Refer [**Fig 7.30b**] to recall the method.

[b] Two or more actuators can also be interlocked, by using a common direction control valve [**Fig. 7.39**], before using individual direction control valves. The interlocking D.C. valve will allow the pressurized oil to flow to only one of the actuators, thereby, preventing the operation of the other. This can ensure that the horizontal actuator will not function when the vertical actuator is in motion, and vice-versa.

In an automated drilling machine, drill-feeding ram will not function when the clamping [or workpiece feeding] ram is moving. Each interlock will require an additional direction control valve.

7.3.3.4 Hydraulic copying [Fig. 7.40]

Hydraulic systems can be conveniently used for precise copying, from templates or master models.

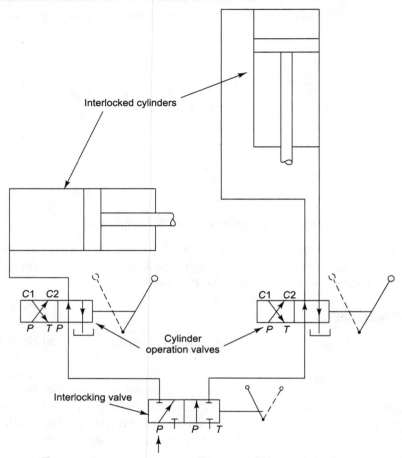

Fig. 7.39: Interlocking 2 cylinders with a direction control valve

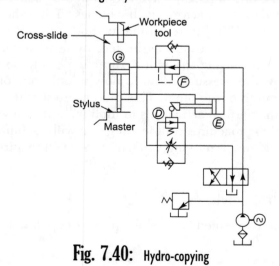

Fig. 7.40: Hydro-copying

A deceleration valve 'D' provides for a changeover from a rapid approach to a slow-cutting feed. The cam attached to the longitudinal travel ram of cylinder E, pushes the roller down, operating deceleration valve, thereby blocking the free flow passage through the valve. As the passage through the deceleration valve is completely blocked, the hydraulic fluid is compelled to pass through the flow control valve, which can be adjusted to get the desired speed for the ram. This is used to control the feed rate of the tool.

The tracing stylus attached to the ram of the cross-slide cylinder G, is held pressed against the master, by the pressurized fluid, from a pressure-reducing valve F. When the pressure on the stylus reaches the setting of the pressure-reducing valve [F], the valve closes the passage between the pump and the closed end of the cylinder G. Closure of the valve holds the stylus in position.

When there is a depression in the master, the stylus ram moves in an outstroke. This moves the tool towards the workpiece, copying the depression. There is a drop in the pressure on the closed side of the stylus cylinder. The pressure-reducing valve [F] opens, allowing pressurized fluid into the cylinder. This reduces the flow to the longitudinal cylinder, and its speed.

A rise in the master profile pushes the stylus in an instroke. The fluid on the closed side [of G] flows out through the bypass check-valve, into the pressure-reducing valve [F]. The tool moves away from the workpiece, copying the rise onto the master.

The longitudinal feed rate can be varied by adjusting the flow control in valve [D]: meter out. The check valve in D permits full flow to the ram side of cylinders E and G, to facilitate rapid return during instrokes of both the cylinders.

We will now see some applications of the hydraulic automation principles, in some sophisticated machine tools.

7.3.4 Applications of hydraulic automation

7.3.4.1 Pull-type broaching machine

Figure 7.41 shows the hydraulic circuit for a pull-type broaching machine. The system is powered by two pumps A and B, which provide three flow rates A, B, [A + B]. These provide three basic speeds for the machine. These speeds can be further varied by flow control valves C and D. Valve 'C' bleeds off part of the flow directed towards the closed end of the cylinder, whereas valve 'D' bleeds off the flow to the rod end of the cylinder. These valves facilitate variation of flow to the cylinder, and its speed.

The counter-balance valve E ensures a positive pressure in the closed-end side of the cylinder, before the ram starts moving in an instroke. This is because the passage of the counter-balance valve is normally held closed, blocked by a spring. However, the rise of pressure in the closed-end side is piloted to the valve. On reaching the pressure setting, the valve opens up, permitting oil from the closed end to be discharged through the counter-balance, and D.C. valves into the tank. There is a non-return valve, for free flow in the opposite direction. The machine is operated by D.C. valve F, which connects the closed, and rod ends of the cylinder, to the pressurized fluid, to effect the outstroke and instroke.

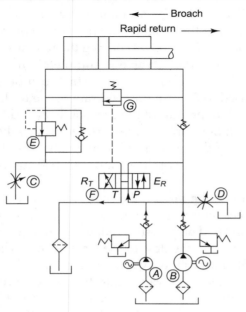

Fig. 7.41: Hydraulic circuit for pull-type broaching machine

The instroke is used as a working stroke in pull broaching machines. Instroke speed is adjusted by the flow control valve D.

During an idle return outstroke, the speed of the ram is increased by complementing the flow from the pumps, with the fluid coming out of the rod end of the cylinder. This is accomplished by using **regenerative action**.

When during an outstroke, the closed end of the cylinder is connected to the pressurized oil, a pilot from the closed-end side operates the pilot-operated valve [G]. Valve G is normally closed. However, the application of pilot pressure from the pressure line opens the valve 'G'. This connects the rod and the closed ends to each other, and to the pressure source. This commences regenerative action.

As the closed end piston area is 1½ to 2 times bigger than the area on the rod-end side, the ram moves in an outstroke. The oil coming out of rod end has no place to go except join the flow to the closed end side, through valve G. With the addition of flow from the rod side to the pump flow, the ram moves much faster, 1½ to 2 times the possible speed of the flow from the pump alone. Thus, the return, idle, outstroke speed is increased by 1½ to 2 times, through regenerative action.

$$\text{Outstroke speed} = \frac{\text{Pump flow}}{\text{Rod area}}$$

7.3.4.2 Cross feed circuits in machine tools [Fig. 7.42]

Many machine tools such as shaping, planing, and grinding machines, advance the tools in a transverse direction, at one or both the ends of the longitudinal travel, to effect the cross feed.

Machine Tool Operation

At the end of the stroke, the tool/workpiece is moved in a direction that is at right angles to longitudinal travel, to take the next cut.

The longitudinal reciprocation [**Fig. 7.42**] is accomplished by direction control valves 'A' and 'B'. 'A' is a small valve, which is operated by dogs/cams at the end of the stroke. At each end of the stroke, a cam strikes the handle of valve 'A', to shift its spool to another position, for a stroke in the opposite direction.

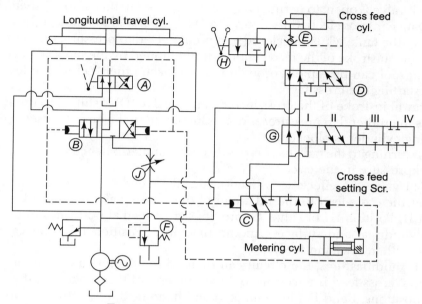

Fig. 7.42: Hydraulic circuit for cross feed

Valve 'A' acts as a control [pilot] valve, while the cylinder is reciprocated by a bigger, pilot-operated direction-control valve 'B'. It is operated by the pilot oil, supplied by valve 'A', and reciprocates the table. The cams can be shifted, to alter the longitudinal stroke. The speed of the longitudinal stroke can be varied by adjusting the outflow-control valve 'J'.

Control valve 'A' is also connected to the pilot-operated cross feed valve 'C'. So both the D.C. valves 'B', and 'C change spool positions simultaneously, at each end of the stroke.

In the position shown in [**Fig. 7.42**], the rod end of the cross-feed cylinder E is connected to the L.H. side of the metering cylinder. The rodless piston of the metering cylinder moves towards the right till it touches the setting screw. The setting screw of the metering cylinder permits changing of the R.H. end position, and the stroke of the piston, and hence, the amount of oil the metering cylinder can admit. As the oil coming out of the rod-end of the feed cylinder has no other place to go except the metering cylinder, the amount of oil, which can come out of the rod-end of the cross-feed cylinder and the stroke of the cross-feed cylinder, is limited. Thus, the setting screw facilitates variation of the cross-feed rate per stroke.

Cross-feed takes place at the right-hand end of the stroke, when the spool of the valve is shifted to L.H. position, shown in [**Fig. 7.42**]. At the L.H. end of the longitudinal stroke, the valve 'C' shifts to the R.H. position. The port connecting the rod-end of the cross-feed cylinder is blocked. The oil from rod end can't flow out to the metering cylinder. The piston can't move and there is no cross-feed at the L.H. end of the stroke. However, valve 'G' connects the pressurized oil supply to the R.H. end of the metering cylinder, while the L.H. end is connected to the tank. This moves the metering piston to the left, till it touches the L.H. end of the cylinder. The metering cylinder is now ready for cross-feed, at the R.H. end of the stroke.

Normally, D.C. valve D connects the pressurized oil to the closed end of the feed cylinder, as shown in **[Fig. 7.42]**. In this position, piloted oil from the closed end, holds the pilot-operated check valve E open, to permit oil from the rod-end to flow into the metering cylinder. For rapid advance in cross-feed, direction valve 'H' is operated to connect the rod-end of the feed cylinder to the tank, while the closed end is connected to pressure, through valve 'D'. In this position, the outstroke of the feed cylinder is not limited by the metering cylinder. So, the workpiece and tool can be rapidly brought close to each other, before the lever valve H is released, for starting the cross-feed.

For withdrawal instroke of the cross-feed cylinder valve 'D' is shifted to the R.H. position. This connects the rod-end to pressure, and the closed end to the tank. Consequently, the path to the metering cylinder is blocked.

All the oil returning to the tank has to pass through counter-balance valve 'F', which ensures a certain backpressure in the circuit.

Position I of valve 'G' effects cross-feed at the R.H. end of the longitudinal stroke. For changing over the cross-feed to the L.H. end of the stroke, valve 'G' must be shifted to the next position II. Position III of valve 'G' provides cross-feed at both ends of the stroke, as in the case of cylindrical and surface grinding machines. Position IV of valve 'G' stops the cross-feed altogether, for hand feed.

With minor modifications, the circuit can be used for shaping and planing machines. In these machines, cross-feed is necessary only at one end of the stroke. So, the valve 'G' will have only 2 positions, I and IV, for auto feed and hand feed, respectively. Furthermore, the cylinder for the longitudinal travel will have the rod only on one side [L.H.] of the piston.

Air in hydraulic system

The presence of air in a hydraulic system leads to a jerky operation of the actuators. This is because air bubbles get compressed when the pressure rises, and they expand when the oil pressure drops. The entry of air into hydraulic systems can be prevented by:

1. Keeping the **suction pipe** of the pump submerged, well below the oil level, so that air cannot enter the pump.
2. The pipe **delivering** the return oil into the tank, should also remain submerged, well below the oil level.
3. The **oil level in the reservoir** should not be allowed to fall below the minimum level marked on the oil level window, so that suction and return oil pipes remain submerged at all times.
4. All **pipe joints** and connections should be tightened properly to prevent infiltration of air in the joints.

When oil is filled for the first time or changed, some air will always infiltrate into the oil. This can be removed by a repeated operation of the hydraulic system without the work-load. When all the air is expelled from the system, the actuators will operate smoothly, without any jerks. The system can then be operated under full load.

7.4 NUMERICAL CONTROL [NC]

In NC machines, the tool/table is moved automatically, according to the program fed into the control computer. Instead of approximately set stoppers, numerals and letters are used to move the tool/table precisely, through computerized actuators. Selection of speed and feed, switching on and off, and tool/table motions, are automated by a program, which depends upon the workpiece material, shape, and operation.

Program

Numerals are converted to impulses, whose number depends upon the least count of the actuator [say stepping motor], and the workpiece size. For example, if the least count of the stepping motor is 0.1, the program should generate 20 impulses, for moving a tool/table by 2.0 mm.

In the simplest version, the program is recorded on paper [or plastic] tapes, by punching the number of holes, equal to the impulses necessary. The tape has a number of tracks to cover the axes and direction [+ for away from the spindle, and − for moving towards the spindle]. **Figure 7.43** shows the standard axes and directions [+/−] for CNC machines. In lathes, the spindle height [Y co-ordinate] is fixed. So, the program moves the tool along X [radial] and Z [axial] axes. In milling machines too, the Z axis coincides with the machine spindle, and X and Y co-ordinates determine the position of the spindle on the work-table.

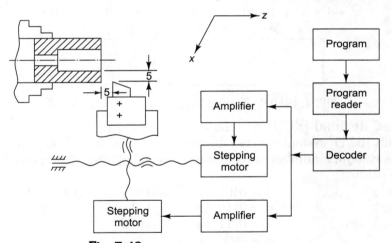

Fig. 7.43: System and axes for CNC lathes

Coding systems

The numerals representing motions must be converted to program codes: holes on the tape. Two systems have been developed: **the decimal system,** and **the binary system**. In both these, the coded number is split into constituents, representing powers of 10 or 2.

The decimal system

The decimal system uses powers of ten, i.e. $1 = 10^0$; $10 = 10^1$; $100 = 10^2$, $\cdots$ $1000 = 10^3$

$$418 = 4 \times 10^2 + 1 \times 10^1 + 8 \times 10^0$$

For covering numbers up to 1000, four tracks with 10-hole options on each track [40 holes options], are necessary. **Figure 7.44a** shows the tape for No. 418 in the decimal system.

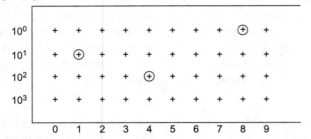

Fig. 7.44(a): Decimal coding $4 \times 10^2 + 1 \times 10^1 + 8 \times 10^0 = 418$

The binary system can cover the same range of 1000 numbers in 10-hole options, on a single track. But it involves more calculations.

For $2^0 = 1$, $2^1 = 2$, $2^2 = 4$, $2^3 = 8$, $2^4 = 16$, $2^5 = 32$, $2^6 = 64$, $2^7 = 128$, $2^8 = 256$, $2^9 = 512$

$418 = 2^8 + 2^7 + 2^5 + 2^1 = 256 + 128 + 32 + 2$

Figure 7.44b shows the single-track punched tape for the binary system. Note that multiples for the powers of 2 are either 1 or zero, i.e. hole or no hole.

$418 = 0 + 2^9 + 1 \times 2^8 + 1 \times 2^7 + 0 \times 2^6 + 1 \times 2^5 + 0 \times 2^4 + 0 \times 2^3 + 0 \times 2^2 + 1 \times 2^1$

Fig. 7.44(b): Binary coding $2^8 + 2^7 + 2^5 + 2^1 = 418$

Binary coded decimal [BCD] system, used in modern CNC machines, combines both the systems. This [BCD] system covers numbers up to 1000 in 12 holes options, in 3 rows of 4 tracks. In this system, a number is written in vertical columns. Numerals 1–9 and zero are represented as below:

1 = 0001 i.e. 2^0
2 = 0010 i.e. 2^1
3 = 0011 i.e. 2^1 and 2^0
4 = 0100 i.e. 2^2
5 = 0101 i.e. 2^2 and 2^0
6 = 0110 i.e. 2^2 and 2^1
7 = 0111 i.e. 2^2 2^1 and 2^0
8 = 1000 i.e. 2^3
9 = 1001 i.e. 2^3 and 2^0
0 = 0000

Figure 7.44c shows No. 418 punched on the tape.

Fig. 7.44(c): Binary decimal coding [BCD] 4 = 0100   1 = 0001   8 = 1000

Machine Tool Operation

Other mediums used for programing are **punched cards** and **magnetic tapes**.

Punched cards, though easy to correct [replace], have limited capacity. Also, the transportation devices and readers for these cards are complicated, in comparison with tape movers and readers. Cards are now used only in very simple program, in positioning N.C. systems.

Magnetic tapes are plastic ribbons, coated with iron oxides. They have much more capacity than punched tapes. They can be erased and used again. This makes them vulnerable to unintended accidental erasion. Also, electrical and magnetic fields in their vicinity, can distort the magnetic signals. Magnetic tapes are encoded by magnetizing small areas, by a computer. The magnetized spots attract damaging ferrous dust. What is more, the damage is not visible, hence, not easily detectable. But their high capacity makes them suitable for the continuous path that N.C. machines use to generate complex profiles. Advent of sealed tape magazines will increase the use of magnetic tapes.

Punched tapes still widely used, can also be made of plastic or plastic/aluminum sandwich. The tapes and their readers are cheaper than magnetic readers. They can be punched manually. The damage is visible, and easily noticeable.

Standard punch tapes [Fig. 7.45] are 1 inch [25.4 mm] wide. They have 8 tracks. In an EIA system, only 6 tracks are used for coding letters and numerals. The fourth [un-numbered in [Fig. 7.45] track is used for parity check for tape/damage, while the eighth track is used for statements such as END OF BLOCK [EOB].

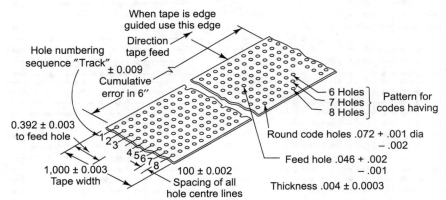

Fig. 7.45: Standard 8-track punch tape

The parity check track adds a hole to each cross row. Binary system requires an even number of holes for each numeral. An additional hole on the parity track makes the total number of holes per row, odd. Damage to the punched tape will add an additional hole, making the number of holes in the damaged row, even. Noting this, the system will initiate the command to stop the reader, and alert the operator about the tape damage.

The latest ISO and American ASCII systems use 7 tracks for coding data, and the 5th track for parity check. In this system, every character code comprises an odd number of holes. The parity check track adds one hole to each row, making the total holes per row, even. Damage to the tape will add a hole, making the total odd. This will stop the reader and alert the operator about the tape damage.

Figure 7.46 shows both EIA and ISO/ASCII codes for numbers, letters, punctuations, mathematical signs, etc.

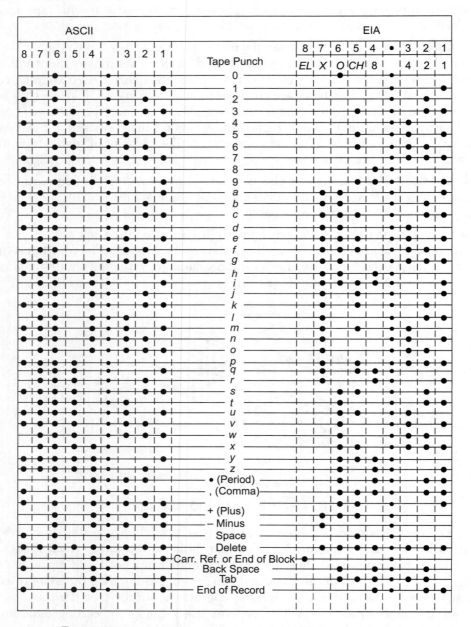

Fig. 7.46: EIA and ASCII [ISO] codes for numerals, symbols, and letters

Punched tape readers [Fig. 7.47 a, b, c]

Tape readers transform the hole code into electrical signals. In the **electro-mechanical reader,** there is a fixed, and a movable contact for each track. Moving contacts enter the punched holes to complete the electrical circuit. If there is no hole, the circuit continues to remain off. Combination of on and off circuits in the tracks, determines the character [numeral, letter, sign, etc.]. The tape is moved forward to the next row. The tapes move at 18 m/min speed, reading 20 to 120 characters per second. Electro-mechanical readers are widely used in positioning systems. There is a risk of tape damage at high speeds.

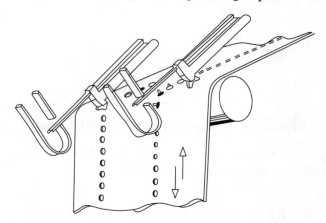

Fig. 7.47(a): Schematic diagram of an electro-mechanical tape reader

In a photo-electric reader [**Fig. 7.47b**], a beam of light falling on a photo-diode, generates a signal to close the electrical circuit. The tape passes between the source of illumination, and the photo-electric cell. A punched hole allows the passage of light beam to close the circuit. If there is no hole, the circuit continues to remain off. Combination of 8 circuit on/off options, identifies the character. A photo-electric reader can read 100 to 1000 characters per second. They can be conveniently used in continuous path, N.C. machine tools.

In one design of a pneumatic reader [**Fig. 7.47c**], the movable contact is moved using compressed air pressure. In another version, fall in air pressure due to release of air through the punched hole, generates an electrical signal to close the circuit.

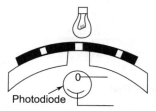

Fig. 7.47(b): Schematic diagram of a photo-electric tape reader

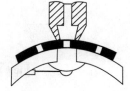

Fig. 7.47(c): Schematic diagrams of pneumatic tape readers

Readers for magnetic tapes cost 10 to 20 times the electro-mechanical readers. Magnetic coders and readers, both use magnetic heads [**Fig. 7.48**] Passage of AC/DC electric current impulse through the coil, generates a magnetic flux in the annular core. This flux magnetizes a small spot of iron oxide, coating the tape. During reading, this magnetized spot induces e.m.f. in the winding, to close the circuit.

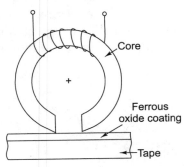

Fig. 7.48: Schematic diagram of magnetic tape reader

For erasing the tape, we only have to pass a high-frequency current through the magnetic head coil, and pass the tape under the magnetic head.

Decoding

Signals from the track are fed to relays. A hole in the tape opens the relay to make a circuit. **Figure 7.49a** shows a pyramid decoder. Combination of on/off circuits identifies the character. Numerals are usually used for moving the table/tool. Currents for the numerals should hence, be proportional to their values. The current for the numeral 8, should be 8 times the current for numeral 1. This is accomplished by using proportional resistances in the circuits. Referring to [**Fig. 7.49b**] it should be noted that $R_2 = 2R_1$, $R_3 = 3R_1$..., $R_7 = 7R_1$ and $R_0 = 0$.

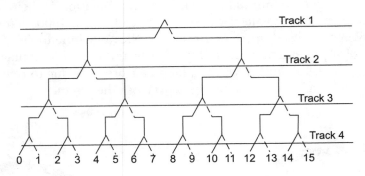

Fig. 7.49(a): A decoder for 16 electrical circuits

Buffer storage

When the tape reading speed is slower than effecting the related motion, the machine operation will stop for signals from the reader. This will cause a delay, which leaves dwell marks on the machined surface. Such marks are unacceptable in straight-cut and continuous path N. C. machines.

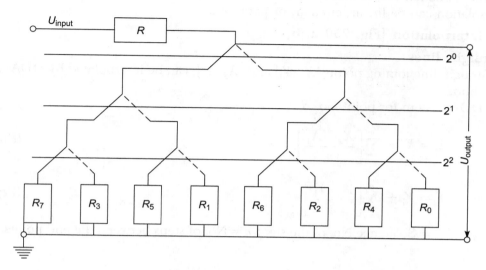

Fig. 7.49(b): A decoder for numerals 0 to 7

The accuracy of a profile depends upon the length of the approximating lines or arcs. If the length is 0.01 and feed rate is 100 mm/min,

$$\text{Machining time} = \frac{0.01 \times 60}{100} = 0.006 \text{ secs}$$

For 100 characters/sec speed of electro-mechanical reader,

$$\text{Reading time} = 0.01 \text{ secs.}$$

There will be a dwell of [0.01 − 0.006 = 0.004] seconds after machining of the approximating straight line. Even a photo-electrical reader [1000 characters/sec] will be slower than the machining time [0.001 − 0.0006 = 0.0004].

The dwell is eliminated by providing the NC machine with buffer storage. Data for the following motion is kept ready in the buffer, while the machine operates on the active storage. As soon as machining is over, the buffer storage data is transferred to active storage, almost instantly, i.e. in milli-seconds.

Interpolator

Interpolation is a process of drawing a smooth curve through a number of points. Interpolation coordinates two or more simultaneous cutting motions, to arrive at the required profile.

Digital interpolator computes the intermediate points',coordinates, according to fixed algorithms. Even straight lines not parallel to the standard axes, call for interpolation. There are two methods of computation of co-ordinates for intermediate points:

1. Digital Differential Analysis [DDA]
2. Direct Function Estimation [DFE]

Interpolation can be linear, circular, or parabolic.

Linear interpolation [Fig. 7.50 a, b]

a. Straight lines

A straight line joining points $[X_1, Y_1]$ and $[X_2, Y_2]$, can be interpolated by DDA or DFE methods.

In DDA system for point No. S,

$$X_s = X_1 + \sum_{s=1}^{n} \left[\frac{X_2 - X_1}{n} \right] \qquad \textbf{(Eqn. 7.3)}$$

$$Y_s = Y_1 + \sum_{s=1}^{n} \left[\frac{Y_2 - Y_1}{n} \right] \qquad \textbf{(Eqn. 7.4)}$$

$s =$ Sequence No. of the step; $n =$ No. of steps between the end points.

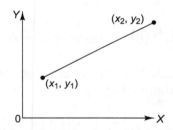

Fig. 7.50(a): Straight line to be interpolated

In DFE method, for all the points on the curve, $F[X,Y] = y - f[X] = 0$. All the intermediate points, with $F[X, Y]$ positive, will lie above the line [**Fig. 7.50b**], while the points with $F[X, Y]$ negative, will be below the line. For the negative points, the following step will be positive along the Y axis. For positive points, the following step will be positive along the X axis.

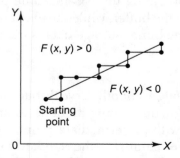

Fig. 7.50(b): Interpolation of the coordinates of intermediate points by DFE method

b. Curves [Fig. 7.50c]

Curves are represented as a series of straight lines. Their increments depend upon the tolerance [T], profile, and curve radius [R].

$$\text{Increment }[1] = 2.8\sqrt{RT} \qquad \text{(Eqn. 7.5)}$$

It will be necessary to find the co-ordinates of the ends of the approximating straight lines.

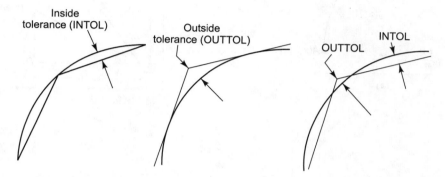

Fig. 7.50(c): Different methods of approximating a curved profile by straight lines

Circular interpolation [Fig. 7.51]

This requires only five program commands: two end points of the arc, radius, its center, and the direction of the motion. Even arcs are interpolated by breaking them into straight lines. Their co-ordinates are computed by the interpolator, which also generates the controlling signals. Either DDA or DFE method is used. By DDA method,

$$X_s = X_1 - \sum_{S=1}^{S=\mu} \frac{Y[s-1] - Y_m}{n} \qquad \text{(Eqn. 7.6)}$$

$$Y_s = Y_1 + \sum_{S=1}^{S=\mu} \frac{X[s-1] - X_m}{n} \qquad \text{(Eqn. 7.7)}$$

s = Sequence No; n = Total No of steps between the curve ends

μ = Instantaneous value of step-wise change in s

For positive points above the arc and the points on the arc, the following step is negative in the X direction. For the negative points below the arc, the next step is positive, along the Y-axis.

For the sake of simplicity and economy, interpolation is restricted to a single quadrant, since when the curves spread over more quadrants, it is necessary to determine a separate set of commands for each quadrant.

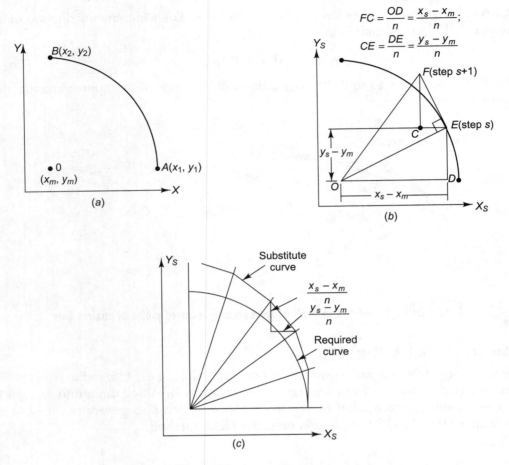

Fig. 7.51: Circular interpolation

Parabolic interpolation

Flexibility of the parabolic equation facilitates its adoption for any free-form/mathematical curve. Parabolic interpolation can be tangential [**Fig. 7.52a**], or 3-point [**Fig. 7.52b**]. The 3-point method gives greater accuracy. Parabolic interpolation requires only 1/50 of the number of points required for linear interpolation. It gives more accuracy, but is very costly. Consequently, most of the continuous path NC machines have to do with linear and circular interpolators.

Complicated profiles often call for simultaneous usage of linear, circular, and parabolic interpolations. Transitions between 2 different types of interpolations are smoothened with the help of software with function routines.

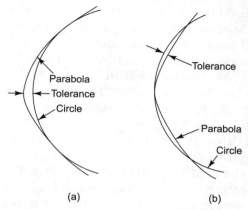

Fig. 7.52: Parabolic interpolation by: (a) tangent parobola, (b) three-point parabola

Drives

Spindle rotary drives have already been discussed earlier. Feed motion drives, in numerically controlled machines, can be broadly classified as:
1. Point-to-point systems and
2. Continuous path systems

1. Point-to-point systems

These are usually powered by stepping motors [**Fig. 7.53**]. These have a toothed stator and rotor. There are 'n' assemblies along the rotor length. Each assembly is staggered by an angle equal to:

$$\frac{\text{Tooth pitch}}{\text{No. of assemblies}}, \text{ i.e. } \frac{P}{n}$$

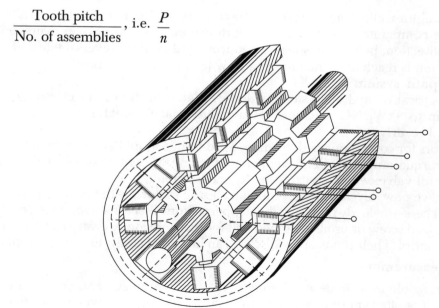

Fig. 7.53: Isometric section of a stepping motor

Each stator winding is connected to the source of the electrical pulses. Excitation of stator assemblies by a series of pulses, turns the rotor through angle p/n with each pulse. Usually, there are 200 pulses. Each pulse turns the rotor through 1.8°, i.e. 1/200 of a circle. The pulse frequencies range from 80–1600 Hz. So, the stepper motor RPM ranges from 24–480 RPM. The torque is still limited to 2000 kg mm only.

Point-to-point systems use 3–12 meters/min speed for moving members. This speed is reduced to 5–10mm/min to prevent an overshoot. Pulse rate is reduced to decrease the feed. This is accomplished by using electro-magnetic clutches. **Figure 7.54** shows an electro-magnetic feed drive. Powered by a 3-phase induction motor, the drive uses a lead screw for moving the machine table. Electro-magnetic cone clutch Ml is engaged for rapid motion, and clutch M2 is used for slow homing. The table motion is monitored by the transducer. Nearing the destination, the transducer signal to the comparator disengages clutch Ml, and engages clutch M2, to slow down the speed.

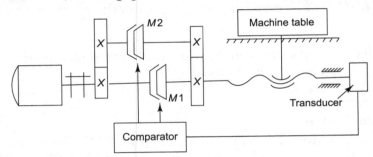

Fig. 7.54: Schematic diagram of an electro-magnetic feed drive

Some NC machines allow an overshoot, beyond the specified distance. After the overshoot, the comparator reverses the travel direction. There is also an overshoot in the opposite direction, before the speed is reduced, and another reversal effected. The required position is reached by inching at a slow [3–30 mm/min] speed.

2. **Continuous path system**

    These use electrical or hydraulic stepless drives. Electrical drives are economical for low power [up to 5kW] NC machines. Heavy-duty machines [above 5 kW] call for hydraulic stepless drives.

    Stepping motors for continuous path feeding, need torque amplifiers. Hydraulic motors serve as the torque amplifiers. The stepper motor only turns the spool of the rotary direction control valve, which controls the hydraulic motor.

    Researchers have now developed conventional electric motors, with better response and finer control. These are slowly replacing the stepper motors. Electric servomotors, coupled directly to the lead screw or using a toothed belt for transmission, are very convenient for slide motion control. Their rotary speed [RPM] can be varied easily, to alter the feed rate.

Displacement measurement

As tool wear is negligible in a single machining cycle, and workpiece and cutting tool are accurately located, it is often more convenient to measure tool/table travel, rather than the workpiece size.

Machine Tool Operation

Absolute measurement mode specifies distances [measurements] from fixed points [zero point] or axes [X, Y, Z]. They are suitable for positioning, and straight line and continuous path controls, of large machines.

Incremental or sequential mode measures distances from the preceding point or line [axis]. They are convenient for small and medium displacement NC machines, and continuous path machines, with digital interpolators.

Measuring devices can be:
(1) Digital type
(2) Analog type

Digital devices [Figs. 7.55, 7.56] use a feedback transducer, which monitors the position of the moving member. The increments are converted to numerical values.

A digital measuring device comprises a transparent calibrated scale and a scanner. Either the scanner or the scale is mounted on the moving member, while the other remains stationary. It is usually convenient to attach the scale to the moving member, and keep the scanner fixed **[Fig. 7.55]**. Photo-diode sensor of the scanner produces a pulse, when a scale-mark is passed over it. Electronic counter adds the pulses to indicate the distance moved.

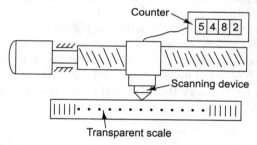

Fig. 7.55: Schematic diagram of a digital displacement-measuring system

Simple digital scales with 0.01 graduations can be very costly. It is more economical to use a combination of two scales, inclined at angle α with each other **[Fig. 7.56]**. The optical interaction between the two scales produces a Moiré pattern grating, whose width depends upon the graduation pitch and the inclination angle α. The resultant, manifold magnification allows the usage of more economic scales, with a wider marking pitch.

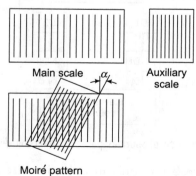

Fig. 7.56: Displacement-measuring system consisting of a main scale and an auxiliary scale

Instead of direct linear measurement, we can measure the rotation angle of the lead screw, to determine the axial displacement. This facilitates the usage of wider spacings for graduations. A graduated disk, mounted on a lead screw, passes through a photo-electric scanner. Although cheaper, rotary feedback devices are subjected to inaccuracies resulting from pitch error, backlash, and lead screw compression.

Elimination of reading errors by direction discriminator

Vibrations and spurious voltages from the machine electrical circuitry, often generate false signals, leading to errors. False signals due to vibrations can be eliminated by using **direction discriminator**.

Direction discriminator [Fig. 7.57] comprises:

1. An uncoded binary scale with one pitch $[S]$ divided into two equal parts of $[S/2]$ width. One part $[S/2]$ is dark, and the other $S/2$ transparent.
2. Two photo-diode sensors $[P_1$ and P_2 at $S/4$ distance from each other, i.e. electrically at 90°]. The signal from P_1 triggers T_1 to produce two square waves, 180° out of phase with each other. Differentiating elements D_1 And D_2 convert the signals to short duration pulses, which pass onto AND logic elements I and II.

The signal from P_2 is 90° out of phase with the signal from P_1. The P_2 signal triggers T_2 to break into 2 in-phase square waves, which are passed onto AND logic elements. Depending upon the displacement, direction of the two signals to AND elements, either add up, or cancel each other. The outputs from AND elements A and B are wired to the reversible counter in such a way that a signal from one is added, while the signal from the other is subtracted.

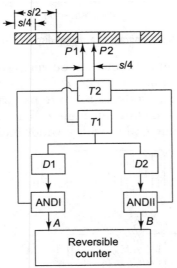

Fig. 7.57: Elimination of spurious pulses by direction discriminator circuit

Vibrations invariably involve 2 motions, in opposite directions. So, the spurious signal from a vibration is first added by A, and then subtracted by B, to eliminate it altogether.

Evaluator circuit comes in handy for eliminating false pulses induced by ambient voltage in the machine electrical circuitry. An evaluator circuit also comprises an uncoded binary scale, and 2 photo-electric sensors. But the sensors are placed at $S/2$ distance from each other. Consequently, the signals from them are 180° out of phase. The signals from the 2 sensors are fed to an OR logic element. OR generates an output pulse only if two 180° out-of-phase pulses are fed to it. A false pulse due to machine circuitry, will have the same polarity at both the sensors. So, it will not be counted at all.

For absolute displacement measurement, coded binary scales or disks are used [**Fig. 7.58**]. Top scale for 2° will have the least pitch, while the pitches for adjacent lower scales will be double the pitch of the adjacent upper scale. The measurement precision will naturally depend upon the uppermost scale with the minimum least count. A black zone on the coded scale reads zero [off], while a bright zone reads one [on]. Coded binary scales have a close resolution of about 2.5 microns.

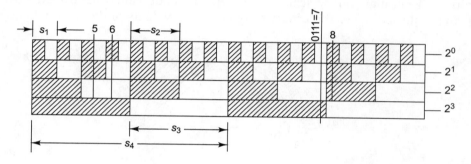

Fig. 7.58: Binary coded scale

A coded binary scale can give a false reading during transition from one digit to another. For example,

For 5 — Binary code 0101
For 6 — Binary code 0110

While passing from 5 to 6 on the coded binary scale, the readings will pass through the following sequence:

0101, 0111, 0110

The intermediate incorrect number 0111[7] may actuate a wrong feedback. This error can be eliminated by:

1. **Switching off the signals** when the transducer passes from one digit to another. This can be done by using a sprocket and a spring-loaded indexing pin, which enters the tooth space of the sprocket. The reader is switched off when the pin is not in the sprocket tooth space.
2. **Using 2 vee scanners** placed at angles, to form a Vee [**Fig. 7.59**]. For zero reading, scanner II is switched on to the next higher scale. For reading 1 [on], scanner I is switched on to the next higher scale.

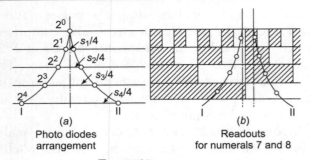

(a) Photo diodes arrangement   (b) Readouts for numerals 7 and 8

Fig. 7.59: V scanning

Coded binary disks

Coded binary scales are very costly. Coded disks [**Fig. 7.60**] are comparatively cheaper. They comprise 4 circular binary scales with a special code called **Modified Gray Code**. The following table [7.3] gives the binary gray code equivalents of the decimal code.

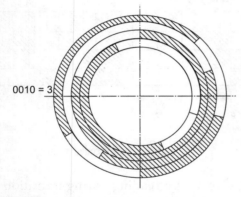

Fig. 7.60: Binary coded ring using modified gray code

Table 7.3: Modified gray code

Decimal Code	0	1	2	3	4	5	6	7	8	9
Modified Gray Code	0000	0001	0011	0010	0110	1110	1010	1011	1001	1000

A number of disks are coupled in succession, through a gearing with a 10:1 ratio. The first disk gives readouts of units, the next one are readouts of tens, the next of hundreds, and so on. Transit errors are avoided by the following rules:
1. If 0 is read on any of the disks, the sensors of a higher order scale on the next disk, is switched on.
2. If 1 is read, the sensors of the next lower order scales are switched on.

Analog displacement measurement

The displacement is converted to mechanical, magnetic, electrical, or photo-electrical analog. Electrical analog can be monitored through voltage, frequency, or phase.

Machine Tool Operation

Most widely used phase-sensitive transducers comprise 3 coils [**Fig. 7.61**]. In a **resolver**, two phase-wound stator coils 'B' and 'C' are placed at 90° to each other. The single phase wound rotor A is connected to the single-phase AC voltage. Tapping the coils U_B and U_C gives the sine and cosine components of the amplitude. If rotor A is coupled to the lead screw of the machine tool, the voltages from coils B and C will indicate the rotation angle of the lead screw, and the resultant axial displacement of the mating nut. Thus, resolver converts the angle of rotation into sine and cosine components.

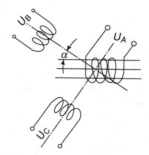

Fig. 7.61: Schematic diagram of a resolver

The Selsyn transducer is the reverse of the resolver. It places 3 stator windings, at 120 to each other. Two-phase voltage is applied to coils U_B and U_C. The single-phase voltage of the coil U_A, interacts with the magnetic fields of coils U_B and U_C. This induces e.m.f. in coil U_A. When coil U_A is coupled with the machine lead screw, the induced voltage indicates the angular position of the lead screw. Selsyn transducer is also used in synchronizing circuits, and hence, known as **Synchro**. Synchros are used more widely than resolvers.

A single synchro has a very narrow range of 4–6 mm linear measurement. The range is increased manifold, by using a number of synchros with precision 10:1 gearing [**Fig. 7.62**]. During travel of the moving member, coarser measurement transducers are automatically, successively switched off in decreasing order, and the finer ones switched on. Only a fine control transducer works during the final stage of displacement. Use of precision ground gears, makes synchros very dear [costly].

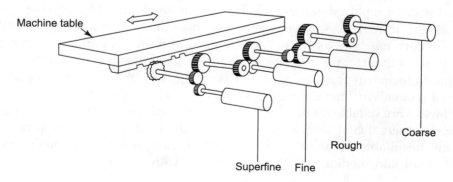

Fig. 7.62: Measuring arrangement using a number of synchros

For large displacements **Inductosyn** [**Fig. 7.63**] is more useful. It comprises a scale and a non-magnetic slider. The scale is printed with rectangular winding, while the slider has two displaced windings.

$$\text{Displacement} = ns + \frac{s}{4}$$

n = No. of steps; s = No. of the particular step

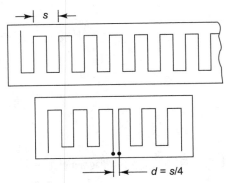

Fig. 7.63: Linear inductosyn

The slider windings are 90° out of phase with each other. The scale and the slider windings act like the rotor and stator windings of the resolver [**Fig. 7.61**]. The motion of the slider along the scale, changes the induced e.m.f. in the windings. The change in e.m.f. gives the rotation angle of the lead screw, and the resultant linear displacement.

$$\text{Displacement} = \frac{s\alpha}{360}$$

s = Pitch of the scale = 2.0 [usually]

α = Rotation angle

The standard 200 long scales, with 2.0 [s] markings, are readily available. For longer distances, the standard scales are placed end-to-end. The gap between the slider and the scales ranges from 0.05 to 0.15 mm. The A.C. voltage, supplied to the scale winding, has a 10–12 kHz frequency. Modern machine tools use combinations of resolvers and inductosyn : resolvers for coarse measurements, and inductosyn for finer ones. Resolvers use 400 Hz frequency power, while inductosyns function at 10–20 Hz frequency. A.C. voltage, supplied to both resolvers and inductosyns, have a somewhat mean 2 kHz frequency.

Analog devices are suitable for measuring **absolute displacements** only. Although simpler, robust, and cheaper, digital devices are suitable only for continuous measurements. This renders them unsuitable for position displays of moving members. Nonetheless, they are suitable for small and medium-size N.C. machines, with positioning and continuous path operation.

The moved distance should be checked by a comparator.

Comparators

Choice of the comparator depends upon the type of transducer used:
1. Discrete-function comparator for digital transducers
2. Continuous function comparator for analog transducers

1. Discrete function comparator

Pre-selection counter is the simplest type of discrete function comparator. The standard electronic preselect counter comprises decade switches, setable manually or automatically, through a punched program. The setting is an integer multiple of the least count [pulse]. The output signal can be generated in the following ways:

(a) By subtraction circuit
(b) By addition circuit
(c) By detection circuit

[a] Subtraction circuit [Fig. 7.64]

Suppose the desired displacement is 240.28 mm, and the least count is 0.01.

Then, $\dfrac{240.28}{0.01} = 24028$ pulses are required. This number is set on the decade 0.01 switches. The feedback pulses from the digital displacement-measuring device are fed to the units counter. Counting backwards, the unit counter reaches 0, after 8 pulses. This automatically switches on the tens counter. When after 2 pulses it reaches 0, the hundreds counter is automatically switched on. When all the counters reach zero, the output signal is produced by AND logic element.

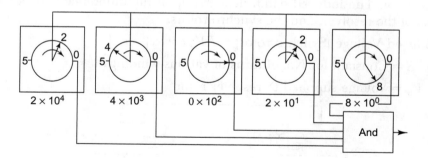

Fig. 7.64: A subtraction circuit pre-selector

[b] Addition circuit [Fig. 7.65]

In this, the counting is forward. The preselector is set to the complement of the required number of pulses, i.e. 100000−24028 = 75972. The counting starts from this number. When 24028 pulses are added, the preselector display shows 100000, and the output signal is produced by AND logic.

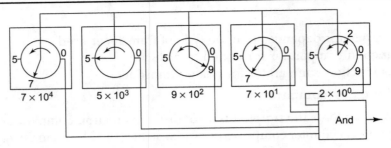

Fig. 7.65: An addition circuit pre-selector

[c] Detection circuit

In this method, the preselector is set for the desired number, say 24028. Initially, all the decade switches show a zero reading. On receiving the pulse from the feedback device, the indicators begin to move forward. Reaching the set number, it sends a signal to AND logic element. After receiving signals from all the decade switches, the logic element produces the output signal.

2. Continuous function comparator

In this, first the digital values of the program are transformed to analog [voltage]. For this, a digital-analog converter [**Fig. 7.67**] is used. It is a transformer with a number of secondary windings, connected serially. Each secondary winding has a tapped step, 10 times bigger than the preceding. No. 422 becomes 4.22 V.A.C. Voltage corresponding to the required displacement, is fed to the rotor of resolver [**Fig. 7.66**]. The resolver divides it into sine and cosine constituents. These are fed to 2-phase windings of the Synchro stator. The induced e.m.f. depends upon the difference between the angular positions of the resolver, and the synchro rotors.

Induced Voltage [Synchro rotor] $= KV_{rr} \cos[\alpha_1 - \alpha_2]$ (Eqn. 7.8)

$K =$ Constant for No. of turns of various windings

$V_{rr} =$ Voltage supplied to resolver rotor

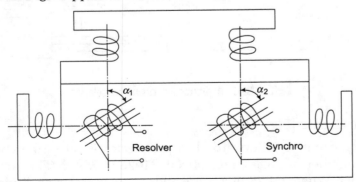

Fig. 7.66: A resolver-synchro coupling as a continuous-function comparator

The induced voltage will be zero when $[\alpha_1 - \alpha_2] = \pi/2$. It will be maximum when $[\alpha_1 - \alpha_2] = 0$, i.e. $\alpha_1 = \alpha_2$.

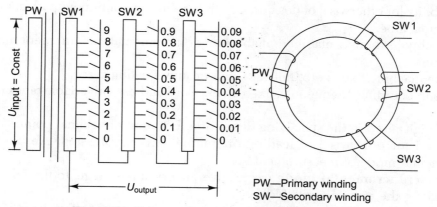

Fig. 7.67: Schematic diagram of a digital-analog converter

Figure 7.68 shows the analog comparator. The primary coil of transformer I is supplied with constant voltage, corresponding to the required displacement on the program. The voltage, varying due to the position of the moving member [analog feedback], is fed to the primary coil of transformer II. When the rotors of the programed and feedback transducer are perpendicular, the comparator output is zero. When the rotors are not perpendicular, the comparator generates a D.C. signal. Its polarity depends upon the sign of the phase difference. Simplicity, reliability, robustness, and low cost of continuous function comparators, make them a popular choice.

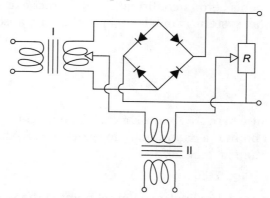

Fig. 7.68: Analog comparator

Programing

Program depends upon the workpiece and the machine program. It can be:
1. Manual
2. Computer-aided

The same workpiece will have different program for different numerical control machines. The machine tool format depends upon the following machine considerations:
(a) Availability [or otherwise] of positioning, straight-line, or continuous path control system, on the machine.
(b) Availability [or non-availability] of absolute or incremental measurement modes on the machine.
(c) Has the machine fixed zero, full-zero shift or fully-floating zero?
(d) What sort of format ? Sequential, fixed block, word address, or variable block programing format?
(e) No. of digits used in the specifying dimension. Location of decimal point.
(f) Provision [or otherwise] of leading or trailing zero suppression. Absolute mode or incremental mode of measurement

1. **Error: Incremental mode** carries over errors at any stage, to the following stage, making the error cumulative.
   An **absolute mode** error, at an intermediate stage, does not affect the following stages:
2. **Interruptions**: Tool breakage calls for going back to the very beginning of the program in the incremental system.
   In an **absolute system,** rewinding to the beginning of the concerned block is enough. We do not have to go to the beginning of the entire program.
3. **Modifications of program** are **irksome in an incremental** mode, since an error in any block affects all the following blocks.
   In **absolute mode,** change of co-ordinates of an intermediate point, does not call for a change of co-ordinates for the following point.
4. **Cross check**: In an incremental mode, going back to the starting point makes the sum of the increments zero: a confirmation of correct execution of the increments.
5. **Cost: Incremental system** is much cheaper than the absolute system.

Zero disposition

1. Fixed zero
2. Full shift zero
3. Full float zero

Machine zero is a point where X, Y, and Z axes of the machine meet.

Workpiece zero or set point is a datum point for workpiece/fixtures. It should be located precisely, with respect to the machine zero.

1. Fixed zero system [Fig. 7.69]

The program gives some co-ordinates to the set point of the workpiece, say X_m, Y_m. In a fixed zero machine, the workpiece must be positioned on the machine table in such a way that co-ordinates of the set point are X_m, Y_m; and the machine axes are parallel to the workpiece axes: edges in a rectangular workpiece. This is a skillful and time-consuming operation. After setting the workpiece, the assigned co-ordinates X_m, Y_m are entered in the MCU console and the 'Start' button is pressed. The table/spindle moves to the set point at the workpiece, to coincide the spindle axis and the set point.

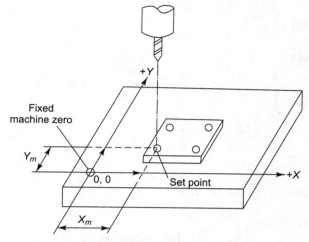

Fig. 7.69: Fixed-zero system

2. Full shift zero [Fig. 7.70]

The system allows shifting of the machine zero, by manipulating zero shift dials, provided in the machine. The workpiece can be clamped in any convenient position on the machine table. But the axes [edges] of the workpiece should be parallel to the machine axes [table edges]. After setting the workpiece, the zero shift dials of the machine are adjusted in such a way that the co-ordinates of the workpiece set point [zero] are X_m and Y_m. In rectangular workpieces, it is convenient to shift zero to the L.H. bottom corner of the workpiece.

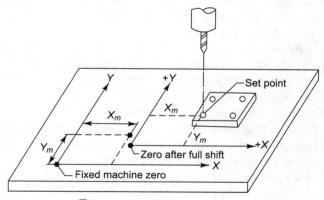

Fig. 7.70: Full-zero shift system

3. **Fully floating zero [Fig. 7.71]** system has no fixed machine zero. The programer can select any point as zero. Naturally, the axes of the workpiece and the machine must be parallel. Co-ordinates of the workpiece set point can even be minus, as shown in [Fig. 7.71]. If the zero point is placed at the geometrical center of the workpiece, as shown in the figure, the workpiece set point will have co-ordinates : $-X_m - Y_m$.

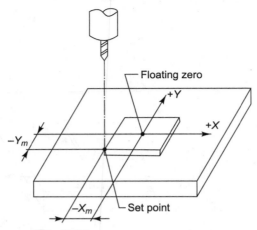

Fig. 7.71: Fully floating zero system

Programing formats can be broadly classified into the 4 categories given below:

1. **Fixed block format** has a fixed sequence. The number of characters in every block is the same. The sequence number precedes the coordinates, which are followed by the end of the block [EOB] statement. As in fixed zero and full-shift zero systems, coordinates are always positive, and there is no need for signs [+/−]. In a fully-floating zero system, the coordinates must be preceded by the plus/minus sign. X, Y coordinates must be repeated in the next block, even if the location does not change.

2. **Tab sequential format** uses tab character to separate the words in a block. The fixed sequence of the sequence number, $X - Y$ coordinates, and the end of the block statement, must be followed. But the repeated word [or dimension] need not be punched again in the following block. For a tab sequential format all the numbers are retained in the MCU memory.

3. **Word address format** follows no fixed sequence, since the displacement information is prefixed by an identifying word [X/Y] acting as the address. Naturally, the program is longer. Except commands for program end, stopping machine, program rewinding [M02 in Table 7.6]; and tool change [M06], all other commands [X, Y, Z, G from Table 7.5] remain in effect till they are replaced by new ones. So, a repeated word need not be punched again in the following block. The exceptions [M02 and M06] are cancelled automatically, after implementation. So, they [M02 and M06] must be repeated if necessary, in every following block.

4. **Variable block system** combines word address and tab sequential formats. Every displacement information must be addressed by a prefix [X, Y, Z, etc.]. Word address pairs are separated by a tab. Repeated words need not be punched again in the following blocks.

Specifying dimensions

Five to eight digit words can be used to specify dimensions. Most N.C. machines use a 3.3 format with 6 digit words for mm dimensions and 2.4 format for inch dimensions. In the 3.3 format, the decimal point is placed after 3 integers, while the 2.4 format places the decimal point after two integers. No. 76.2 will be written as 076200 in the 3.3 format, and 11.15 inches will be 111500 in the 2.4 format. In **leading zeros suppression,** the machine reads from right to left, and 076200 will become 76200. In **trailing zero suppression,** the machine reads from left to right. So, 111500 will become 1115. Zero suppression is not applicable to digital codes for other functions.

Displacement [Point-to-Point] program

Figure 7.72 shows a workpiece for the displacement program. First, let us find the X, Y co-ordinates, as per various zero systems. The first step will be to number the holes, for the sequence to be followed. Taking tentatively, $X = 20$ and $Y = 29$ for fixed zero machine, the co-ordinates will be as below.

	Point No.		Zero
	1	2	at
[1] Fixed Zero			
X	20	50	m/c
Y	29	55	zero
[2] Full Shift Zero			
X	8	38	L.H. Bottom
Y	8	34	Corner
[3] Fully Floating Zero			
X	−15	+15	Center of
Y	−21	+5	workpiece

The 2.4 format can also be used for mm. Similarly, the 3.3 format can also be used for inches. This gives us many combinations of formats. We will work out program for the 4 formats in the following example, for the workpiece shown in **[Fig. 7.72]**.

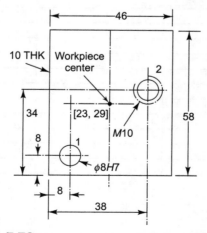

Fig. 7.72: Workpiece for point-to-point program

End of the block is represented by letters: EOB. The program begins with EOB.

Format	Fixed Block	Tab Sequential	Word Address	Variable Block
Zero:	Fixed	Full Float	Full Shift	Full Float
Dimensions	X = 3.3	X = 2.4	X = 2.4	X = 3.3
Format	Y = 3.3	Y = 2.4	Y = 3.3	Y = 3.3
'0' Suppression	None	Trailing	Trailing	Leading
	EOB	EOB	EOB	EOB
Sequence	0	0	N	N
No	0	0	0	0
	1	1	0	0
	0	TAB	1	1
X	2	− [X]	X	TAB
	0	1	0	X
	0	5	8	−
	0	TAB	Y	1
	0	− [Y]	0	5
		2	0	0
	0	1	8	0
Y	2	EOB	EOB	0
	9	0 [Seq. No]	N	TAB
	0	0	0	Y
	0	2	0	−
	0	TAB	2	2
	EOB	1 [X]	X	1
Seq.	0	5	3	0
No.	0	TAB	8	0
	2	0 [Y]	Y	0
	0	5	0	EOB
	5	EOB	3	N
	0		4	0
	0		EOB	0
	0			2
	0			TAB
	0			X
	5			1
	5			5
	0			0
	0			0
	0			0
				TAB
				Y
				5
				0
				0
				EOB

Function codes

Even functions like changing the tool, switching on the spindle, setting the RPM of a spindle and the feed, and starting the coolant pump, can be controlled numerically, in sophisticated NC machines. Loading and unloading workpieces is however, invariably done manually.

Sequence number is prefixed by 'N' and represented by a 3-digit number.

Tool change is usually programed on machines equipped with automatic tool changers and programable turrets. Prefixed by letter T, tool change is represented by a 2 to 5 digit number.

Speed is related to the tool. So, machines provided with tool changers, also have a provision for automatic speed change. The speed function is represented by 3 digits, prefixed by 'S'. If a machine has 12 stepped speeds, they might be assigned a number between 001–012. Infinitely variable speeds can be directly represented by 3 digits, giving the RPM. The speeds can also be represented by **magic three code equivalents,** given in Table 7.3. In these, the RPM is rounded to two digits, which make digits two and three of the magic code. The number of digits before the decimal point are added to No. 3, to obtain the first digit of the magic code. The following table shows the conversion of some speeds [RPMs] to the magic 3-digit code.

Table 7.3: Magic three code

R.P.M.	First digit	Rounded last 2 digits	Magic equivalent
16	2+3 = 5	16	S516
112	3+3 = 6	11	S611
1500	4+3 = 7	15	S715

Feed function

Feed rates are identified by the prefix 'F'. The rates [mm/rev or mm/min] are converted to the magic 3-digit code. However, feed values are usually smaller, with few, if any, integers on the left of the decimal point. When there is no integer at all, the first digit for the magic three digit code is obtained, by subtracting the number of zeros after the decimal point from the number three. The second and the third numbers are obtained by rounding the feed value to two figures. The following table gives some examples of conversion of feed rates to magic three code.

Feed rate	First digit	Rounded 2 digits	Magic three equivalent
1.25	1 + 3 = 4	12	F412
0.762	3 − 0 = 3	76	F376
0.0256	3 − 1 = 2	25	F225
0.00508	3 − 2 = 1	50	F150

Prefixes for dimensions

In addition to X and Y coordinates workpiece specifications also comprise angular dimensions from the standard X, Y, Z axes; secondary dimensions parallel to X-Y-Z axes, and interpolation parameters parallel to the std. axes. These are identified for the type, by the following prefixes:

Prefix	For	Prefix	For
X	Primary X Dim	U	Secondary Dim parallel to X axis
Y	Primary Y Dim	V	Secondary Dim parallel to Y axis
Z	Primary Z Dim	W	Secondary Dim parallel to Z axis
Prefix	For	Prefix	For
A	Angular Dim about X axis	I	Interpolation parameter parallel to X axis
B	Angular Dim about Y axis	J	Interpolation parameter parallel to Y axis
C	Angular Dim about Z axis	K	Interpolation parameter parallel to Z axis

Codes for preparatory functions

Codes for preparatory functions immediately follow the sequence number. Codes for preparatory functions are prefixed by the letter 'G'. The function is represented by a 2 digit number. The following table gives the codes for some of the often-used preparatory functions.

Table 7.5: Codes for some important preparatory functions

Code	Function
G00	Rapid Traverse Positioning
G01	Linear Interpolation
G02	Clockwise Circular Interpolation
G03	Anti-clockwise Circular Interpolation
G04	Dwell
G08	Suppress deceleration: Ramp down inhibit
G09	Cancel G08
G17	XY Plane Selection for Interpolation
G18	XZ Plane Selection for Interpolation
G19	YZ Plane Selection for Interpolation
G28	Return to reference point
G30	Cancel mirror image
G31	Mirror image with Axis command
G39	Tool Radius Vector Setting prior to Change of Direction
G40	Cancel Cutter Dia/Tool Nose Radius Compensation
G41	Cutter Dia/Tool Nose Radius Compensation, Left
G42	Cutter Dia/Tool Nose Radius Compensation, Right

Contd.

Machine Tool Operation

Code	Function
G45	Tool Offset Increase
G46	Tool Offset Decrease
G49	Cancel Tool Length Compensation
G50	Programming of Absolute Zero Point in Turning/Cancellation of Do Loop
G51	Do Loop
G55	Full Zero Shift in CNC Milling/Drilling Machines
G70	Inch Data Input
G71	Metric Data Input
G78	Milling Cycle Stop
G79	Milling Cycle
G80	Cancel Cycle
G81	Drilling Cycle
G82	Facing Cycle
G84	Tapping Cycle
G85	Boring Cycle
G90	Absolute Dimension Input
G91	Incremental Dimension Input
G92	Programing of Absolute Zero Point in milling
G94	Feed in mm/min [Inch/min]
G95	Feed in mm/rev [Inch/rev]
G96	Constant Surface Speed
G97	Spindle Speed in Revs/min [RPM]
G98	Return to Initial [I] Level in Canned Cycle
G99	Return to R level in Canned Cycle

Important functions, which are not preparatory, nor a part of the cutting program, are categorized as miscellaneous functions. Their code comprises a two digit number, prefixed by the letter 'M'. The following table [Table 7.6] gives the codes for often-used miscellaneous functions.

Table 7.6: Important miscellaneous functions

Code	Function
M00	Stop program: all slide motions, spindle rotation, and coolant pump
M01	Optional stop, activated manually, from console
M02	End of program. It stops machine and rewinds the program
M03	Start clockwise [Looking down towards workpiece] spindle rotation
M04	Start anti-clockwise spindle rotation

Contd.

Code	Function
M05	Spindle stop [RPM]
M06	Tool change
M08	Coolant pump on
M09	Coolant pump off
M10	Clamp [Spindle, Quill, Fixture, slide, etc.]
M11	Unclamp [Spindle, Quill, Fixture, slide, etc]
M13	Start spindle clockwise and switch on the coolant pump
M14	Start spindle anti-clockwise and switch on the coolant pump
M19	Stop spindle in a pre-specified angular position: oriented stop
M21	Mirror image on X axis
M22	Mirror image on Y axis
M23	Mirror image off
M30	End of program; stop m/c and rewind memory [tape]
M41-45	Gear change
M51-59	Z Axis setting with cams and switches on 2½ axis N.C. M/Cs
M98	Go to sub-routine
M99	Return from sub-routine [to main program]

Comprehensive point-to-point program

The program for the workpiece, shown in **[Fig. 7.72]**, can be completed now. The operations at the two positions are:
(1) Drilling and Reaming $\Phi 8H7$
(2) Drilling Φ 8.5 and Tapping M10 X 1.5

The machine tool format will be: N3. G2. X3.3. Y3.3. M2 EOB. The dots indicate the **Tab Sequential Format**. The suffixes after N, G, and M specify the number of digits in the corresponding codes.

The machine is equipped for absolute and incremental dimensioning in mm and inches, and point-to-point control. It has a full-range zero shift, and suppressing trailing zeros. During workpiece loading/unloading and tool change, the spindle is positioned at point X = 150, Y = 150, clear of loading/unloading area. The zero is shifted to the L.H. bottom corner of the workpiece.

The coordinates of the two points will be

Point No.	X		Y	
	On Drg	On Tape	On Drg	On Tape
1	8	008	8	008
2	38	038	34	034

In the tabulated program for the part in **[Fig. 7.72]**, the following points should be noted.

Machine Tool Operation

Notes:
(1) 'T' indicates Tab, while 'E' represents End of Block [EOB]
(2) EOR specifies the point up to which the tape/memory is re-wound after the M02/M30 command. Some older machines use % symbol instead of EOR.
(3) Setting zero at the bottom L.H. corner of the workpiece, is indicated by N000 block instructions.

The following table gives the full program manuscript.

Program manuscript for part in [Fig. 7.72] (p 591)

N	T/E	G	T/E	X	T/E	Y	T/E	M	T/E	Remarks
EOR 000				0		0				Set Machine Zero
005	T	90	E							Abs. Dim. Selection
010	T	71	E							Metric Dim. Selection
015	T	00	E							Rapid Positioning
020	T	80	T	15	T	15	T	06	E	Clamp Φ7.5 Drill
025	T	81	T	008	T	008	T	51	E	Drilling Cycle
030	T		T		T		T	03	E	Start Spindle RPM Clockwise
035	T		T		T		T	05	E	Stop Spindle RPM
040	T	80	T	15	T	15	T	06	E	Clamp Φ8 Reamer
045	T	81	T	008	T	008	T	52	E	Reaming Cycle
050	T		T		T		T	03	E	Start Spindle RPM Clockwise
055	T		T		T		T	05	E	Stop Spindle RPM
060	T	80	T	15	T	15	T	06	E	Clamp Φ 8.5 Drill
065	T	81	T	038	T	034	T	53	E	Drilling Cycle
070	T		T		T		T	03	E	Start Spindle RPM Clockwise
075	T		T		T		T	05	E	Stop Spindle RPM
080	T	80	T	15	T	15	T	06	E	Clamp M10 Tap
085	T	84	T	038	T	034	T	54	E	Tapping Cycle
090	T		T		T		T	03	E	Start Spindle RPM Clockwise
095	T		T		T		T	05	E	Stop Spindle RPM
100	T	80	T	15	T	15	T	02	E	End of the Program

Let us consider an example of a positioning-cum-straight cut program for the workpiece shown in **[Fig. 7.73]**, for a Φ 10.5 hole and a 12 wide X 3 Dp milled slot.

The machine constraints/facilities are:
(a) Fully-floating zero: calls for workpiece central axes
(b) Only absolute dimensions, in mm only

(c) Φ 12 Endmill will be used for 12 wide slot. The milling cutter should be clear of the workpiece, before and after cutting.

(d) Fixed Block Format, with no zero suppression at all. This format has a fixed sequence: Seq. No., X Dim, Y Dim, End of Block [EOB].

With the floating zero placed at the geometrical center of the workpiece, the coordinates of the drilled hole will be X = 000.000, Y = 030.000. If we provide a 2 mm clearance between the milling cutter and the workpiece, the coordinates of the cutter before beginning the cut will be:

$$X = -\left[\frac{160}{2} + 2[\text{clearance}] + \frac{12}{2}[\text{cutter radius}]\right] = -88 = -088.000$$

$$Y = -025.000$$

At the end of the cut, the coordinates of the cutter will be:

$$X = 088.000, \text{ and } Y = -025.000$$

The machine tool fixed block format is: N3, G2, X3.3, Y3.3, M2, M2, EOB.

The availability of two M2 statements in the fixed block, allows the coding of 2 miscellaneous functions in a single block.

The following table shows the complete program manuscript for the part in [Fig. 7.73]. Note that the number of digits in every column of the fixed block should be the same, even if it means giving the same command twice, as M06 in EOR 005. The tool change will be done at point X = 90; Y = 60, i.e. clear of the workpiece loading/unloading area.

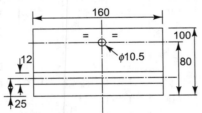

Fig. 7.73: Workpiece for positioning cum straight-cut program

Program manuscript for part in [Fig. 7.73]

N	G	X	Y	M	M	E	Remarks
000		−080000	−050000				Setting Machine Zero
EOR 005	80	+090000	+060000	06	06	E	Cancel Cycle and Clamp Φ 10.5 Drill
010	81	+000000	+030000	51	13	E	Drilling Cycle, Start spindle Clockwise
015	80	+090000	+060000	05	09	E	Stop spindle, Stop Coolant
020	80	+090000	+060000	06	06	E	Clamp Φ 12 Endmill Cutter
025	00	−088000	−025000	52	52	E	Position Cutter to Start Cycle
030	79	+088000	−025000	52	13	E	Milling Cycle, Coolant Start
035	78	+088000	−025000	05	09	E	Stop Milling Cycle, Spindle RPM and Coolant
040	80	+090000	+060000	02	02	E	End of the program, Rewind

Continuous path contouring program

Contours can be approximated by circular / parabolic arcs and straight lines. We will consider only circular arcs and straight lines interpolating program for parabolic interpolators are costlier, and hence, rare.

Let us consider contouring on lathes. While programing NC lathes, we should remember the following features of N.C. lathes:

N.C. lathes: peculiarities

1. Incremental system of dimensioning
2. Axes – Spindle – Z axis
   Radial – X axis
3. Angle between [tool] motion direction and Z axis is represented by the sine of the angle and identified by the prefix I.
4. For the facing motions, square with the axis, i.e. angle 90°,

    $I = \sin 90° \approx 0.99999$

    As sine of any angle is less than one, the decimal point, and the zero preceding it, is omitted.

    $I = 99999$

K coordinate gives value of the cosine of the angle $[\theta]$ between the spindle axis and tool motion. For parallel turning, the angle is zero.

$K = \cos 0° = 1 \approx 0.99999 \approx 99999$

Contouring program for turning a cone [tool motion in a straight line inclined to X and Z axes]

The inclination of the straight line can be specified in two ways:
  a. Angle $[\theta]$ between the line and Z axis.
  b. The lengths of the tool paths: distances of the line ends, from X and Z axes.
  a. Angle $[\theta]$ of the angular cutting path line with the spindle, i.e. Z axis is designated by a 5 digit number, with the prefix I, for sine of the angle, and prefix K for cosine of the angle, as explained earlier. The following table gives some examples of converting sine/cosine values to 5 digit numbers.

Angle θ	Sine Value	Converted Value	Cosine Value	Converted Value
30°	0.5	I 50000	0.8666	K 86660
45°	0.7071	I 70710	0.7071	K 70710
60°	0.8660	I 86660	0.5	K 50000
90° [Facing]	1.0	I 99999	0	K 00000
0° [Turning]	0	I 00000	1.0	K 99999

Naturally, if there is a provision for suppressing the trailing zeros, I 5000 will become I 5, and K 70710 will become K 7071.

b. Tool path lengths [Fig. 7.74b]

In this system, I and K values are replaced by Feed Rate Number [FRNJ. Prefixed by the code F, FRN is a five digit number: three to the left of the decimal point, and two to the right of the decimal point.

$$FRN = \frac{10 \text{ Feed Rate [mm/min]}}{\text{Length of tool path [mm]}} \qquad [1'' = 25.4 \text{ mm}]$$

When the tool motion is controlled by simultaneous motions along the X and Z axes, only FRN is stated, and values of I and K are omitted.

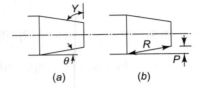

Fig. 7.74: Specifying angular path

Effect of tool nose radius [r]

Generally, all cutting tools have radiused noses. This radius increases/decreases the tool travels, beyond the workpiece dimensions.

Figure 7.75a shows the effect of the nose radius on the tool path length. It should be noted that Tool Travel $P'Q'$ will be less than the workpiece length PQ by '$r \tan \frac{\theta}{2}$' while tool travel $R'S'$ will be '$r \tan \frac{\theta}{2}$' more than the workpiece dimension RS. In grooves [**Fig. 7.75b**], tool travel will be reduced by '$2r$', while tool travels preceding and following the grooves, will increase by 'r'.

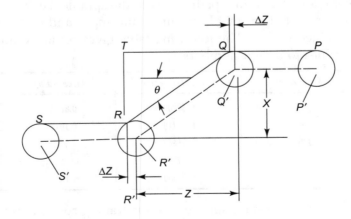

Fig. 7.75(a): Calculation of tool offset for a tapered bend

Thus, referring to [**Fig. 7.75b**],
$$S'R' = SR - 2r, \quad P'Q' = PQ + r, \quad T'U' = TU + r$$

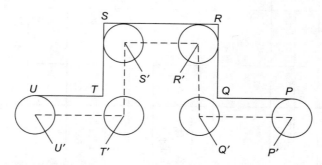

Fig. 7.75(b): Calculation of tool offset for a rectangular bend

Lengths ST and QR will however, coincide with the tool travels: $S'T' = ST$; $QR = Q'R'$
Let us work out tool travels for the workpiece in [**Fig. 7.76**]. For tool radius $[r]$ of 2.0 mm,

$$P'Q' = PQ - r\tan\frac{\theta}{2} = 25 - 2\tan\frac{30}{2} = 24.464$$

$$R'S' = RS + r\tan\frac{\theta}{2} = 19 + [2 \times 0.2679] = 19.536$$

Lengths TR, TQ, QR in the workpiece will however, remain unchanged and will be equal to tool travels $T'R'$, $T'Q'$, $Q'R'$, respectively. [Ref **Fig 7.75a**]

Let us program the workpiece in [**Fig. 7.76**], for a 2 mm radius tool. Let us place zero at 5 mm, clear from the R.H. end of the workpiece, i.e. at 100 from the L.H. center of the lathe. The X co-ordinate from Z axis should also place the tool clear of the workpiece, to permit tool change, i.e. replacement of the wornout tip/tool; say at $X = 50$.

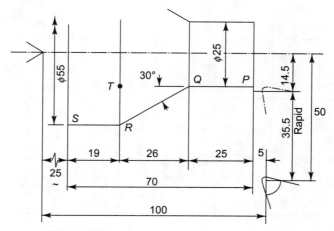

Fig. 7.76: Workpiece for program in Table 7.7

Table 7.7: Word address format program for workpiece in [Fig. 7.76] on lathe X3.3, Y 3.3: Feed Function 0000.0

N 000	G		X 050000	Z 100000 [from L.H. center]	I	K	F Feed	S Speed RPM	M	M	E
EOR N005	G90	Abs. Dim Mode for zero setting									
10	G71	Dims in mm									
15	G95	Feed in mm/Rev									
20	G04	Dwell [for feed rate change & speed selection]	X02				F440	S650			E
25	G01	Linear interpolation	050000	Z100000							E
30	G91	Incremental dimensioning code							M06	Mount/Replace Turning Tool	E
35	G01	Linear Interpolation				K99999					E
40	G01	Linear Interpolation	−035500		I 99999						E
45	G04	Dwell for Feed Change	X02				F320		M03	Start Spindle RPM Clockwise	E
50	G01	Linear Interpolation		−029464 [5 Approach]		K99999					E
55		30° Angle Turning Dia 55	015000	−026000 −019536	150000	K86600 K99999					E
60	G04	Dwell for Feed Change	X02				F440		M09	Coolant off	E
									M08	Coolant on	
65	G01	Linear Interpolation Tool Withdrawal	022500		I 99999				M05	Spindle stop	E
70		Back to Zero position		075000		K99999			M02	Program End	E

Machine Tool Operation

Remember that the tool travels should be specified from the center of the nose radius. At zero point, the radius center coincides with the zero. So the X coordinate for turning Φ25 will be: { 50 – 25/2 – 2[nose rad] } = – 35.5. The tool can approach this starting position at a rapid feed of 4 mm/Rev [F440]. The feed can be changed to the turning feed of 0.2 mm/Rev [F320] with 2 seconds [X02] dwell. The speed can be 500 RPM, i.e. as per the 3 digit code: S650. We have already worked out the tool travels P'Q' and R'S' as **24.464** and **19.536,** respectively.

We can now tabulate the program in a word address format. It should be remembered that in this format, it is necessary to provide an **identifying code** [X/ Z] for the dimension address. The repeated words need not be punched again, for the commands remain in force till they are replaced by new ones, except of course, M02 and M06. There is no need to follow any **fixed sequence**.

Programing convex and concave radii

Circular arcs are specified by 4 variables:

  a. Direction of tool [cutting] motion: Clockwise/Anti-clockwise
  In some machines like CNC turning centers [p 629], this direction can be −Y axis, i.e. look upward towards the Z axis. In other type of machines, the direction of looking is +Y axis, i.e. downward, from above the Z axis. This is the direction assumed in the following example on programing radiused workpieces.

  b. Incremental departures: The distance traversed by the tool along the X and Z axes during contouring. The departures define the extent [length] of the curve. **Figure 7.77** shows the X and Z departures for various circular arc configurations.

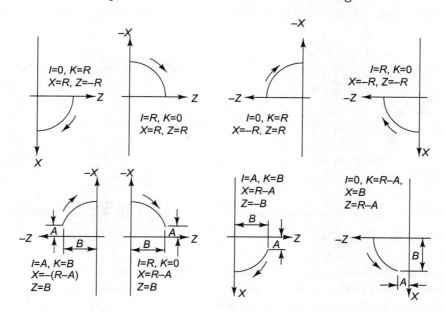

Fig. 7.77: Method of determination of *I* and *K* values, and the *X* and *Z* departures for circular arcs

c. **Arc center** is specified by *I* and *K* prefixes. I coordinate is the [absolute i.e. always +] distance of the curve starting point from its center along the *X* axis. *K* represents the absolute distance along the *Z* axis. **Figure 7.77** shows the typical shapes of circular arcs with the *I* and *K* coordinates.

If the arc passes through more quadrants, separate blocks of instructions are necessary for each quadrant.

d. **Feed Rate No.** [FRN] is represented by a five digit number: 000,00.

$$\text{For Circular Arcs FRN} = \frac{10 \text{ X desired feed rate [mm/min]}}{\text{Radius of Arc [mm]}} \quad \text{(Eqn. 7.9a)}$$

Even in circular contours, the tool nose radius changes the tool travels [**Figs. 7.78 a, b**]. In a concave profile [**Fig. 7.78a**], the tool travel along the *Z* axis, decreases by the tool radius '*r*', while the length of the following straight travel increases by length '*r*'.

In the convex profile [**Fig. 7.78b**], the travel for the straight length [*PQ*], preceding the arc will decrease, while the tool travel for the following straight length [*RS*], will increase.

In [**Fig. 7.78b**],

$$P'Q' = PQ - \qquad \text{(Eqn. 7.9b)}$$

$$Q'T' = QT + \left\{ r - (R - \sqrt{R^2 - r^2}) \right\} \qquad \text{(Eqn. 7.9c)}$$

The tool travels along *X* axis will remain equal to the corresponding workpiece dimensions.

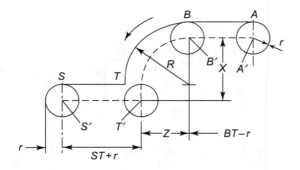

Fig. 7.78(a): Calculation of tool offset for an anti-clockwise circular bend

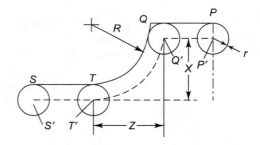

Fig. 7.78(b): Tool offset for a clockwise circular bend

Let us program the workpiece shown in [**Fig. 7.79**]. We will set zero at 5 mm from the right-hand end of the workpiece. It will be 112 from the lathe L.H. center. The X coordinate of the set point can be 55, tentatively. Tool nose radius is 3 mm. We can use **Word Address Format**. In this, **dimensions** need to be **addressed** [**X/Z**]. Repeated words need not be repeated in the program. The machine has a provision for **suppressing the leading zeros**.

Let us calculate the tool travels for the straight lengths preceding and following the curves [in **Fig. 7.79**].

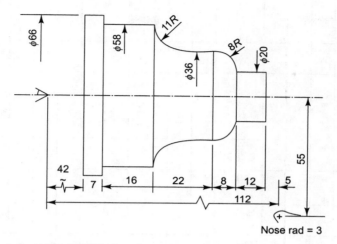

Fig. 7.79: Workpiece for program in Table 7.8

$$P'Q' = PQ - \left\{ r - (R - \sqrt{R^2 - r^2}) \right\}$$

$$= 12 - \left\{ 3 - (8 - \sqrt{8^2 - 3^2}) \right\}$$

$$= 12 - \{ 3 - [8 - 7.42] \}$$

$$= 9.584 \text{ [Entry N045 in Table 7.8]}$$

Motion [Departure] along Z axis for the arc

$$= R + \left\{ r - (R - \sqrt{R^2 - r^2}) \right\}$$

$$= 8 + \{ 3 - 0.58 \}$$

$$= 10.42 \text{ [N050 in Table 7.8]}$$

Departure along the X axis will be the same as the corresponding dimension of the workpiece, i.e. $R = 8$

For the concave 11R portion, the tool travel along the Z axis will be reduced. It will be $R - r = 11 - 3 = 8$ [N050].

The straight tool travel for Φ 58 will increase by 'r', i.e. 3 mm

$$T'S' = TS + 3 = 16 + 3 = 19 \text{ [N065]} - 3 \text{ [shoulder corner rad]} = 16$$

We can now tabulate the program manuscript [Table 7.8]. Note the 2 second dwells [X02] for feed changes, and usage of rapid feed for tool approach and withdrawals.

Table 7.8: Programed workpiece with arched profile [Fig. 7.79]

Word address format with Abs./Incremental programing, X 3.3, Z 3.3 Full Shift Zero; Suppression of leading zeros. Two Misc Functions; Feed 100 /Rev] i.e. 0.1 mm/Rev; Speed = 125 RPM, i.e. S613

N	T/E	G		X	Z	I	K	F	S	M	M	E
000				55000	112000							E
EOR		G90	Abs Dim									
N005												
N010		G71	mm Dim									
N015		G95	Feed mm/Rev									
N020		G04	Dwell for Feed speed change	X02				F440	S613			
N025		G01	Linear Interpolation	55000	112000					M06	Clamp tool	E
N030		G91	Incremental Dim									E
N035		G00	Rapid positionning for ϕ 20	−32000		199999						E
N040		G04	Dwell for Feed change	X02				F310		M03	Spindle start clockwise	E
N045		G01	Linear Interpolation [2 approach]		−11584		K99999					E
N050		G02	Circular Interpolation Clockwise	8000	−10420	10	K8000					E
N055		G01	Linear Interpolation		−11000		K99999					E
N060		G03	Circular Interpolation Anti-clockwise	8000	−8000	18000	K0					E
N065		G01	Linear Interpolation		−16000		K99999					E
N070			Face shoulder	−4000		199999						E
N075			turn ϕ66 [1 over travel]		−11000		K9999					E
N080		G04	Dwell for Feed change	X02				F440		M09	Coolant off	E
N085		G01	Linear Interpolation	19000		199999				M05	Spindle stop	E
N090					71000		K99999			M02	End of the Program	E

| | | | | | | | | | | M08 | Coolant on | |

Machine Tool Operation

Computer-aided programing

Approximating a curve by many, small straight lines, involves computing the end co-ordinates of a multitude of approximating lines. Usage of a computer for the calculations saves time, and minimizes human errors.

Computers too recognize only two types of signals: on/off [1/0]. But computers use more compact, straight binary codes instead of binary decimal codes used in NC machines. A computer uses special languages and assembly programs for coding words, signs, and numerals in 1/0 system. In addition to translating the simplified programs into computer language, the assembly programs also perform mathematical operations: adding, subtracting, multiplying, dividing, etc. FORTRAN [Formula translation] compiler [assembly] program is widely used in the engineering industry. Banking, trading, etc. also have numerous specialized programs at their disposal.

The NC software consists of two programs:

[a] GENERAL PROCESSOR for machine tool applications such as LINE, POINT, CIRCLE, RADIUS, TOLERANCE.

[b] PART PROGRAMER for a particular workpiece. The program is written in the language, comprehensible to the general processor. This calls for control statements through provision of values for N, G, F, S, M, and T functions, in the format compatible with the machine tool. An additional set of instructions, a POST PROCESSOR, adapts co-ordinates, tool departures, etc. to suit the machine tool/control unit combination. Programming and Post-processing instructions of a machine tool must be studied and fully understood, before undertaking programing for a particular part.

Automatically programed tools [APT]

Apt-IV version comprises 400 words, convenient to tackle sophisticated, 3-dimensional program, APT language, however, calls for a large computer, with a very high capacity.

ADAPT, a subset of APT, comprises 160 words and punctuations, suitable for 2-dimensional programs, needing only a medium size computer.

AUTOMAP, a 50 words version of APT, is convenient for programing 2-dimensional lines and circles.

EXAPT, a German extension of APT with versions suitable for:

[a] Positioning NC machines: EXAPT I
[b] Lathe operations: EXAPT II
[c] Drilling and milling with continuous path control: EXAPT III. X-Y plane and only depth along Z axis [2.5 Dim control]

Other programs unrelated to APT are as follows:

[p] SNAP [Simplified Numerical Automatic Programer] used mainly for point-to-point work; requires a very small computer.
[q] Autospot [Automatic System Positioning Tools] is a 100 words version, for point-to-point programs.
[r] SYMAP [Symbolic Language for Machine Programing] with German language, has 6 versions for different kinds of work:

1. SYMAP [P] for positioning control
2. SYMAP [S] for straight line control
3. SYMAP [B] for 2.5 axes Dim control
4. SYMAP [PS] combination of [P] and [S]
5. SYMAP [DB] for lathes with continuous path
6. SYMAP [DS] for lathes with straight-line path

APT programs

These comprise 4 statements:
1. Motion Statements: Path of the cutting tool
2. Geometry Statements: Surface configurations to be followed by the tool
3. Post Processor Statements: Entries for functions M, F, S, T, and G
4. Auxiliary Statements: Cutter Dia, Tolerances, etc. not covered by statements 1, 2, 3

1. Motion statements

a. Point-to-point programs call for motion from one point to another. There is no constraint for following any particular path. The points are identified by the prefix P. For the workpiece in [Fig. 7.72] [p 591], the program will be as below:
FROM/P1
GO TO/P2
GO TO/P3
Here FROM and GO TO are motion statements, while P1, P2 are geometry statements. Incremental departures such as, X = 20, Y = 10, Z = 30 will be stated as below:
GODLTA/20, 10, 30

Programing a pattern

Only a few characteristic point coordinates need to be specified. For the workpiece shown in [Fig. 7.80], the geometrical statement will be as below:
PAT 1 = PATERN/LINEAR, P1, P2, 10
PAT 2 = PATERN/LINEAR, P1, P3, 16
PAT 3 = PATERN/GRID, PAT 1, PAT 2
The motion statements for the part [Fig. 7.80] will be as below:
FROM/SETPT
CYCLE/DRILL, CAM, 1
GO TO/PAT 3
CYCLE/OFF
GO TO/ PATERN can be added with a modifying word OMIT, RETAIN, etc. OMIT n3, n5, n6 will enable omitting points 3, 5, 6, or whatsoever desired. RETAIN n7, n11 will ensure that the tool visits only the points after the word: retain.

The co-ordinates of the set-up point [SETPT] must be specified.

Words DRILL and CAM1 are translated into G81 [Drilling Cycle] and M52 [Manually set depth cam].

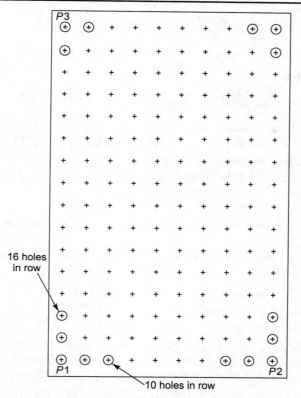

Fig. 7.80: Programing a pattern/grid

Continuous path program

The cutter is guided by three planes [Fig. 7.81]
1. PS [P SURF] at bottom
2. DS [D SURF] at side, i.e. drive surface
3. CS [C SURF] at end, to stop the cutter change its direction

Figure 7.81a shows a tool moving in a straight line, while [**Fig. 7.81b**] shows a tool moving along a curved path. The stop [CS] surface can be at the forward end [GO TO], at the cutter center [GOON], or at the rear end of the cutter [GOPAST], as shown in [**Fig. 7.81c**].

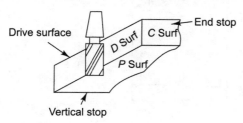

Fig. 7.81(a): Straight drive surface

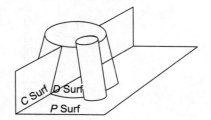

Fig. 7.81(b): Curved inclined drive surface

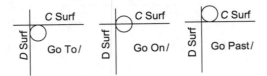

Fig. 7.81(c): Stopping 'To', 'On', and 'Past' C surf

The cutter position, with respect to the drive surface can also be specified, as on the left of the drive surface [TLLFT], to the right of the drive surface [TLRGT]. The center of the cutter may move along the drive surface [TLON].

The **cutter axis** is specified by the direction vector components I, J, K. The motion statement will be:

TLAXIS/I, J, K

For making the tool axis normal to the workpiece surface, the motion statement will be:

TLAXIS/NORMPS

After pre-positioning the cutter [TO/ON/PAST], its motion direction is specified relative to the prior motion.

GOLFT	Go left	GORGT	Go right
GOFWD	Go forward	GOBACK	Go backward
GOUP	Go upward	GODOWN	Go downward

GOFWD and GOBACK commands are used at the point of tangency. Deviations up to 4° from the original direction can be obtained by the GOFWD command.

Command GOUP for withdrawal of the tool, and GODOWN for plunging the tool towards the workpiece, should generally be avoided. Oblique '/' sign should be followed by drive, part, and check [stop] surfaces of the operation, in the required sequence.

Direction of turn is relative with the previous motion, facing in the direction of the previous motion. For milling a slot in **[Fig. 7.82]**, path 1, 2, 3, 4, the statement will be as below:

FROM/SP
GO/TO, A SURF, TO, P SURF, TO, H SURF
GOFWD/A SURF, PAST, B SURF
GOLFT/B SURF, TO E SURF
GORGT/E SURF, TO, F SURF
GORGT/F SURF, PAST, G SURF

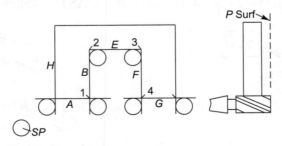

Fig. 7.82: Example describing the directions of turns

Machine Tool Operation

A block of motion statements can be stored for later use by prefixing the block with command MACRO, and ending the block with the TERAC command. The macro statement should be assigned a symbol for easy search later, e.g.,

SITA = MACRO

For later retrieval of the stored block, we only have to type:
CALL/SITA
Usage of MACRO command reduces the number of statements.

Vectors

For motion in the direction of a vector, commands will be:
FROM / SP
INDIRV / X, Y, Z components of the vector

APT geometry statements

A geometry statement specifies the configuration of the surface to be generated by cutting tool. The prefix [say P1] represents the symbol. The middle part, separated from the prefix by an equal sign, gives a general description [say point]. The suffix separated from the middle by an oblique [/] sign, provides specific information [say co-ordinates] of the geometrical element [say the point].

Specifying a point

A point can be specified in a number of ways: intersection of two lines, intersection of a circle and a line, intersections of two circles, and the center of a circle.

POINT specifying methods
[1] Co-ordinates: P1 = POINT/X, Y, Z
[2] Intersection of two lines L1 and L2: P2 = POINT/INTOF L1, L2
[3] Intersections of a line L1 and a circle C1 [**Fig. 7.83**]:

P3 = POINT / Y SMALL, INTOF, L1, C1

or

P3 = POINT/X SMALL, INTOF, L1, C1
P4 = POINT / Y LARGE, INTOF, L1, C1

or

P4 = POINT/X LARGE, INTOF, L1, C1

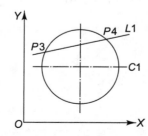

Fig. 7.83: Points obtained by the intersection of a line and circle

[4] Intersections of two circles C1 and C2 [**Fig. 7.84**]:

    P5 = POINT / X SMALL, INTOF, C1, C2

or

    P5 = POINT / Y LARGE, INTOF, C1, C2
    P6 = POINT / X LARGE, INTOF, C1, C2

or   P6 = POINT/Y SMALL, INTOF, C1, C2

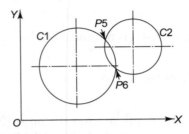

Fig. 7.84: Points obtained by the intersection of two circles

[5] Center of a circle C1:

    P7 = POINT/ CENTER, C1

Specifying line

LINE can be specified in 6 ways, as shown below:

[1] Joining two points P1 and P2: L1 =LINE/P1, P2
[2] Passing through point P1, and making a 30° angle with X axis:

    L2 = LINE /P1, ATANGL, 30

[3] Passing through P1 and making a 60° angle with line L2:

    L3 = LINE /P1, ATANGL, 60, L2

[4] Passing through P1 and parallel to line L10:

    L4 = LINE/P1, PARLEL, L10

[5] Lines parallel to line L10 [**Fig. 7.85**]:

    L5 = LINE / PARLEL, L10, Y LARGE, 12
    L6 = LINE /PARLEL, L10, X SMALL, 16

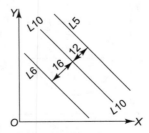

Fig. 7.85: Lines parallel to a given line

[6] Lines passing through P1 and tangential to circle C1 [**Fig. 7.86**]:

   L7 = LINE /P1, LEFT, TANTO, C1
   L8 = LINE /P1, RIGHT, TANTO, C1

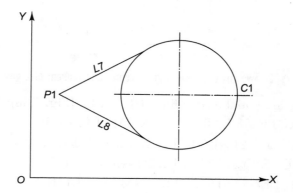

Fig. 7.86: Two lines tangent to a given circle and passing through a given point

Specifying planes

Planes can be specified in 4 ways as stated below:
(1) Plane through 3 points P1, P2, P3: PL1 = PLANE / P1, P2, P3
(2) Planes through P1, parallel to plane PL7 at distances 20 and 8 above PL7:

   PL2 = PLANE / PARLEL, PL7, Z LARGE, 20
   PL3 = PLANE / PARLEL, PL7, Z SMALL, 8

(3) Plane parallel to XY-plane 16 above XY-plane [Z+]:

   PL4 = PLANE/0, 0, 1, 16

(4) Plane parallel to XY-plane 25 below XY plane [Z–]:

   PL5 = plane/0, 0, 1, –25

Specifying circles

Circles can be specified in 4 ways:
(1) Center coordinates X, Y, Z and radius R:

   C1 = CIRCLE/X, Y, Z, R

[2] Center P1 and radius R:

   C2 = CIRCLE/CENTER, P1, RADIUS, R,

[3] Center P1 and tangential to other circle C7 [**Fig. 7.87**]:

   C3 = CIRCLE/CENTER, P1, SMALL, TANTO, C1
   C4 = CIRCLE/CENTER, P1, LARGE, TANTO, C1

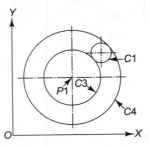

Fig. 7.87: Two circles having a given center and tangent of a given circle

[4] Circle of radius R tangential to two lines L1 and L2 [**Fig. 7.88**]:

    C5 = CI RCLE / YLARGE, L2 X SMALL, L1, R
    C6 = CIRCLE/ YLARGE, L2, XLARGE, L1, R
    C7 = CIRCLE/ YSMALL, L2, XLARGE, L1, R
    C8 = CIRCLE/ YSMALL, L2, XSMALL, L1, R

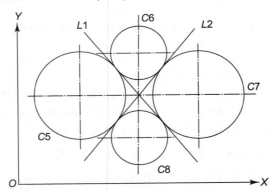

Fig. 7.88: Four circles tangential to two given lines

Post-processor statements

Post-processor listing of the programed machine tool should be referred to, before drafting the statements. Some common commands stated below are widely used in CNC machine tools.

1. COOLANT ON can have further options of mist [COOLANT/MIST], and flooding [COOLANT/FLOOD], in some sophisticated machine tools.
2. COOLANT OFF.
3. Spindle speed can be in revolutions per minute [SPINDL/800RPM], or linear [surface] speed per minute [SPINDL//25SMM]. Speeds in feet per minute will be suffixed SFM, i.e. 80SFM. The computer automatically selects the nearest RPM when the speed can be varied only in steps.
4. SPINDL/ON.
5. SPINDL/OFF.
6. Feed rates in mm/min are not provided with any suffix [say FEDRAT/70]. The rates in mm/rev, however, are suffixed MMPR [say FEDRAT/0.2, MMPR].

7. RAPID command for quick non-cutting approach can be used only once, before commencing a slower feed cutting operation.
8. In inch system, No. of threads per inch [TPI] is specified without any suffix: PITCH/6 means 6 TPI.
9. THREAD/ Command starts the threading cycle.
10. The post-processor type is specified with the prefix MACHIN and an oblique [/], e.g. BXCIN post-processor in a Cincinnati machine will be MACHIN/BXCIN.
11. Command TURRET/ specifies the particular turret position, say TURRET/1.
12. SELECTTL/ is used for selecting the next tool. It also specifies the tool length.
13. LOADTL loads the tool of the given Tool No. [TL].
14. CYCLE/ can be drilling, milling, boring, or facing. Preparatory functions corresponding to codes G80-82, G84-85 will be represented by the following APT commands. CYCLE/OFF [instead of G80], CYCLE DRILL [G81], CYCLE FACE [G82], CYCLE TAP [G84], and CYCLE/BORE [G85].
15. END represents completion. It stops the machine and closes the Machine Control Unit [MCU].
16. At the end of the program, the suffix **FINI** initiates tape rewinding.
17. Command SEQ NO/ is used for allotting the sequence numbers. It also specifies the increment in numbers and the starting point.
    SEQ NO/100, INCR, 5 instructs that the sequence should commence from No. 100 and increment by 5, i.e. Nos. will be 100, 105, 110, 115...
18. SEQ NO/OFF terminates sequence numbering.
19. DELAY/T provides delay of 'T' seconds.

APT auxiliary statements

Some of the commonly used statements are listed below.
1. PARTNO specifies part name in words without any commas,
   e.g. PARTNO DRIVE SHAFT.
2. CLPRNT command prints out the coordinates of the points specifying the tool/cutter path.
3. Cutter diameter is specified as CUTTER/Φ, e.g. CUTTER/125. From the cutter Φ, the computer calculates the tool center offsets, and works out the cutter center path for the tape instructions.
4. MCHTOL/X specifies X [mm/inch] limit of the tool overshoot before taking a turn.
5. INTOL/ specifies the inside tolerance limit of the approximating lines for a curve, e.g. INTOL/0.01.
6. OUTTOL/ gives the outside tolerance limit for the approximating lines for a curve, e.g. OUTTOL/0.015.
7. UNITS/MM [or INCH] records the choice of the units.
8. $ indicates end of a line for the statement to be continued.
9. '$ $' separates the APT statement to be punched on the tape, from a remark only to be printed out,
   e.g. CUTTER/250 $$ 250 MM FACE MILL. The part following $$ is not included in the program, but it is printed in the program listing.

10. **REMARK/** is not a part of the program. It elaborates a statement. REMARK begins in the first column. There are no spaces between the words,
   e.g. REMARK/T24 tells the operator to use Tool No. 24.

APT computation statement

The following table gives APT operators for often used arithmetic and trigonometric functions.

Table 7.8: APT operators for important arithmetical and geometrical functions

Arithmetic or Geometric Function	APT Operator	Arithmetic or Geometric Function	APT Operators
+, X + Y	+, X + Y	sin X	SINF[X]
−, X − Y	−, X − Y	cos X	COSF[X]
x, X × Y	*, X * Y	ln X	ALOGF[X]
÷, X ÷ Y	/, X / Y	arc tan X	ATANF[X]
X^2	X ** 2	tan X	TANF[X]
$\sqrt{X}$	SQRTF [X]	$\log_{10} X$	ALOGIO[X]
e^x	EXPF [X]		

Drafting APT Program

All important points, lines, planes, circles, etc. must be labeled first.

   Points: P1, P2, P3...

   Lines : L1, L2, L3...

   Planes: PL1, PL2, PL3...

   Circles: C1, C2, C3...

The part [to be programed for] and the post-processor should be identified. Then the auxiliary statements for co-ordinates, listing [CLPRINT] tolerances [INTOL / OUTTOL], Cutter Φ [CUTTER / Φ], unit system [UNITS / MM]. Write post-processor feed rate [FEDRAT/] and spindle speed [SPINDL/] statements. The next stage involves geometrical statements, specifying all points, lines, circles, etc. Then come the motion statements.

Usually, for convenience, auxiliary and post-processors statements are interspaced with the motion statements. Naturally, lines, points, planes, circles, etc. must be defined before stating the motions. The sequence can be changed suitably for easy comprehension.

Let us draft a program for drilling two holes in the workpiece, shown in **[Fig. 7.89]**, with PD.l 1 post-processor. **Figure 7.89** also shows the layout on the machine table with the tool.

The Z co-ordinates of the set point should reckon the tool [drill length]. The clearance below the drill point should be big enough to allow change of the tool, at the set point.

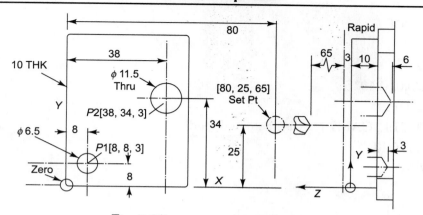

Fig. 7.89: Point-to-point APT program

APT Point-to-Point Program for Part in [Fig. 7.89]

PARTNO   COVER PLATE
MACHIN/P, 11
UNITS/MM
  SET PT = POINT/80,25,65
   P1 = POINT/8,8,3 [3 mm is the approach from the workpiece]
   P2 = POINT/38,34,3
 REMARK/SET, FEDRAT, 0.15MMPR
 REMARK/SET, SPINDL, 800RPM
 REMARK/DRILL, 6.5MM
 SPINDL/ON
 COOLANT/ON
 FROM SETPT
 RAPID
 GOTO/P1
 GODLT A/0,0,–16*
 GODLT A/0,0,16*
[* CYCLE/DRILL, CAM 1 for cam actuated feed]
 SPINDL/OFF
 COOLANT/OFF
 RAPID
 GOTO/SETPT
 REMARK/SET, FEDRAT, 0.25MMPR
 REMARK/SET, SPINDL, 400
 REMARK/DRILL, 11.5MM
 SPINDL/ON
 COOLANT/ON

```
    FROM SETPT
    RAPID
    GOTO/P2
    GODLTA/0,0, – 19*
    GODLTA/0,0,19*
[* CYCLE/DRILL, CAM2 for cam actuated feed]
    COOLANT/OFF
    SPINDL/OFF
    RAPID
    GOTO/SETPT
    REWIND
    FINI
```

If the drill length is controlled by a cam [CAM 1] CYCLE/DRILL, CAM, 1 command will replace the GODLTA commands, along with the coordinates [0, 0, –16], [0, 0,16], for Φ 6.5 hole at point P1. The cam setting will take care of the drilling stroke as well as the drill withdrawal.

APT continuous path program

Let us draft a program for milling the profile of a cam, shown in **[Fig. 7.90]**, with ABM8 processor, using 16 mm HSS endmill cutter, with 4 teeth Feed 0. 1/Tooth [0.4/Rev]; Spindle speed = 500 RPM.

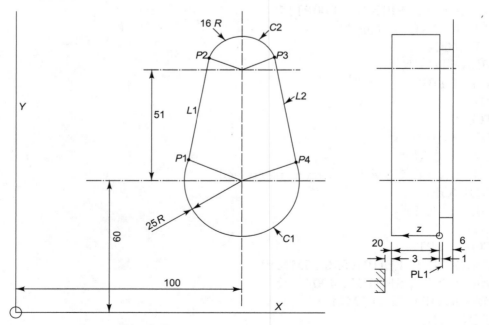

Fig. 7.90: Workpiece for profiling lines and circles

The part program will be as below:
PARTNO CAM
MACHIN/ABM, 8
UNITS/MM
CLPRNT
SETPT = POINT/0,0,23 $$ USE 6 MM SHIM TO LOCATE TOOL
P1 = 75.4, 64.4, 23
P2 = 84.2, 113.8, 23
P3 = 115.75, 113.8, 23
P4 = 124.6, 64.4, 23
L1 = LINE/P1, P2
L2 = LINE/P3, P4
C1 = CIRCLE/100, 60, 25
C2 = CIRCLE/ 100, 111, 16
PL1 = PLANE /0, 0, 1, 0
CUTTER/ 16
FEDRAT/0.4, MMPR
REMARK / SET, SPINDL, 500, RPM
INTOL / 0.012, $$ CUT CHORDS WITHIN 0.012 MM OF THE PERFECT CIRCLE
OUTTOL/ 0.012, $$ TANGENT WITHIN 0.012 OF THE PERFECT CIRCLE
FROM / SETPT
RAPID
GOTO, P1, TO, PL1, $$ START UP STATEMENT
SPINDL/ON
COOLANT/ON
GORGT/L1, TANTO, C2
GOFWD / C2, TANTO, L2
GOFWD/L2, TANTO, C1
GOFWD/C1, TANTO, L1
GOTO P1
SPINDL/OFF
COOLANT/OFF
RAPID
GOTO / SETPT
FINI

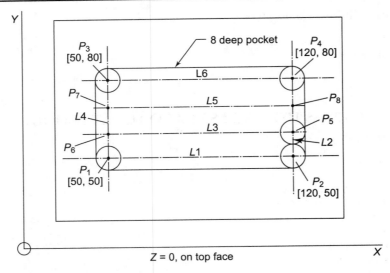

Fig. 7.91: Programing a milling pattern

APT PROGRAM for milling, in a pattern for the pocket in the workpiece, in [Fig. 7.91]:
The cutter should move in the following path.

SETPT-P1, PLUNGE TO DEPTH-P2-P5-P6-P7-P8-REPEAT CYCLE for the next cut.

If the maximum permissible depth is 4 mm, the pocket milling will require two cuts.

PART NO FELTHOLDER [Fig. 7.91]

SETPT = POINT / 0, 0, 6
P1 = POINT / 50, 50, 0
P2 = POINT/120, 50, 0
P3 = POINT / 50, 80, 0
P4 = POINT/ 120, 80, 0
P10 = POINT/50, 50, 6
L1 = LINE/P1, P2
L2 = LINE / P2, P4
L3 = LINE / PARLEL, L1, Y LARGE, 10
L4 = LINE/P1, P3
L5 = LINE / PARLEL, L1, Y LARGE, 20
L6 = LINE / P3, P4
PL1 = PLANE/0, 0, 1, −4 [zero of Z axis on workpiece top face]
PL2 = PLANE / 0, 0, 2, −8
CUTTER/ 20

REMARK / SET, FEDRAT, 0.4, MMPR
REMARK / SET, SPINDL, 800, RPM
FROM SETPT
RAPID
GOTO / P10 $$ RAPID APPROACH TO THE POINT ABOVE P1
SPINDL/ON
COOLANT/ON
GODLTA / 0, 0, −10 $$ PLUNGE CUT TO DEPTH OF 4 MM
GO / ON, L1, TO, PL1, ON L2
PRAKASH = MACRO STATEMENT
GOLFT/ON, L2, ON, L3
GOLFT/ON, L3, ON, L4
GORGT / ON, L4, ON, L5
GORGHT / ON, L5, ON, L2
GORGHT / ON, L2, TO, P5 [TO REMOVE PROJECTION BETWEEN P5, AND P8]
GO[BACK] / ON, L2, ON, L6
GOLFT / ON, L6, ON, L4
GOLFT / ON, L4, ON, L1
TERMAC $$ THE MACRO LOOP ENDS HERE
GODLTA / 0, 0, 10 $$ TOOL IS RETRACTED
LOOPND $$ SIGNIFIES END OF LOOPING
GODLTA / 0, 0, −14 $$ PLUNGE CUT TO DEPTH OF 4 MM
GO / ON, L1, TO, PL2, ON L2
CALL PRAKASH $$ EXECUTE MACRO CYCLE
GODLTA / 0, 0, 14 $$ TOOL IS RETRACTED
LOOPND $$ SIGNIFIES END OF LOOPING
SPINDL/OFF
COOLANT/OFF
RAPID
GOTO / SETPT
FINI

The Macro command with the identifying label PRAKASH, allows us to repeat the line pattern L2-L3-L4-L5-L2-L6-L4 with the increased depth of the cut.

Turning program

Let us program a turned workpiece, shown in **[Fig. 7.92]**.

In lathes there are only two axes: Z axis for the spindle, and X axis for radial position of the tool, with respect to the Z axis.

RPM = 125, Tool nose Rad = 2, feed = 0.1/Rev

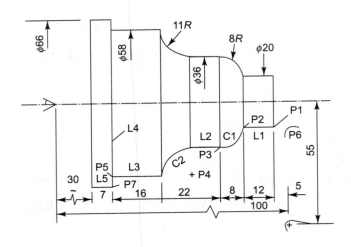

Fig. 7.92: Workpiece for program in Table 7.8

The machine has PLATH, 10 post-processor. Let us take set point at 5 mm from the R.H. end face of the workpiece, i.e.[65 + 5 + 30 = 100] from the L.H. center point. So Z = 100. The X coordinate of set point = 66/2 + 20 = 53 ≈ 55

```
PART NO   SHAFT
MACHIN/PLATH, 10
CLPRNT
SET PT = POINT/100, 55
P1 = POINT/95, 10
P2 = POINT/83, 10
P3 = POINT/75, 18
P4 = POINT/64, 29
P5 = POINT/37, 29
P6 = POINT/100, 12
P7 = POINT/37, 33
L1 = LINE/PI, P2
C1 = CIRCLE/75, 10, 8
```

L2 = LINE, RIGHT, TANTO C1
C2 = CIRCLE/64, 29, 11
L3 = LINE/P5, P4
L4 = LINE/P5, RIGHT PERPTO, L3
L5 = LINE/P7, RIGHT, PERPTO L4
CUTTER/4 [double of tool nose radius]
FEDRAT/0.1, MM PR
SPINDL/125, RPM
INTOL/0.012
OUTTOL/0.012
FROM/SETPT
RAPID
GOTO/P6 $$ RAPID APPROACH TO POINT 5 CLEAR FROM P1
COOLANT/FLOOD
GO/TO, L1, $$ ONE SURFACE START STATEMENT
GOLFT/L1, TO, C1
GOFWD/C1, TO, L2
GOFWD/L2, TO, C2
GOFWD/C2, TO, PAST, L3
GORGT/L3, TO, L4
GOLFT/L4, PAST, L5
RAPID
GOTO/SETPT
SPINDL/OFF
COOLANT/OFF
REWIND
FINI

Distributive numerical control [DNC-1]

Before the advent of compact, cheap microchips, computers were bulky and costly. During those times, a single, big mainframe computer served many machine tools. It stored all the part programs. It gave individual machine tools the required program during manufacture. This system of one big program computer, serving many programed numerical control machine tools, is called **Distributive Numerical Control [DNC1].**

Computerised numerical control [CNC]

Development of compact, cheap microchips reduced the prices of computers substantially. Modern CNC machines are equipped with captive computer, loaded with the control program which can usually be used for lathes as well as milling machines.

The machine computers are provided with large memories. We only have to run the punched part program tape, through the computer once, and the program is stored in its memory. We do not need the tape again to run the part program. The tape is however, retained as a back-up.

Latest CNC machines do not require a punched tape at all. They are provided with Manual Data Input [MDI] consoles with keyboards, which facilitate the writing of sophisticated part program, directly into the memory of the console.

Other developments in the field are as below.
1. Diagnostics, Monitoring, and Back-up systems, which give us alerts about breakdown, deploy back-up, record down time and their reasons.
2. In-process inspection of workpiece, and correction of deviations, due to tool wear, nose radius variation, etc.
3. Program editing facilities, with graphic display for tool position and trajectory.
4. Choice of circular, parabolic, cubic interpolations.
5. Macros for storage and repetition of patterns.
6. Axis inversion and mirror image facilities for saving part programing time for workpieces with symmetrical shapes.

Machining centers

Usually, standard machine tools like lathes and milling machines, call for considerable skillful manipulation. This non-cutting, idle time can be reduced considerably, by resorting to automation. Furthermore, more number of operations cause more dimensional and formal deviations, for in every loading and setting, there might be some variations in dimensions from datums. Locators may also have some permissible variations, necessary for any manufacture.

Machining centers aim to finish all operations in a single [at most two] loading. They combine a number of machine tools in a single center. After loading a workpiece in a fixture, it is transferred or indexed, to position it to individual machining stations.

In a transfer machine I designed for a pressure gauge body, hydraulically-operated slides took the job to milling, drilling, and tapping stations, to finish all machining in a single loading. Even this time can be overlapped on machining time, by using an indexing table to load/unload a workpiece, while machining is going on. Machining centers ensure better precision and consistency, with little variation, if at all, in a batch.

Machining centers reduce machining costs substantially. A single unskilled operator replaces many skilled machinists, for the operator only has to load/unload a workpiece onto the fixture. Even the change of a tool is automated, by using preset tools and a tool changer [**Fig. 7.93**], with a bank on drum or chain-type magazine [**Fig. 7.94**].

Turning centers use 12–50 stations turret, instead of a tool-post. Some centers use two turrets: one for presetting the tools, while the other turret carries the set tools, being used in machining. Unlike the turrets in capstan and turret lathes, machining center tool turrets are used only for storage of tools. Before usage, the indexing tool-changer [**Fig. 7.93**] brings the tool in a cutting position. When the number of tools exceeds 50, a chain type magazine [**Fig. 7.94**] is necessary.

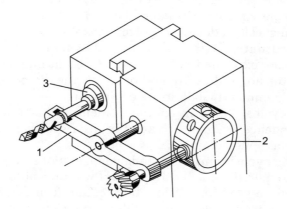

Fig. 7.93: Changer arm mechanism used with tool magazines

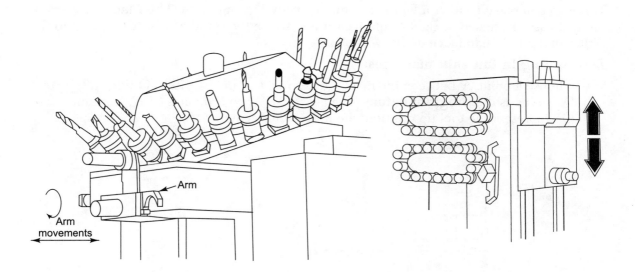

Fig. 7.94(a): Drum-type tool magazine

Fig. 7.94(b): Chain-type tool magazine

Spindle speed [rpm] can be the exact recommended value in an infinitely variable drive. If there is a stepped variation [gear box], a proper range is selected by M40–M45 commands. The RPM is prefixed by the letter 'S'.

Diameter programing

X co-ordinates of the cycle starting point [A] [**Fig. 7.95**] and the tool motions are stated in diameters from the workpiece zero: the center of the R.H. end face of the workpiece. Diameter programing facility automatically reduces the tool motions along the X-axis, to half the difference between the X co-ordinates. Circular interpolation is executed by using X, Z co-ordinates, and 'I' value, explained before. Some CNC turning centers can use X, Z co-ordinates of the end of the arc, radius [R], and of course the direction: clockwise [command G02] or anti-clockwise [command G03], to specify the arc.

Canned [Standard] **cycles**, stored in computer memory [G codes 81–89] can be either fixed with a definite sequence of steps, or **variable** [user specific], such as for the holes' of required size, at a specified pitch diameter.

Fixed canned turning cycles comprise 4 straight line motions: Rapid approach to required depth, Two motions at required feed, and Rapid tool withdrawal to the cycle starting point. Fixed canned cycles can be used for turning, boring, and facing [G86]; and taper turning/facing [command G87/88] with I / K vectors with proper signs.

It should be noted that X and Z coordinates are from point A, the chosen starting point for the cycle. We will now see how point 'A' is chosen. For that we must first set zero.

Setting zero and cycle starting point [A]

Turning centers have either a Full Zero Shift, or Fully Floating Zero. **The Machine Zero** in the first case is situated at right of the headstock, above the spindle axis. It is shifted to the center of the right hand face of the workpiece.

Zero setting in full shift machines:

1. Choose a point 'A' between the machine Zero O' and the part Zero O [**Fig. 7.95**]. Point 'A' serves as the **program datum**. The $[-Z^1_A, -X^1_A]$ co-ordinates of 'A', from m/c Zero O', represent the tool tip position: the nose radius center.

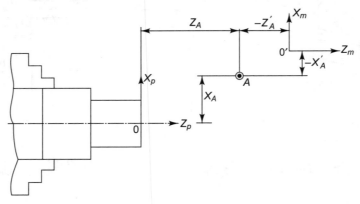

Fig. 7.95: Zero setting on turning center with full zero shift

Machine Tool Operation

2. Coordinates $[Z_A, X_A]$ of 'A', with respect to part zero [RH end center] O are practically found by moving the tool from point 'A' to the workpiece [part] Zero [0].

The m/c Zero is shifted to the workpiece [part] Zero by using G50 command in the following sequence:

Seq. No. N–G50 X 0 Z0

Seq. No. N–G00 $-X^1_A$ $-Z^1_A$ [Note that X^1_A is the actual radial distance]

Seq. No. N–G50 X_A Z_A [X_A is the diameter dimension: twice the radial distance]

These commands place the tool at point A. For tool motions, we will have to use coordinates from the part zero point O.

For the part shown in [Fig. 7.96], the coordinates of point A are as below:

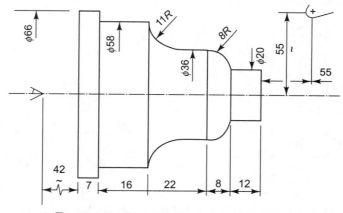

Fig. 7.96: Workpiece for program in Table 7.8

From m/c zero [O'] – X^1_A = 125, $-Z^1_A$ = 125
From part zero [O'] X_A = 110, Z_A = 55 [found practically]

So, for shifting the machine zero to part zero, the commands will be as below:

N–G50 X0 Z0

N–G00 X –125 Z–125

N–G50X 110 Z 55

Zero setting for fully floating zero machines

In the system Reference Tool TOO is used only for setting zero datum. For that, tool TOO is touched to the center of the right-end face of the workpiece. The STOP button is pressed to set Z axis zero at the touched point. Then accurately measure a convenient diameter of the workpiece, and touch the tool on the measured diameter. Press keys for the measured diameter as the X coordinate [diameter programming]. Press the setting [STOP] button.

In fully floating zero machines, point A is placed in the first quadrant. Its coordinates are specified in the main program with the following command [**Fig. 7.96**].

N– G50 X_A Z_A [Note that X_A is the of A diameter coordinate from O]

For the position in [**Fig. 7.96**], the coordinates of point A will be specified as below.

N–G50X 110 Z 55

Offsets for tool nose radius and tool position

The tool nose radius is entered into the computer register. The register also records the tool number, offset register number, X and Z values of offsets with appropriate sign, and TNV number, indicating the possible directions of the cutting motions [**Fig. 7.97**].

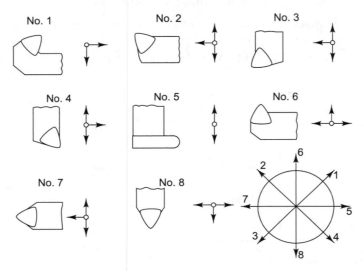

Fig. 7.97: Standard tool nose vector (TNV) numbers of tools used on turning centers, the suggested directions of cutting motions

The X and Z offset values of the tools are obtained by touching the workpiece with the concerned tool. Touching the tool on the R.H. end face of the workpiece, gives on the machine display, the Z offset of the tool. For X offset, the tool is touched on an accurately measured diameter. The difference between the displayed X coordinate, and the measured workpiece diameter, gives the X offset for the tool. For example, if we touch on 58 of the workpiece in [**Fig. 7.96**], and the display shows X = 62, the X offset will be = 62 – 58 = 4. The Z offset, the Z display reading on touching R.H. face, is 5.

The following table shows the Tool Length Offset Register for 3 tools.

Tool length offset register

Tool No	Offset RegNo	X offset	Z offset	TNRNoseRad	TNV No
01	01	−3	7	1.5	3
02	02	4	−6	2	3
03	03	4	5	3	3

After registering the tool nose radii and offsets in the offset register of the machine computer, it will automatically compensate for the tool nose radius, and take care of the offset. The program can be drafted, assuming zero nose radius on the tool.

Left and right hand coordinates systems

The CNC turning centers usually use the **Right-Hand [G42] Coordinates System**. In this, the + X axis points upwards, away from the operator. External turning and facing outwards from the center can be executed by the command G42 [cutter ϕ/ tool nose radius compensation, right]. For boring and facing from outside, towards the center, will call for the command G41: cutter ϕ/ TNR compensation; left.

Some CNC lathes use the Left-Hand [G41] Coordinates System. In this, the +X-axis points toward the operator. Consequently, command G41 is used for external turning, and facing outwards from the center. Facing towards the center and boring, call for the G42 command.

Direction of contouring by circular interpolation [clockwise / anti-clockwise] are also opposite, in the right-hand and left-hand coordinates systems. These differences should be reckoned while drafting the part program.

Every part program commences by the choice of units [metric/inch] canceling previous cutter Φ TNR compensation [G40], and canceling cycle [G80].

Turning center program for the workpiece in **[Fig. 7.96]** with full zero shift and R.H. coordinates system.

No	Program	Explanation
N010	G71 G40G49	Metric Units, Cancel TNR compensation, Cancel Tool Length
	G80	Compensation, Cancel Cycle
N020	G50 X 0Z0	Setting Abs. Zero point: synchronizing m/c zero 0' with digital display zero
N030	G00X-125Z-125	Rapid positioning to point A
N040	G50X110 Z 55	Shifting zero from 0' to 0 [workpiece]
N050	T0101 M06	Clamp Tool No T0101 [Ref offset register]
N060	G96 S 25	Set Constant Surface Speed 25 m/min
N070	G90 G00 X 00Z3	Abs. Dimensions, Rapid Positioning for facing with 3 mm face clearance
N080	G01 G42X00Z0	Linear Interpolation, plunge into face at Feed 0.1/ Rev
	F0.1	
N090	X 20 M08	Face outward to Φ [or Dia] 20, Coolant on
N100	Z–12	Turning ϕ 20 x 12
N110	G03 X 36 Z-20	Circular interpolation anticlockwise, Radius form to 36 Φ [or Dia]
	R8	Radius 8
N120	G01 Z-31	Linear Interpolation, turn ϕ [or Dia] 36 till 11R start

Contd.

No	Program	Explanation
N130	G02 X 58 Z-42	Circular Interpolation, clockwise, Form 11R to 58 Φ
	R11	
N140	G01 Z-58	Linear Interpolation, turn ϕ 58 X 16
N150	X66	Face Collar to ϕ 66
N160	Z-70	Turn ϕ 66 X 70: (51g extra for parting)
N170	X75	Move clear from workpiece
N180	G00G40X 110	Rapid positioning, Cancel TNR compensation, Move to
	Z55	starting point A
N190	M05 M09	Stop spindle, Stop coolant
N200	M06 T0202	Replace T0101 by T0202 [4 wide Parting Tool]
N210	G00 X 70 Z-69	Rapid positioning [for parting tool]
N220	G01 X-2F0.1	Linear Interpolation, Part with 1 mm over travel beyond
	M08	center at Feed 0.1 / Rev, Coolant on
N230	G00 X 110Z55	Rapid positioning to point A, Coolant off
	M09	
N240	G50 X 125	Shift 0 to 0'
	Z 125	
N250	G28 X 0Z0	Return to reference point, Stop spindle
	M05	
N260	M30	Program End. Machine stops, Program rewinds

Canned cycles usage

If the workpiece does not have curved profiles, we can use canned cycles [G81–89], stored in the computer memory. **Figure 7.98a** shows canned cycles for turning, facing and tapering diameters, and faces. **Figure 7.98b** shows a tapered workpiece, suitable for a canned cycle.

The material to be removed is divided into separate blocks, suitable for the canned cycles. If we set the fully floating zero at 2 mm from the workpiece [**Fig. 7.98b**] R.H. end, the block will be as below.

a. Block P with turning cycle starting from point P 1 [X 47 Z 2]
b. Block Q with taper turning cycle starting from point Q1 [X 47Z-25]
c. Block R by facing canned cycle commencing from point R1: X30 Z2

The starting point of the cycle or the program datum should be clear of the workpiece, to allow easy change of the tools: say X50 Z 30. The CNC program will be as below.

Canned cycles program for fully floating zero and R.H. coordinates, for the workpiece in [**Fig. 7.98b**]

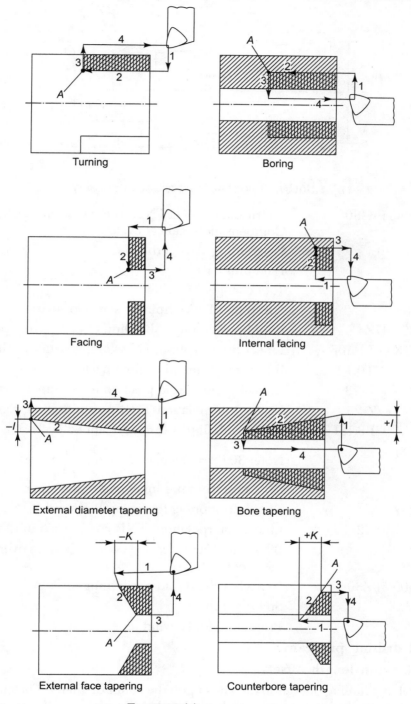

Fig. 7.98(a): Canned cycles

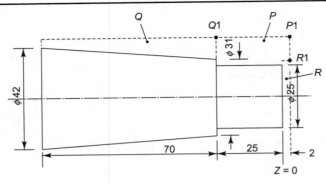

Fig. 7.98(b): Tapered workpiece suitable for canned cycle

N010 G71G40G49G80	Metric data, cancel TNR compensation, cancel Tool Length Compensation, cancel cycle
N020G50X50Z30	Programing Abs zero
N030G95	Feed mm / Rev
N040M06	Tool change, TO 101
N050S600M03	Speed 600 RPM, Spindle start clockwise
N060G00G90X47Z4	Rapid positioning, Abs Dim, to cut starting position
N070G01G42X47Z2M08	Linear Interpolation, TNR compensation, right; Coolant on
N080G86X25Z-27F0.1	Blk P with Canned turning cycle, Feed 0.1/Rev
N090G80G00X47Z-23	Cancel cycle, Rapid positioning for taper turning
N100G01G42X47Z-25	Linear Interpolation TNR compensation, right
N110G87X42Z-97I-8.16F0.1	Blk Q with Taper Turning Canned cycle
$I = \dfrac{(47 - 30.68)}{2}$	Note: 30.68 = calculated dia for $\dfrac{11}{70}$ taper at Z = -25
Feed 0.1/Rev	i.e. 2 mm from the small end of the taper
N120G00X28Z4	Rapid positioning for facing R.H. end
N130G01G42X28Z2	Linear Interpolation, TNR compensation to right
N140G86X-2Z0	Blk R with Canned facing cycle with 1 mm overtravel beyond the center
N150G80G00X50Z30	Cancel cycle, Rapid positioning to Pt 'A'
N160M05	Spindle stop
N170 M30	Program end, Rewind tape

Milling and drilling program

Tool diameter and length offsets

The tool travel in a milling operation depends upon the cutter diameter. The cutter should be clear of the workpiece, before commencing feed, and after completion of the operation. Usually,

Cutter Travel = workpiece length + cutter Φ [or Dia] + 2 clearance [2–5 mm]. So, if the cutter Dia is changed, the necessary tool travel must also be changed.

CNC machines are provided with automatic cutter diameter compensation. The part program is written for a zero diameter cutter. However, the cutter size must be entered into the machine tool offset register [usually D register], already discussed earlier, in the coverage on CNC lathes programing. The following table shows some typical entries.

Tool Diameter Offset Register [D-register]

Tool No.	Register No.	Cutter Diameter
01	01	25
02	02	100

We only have to enter the tool register No. in the part program, and the computer will take care of the cutter offset. Like the lathe tool, the cutter diameter offset can be **right** [G42] or **left** [G41]. For determining the offset direction, the programmer should face in the direction of the cutter travel **[Fig 7.99a]**. If the starting point of the cutting travel is on the left of the workpiece, the cutter offset should be on the left [G41]. If the cutter starting point is on the right of the workpiece, the tool offset will be on the right [G42]. The offsets are canceled by the command G40.

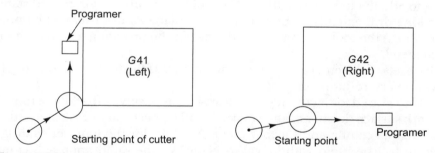

Fig. 7.99(a): Tool diameter compensation left (G41) and right (G42) on machining centers

Tool length offset

It is an addition / subtraction operation, programed by codes G45 / G46. Differences in tool lengths are entered as offset values in the H register of the controlling computer. Various spindles of the machining center can use tools of different lengths. Changes in tool lengths [due to wear / replacement by a new one] can be taken care of, by simply entering the differences in lengths in the H register, and using the tool register number in the part program. There is no need to change the program, if the tool length is changed. We only have to change the tool length offset.

To determine the tool length offset, we must first decide the position of zero along the vertical Z axis. It is usually convenient to keep zero at 3 to 5 mm from the workpiece top surface **[Fig. 7.99b]**. The horizontal plane [A], through the zero, is the surface up to which a cutting tool end can approach in a rapid motion [G00]. It is convenient to keep the longest cutting tool T02, $\phi 12 \times 97$ 1g cutter in **[Fig. 7.99b]**, touching plane A, at the end of the rapid approach position. Then, T02 will have zero length offset.

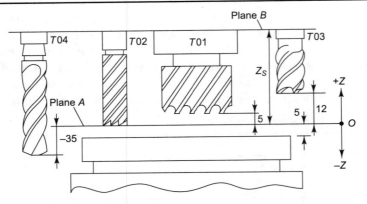

Fig. 7.99(b): Tool length offset on machining centers

The offsets of the other tools will be the distance of their cutting end faces from the plane A. The shell Endmill [T01] will have 5 mm offset, while the spot-facing cutter [T03] will have 12 mm length offset.

Let us suppose that the stock of the standard length ϕ 14 drills in the workshop has been exhausted, and we have to use a long series drill as the only immediate alternative available. We only have to add the long series drill in the tool length offset register, and state the length offset. As the long drill extends 35 beyond the zero plane A, its offset will be **minus 35**. We only have to change the tool register No. and offset in the program. The remaining program remains unchanged.

The length offsets of the end cutting tools will change as the tools wear out and their length decreases, with every resharpening.

Command G45 is used for increasing the offset, i.e. for positive offset. If the tool is projecting beyond the work surface due to negative offset, it will be necessary to take back the tool along the Z axis, so that its cutting end touches plane A. Thus, for the long series drill, T04 rapid approach will in fact be rapid withdrawal to plane A. The rapid approach command G00 G45 Z12 H04 will precede the slow feed command.

H04 is the tool offset H register No. The following table gives the H register entries, for the tool offsets in **[Fig. 7.99b]**.

Tool No	H Register No.	Length Offset mm	Tool Diameter mm
01	01	5	40
02	02	0	12
03	03	12	16
04	04	−35	14

Setting zero of milling/drilling machining centers [G55]

CNC machine with Full Shift Zero or Fully Floating Zero use the following procedures for setting zero.

First, a workpiece with a hole [A] of known size is clamped on the machine table, after making the workpiece edges parallel to the m/c table edges, with a dial gauge.

Setting Full Shift Zero

The m/c spindle is fitted with a roller, close running fit [H7/g6], with hole A. The spindle is moved to align its center with hole A. This is checked by passing the roller into hole A. The X and Y coordinates on the display in this position, are noted as A_x, A_y. The zero is shifted to position A by following commands, manually. That the commands are to be carried out manually, is indicated by underlining the commands.

G55 ENTER X A_x ENTER EOB

G55 ENTER Y A_y ENTER EOB

The zero is however, usually shifted to the L.H bottom corner B of the workpiece [**Fig. 7.100**]. For that, it is necessary to use a roller of a known radius R. Fitting the roller to the spindle, the roller is moved manually to the touch edge V of workpiece and the display reading for the X coordinate, say B_x is noted. Similarly, the roller is touched to workpiece edge H, and the Y coordinate B_y is noted. The zero is then shifted to Point B, by the following commands.

G55 ENTER X $[B_x + R]$ ENTER EOB

G55 ENTER Y $[B_y + R]$ ENTER EOB

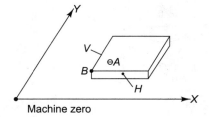

Fig. 7.100: Zero setting in machining center full zero shift

Setting Fully Floating Zero

For shifting fully floating zero to point A, the following commands will be used. G00 XAx YA$_y$ G56

Note the command No. G56 is different than the command G55, used for full shift zero. For shifting the zero to corner point B of the workpiece, the command will be

G00X$[B_x + R]$ Y$[B_y+R]$G56

Co-ordinates of point A are obtained by using roller diameter, good fit in hole A, whereas the co-ordinates of point B are obtained by using a roller of known radius R, and following the procedure for shifting full shift zero, explained above.

The roller diameter can also be reckoned by using the tool [cutter] diameter offset. While shifting zero to point B, if we use a roller of 25 diameter [R = 12.5], the center shift to B can be effected by using the G45 command, and D tool register No. for the roller [treated as a tool]. Then the zero shift commands will be as below.

G00 X B_x G45 D01

G00 Y B_y G45 D01

The full float allows the zero to be shifted to any convenient point.

Special features of milling and drilling CNC machines

Tool radius vector setting facility

When the change in the direction of the cutter travel is angular [**Fig. 7.101a**], damage to sharp corner A can be avoided, by rotating the cutter about point A, by the angle dependent on the angle of direction change. The angle is specified by dimensions J and I in the direction of the Y and X axes. For the angle in [**Fig. 7.101b**], the direction change command will be as below:

G39 I16 J-30

In the latest models of CNC machining centers, the cutter overtravels past point A, to a point tangential to the following angular surface. The overtravel is calculated by the machine control computer. There is no need for the G39 command at all [**Fig 7.101**].

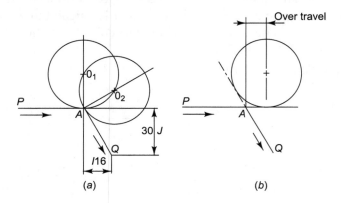

Fig. 7.101: Tool radius vector setting in machining centers of tool

Do Loop [G51]

When the program features repeat themselves, the program repetitions can be eliminated by using the Do Loop facility. The number of repetitions are stated between the commencement command **G51**, and the termination command **G50**. The tool departures are stated in an incremental dimension mode.

Sub-routine Program [M98]

An independent program, sub-routine is placed at the end of the main program. It is identified by address P, and a 3-digit number. Sub-routine is programed as a fresh sequence number. Execution of the main program can be transferred to sub-routine, by the command M98. The statement will be as under:

N ⋯ M98P ⋯ [Sub-routine No.]

The first block of the sub-routine is a labeling statement.

P ⋯ N010 [1st step of the sub-routine]

After the end of the sub-routine command, M99 is used to return to the main program.

Mirror imaging [M21, M22]

Mirror imaging comes handy while programing workpieces with symmetrical features. **Figure 7.102 a** and **b,** show mirror images of cut-outs about X and Y axes, respectively. Configuration No. 6 in **[Fig. 7.102c]**, also can be considered as a succession of 2 images of No. 5, about Y and X axes. Command **M21** gives a mirror image about the **X axis,** while mirroring about the **Y axis** can be effected by command **M22**.

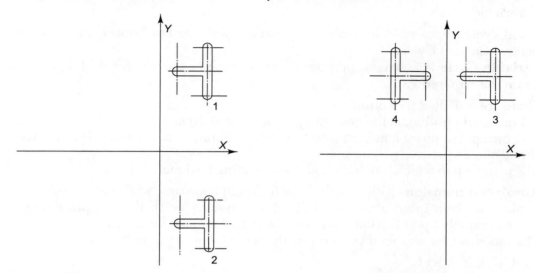

Fig. 7.102(a): Mirror imaging about X axis   **Fig. 7.102(b):** Mirror imaging about Y axis

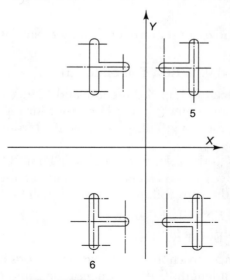

Fig. 7.102(c): Mirror imaging

The program for the copied configuration can be stored as a sub-routine at the end of the main program. Before repeating the mirror imaging command, it is necessary to cancel the preceding mirror imaging command [M21 / M22], by command M23.

Canned cycles

Canned cycles, stored in the memory of the computer, under G code can be:
1. Fixed, or
2. Variable

Fixed cycle, repeated at a number of locations, with the **Do Loop** command, is called Repetitive Canned Cycle.

Variable Cycle is a general purpose program, such as the one for drilling equi-spaced holes on the required P.C.D.

There are usually 3 tool positions:
1. I or initial position, at the beginning of the canned cycle
2. R or rapid approach, at high feed to within 1–3mm, from the workpiece surface to be machined
3. Z plane, up to which tool travels at a slow-cutting feed rate.

Absolute dimensions [G90] are used for the rapid approach, while cutting stroke from R to Z plane is given in an incremental dimension system [G91]. From Z plane, the tool is withdrawn rapidly up to R [G99], but more often to the initial position I [G98].

The statement for a canned cycle is usually in the following format:

N–G--G--- X Y Z RFL

N = Sequence No, G = Cycle code, G = Withdrawal Point Code G98/99, X, Y = Co-ordinates of the point to be machined, Z = Level at operation completion, R = End of Rapid, F = Feed Rate, L = No. of Repetitions

The following example shows a program for drilling 8 Nos. of $\phi 9$ holes, in the part shown in [**Fig. 7.103**].

RPM = 1000, Feed 0.12/Rev, Full Shift Zero

The holes are placed at a pitch of 11, in X direction, and 12 in Y direction. Let us take the tool length offset as 12 [I = 25], and enter it in the H register for register No. H01.

Plane A, for determining tool length offset, will be at 13 from the workpiece, and level R, 3 mm from the workpiece.

First, let us shift the zero to the L.H. bottom corner Point 'C in [**Fig. 7.103**]. If 10 Radius roller is used for touching edges V and H, and co-ordinates on the display, in touched position are $C_x = 40$; $C_y = 30$, and the zero shifting commands will be:

G55 ENTER X [40+10] ENTER EOB and

G55 ENTER Y [30+10] ENTER EOB

The CNC program commences by choosing Metric dimensions [G71], and canceling previous tool diameter [G40] and tool length [G49] compensations, and of course, the previous cycle [G80].

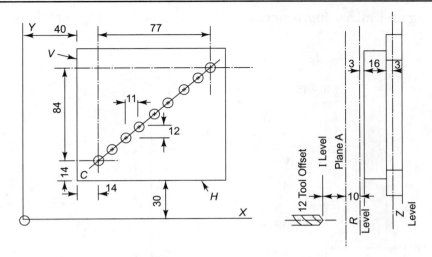

Fig. 7.103: Programing canned drilling cycle

Drilling Program for Plate in [Fig. 7.103]

Program	*Explanation*
N010G71G40G49, G80G91	Metric Dims, Cancel cutter Dia compensation, Cancel Tool Length Compensation, Cancel previous cycle, Incremental dimensions
N020 G92X0Y0Z0	Start synchronization of program zero and display zero
N030S1000M03	Set Speed = 1000 RPM, Start Spindle Clockwise
N040G95F0.12M08	Feed mm/Rev, 0.12/Rev, Start Coolant
N050 G19 G45 G00Z-10 H0 1	YZ plane selection, Tool Offset increase, Rapid positioning: 'A' to 'R'[3 mm from workpiece top] Actual travel = –[10 + 12 offset] = –22
N060 G81 G99 X14Y14 Z-22 R-10 F0.1	Drilling Cycle, Return to R in canned cycle, R = 25 – [12[length offset] +10[Rapid Approach] = 3 from workpiece top Distance between R and A levels = 10; R to Z = 3 + 16 + 3 = 22; Drill No. 1 hole
N070 G91 X11 Y12L8	Incremental Dim., Repeat cycle to drill 8 more holes at X = 11 and Y = 12 increments.
N080G80M09	Cancel cycle, Coolant off
N090 G28 X0Y0Z0M05	Return to reference point, Stop spindle
N100 M30	Program End, Rewind Memory

Contouring and mirroring program for the workpiece in [**Fig. 7.104**] for m/c with full shift zero.

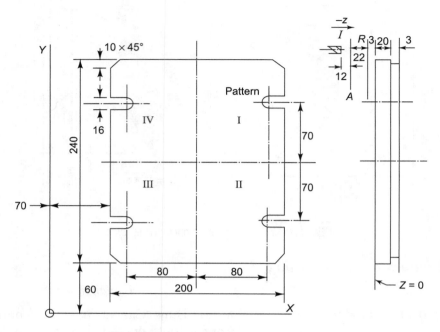

Fig. 7.104: Workpiece for contour milling

The first step will be to select the contouring cutter, say $\phi\,12$ endmill. We must enter the diameter and length offsets of the cutter, in the D and H registers of the m/c computer.

	H Register			D Register	
Tool No.	Reg No.	Length offset mm	Tool Dia mm	Register No	Offset
T01	H01	12	12	D01	6

Approach [clearance plane R] and overtravel [Z level], will both be taken 3 mm above and below the workpiece top and bottom surfaces.

Shifting zero to the center of the workpiece will allow mirroring the program in the first quadrant, to the other 3 quadrants. If we use $\phi\,20$ roller [R = 10] for touching the left and the bottom [in **Fig. 7.104**] edges of the workpiece, and get display readings as X = 60 and Y = 50, respectively, we will have to add half the part length [200/2] and width {240/2} along with the touching roller radius [R = 20/2], to the displayed co-ordinates, to shift the center to the part center with the command G55.

    G55 ENTER X [60+ 10 + 100 = 170] ENTER EOB
    G55 ENTER Y [50 + 10 +120 = 180] ENTER EOB

Machine Tool Operation

We will machine all the edges first. Milling of 10 × 10 chamfer and 16 wide slot in the first quadrant, will be done next. Mirror imaging the program for the chamfer and the slot, to other three quadrants will finish the work.

Seq. No. Program	Explanation
N010 G71G40G49G80	Metric dim, Cancel compensations for cutter Dia and length, Cancel previous cycles
N020 G92X0Y0Z0	Synchronization of display zero with the program zero
N030 G94 S650 M03	Feed mm/min, RPM 650, Spindle start clockwise
N040 G90 G17G00 G41 X-100 Y-120 J240 D01	Abs Dim, XY plane selection, Rapid positioning, Cutter Dia compensation, left; Next movement in Y direction i.e. J = 240, cutter Dia [compensation] in register D01
N050G19G00Z-22	YZ plane selection, Rapid positioning to R level
N060 G91 G19 G01 Z-24F70	Incremental Dims, YZ plane selection, Linear Interpolation, Feed to level Z at 70/min [1 over travel]
N070 G17 G01 Y240 M08	XY plane selection, Linear Interpolation, Coolant on, Machine L.H. edge
N080 G39I 200	Setting Tool Radius Vector before direction change
N090 G01 X200	Linear Interpolation, Machine 200 long top edge
N100 G39 J-240	Set Tool Rad. Vector before direction change
N110 G01 Y-240	Linear Interpolation, Machine R.H. edge
N120 G39 I-200	Set Tool Rad. Vector for direction change
N130 GO1 X-200	Machine 200 1g bottom edge
N140 G19 G00Z46	Select YZ plane, Rapid positioning, Cutter withdrawal to I level
N150 G79 G99 Z-24 M08	Milling Cycle, Return Rapidly to R level 22 from A level, Move at slow feed rate to level Z, Coolant on [1mm over travel below workpiece bottom]
N160 G90	Abs Dim
N170 P280 M98	Call subroutine Milling Pattern I
N180 M21	Mirror Image about X-axis

N190 P280 M98	Mill Pattern II
N200 M22	Mirror Image of Pattern II about Y axis
N210 P280 M98	Mill Pattern III
N220 M23	Cancel Mirror Image
N230 M21	Mirror Image Pattern III about X axis
N240 P280 M98	Mill Pattern IV
N250 G80 G49 M09	Cancel Cycle, Cancel Tool length compensation, Stop coolant
N260 G28 X0 Y0Z0 M05	Return to Ref Pt., Stop spindle
N270 M30	End of Program, Rewind memory [tape]
N280 N010 G41 G90 G00 X100Y110	Subroutine starts, Cutter Dia Compensation, Left; Abs. Dim, Rapid positioning
N020 G39 G91I-10 J 10	Tool Radius Vector for direction change, Incremental dims, I [X] J[Y] co-ordinates for new direction
N030 G01 X-10Y 10	Linear Interpolation, Mill chamfer
N040 G00 Z24	Rapid Vertical Positioning for 16 wide slot
N050G90G41G00X100Y78	Ab Dim., Cutter Dia Compensation Right, Rapid positioning
N060 G01 Z-24	Linear Interpolation, vertical cutter positioning at feed rate
N070 G39G01G79	Incremental dims, Linear Interpolation, Milling cycle
N080 X-20	Mill slot top straight edge
N090 G03 Y-16X-8	Circular Interpolation anti-clockwise, Mill 8 Rad
N100G01 X20	Mill slot bottom edge
N110G00Z24	Withdraw cutter rapidly to R level
N120M99	Return to the main program

CHAPTER 8

TOOL ENGINEERING

Machine tools are machines used for maneuvering tools to suit the operation at hand. Machine tools are, therefore, incomplete without tools.

The dictionary defines a **tool** as a 'mechanical implement for working on something'. As a verb, tooling implies dressing [forming-chipping] stone, or using an object or a person for some purpose. In engineering or metal working industry, the word tool is broadly used for implements/devices used for cutting or shaping metal. Press dies used for bending, forming, and drawing [cupping] are also called tools, with the appropriate prefix: for example, bending tool, drawing tool, etc.

Tooling however, has a much wider meaning. It includes gauges, jigs and fixtures, even die-casting and plastic moulds. As a verb, tooling now means providing all the auxiliary equipments [except machines], necessary to manufacture a machine/consumable part. The following pages will look at the design, manufacture, and usage of tooling.

8.1 WHAT TOOLING COMPRISES

- **Cutting, or shaping tools:** Turning/shaping/planing tools, drills/reamers/taps, milling cutters, broaches, press/forging dies.
- Work piece holders and manipulators: Jigs and Fixtures, conveyors.
- Workpiece measuring devices: Gauges, inspection fixtures.
- Cutting tools

Tool materials

Tool material choice depends upon the material to be cut. Tools for wood work can be made from carbon steel. But for machining harder, tougher steels, heat-resistant [red-hard], High Speed Steel [HSS] tools are necessary. Carbide tools can be used at 4 times higher cutting speeds than HSS tools. Very hard workpiece materials call for much harder Ceramic or Diamond tools.

We will first study the tools materials. Workpiece materials will be discussed later.

8.2 REQUISITES OF TOOL MATERIALS

1. **Hardness:** Should exceed that of the material being cut, to ensure good wear resistance.
2. **Toughness:** To withstand impact and vibrations, inherent in metal-cutting.
3. **Heat resistance [red hardness]:** Can retain shape and hardness at high temperatures that are encountred in cutting.

4. **Workability:** Is particularly important when the tool must be shaped [formed] to suit the workpiece. High speed steels can be ground precisely, while carbides must be moulded [compacted] to the required shape and lapped by diamond wheels, as grinding is un-economical, and machining impossible.
5. **Economy:** Even the most ideal tool material cannot be availed, if it is too expensive compared to the advantages it offers.

8.3 COMMON TOOL MATERIALS

8.3.1 High carbon steels [T60, T70, T85, T118]

These contain 0.85 to 1% carbon and can be hardened up to 67 RC. The cutting speeds for carbon steel tools are about half the speeds used for high-speed steel. So, although cheaper, carbon steel tools are used only for cutting soft materials such as aluminum, magnesium, and wood [Refer to Table 8.1], and hand tools.

T70 (0.7% C) is convenient for cold working tools such as chisels, scissors, and knives. Metal cutting tools such as threading dies and taps, milling cutters, and planing tools, are made from T118 (1.18% C).

8.3.2 Carbon-chromium steels

Carbon-chromium steels T 105 Cr 1 and T 105 Cr 1 Mn 60 (1.05% C 1% Cr 0.6% Mn) have excellent compressive strength and wear resistance, and are used for forming and knurling tools and press tools.

8.3.3 High carbon high chromium

High carbon (1.6–2.15%), **high chromium** (12%) [HCHC] **steels** make excellent material for press tools and cold forming tools, subjected to impact load.

8.3.4 Non-shrinking tool steels

Non-shrinking tool steels T 110 W 2 Cr 1 and T 90 Mn 2 W50 Cr 45 are convenient for making parts which cannot be finished after heat-treatment. These steels are suitable for fine engraving tools, slender punches, and un-ground taps.

8.3.5 High Speed Steels

High Speed Steels [IS: T 75 W 18 Co 6 Cr 4 V1 Mo 75 and T 83 Mo 600 W 6 Cr 4 V 2]. The first is called 18–4–1 Tungsten tool while the later is called 6-6-2 Molybdenum tool.

Tool Engineering

Hardened to 65 RC, HSS tools possess commendable heat resistance (red hardness), which makes them eminently suitable for high-speed metal cutting. Most of the standard metal cutting tools such as drills, reamers, and milling cutters are made from High-Speed Steels [HSS]. Ease in grinding makes it a convenient material for form tools. HSS tools are widely used for turning. However, rigid powerful machines can be run at much higher speeds using carbide tools.

8.3.6 Cast cobalt (stellite) tools

Also called white bits, these contain costly materials like cobalt (50%), chromium (28%), tungsten (20%), and (2%) carbon. Although very hard and convenient for cutting forgings and castings, 'Stellite' can be shaped only by casting, and is very difficult to machine and grind. This limitation of workability has rendered cast cobalt tools un-economical and obsolete.

Table 8.1 gives the composition and important properties of tool steels. These often contain chromium [Cr], Molybdenum [Mo], Vanadium [V], and Tungsten [W], in addition to Carbon [C].

International equivalents of the above Indian Standard Tool Steels are given in Table 8.9.

8.3.7 Carbides

Carbide tools are both, very hard and brittle, and have excellent wear resistance. The addition of Tantalum and Niobium carbides improves their wear resistance, making the tool more suitable for steel and other materials that produce long, continuous chips. Table 8.3 shows the carbide grades for various materials. ISO grades with the prefix **P** have a high wear resistance while prefix **K** indicates higher toughness.

Coating carbide tips with Titanium, triples the tool life. **Figure 8.1** shows the effect of coating carbide tips on its wear. The coat enhances surface hardness, without compromising the toughness of the core. Multi-layer coating further boosts the wear resistance. Some modern coated grades such as **Widialon,** are deposited with as many as 13 layers of harder materials. Table 8.4 lists Widia-coated carbide grades with applications, and the ISO equivalents.

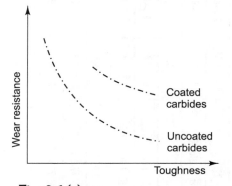

Fig. 8.1(a): Effect of coating carbides

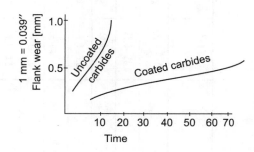

Fig. 8.1(b): Flank wear for coated and uncoated carbide tips

Table 8.1: Composition of tool steels

Designation	%C	%Si	%Mn	%Ni	%Cr	%Mo	%V	%W	Max. Hrd. HRC **
T 118	1.1–1.25	0.1–0.3	0.2–0.35	–	–	–	–	–	–
T70	0.65–0.75	0.1–0.3	0.2–0.35	–	–	–	–	–	35–48
T60	0.55–0.65	0.1–0.35	0.5–0.8	–	–	–	–	–	–
T85	0.8–0.9	0.1–0.35	0.5–0.8	–	–	–	–	–	–
T 105 Cr 1 Mn 60	0.9–1.2	0.1–0.35	0.4–0.8	–	1.0–1.6	–	–	–	62
T 105 Cr 1	0.9–1.2	0.1–0.35	0.2–0.4	–	1.0–1.6	–	–	–	–
T 215 Cr 12	2.0–2.3	0.1–0.35	0.25–0.5	–	11.0–13.0	0.8 Max	0.8 Max	–	54–61
T 160 Cr 12	1.5–1.7	0.1–0.35	0.25–0.5	–	11.0–13.0	0.8 Max	0.8 Max	–	54–61
T 140 W 4 Cr 50	1.3–1.5	0.1–0.35	0.25–0.5	–	0.3–0.7	–	–	3.5–4.2	–
T 110 W 2 Cr 1	1.0–1.2	0.1–0.35	0.9–1.3	–	0.9–1.3	–	0.25 Max	2.25–1.75	58–64
T 90 Mn 2 W 50 Cr 45	0.85–0.95	0.1–0.35	2.25–1.75	–	0.3–0.6	–	–	0.4–0.6	–
T 50 Cr 1 V 23	0.45–0.55	0.1–0.35	0.5–0.8	–	0.9–1.2	–	0.15–0.3	–	45–63
T 55 Ni 2 Cr 65 Mo 30	0.5–0.6	0.1–0.35	0.5–0.8	2.25–1.75	0.5–0.8	0.25–0.35	–	–	–
T 30 Ni 4 Cr 1	0.26–0.34	0.1–0.35	0.4–0.7	3.9–4.3	1.1–1.4	–	–	–	–
T 35 Cr 5 Mo 140 V 1	0.3–0.4	0.8–1.2	0.25–0.5	–	4.75–5.25	1.2–1.6	1.0–1.2	–	38–53H*
T 35 Cr 5 Mo 140 W 1V 30	0.3–0.4	0.8–1.2	0.25–0.5	–	4.75–5.25	1.2–1.6	0.2–0.4	1.2–1.6	38–55H*
T 75 W 18 Co 6 Cr 4 V 1 Mo 75	0.7–0.8	0.1–0.35	0.2–0.4	–	4.0–4.5	0.5–1.0	1.0–1.5	17.5–19.0	62–65
T 83 Mo 600 W 6 Cr 4 V 2	0.75–0.9	0.1–0.35	0.2–0.4	–	3.75–4.5	5.5–6.5	1.75–2.0	5.5–6.5	60–65
T10 Cr 5 Mo 75 V 23	0.15 Max	0.1–0.35	0.25–0.5	–	4.75–5.25	0.5–1.0	0.15–0.3	–	–

Underline indicates percentage in decimals.

* Hot working S+I.

** Working hardness

Table 8.2: Applications of tool steels

Type	Designation	Uses
Carbon tool steels	T 118 T 70 T 60 and T 85	Engraving tools, files, razors, shaping tools, drills, shear blades, chisels, press tools, etc. Die blocks, hand tools, garden and agricultural tools
Carbon-chromium tool steels	T 105 Cr 1 Mn 60 T 105 Cr 1	Cold forming rolls, lathe centers, knurling tools, press tools
High carbon, high chromium tool steels [HCHC]	T 215 Cr 12 T 160 Cr 12	High-quality press tools, drawing and cutting dies, shear blades, thread rollers, cold rolls, etc.
Fast finishing tool steels	T 140 W 4 Cr 50	Finishing tools with light feeds, marking tools, etc.
Non-deforming tool steels [OHNS]	T 110 W 2 Cr 1 T 90 Mn 2 W 50 Cr 45	Engraving tools, press tools, gauges, taps, hand reamers, milling cutters, cold punches, knives, etc.
Shock-resisting tool steels	T 50 Cr 1 V 23	Chisels, rivets shaping tools, shear blades, scarfing tools, trimming dies, heavy duty punches, pneumatic chisels, etc.
Nickel-chromium molybdenum tool steels	T 55 Ni 2 Cr 65 Mo 30 T 30 Ni 4 Cr1	Cold punches, trimming dies, shear blades, scarfing tools, pneumatic chisels
Hot work tool steels	T 35 Cr 5 Mo 140 V 1 T 35 Cr 5 Mo 140 W 1 V 30	Dies for extrusion, stamping dies, casting dies for light alloys, forging dies, etc.
High-speed tool steels [HSS]	T 75 W 18 Co 6 Cr 4 V 1 Mo 75 T 83 Mo 600 W 6 Cr 4 V 2	Drills, reamers, broaches, form cutters, milling cutters, deep hole drills, slitting saws, and other high-speed and heavy-cut tools
Carburizing steels	T 10 Cr 5 Mo 75 V 23	Used for case-hardened molds for plastic material

Underlin indicates percentage in decimals

Re-sharpening of the carbide tips involves lapping on expensive diamond wheels. Consequently, it is much more economical to clamp the carbide tip insert in a steel-holder [**Fig. 8.2**]. The inserts are provided with more than one cutting edge. After wear, the insert can be indexed to bring a new cutting edge in the operating position. These inserts can be thrown away after wear, instead of resorting to expensive lapping on diamond wheels.

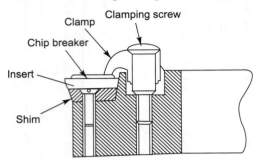

Fig. 8.2: Use of shim in clamping insert

Table 8.3 ISO carbide grades

Grade	Workpiece material
P	Long chipping materials, e.g. steel.
M	Material between P and K, e.g. Stainless steel castings and High-temperature alloys.
K	Short chipping materials, e.g. Cast iron

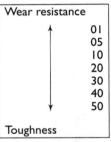

	operation and working conditions
P01	Finishing of steel and steel castings at very high-cutting speeds, and light feeds under stable working conditions.
P10	Finishing and light roughing of steel and steel castings at high-cutting speed, and moderate feed under favouable working conditions.
P20	Light and rough machining of steel and stainless steels at moderate-cutting speeds, and feeds under less favourable working conditions.
P30	Roughing of steel and steel castings at moderate-cutting speeds, and feeds under unfavourable working conditions.
P40	Roughing of both carbon and stainless steels and steel castings at low-cutting speeds under difficult working conditions.
M10	Finishing and light roughing of high-strength, thermal-resistant materials. High resistance to notch-wear. Suitable for machining at comparatively high speed.
M40	Machining of austenitic stainless steel castings. Permits the use of large positive rakes and low cutting speeds under unfavourable conditions.
K01	Finishing of cast iron. The most wear resistant 'K' Grade.
K10	Finishing of both steel and cast iron, when high-edge sharpness is required.
K10	Finishing and roughing of high and low-alloy cast iron. An extremely fine-grained grade, combining toughness with very high wear-resistance.
K10	An alternative grade for cast iron and other short-chipping materials. Suitable as a finishing grade for hot-strength alloys.
K20	Roughing of cast-iron and hot-strength alloys at heavy feeds and low-cutting speeds, under unfavorable working conditions.

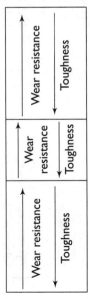

Table 8.4: Widia carbide tip grades

WIDALON/ WIDADUR/ WIDIA grade	Range of ISO application group	Field of application
WIDALON coated grades		
TK 15	P05–P30 and M15–M30 conditions.	Finishing and semi-finishing of moderately hard steel, at higher-cutting speeds, under favorable machining
HK 15	M10–M25 and K5–K30	Finishing and semi-finishing of cast steel, grey malleable and nodular cast iron, under favorable machining conditions.
HK 15 M	M10–M15 and K10–K20	Milling of cast steel, grey and nodular cast iron for light to moderate chip cross-sections, at high-cutting speeds.
WIDADUR coated grades		
TG	P10–P30, M10–M20 and K05–K20	For light and, medium rough turning applications, for steel, as well as cast iron.
TN	P01–P25 and M10–M15	Precision turning of steel at very highcutting speeds, under favorable machining conditions.
TN 25	P10–P20 and M10–M20	Finishing and semi-finishing of steel at highcutting speeds, under favorable as well as moderately unfavorable conditions.
TN 35 M	P 20–P30 and M20–M30	Milling of steel and cast steel under moderate chip load and high-cutting speeds.
TR	P20–P40 M30–M40 and K30–K40	Turning of steel and cast iron with high chip cross-section, or under unfavorable machining conditions.
HC 15	K10–K20 and M10–M20	Turning of cast iron and other short-chipping materials, at high speeds.
WIDIA hard metal grades		
TT03 S	P01–P15	Precision finishing and roughing of steel, cast steel, stainless steel, acid and heat-resistant alloys, under very favorable conditions, with light-chip cross-section.
TTX	P10–P20	Finish turning of steel at high-cutting speeds and feeds. Also useful for grooving, threading, and deep-hole drilling.
TTS	P20–P30	Rough and finish turning of long chipping material with large chip cross-section at medium-cutting speeds. Also useful for grooving and threading, under unfavorable conditions.

Contd.

WIDALON/ WIDADUR/ WIDIA grade	Range of ISO application group	Field of application
TTR	P30–P40 and M30–M40	Rough turning and milling of steel with high chip cross-section and under unfavorable conditions.
AT10	M10	Turning of grey cast iron, alloy cast iron, spheroidal graphite iron, and white hearth iron. Also useful for turning of aluminum with high silicon content and hardened steel.
ATM	M10–M30	Rough turning and milling of grey cast iron, alloyed iron, spheroidal graphite iron, white hearth iron, and stainless steel. Also useful for turning of aluminium with high silicon content and hardened steel.
AT15 S	M15–M30	Rough turning of steel and cast iron with heavy cuts. Also useful for rough machining of cast steel, molybdenum alloy steel, and alloyed grey cast iron.
TTM	P15–P30	General purpose grade for rough and finish milling of all types of steel.
TH 03	K01	Precision turning of hard cast iron, hard steel and other tough materials, under very favorable machining conditions with light-chip loads.
TH 05	K01–K10	Precision turning of hard cast iron, hard steel and other tough materials under favorable machining conditions with light-chip loads.
TH 10	K10	Turning, milling of cast iron, non-ferrous materials, plastics, etc. with moderate chip loads, under favorable and unfavorable machining conditions. Also useful for thread-cutting, thread-whirling of cast iron, and deep-hole drilling of all types of steel.
TH 20	K20	General purpose grade for short-chipping materials, non-ferrous materials, and plastics. Also useful for threading cast iron; reaming, scraping, broaching, and deep-hole drilling in all types of steel.
THM	K10–K20	Grade with extremely high-cutting edge stability for turning and milling short-chipping, non-ferrous and non-metallic materials. Also useful for deep-hole drilling, reaming, scraping, and broaching in all types of steel.

8.3.8 Ceramics

Ceramic tips are cheaper than tungsten carbide tips. Although very brittle, ceramic tools can be operated at twice the carbide speed for machining cast iron and hard iron steels.

8.3.9 Diamond tips

The strength of natural diamond can be increased two to three times, by sintering diamond powder under high pressure, at a high temperature. Diamond inserts can be used for high-speed (800 m/min) machining of work materials, with high abrasion-resistance, such as silicon-aluminum pistons, and engine blocks.

8.3.10 Borazon tips

Also known as synthetic diamond, Borazon tips are used for machining non-ferrous alloys and non-metals such as fibre, glass, and plastics. They are unsuitable for machining ferrous materials.

8.4 COMMON WORKPIECE MATERIALS

Work materials can be broadly classified as metals and non-metals. Metals can be further categorized as ferrous and non-ferrous metals. Ferrous metals contain iron. Metals, which do not contain iron, are called non-ferrous metals.

Engineers also encounter non-metals such as timber, plastics, and elastomers in their field of work.

8.4.1 Steels

Many components of machines are made from a variety of steels. The iron content in steel ranges from 70%–98.5%. Other elements when added to iron, change its properties substantially. The following table states the properties and percentage range of the alloying elements.

Table 8.5: Alloying elements

Sr. No.	Element	Percentage	Effect
1.	Carbon, C	0.1–1.6	*Increases* strength, propensity to harden. *Reduces* toughness, ductility, fluidity.
2.	Manganese, Mn	0.2–2	*Increases* toughness, forgeability, wear and impact-resistance. *Reduces* machinability.
3.	Chromium, Cr	0.2–13	*Increases* strength, hardness, propensity to harden, wear and corrosion-resistance, cutting and magnetic properties. *Reduces* forgeability, machinability.

Contd.

Sr. No.	Element	Percentage	Effect
4.	Nickel, Ni	1.25–4	*Increases* toughness without affecting hardness, removes magnetic properties. *Reduces* linear expansion up to 100°C.
5.	Tungsten, W	0.5–25	*Increases* toughness, hardness, and cutting strength.
6.	Vanadium, V	0.15–4	*Increases* strength and resistance to impact, heat and improves cutting properties.
7.	Molybdenum, Mo	0.2–10	*Increases* resistance to fatigue and corrosion, and imparts hardness-retention.
8.	Silicon, Si	up to 0.45	Increases strength, hardness, propensity to harden, and elasticity. *Reduces* forgeability, and causes black short in forging.

8.4.1.1 Types of steels

Mild steels (En 2A) Mild steels contain less than 0.3% carbon, 0.1–0.8% manganese, and 0.3% silica. The exposed surface of mild steel can be case-hardened up to 1 mm skin depth. Parts which do not need hardening, can be made more machinable by restricting the carbon content to below 0.15%. Called 'Free cutting steel' (EN 1A), it is the cheapest steel available. However, free cutting steel cannot be case-hardened. The Indian Standards Institution specifies carbon steel by the prefix 'C', and carbon percentage by the following numerals (see Table 8.6). ISI steels C 7 and C 10 fall in the free cutting steel category, while ISI steels C 20 and C 25 fall in the mild steel category. Table 8.6 gives the composition and physical properties of carbon steels. The tensile strength of mild steels varies from 32–54 kgf/mm^2. The tensile strength increases with an increase in carbon percentage.

High-tensile steels Medium carbon steels contain 0.35 (C 35) to 0.5% (C 50) carbon. The tensile strength ranges from 55–72 kgf/mm^2. High tensile steels are used for machine components, like fasteners, keys, spindles, axles, and gears subjected to higher stresses. English steels En 8 and En 9 fall under high-tensile steel category. ISI high-tensile steels C 40, C 45, C 50 are usually tempered after hardening. Table 8.8 states the composition of alloy steels. English steel En 24 (IS 40 Ni 2 Cr 1 Mn 28) can be heat-treated to 150 kg/mm^2 strength and 440 HB. Nickel chrome steel En 36 (IS 13 Ni 3 Cr 80) is used for highly stressed parts such as gears, which require tough core and high strength, after hardening. Collet steel En 44 (IS 50 Cr 1 V 23) is used as a spring steel. It is usually hardened and tempered to 47 RC.

Table 8.6: Composition and properties of carbon steels

Designation	%C	%Mn	Tensile strength (kg/mm²)	% Minimum elongation (gauge length $5.65\sqrt{a}$* cylindrical test piece)	Yield Stress (kg/mm²)	Izod impact value min. (if specified) (kg/m)	Brinell hardness (if specified) HB
C 07	0.12 max	0.50 max	32–40 [45.5 – 56.9 × 10³] psi	27	20 [28.4 × 10³] psi	—	—
C 10[†]	0.15 max	0.30–0.60	34–42 [48.3 – 59.7 × 10³]	26	21 [29.9 × 10³]	5.5	—
C 14[†]	0.10–0.18	0.40–0.70	37–45 [52.6 – 64 × 10³]	26	22 [31.3 × 10³]	5.5	137
C 15	0.20 max	0.30–0.60	37–49 [52.6 – 69.7 × 10³]	25	24 [34.1 × 10³]	—	137
C 15 Mn 75	0.10–0.20	0.60–0.90	42–50 [59.7 – 71.1 × 10³]	25	25 [35.5 × 10³]	—	163
C 20	0.15–0.25	0.60–0.90	44–52 [62.6 – 73.9 × 10³]	24	26 [37 × 10³] psi	—	156
C 25	0.20–0.30	0.30–0.60	44–54 [62.6 – 76.8 × 10³]	23	28 [39.8 × 10³]	—	170
C 25 Mn 75	0.20–0.30	0.60–0.90	47–57 [66.8 – 81 × 10³]	22	28 [39.8 × 10³]	—	207
C 30[†]	0.25–0.35	0.60–0.90	50–60 [71.1 – 85.3 × 10³]	21	30 [42.7 × 10³]	5.5	179
C 35	0.30–0.40	0.30–0.60	52–62 [73.9 – 88.2 × 10³]	20	31 [44.1 × 10³]	—	187
C 35 Mn 75[†]	0.30–0.40	0.60–0.90	55–65 [78.2 – 92.4 × 10³]	20	32 [45.5 × 10³]	5.5	223
C 40[+]	0.35–0.45	0.60–0.90	58–68 [82.5 – 96.7 × 10³]	18	33 [46.9 × 10³]	4.1	217
C 45[+]	0.40–0.50	0.60–0.90	63–71 [89.6 – 101 × 10³]	15	36 [51.2 × 10³]	4.1	229
C 50[+]	0.45–0.55	0.60–0.90	66–78 [93.8 – 111.9 × 10³]	13	38 [54 × 10³]	—	241
C 50 Mn 1	0.45–0.55	1.10–1.40	72 min [102.4 × 10³]	11	40 [56.9 × 10³]	—	255
C 55 Mn 75[†]	0.50–0.60	0.60–0.90	72 min [102.4 × 10³]	13	40 [56.9 × 10³]	—	265
C 60	0.55–0.65	0.50–0.80	75 min [106.6 × 10³]	11	42 [59.7 × 10³]	—	255
C 65	0.60–0.70	0.50–0.80	75 min [106.6 × 10³]	10	43 [61.1 × 10³]	—	255

*a, area of cross-section
[†] Steels for case carburizing
[+] Steels for hardening and tempering
Mn 75: average content of Mn is 0.75%.

1 kg = 10 N = 2.2046 Lbs   1 kg/mm² = 1422 Lbs/Inch²
Properties given are for bars and forgings in the oil hardened and tempered condition.   1 kg/m = 0.672 Lbs/Ft.

Table 8.7: Applications of carbon steels

Designation	Typical uses
C 07, C 10	Used for cold forming and deep drawing. Rimming quality, used for automobile bodies, cold-heading wires and rivets. Killed quality, used for forging and heat-treating applications.
C 10 and C 14	Case hardening steels, used for making camshafts, cams, light-duty gears, worms, gudgeon pins, selector forks, spindles, pawls, ratchets, chain wheels, tappets, etc.
C 15	Used for lightly-stressed parts. The material, although easily machinable, is not designed specifically for rapid cutting, but is suitable where cold work, such as bending and riveting may be necessary.
C 15 Mn 75, C20, C 25 and C 25 Mn 75	General purpose steels for low-stressed components.
C 30	Used for cold formed levers, hardened and tempered tie rods, cables, sprockets, hubs and bushes, steel tubes.
C 35	Used for low-stressed parts, automobile tubes and fasteners.
C 35 Mn 75	Used for making low-stressed parts in machine structures, cycle and motor cycle frame tubes, fishplates for rails and fasteners.
C 40	Used for crankshafts, shafts, spindles, automobile axle beams, push rods, connecting rods, bolts, lightly-stressed parts, etc.
C 45	Used for spindles of machine tools, bigger gears, bolts and shafts.
C 50	Used for making keys, shafts, cylinders, and machine components requiring moderate wear-resistance. In a surface-hardened condition, it is also suitable for large-pitch worms and gears.
C 50 Mn 1	Rail steel. Also used for making spike bolts, gear shafts, rocking Levers, and cylinder liners.
C 55 and C 55 Mn 75	Used for making gears, cylinders, cams, keys, crank shafts, Sprockets and machine parts requiring moderate wear-resistance, for which toughness is not of primary importance.
C 60	Used for making spindles for machine tools, hardened screws and nuts, couplings, crank shafts, axles, and pinions.
C 65	High-tensile structural steel for making locomotive carriage and wagon tyres. Typical uses of this steel in the spring industry include, engine valve springs, small washers and thin stamped parts.

Table 8.8 gives the composition and important properties of widely used alloy steels.

Table 8.8: Composition and properties of alloy steels

[Note: Bracketed figures gives strength in Lbs/Inch^2]

Designation	% C	% Si	% Mn	% Ni	% Cr	% Mo	Tensile strength kgf/mm^2	Yield strength kgf/mm^2	% Min. elongation $l = 5.65\sqrt{a}$	Min. Izod impact value kg fm	Brinell hardness number HB
40 Ni 1 Cr 1 Mo 15	0.35–0.45	0.1–0.35	0.4–0.7	1.2–1.6	0.9–1.3	0.1–0.2	80–95 [1,13,760–1,35,100psi]	60 [85.3×10^{3}psi]	16	5.5 [40.56 Lb Ft]	229–277
							90–105 [1,28,000–1, 49, 300 psi]	70 [99.5×10^3]	15	5.5 [40.56]	255–311
							100–115 [1,42,200 –1, 63, 500]	80 [113.8×10^3]	13	4.8 [35.4]	285–341
							110–125 [156.4×10^3–177.7×10^3]	88 [125.1×10^3]	11	4.1 [30.24]	311–363
40 Ni 2 Cr 1 Mo 28	0.35–0.45	0.1–0.35	0.4–0.7	2.25–1.75	0.9–1.3	0.2–0.35	80–95 [113.8×10^3–135.1×10^3]	60 [85.3×10^3]	16	5.5 [40.56]	229–277 (En 24)
							90–105 [128×10^3–149.3×10^3]	70 [99.5×10^3]	15	5.5 [40.56]	255–311
							100–115	80 [113.8×10^3]	13	4.8 [35.4]	285–341
							110–125	88 [125.1×10^3]	11	4.1 [30.24]	311–363
							120–135 [156.4×10^3–177.7×10^3]	100 [142.2×10^3]	10	3.0 [22.1]	341–401
							155 min [170.6–192×10^3]	130 [184.9×10^3]	6	1.1 [8.11]	444 min
31 Ni 3 Cr 65 Mo 55	0.27–0.35	0.1–0.35	0.4–0.7	3.25–2.75	0.5–0.8	0.4–0.7	90–105 [220 × 10^3 psi]	70 [99.5×10^3]	15	5.5 [40.56]	255–311
							100–115 [128–149.3×10^3]	80 [113.8×10^3]	12	4.8 [35.4]	285–341
							110–125 [142.2–163.5×10^3]	80 [113.8×10^3]	11	4.1 [30.24]	311–363
							120–135 [156.4–177.7×10^3]	100 [142.2×10^3]	10	3.5 [25.8]	341–401
							155 min [170.6–192×10^3]	130 [184.9×10^3]	8	1.4 [10.32]	444 min
							[220 × 10^3]				

Designation	% C	% Si	% Mn	% Ni	% Cr	% Mo	Tensile strength kgf/mm²	Yield strength kgf/mm²	% Min. elongation $l = 5.65\sqrt{a}$	Min. Izod impact value Kgfm	Brinell hardness number HB
40 Ni 3 Cr 65 Mo 55	0.35–0.45	0.1–0.35	0.4–0.7	3.25–2.75	0.5–0.8	0.4–0.7	100–115 [142.2–163.5 × 10³]	80 [113.8 × 10³]	12	4.8 [35.4]	285–341
							110–125 [156.4–177.7 × 10³]	88 [125.1 × 10³]	11	4.1 [30.24]	311–363
							120–135 [170.6–192 × 10³]	100 [142.2 × 10³]	10	3.5 [25.8]	341–401
							155 min [220 × 10³ psi]	130 [184.9 × 10³]	8	1.4 [10.32]	444 min
55 Si 2 Mn 90	0.5–0.6	1.5–2	0.8–1.0	—	—	—	160–200 [227.5–284.4 × 10³]	150 [213.3 × 10³]	6	—	440–510
50 Cr 1 V 23 (En 44)	0.45–0.55	0.1–0.35	0.5–0.8	—	0.9–1.2	—	190–240 [270.2–341.3 × 10³ psi]	180 [256 × 10³]	4	—	500–580

Table 8.9: Equivalent steels in various international standards

IS (Indian)	CSN (Czech)	DIN (German)	BS (British)	ASTM (American)	JIS (Japanese)	GOST (Russian)	UNI (Italian)	AFNOR (French)
14MnlS14*	11110	15S20	En 1A	C 1113	SUM2	A12	—	1S F2
40Mn2S12*	11140	45S20	En 8M	C 1140	SUM5	A40G	—	35MF 4
C40	12040	C35K	—	—	S35C–D	—	—	—
C45*	12050	CK45	En 8D	1040	S45C	MCr6	C 40	XC45f
C75	12081	2076	En 42	—	—	—	—	—
17MnlCr95	14220	16MnCr5	—	5115	—	20XGA	16MC5	16MC5
T50CrV23	15260	50CrV4	En 47	6150	SUP10	50XFA	50CV4	50CV4
15Ni2CrMo15*	16220	—	En 325	4317	SNCM22	—	17NCD7	18NCD6
40Ni2CrMo28*	16341	36CrNiMo4	En 24	4340	SNCM8	40HXMA	40NCD7	—
13Ni3Cr80	16420	ECN35	En 36	3315	SNC22	12HX3A	15NC11	14NC12
30Ni4Cr1	16640	VCN45	En 30A	—	—	—	—	35NC15
T215Cr12	19436	210Cr46	—	—	SKD1	X12	UX200C13	Z200C12
T110W2Cr1	19710	105WCr6	—	—	SKS31	XBG	U100WC	105WC15.04
T83Mo600W6Cr4V2	19800	DMo5	—	M2	SKH9	—	UX82WD65	Z85WD0606
T123W14Co5CrAV4	19810	EV4Co	—	—	—	P14F4	—	Z125WV15-03
T75W18Co6Cr4	19885	E18Co5	—	T4	SKH3	P18VK5F2	UX80WK185	Z85WK1805
V1Mo75								

* Preferred steels

Tool Engineering

With the advent of free global trade, the import of special steels has become easy. In fact, many standard British [BS], German [DIN], and American steels are readily available in India. Table 8.9 gives the equivalents of widely used Indian standard steels.

$1 \text{kg/mm}^2 = 1422 \text{ Lbs/Inch}^2$ [psi] $1 \text{ kgf m} = 7.3756 \text{ Lbs.Ft.}$

8.4.1.2 Cast irons

Cast irons contain 2–2.5% carbon. Cast iron dampens vibrations, and also has self-lubrication properties. It is therefore, used extensively for machine beds and frames. Easy casting renders intricate and artistic shaping easy and reduces machining costs substantially. The need for a pattern, however, increases the initial cost as well as the manufacturing time. Cast irons have immense compressive strength, but are weak in tension, due to brittleness. Furthermore, smaller cross-sections of the same grade have a higher tensile strength, than bigger sections.

Table 8.10 gives grades of **grey cast iron.** The letters GCI indicate grey cast iron. The number indicates the tensile strength of 15 thk section in kg/mm^2.

Table 8.10: ISI grey cast iron grades

$1 \text{ kg/mm}^2 = 1422 \text{ Lbs/Inch}^2$ [psi]

Grade	Tensile strength kg/mm^2		Brinell hardness HB	Applications
	Ultimate	Design		
GCI 15	15 [21,330 psi]	1 [1422 psi]	149 to 197	Columns, pedestals, beds, covers, housings arms, sliding tables
GCI 20	20 [28,440 psi]	3 [4270 psi]	179 to 223	Beds, columns, guideways, housings, pullies, gears
GCI 25	25 [35,550 psi]	5 [7110 psi]	197 to 241	Gears, chucks, hydraulic cylinders, valves, work tables, flywheels, pump bodies
GCI 30	30 [42,660 psi]	6 [8530 psi]	207 to 241	Gears, chucks, hydraulic cylinders, valves, work tables, flywheels, pump bodies

Spheroidal cast irons are specified by the prefix 'SG'. The suffix number indicates tensile strength (in N/mm^2) in the numerator, and elongation percentage in the denominator. Thus, SG 370/17 casting will have 370N/mm^2 tensile strength, and 17% elongation. The standard grades are: SG 370/17, SG 400/12, SG 500/7, SG 600/3, and SG 800/2. The last grade has the highest tensile strength of 800N/mm^2, but least elongation of 2%.

Meehanite castings [Table 8.11] offer a wide variety of heat, wear, and corrosion-resistant castings.

Table 8.11: Meehanite castings

$1 \text{ N/mm}^2 \approx 0.1 \text{ kg/mm}^2 = 142.2 \text{ psi}$

Type	Tensile strength N/mm²	Yield strength (N/mm²)	Young's modulus N/mm²	BHN
Heat resisting				
HR (carbidic/pearlitic)	280 [39,820 psi]	–	147000 [20.9 × 10⁶psi]	> 300
HS (nodular graph/ferritic)	420–700 [59,720–99540 psi]	310–520 [44,080–73,940 psi]	160000 [22.75 × 10⁶psi]	200
HSV (nodular graph/pearlitic)	700–840 [99540–1,19450 psi]	–	35 to 56 × 10⁴ [49,77–79.63 × 10⁶ psi]	200
HE (flake graph/pearlitic)	170 [24170 psi]	–	700000 [99.54 × 10⁶ psi]	170
Wear resisting				
W1 (Carbidic/pearlitic)	350–420 [49,770–59724 psi]	–	–	500–600
W2 (Carbidic/martensitic)	350–420	–	–	500–600
W4 (Carbidic/austenitic)	420–560 [59,724–79630 psi]	–	–	400–700
WSH (nodular graph/austenitic)	700 [99,540 psi]	520 [73,940 psi]	168000 [23.89 × 10⁶ psi]	350–500
Corrosion resisting				
CC (Flake graphite/pearlitic)	280 [39820 psi]	–	–	200
CR (Flake graphite/Ni austenite)	170 [24,170 psi]	–	–	130–180
CRS (nodular graphite/Ni austenite)	380 [54,040 psi]	210 [29,860 psi]	–	140–200
General				
GM 60 (flake graphite/sorbo pearlitic)	380	170 [24,170 psi]	150000 [21.33 × 10⁶ psi]	>230
GA 50 (flake graphite/pearlitic)	350 [49,770 psi]	140 [19,910 psi]	140000 [19.91 × 10⁶ psi]	>220
GC 40 (flake graphite/pearlitic)	280 [39,820 psi]	90 [12,800 psi]	115000 [16.35 × 10⁶ psi]	160
GE 30 (flake graphite/pearlitic)	210 [29,860 psi]	70 [9,950 psi]	90000 [12.8 × 10⁶ psi]	180
GF 20 (flake graphite/pearlitic ferritic)	140 [19,910 psi]	60 [8,530 psi]	63000 [8.95 × 10⁶ psi]	160
AQ (flake graphite/pearlitic bainitic)	350 [49,770 psi]	–	–	280
SF 60 (nodular graphite/ferritic)	420 [59,720 psi]	310 [44,080 psi]	160000 [22.75 × 10⁶ psi]	160
SP 80 (nodular/pearlitic/ ferritic)	560–700 [79,630–99,540 psi]	500 [71,100 psi]	175000 [24.89 × 10⁶ psi]	200
SH 100 (nodular/pearlitic)	630 [89,590 psi]	450 [63,990 psi]	165000 [23.46 × 10⁶ psi]	240
AQS (nodular/pearlitic bainitic)	560–1260 [59,720–79,170 psi]	500–950 [71,100–1,35,090 psi]	–	225–500

Cast steels have the castability of iron and the strength of steel. High tensile steel castings are specified by the prefix 'CS'. The suffix number indicates the tensile strength (in kg/mm²). Brinnel hardness increases with the tensile strength, but the elongation and reduction, decrease. For example, CS 65 has 65 kg/mm² tensile strength, 17% elongation, and 37% area reduction capacity. But CS 125 has only 5% elongation and 12% reduction, although it has 125 kg/mm² tensile strength.

Carbon steel castings are specified by yield strength and tensile strength. Thus, the 20-40 grade has 20 kg/mm^2 yield strength and 40 kgmm^2 tensile strength. The standard grades are: 20–40, 23–45, 26–52, 27–54, and 30–57. Elongation and reduction decrease as strength increases. For example, the 20–40 grade has 25% elongation and 40% area reduction, while the 30–57 grade has only 15% elongation and 21% area reduction.

Alloy steel castings are available in 7 grades [Table 8.12]. These castings are usually stress-relieved by cooling them down slowly, after heating to 600–650°C.

Table 8.12: Alloy steel castings grades

1 N/mm^2 = 142.2 psi   [1 Nm = 0.73756 Lb Ft]

Properties	Gr 1	Gr 2	Gr 3	Gr 4	Gr 5	Gr 6	Gr 7
Tensile strength (N/mm^2)	550 [78210 psi]	470 [66834 psi]	520 [73944 psi]	490 [69680 psi]	520 [73944 psi]	630 [89590 Psi]	630 [89590 Psi]
Yield strength (0.5% Proof stress) (N/mm^2)	350 [49770 psi]	250 [35550 psi]	310 [44080 psi]	280 [39820 psi]	310 [44080 Psi]	430 [61150 Psi]	430 [61150 Psi]
Elongation (%)	17	17	15	17	17	15	15
Izod impact value (Nm)	–	–	–	34.5 [25.446 Lb.Ft]	–	27.6 [20.357 Lb.Ft]	–
Stress relieving temp. (°C)	600–650 [1110–1200 °F]	600–650 [1110–1200 °F]	690–710 [1270–1310 °F]	600–650 [1110–1200 °F]	660–690 [1130–1270 °F]	600–650 [1110–1200 °F]	600–650 [1110–1200 °F]
Uses	Up to 400°C [750°F]	Turbine castings up to 450°C [840°F]	Turbine castings up to 510°C [950°F]	Valves, pressure vessels up to 510°C [950°F]	Turbine casting up to 580°C [1075°]	Chemical, petroleum industries up to 600°C [1110°F]	Chemical, petroleum industries up to 650°C [1200°F]

Alloy steel castings are available in 5 corrosion-resistant grades, with 49 kg/mm^2 [Gr 4] to 55 kg/mm^2 [Gr 1] tensile strengths, and 15% [Gr 3] to 17% [Gr l] elongation. They have a marked yield point, ranging from 57% [Gr 4] to 53% [Gr 2] of the tensile strength. Heat resistant grades [Gr 6–7] have 63 kg/mm^2 tensile strength and 15% elongation.

8.4.1.3 Non-ferrous metals

Figure 8.3 charts the widely used non-ferrous metals. They do not contain any iron and can be broadly divided as basic metals, and their alloys.

Aluminum

An excellent thermal conductor, and immune to corrosion, aluminum weighs only one-third the weight of iron. It is also easy to cast, machine, and draw. There are 14 standard IS grades of wrought aluminum and its alloys. They are specified by numbers ranging from 19,000 to 74,530. The tensile strength ranges from 6.5 kgf/mm^2 to 49 kgf/mm^2 (65–490 N/mm^2). Cast alloys of aluminum are graded in 16 number (2447 to 4420). Their tensile strength ranges from 11 kgf/mm^2 to 28 kgf/mm^2 (110–280 N/mm^2).

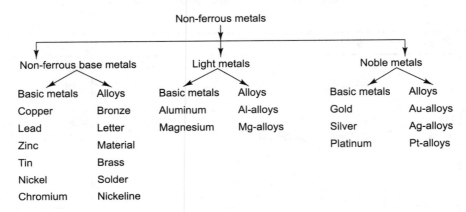

Fig. 8.3: Non-ferrous metals

In non-ferrous metals, the tensile strength is directly related to Brinell hardness (BHN). In aluminum alloys,

Tensile strength = 6t = 0.26 BHN for cast aluminium
= 0.35 BHN for Al–Cu–Mg alloys
= 0.44 BHN for Al–Mg alloys

Copper

Copper is an excellent conductor of heat and electricity. It resists chemical attacks and corrosion and is heavier than steel (Sp. Gr. 8.93). It is soft, tough and easy to mold and work with, in hot as well as cold conditions. Copper is used widely in industries such as electrical cables and food processing.

The addition of tin to copper increases its strength. Copper-tin alloys (called tin bronzes) are immune to corrosion and chemical reactions, and have excellent bearing properties. Tin bronzes (Table 8.13) are used for valves, bearings, worm wheels, and springs exposed to steam. The tensile strength of tin bronzes ranges between 15–55 kgf/mm^2, and Brinell hardness between 60–150 BHN. Alloys of copper with zinc are called brasses. Brasses have excellent bearing properties and corrosion resistance. They have hardness between 70–170 BHN, while their tensile strength ranges between 28–70 kgf/mm^2 (Table 8.13). Brasses and bronzes can be machined at high speeds to a fine surface finish.

Tool Engineering

Bearing alloys

These contain tin, lead, and copper. Lead is both, chemical and corrosion-resistant. Tin is soft, ductile, and air and water corrosion-resistant. The Indian Standards Institution has standardized a wide variety of anti-friction bearing alloys (Table 8.14). Bearing alloys provide a good surface finish and are easy to machine. Table 8.15 gives the applications of bearing alloys.

Table 8.13: Brasses and bronzes

(i) Properties of brasses of different chemical compositions [1 kg f/mm^2 = 1422 psi]

Chemical composition			Mechanical properties		
% Cu	% Pb	% Zn	Tensile strength (kgf/mm^2)	Elongation* (%) length $[5.8 \sqrt{a}]$	Brinell hardness (HB)
54–57	up to 2.5	Rest	>45	10	110
57–59.5	1.0–3.0	Rest	37–68	12 to 25	90–170
59.5–62	up to 0.3	Rest	34–59	23 to 30	80–170
62–65	up to 0.2	Rest	30–55	35 to 45	70–160
66–69	up to 0.1	Rest	29–54	35 to 45	70–160
69.5–73	up to 0.07	Rest	28–53	35 to 44	70–155

(ii) Chemical composition and properties of tin bronzes

Chemical composition					Mechanical properties		
% Cu	% Sn	% Zn	% Pb	% P	Tensile strength (kgf/mm^2)	% Elongation gauge length $(5.8 \sqrt{a}*)$	Brinell hardness (HB)
96	4	–	–	<0.4	32–45	10–25	65–120
94	6	–	–	<0.4	35–50	12–20	70–130
92	8	–	–	<0.4	40–55	20–30	80–150
91	8.5	–	–	0.3	55	15	150
90	10	–	–	–	22	15	60
88	12	–	–	–	24–38	18	80–95
86	14	–	–	–	20	18	85
85	5	7	3	–	15–25	10–12	60–75
86	10	4	–	–	20	10	65

*a, area of cross section, round.
1 kgf = 10 N.

Table 8.14a: Anti-friction bearing alloys

Grade	Alloy	Alloying elements (%)							Impurities (%, max)						Total of indicated impurities per cent
		Sn	Sb	Pb	Cu	Ni	Cd	As	Zn	Fe	As	Al	Bi	Zn	Cd	
90	Sn 90 Sb 7 Cu 3	90 Min	6.5–7.5	03 Max	2.5–3.5	—	—	—	—	0.08	0.10	0.005	0.05	0.008	0.05	0.15
84	Sn 84Sb 10 Cu 5	84 Min	9–11	3 Max	5–6	—	—	—	—	0.08	0.10	0.005	0.05	0.008	0.10	0.17
75	Sn 75 Sb 11 Pb	74–76	10–12	Remainder	2.75–3.25	—	—	—	—	0.08	0.10	0.005	0.05	0.008	0.10	0.17
69	Sn 69 Zn 30	68–70	0.2 Max	0.3 Max	1.0–1.4	—	—	—	29–31	0.10	0.15	0.005	0.08	—	0.10	0.22
60	Sn 60 Sb 11 Pb	59–61	11–12	Remainder	2.5–3.5	—	—	—	—	0.10	0.15	0.005	0.08	0.01	0.10	0.22
20	Sn 20 Sb 15 Pb	19–21	14–16	Remainder	1.25–1.75	—	—	—	—	0.10	0.20	0.005	0.10	0.01	0.10	0.26
10	Sn 10 Sb 14 Pb	9.5–10.5	13–15	Remainder	0.5–1.0	0.8–1.5	0.7–1.5	0.3–0.8	—	0.10	0.20	0.005	0.10	0.01	0.10	0.28
6	Sn 6Sb 15 Pb	5–7	14–16	Remainder	0.8–1.2	—	—	—	—	0.10	—	—	0.10	0.01	—	0.15
5	Sn 5 Sb 15 Pb	4.5–5.5	14–16	Remainder	0.3–0.7	—	—	—	—	0.10	0.25	0.005	0.10	0.01	0.10	0.28
1	Sn 1 Sb 15 Pb	0.75–1.25	15–16	Remainder	0.5 Max	—	—	0.8–1.1	—	0.10	—	0.005	0.10	0.01	0.10	0.28

Table 8.14b: Applications of anti-friction bearing alloys

Grade	Pouring temperature °C	Typical uses
90	340 to 390 [644–734°F]	For lining of petrol and diesel engine bearing, crossheads in steam engine, and other bearings used at high speeds. (As the tin content drops in these alloys, their resistance to shock and load increases.)
84	430 to 460 [806–860°F]	
75	360 to 400 [680–752°F]	Mostly used for repair jobs in mills and marine installations. (Because of its long plastic range, it can be spread as a wipe joint.)
69	500 approx [932°F]	For underwater applications, as a bearing alloy and gland packings.
60	370 to 400 [698–752°F]	For lining of bearings required for medium speeds, such as centrifugal pumps, circular saws, converters, dynamos, and electrical motors.
20 10	370 to 410 [698–770°F]	For low-speed bearings, such as pulp crushers, concrete mixers, and reconveyors.
6	500 to 530 [932°–986°F]	Heavy-duty bearings, rolling mill bearings in sugar, rubber, paper, steel industries, etc. Bearings for diesel engines, cross-heads in steam engines, turbines, etc. Generally for heavy-duty jobs.
5	350 to 390 [662°–734°F]	For mill shaftings, railway carriage and wagon bearings.
1	–	Used as thin-line overlay on steel strips, where the white metal lining material is 0.076 mm thick.

8.5 CHOICE OF MATERIAL

The relevant section on 'Design of various machine tool components', gives recommendations for the materials to be used for the parts. We discuss below, the general considerations for selection of materials for various applications.

Considerations for choice of materials include:
1. Function
2. Load: stress, strain [deflection]
3. Life: wear [necessity for hardening]
4. Economy: cost

1. **Function:** For hardened, replaceable bearing bushes, the material should be softer than the mating shaft. Otherwise, the costlier-to-replace shaft will wear out faster. Cutting tools must be harder than the material they sever, otherwise they themselves will wear out rapidly. Cutting tools subjected to severe wear should be harder than the workpiece material [e.g. carbides], or hardenable by heat treatment [carbon/tool steels]. Parts subjected to corrosion should be made of corrosion-resistant materials [e.g. boiler parts of brass].

2. **Load:** Highly stressed parts should be made of high-strength materials [e.g. fasteners made of medium carbon steels]. Very highly stressed parts call for hardenable steels

[press rams of En24[40Ni2CrlMo28], heat-treated to 160 kg/mm^2 strength]. Parts calling for ease in forming should have very low strength, but high elongation [C08/ C15 usage for deep drawn cups, wrought iron usage for rivets, needing head formation for assembly]. Low stress parts such as handle knobs can be made of plastic/wood or any other cheaper, easily machinable/formable material.

3. **Life:** Wear/breakage calls for replacement, entailing costly m/c idle time. Parts such as anti-friction bearings must be made of materials which can be heat-treated to a high degree of hardness. Parts subjected to fatigue due to high RPM/frequency of stressing and destressing/reverse stressing, should be subjected to very low stresses, to prolong their life. Such parts also call for high-strength materials.

4. **Economy:** Cost of the material, its machining, and heat treatment should be commensurate with its probable life. Parts subjected to low stress, but high wear can be economically made of carburizing steels, which can be case-hardened to 0.8–1.2 deep hard [RC60] skin, e.g., guide pillars of die sets, and drill guide bushes. Wood and plastics can also be used for certain low-stress parts. It should be remembered that fully-hardenable steels are much more difficult to machine than carburizing steels. One can only take 0.5–1 cut while machining high-carbon, high-chromium steels. Machinability is an important consideration for the economic choice of the material.

8.5.1 Machinability of work material

The ease in cutting the material is called machinability. Good machinability implies low cutting-force, less tool wear, and good surface finish. Machinability depends upon the following factors:

- **Chemical composition:** Addition of lead improves the machinability of steels, stainless steels, and copper-based alloys. The presence of free graphite flakes of carbon improves the machinability of ferritic cast iron and addition of zinc and magnesium, enhances the machinability of aluminium alloys.
- **Structure:** Soft, and ductile ferrite work material increases tool life, but affects the surface finish. Pearlite reduces machinability and improves surface finish. The spheroidization of carbon improves the machinability of high carbon (0.7–0.9%) steels.
- **Mechanical properties:** Hard and brittle materials have low machinability, whereas soft and ductile materials are easy to machine. High shear-strength increases the cutting force and reduces machinability.
- **Physical properties:** Thermal properties such as conductivity determine the rate of heat dissipation and tool wear. Thermoelectric tool-wear depends upon electrical properties.
- **Cutting conditions:** Speed, feed, cut depth, cooling, and lubrication have a marked effect on the ease in machining, cutting energy, and tool wear. Cutting conditions are hence, clearly specified and noted while measuring the machinability of various work materials.

Tool Engineering

Measurement of machinability
The following are the criteria for judging machinability:
1. Possible cutting speed (and production rate)
2. Tool life
3. Possible surface finish
4. Cutting force/energy required
5. Temperature rise at cutting point

Most of the above factors are interrelated. Tool life decreases as the cutting speed is increased. Nonetheless, for every workpiece and tool material combination, there is a limit beyond which the speed cannot be increased. Better surface finish calls for higher speed and lesser feed. Cutting force depends upon the area of the cut, i.e. the depth of the cut and feed rate, while energy also depends upon the speed of cutting. The following paragraphs detail the various aspects of metal cutting and their relationships.

8.6 MACHINING

Cutting workpiece materials to the desired size, calls for many considerations: the constraints due to material hardness [speed], cutting tool shape, cutting forces [material resistance] and torques, possible metal removal rates, and resultant power demands.

8.6.1 Cutting speed

Cutting speed depends upon the workpiece hardness [**Fig. 8.4**]. Although an increase in the speed, beyond the recommended level, reduces tool life, it increases productivity. In automatic lathes, tool life is sacrificed to increase productivity. The specialized reference book on metal-cutting, gives formulae for computing the optimum speed for maximizing productivity/profitability, reckoning workpiece loading/unloading time, tool changing time, tool life in number of workpieces/tool, and of course the cutting tool material [HSS/carbide].

The speeds in [**Fig. 8.4**] are applicable to **turning and milling operations**. The roughing cut ranges from 3–8 mm, while the finishing cut ranges from 0.1 to 0.8 mm.

Die steels, used for press tools, must however, be machined with cuts of 0.5–1 mm even during roughing. For overheating and sudden application coolant can lead to hardening before completion of machining. Drilling, reaming, tapping speeds are lesser. Table 8.17 below gives multiples to convert the speed in [**Fig. 8.4**] to the speed convenient for that operation.

Table 8.16: Multiplying factor [M] for converting speeds in [Fig. 8.4] to speeds for other operations

Operation	Turning Milling	Drilling	Reaming	Tapping	Broaching
Multiplying factor	1.0	0.83	0.52	0.42	0.36

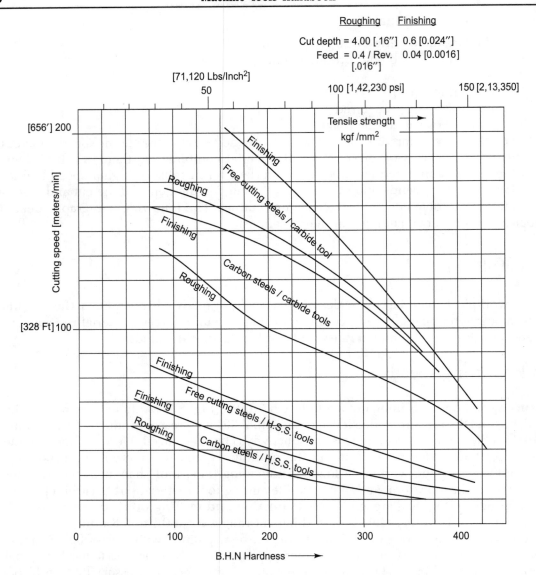

Fig. 8.4: Hardness, tensile strength of workpiece and cutting speed

8.6.2 Feed

Feed depends upon the finish [roughing/finishing]. Roughing is done at 0.4–0.6 mm/Rev feed. Finish turning is done with 0.05–0.2 mm/Rev feed. The **milling feed** is stated in mm/tooth and the linear rate [mm/min] depends upon the type of cutter. Table 8.16 gives the roughing and finishing feed rates per tooth, for various types of milling cutters and workpiece materials.

Tool Engineering

Table 8.17: Feed/tooth for milling cutters

Cutter type	Cast Irons Roughing	Finishing	Steels Roughing	Finishing	Bronzes Roughing	Finishing
Slitting	0.08 [.003"]	0.03 [.001"]	0.05 [.002"]	0.03 [.001"]	0.1 [.004"]	0.05 [.002"]
Slab mills	0.30 [.012]	0.20 [.008"]	0.25 [.01"]	0.15 [.006]	0.55 [.022]	0.40 [.016]
Form relieved cutters	0.15 [.006]	0.10 [.004]	0.10 [.004]	0.05 [.002]	0.25 [.001]	0.15 [.006]
Side and face cutters	0.30 [.012]	0.15 [.006]	0.20 [.008]	0.08 [0.003]	0.45 [.018]	0.25 [.01]
End mills	0.20 [.008]	0.08 [.003]	0.15 [.006]	0.05 [.002]	0.30 [.012]	0.15 [.006]
Face mill up to 45°	0.4 [.016]	0.15 [.006]	0.3 [.012]	0.15 [.006]	0.4 [.016]	0.20 [.008]
Above 45°	0.5 [.020]	0.20 [.008]	0.4 [.0]	0.15 [.006]	0.75 [.03]	0.40 [.016]

Drilling feed ranges from 0.02 to 0.01 of the drill diameter [D].

Drill Φ	Up to 8.5	9–18	18.5–29	> 30
Feed mm/Rev	0.02 D	0.015 D	0.012 D	0.01 D

Reaming feed is double the drilling feed for the same Φ.

Broaching feed [rise/tooth] is also specified per tooth. It depends upon the type of broaching [keyway/round/surface] and workpiece material. Table 8.18 gives the tooth rises for various workpiece materials and operations.

Table 8.18: Rise/tooth 't' (mm) for broaching

Sr. No.	Material	Shape to be broached			
		Keyway	Splines	Round	Surface
1.	Steel	0.05–0.2 [.002–.008]	0.025–0.08 [.001–.003]	0.015–0.03 [.0006–.0012]	0.05–0.08 [.002–.003]
2.	Cast iron	0.06–0.2 [.0024–.008]	0.04–0.10 [.0016–.004]	0.03–0.1 [.0012–.004]	0.06–0.12 [.0024–.0047]
3.	Aluminium	0.05–0.08 [.002–.003]	0.02–0.10 [.0008–.004]	0.02–0.05 [.0008–.002]	0.06–0.2 [.0024–.008]
4.	Brasses and bronzes	0.08–0.20 [.003–.008]	0.05–0.12 [.002–.0047]	0.05–0.12 [.002–.0047]	0.06–0.15 [.0024–.006]

Note: For finishing, Teeth rise/Tooth = t_f = $[0.3–0.5]\,t$

8.6.3 Metal removal rate

Metal removal rate depends upon the cutting speed, feed, and the cutting process: turning, milling, drilling, etc. Table 8.19 gives the formulae for determining the material removal rates for various machining processes.

Table 8.19: Metal removal rate [cm³/min] formulae for various machining processes

[1″ = 25.4 mm] [1 Inch³ = 16.387 cm³] [1 cm³ = 0.061 Inch³]

Process	Formula	Constituents
1. Turning	$df \cdot \dfrac{\Pi D N}{1000}$	d = cut depth [mm]; f = feed mm/Rev D = Workpiece cutting diameter [mm] N = RPM
2. Milling	$\dfrac{b d Z N f}{1000}$	b = Cut width [mm] d = Depth of cut [mm] Z = No. of teeth [from Tool Geometry section] N = Revs./min f = feed/tooth [mm]
3. Drilling	$\dfrac{\Pi D^2 N f}{4}$	D = Drill diameter [mm] N = RPM f = Feed/Rev [mm] from Table 8.17
4. Broaching	$\dfrac{L b d V}{P}$	L = Workpiece length [mm] P = Tooth pitch [mm] b = Broach width [mm] d = Rise/Tooth [mm] from Table 8.18 V = Cutting speed [m/min]

8.6.4 Cutting power, torque, and force

Milling cutters and twist drills are allowed greater wear than turning tools before re-sharpening. Therefore, even if the work/tool materials combination remains the same, milling and drilling will require more power for removing 1 cm³ material per minute, than turning. This is because a blunt tool requires more force/power than a sharp tool.

Table 8.20 gives the Specific Power [P_s] for the removal of one cc of often used materials with different levels of hardness. The values do not take tool wear into consideration, nor the feed rate. Feed factor [C] from Table 8.21 takes into account feed/rev [or tooth].

Table 8.20: Specific power [P_s] for various materials [kilowatt/cc/sec]

[1cc = 0.061 Inch³] [1 Inch³ = 16.39 cm³ [cc]]

BHN	Steels	C.I.	Non-ferrous material
100	1.12		
120	1.12–1.8	0.76	
140	1.15–1.88	0.96	

contd.

[Multiply by 22 for H.P. for 1 Inch³/sec]

BHN	Steels	C.I.	Non-ferrous material
160	1.2–2.02	0.82–1.04	2.24
180	1.31–2.16	0.82–1.42	1.64–2.54
200	1.30–2.27	1.64–1.72	1.97–3.22
240	1.56–2.38	2.48–2.51	2.4–3.0
260	1.69–2.59	2.51	2.54
300	2.18–2.73		3.28
320	2.81		
340	2.62–2.89		
360	2.73–3.11		
400	3.55		

Table 8.21: Feed factor C for various feeds

Feed/rev or tooth	C	Feed/rev or tooth	C
0.025 [.001"]	1.6	0.33 [.013"]	0.98
0.051 [.002"]	1.4	0.356 [.014"]	0.97
0.076 [.003"]	1.3	0.38 [.015"]	0.96
0.10 [.004"]	1.25	0.40 [.016"]	0.94
0.125 [.005"]	1.19	0.45 [.018"]	0.92
0.152 [.006"]	1.15	0.50 [.020"]	0.90
0.178 [.007"]	1.11	0.60 [.024"]	0.87
0.203 [.008"]	1.08	0.70 [.028"]	0.84
0.230 [.009"]	1.06	0.80 [.031"]	0.82
0.254 [.010"]	1.04	1.0 [.0394"]	0.78
0.275 [.011"]	1.02	1.50 [.059"]	0.72
0.305 [.012"]	1.00		

Similarly, efficiencies of various drives vary significantly [60%–90%].

$$\text{Motor power } [P_m] = \frac{P_s CW}{\eta} \times \text{Material removal rate } [\text{cm}^3/\text{sec}] \qquad \textbf{(Eqn. 8.1)}$$

[1 cm³ = 0.061 Inch³]

W = Factor for permissible wear in various machining processes [Table 8.22]

C = Feed factor [Table 8.21]

η = Drive efficiency [percentage/100] from Table 8.23.

Table 8.22: Wear factor for various machining processes

Sr. No.	Process	Wear factor W
1.	Turning—Rough	1.6 – 2.00
	Finish	1.1
2.	Milling—Slab Mill cutter, End Mill	1.1
	Face Mill — Rough	1.3 – 1.6
	Finish	1.1 – 1.25
3.	Drilling	1.3 – 1.5
4.	Broaching	1.2 – 1.3

$$\text{Torque } [T \text{ kgm}] = \frac{975 \, P_m}{N} \qquad [1 \text{ kg m} = .07521 \text{ Lb Ft}] \qquad \textbf{(Eqn. 8.2)}$$

$$P_m = \text{Motor power [kW]; } N = \text{RPM} \qquad [1 \text{ kW} = 1.341 \text{ HP}]$$

$$\text{Turning/Milling tangential cutting force } [F_{kg}] = \frac{6120 \, P_m}{V} \qquad \textbf{(Eqn. 8.3)}$$

$$V = \text{Cutting speed [m/min]} \qquad [1 \text{ m} = 3.281 \text{ Ft}]$$

$$\text{Milling feeding thrust } [F_m] = \frac{3.9 \times 10^6 \, P_m}{DN} \qquad [1 \text{ kg} = 2.2046 \text{ Lbs}] \qquad \textbf{(Eqn. 8.4)}$$

$$\text{Drilling thrust } [F_d] = 1.16 \, P_s \, D \, [100f]^{0.85} \qquad \textbf{(Eqn. 8.5)}$$

$$P_s = \text{Specific power [Table 8.21]; } D = \text{Drill diameter [mm]}$$

$$f = \text{Feed/Rev [mm]} = 0.01 - 0.02 \, D$$

$$\text{Broaching force } [F_b \text{ kg}] = 1.4 \, F_{bu} \, b \, d. \, \frac{L}{p} \qquad \textbf{(Eqn. 8.6)}$$

$$F_{bu} = \text{Broaching force [kgf/mm}^2\text{] from Table 8.24}$$

$$B = \text{broach width [mm]; } d = \text{Rise/tooth [mm]}$$

$$L = \text{Workpiece length [mm]; } P = \text{Tooth pitch [mm]}$$

Table 8.24: Unit broaching force $[F_{bu}]$ kgf/mm² [1 kg f/mm² = 1422 psi]

Workpiece material	Broach type		
	Round	Splines	Keyway
Steel up to 6 t = 75 kg/mm² [1,06,650 psi]	762 [1083560 psi]	230 [3,27,060]	202 [2,87,240 psi]
Cast iron BHN 190	300 [4,26,600 psi]	152 [2,16,140]	115 [1,63,530 psi]

Tool Engineering

8.6.5 Grinding parameters

Grinding differs significantly from metal removal by sharp-edged tools. It is much slower than cutting as grinding cuts are a very small fraction [0.005–0.05 mm] of the cuts in turning, milling, etc. Also, some idle passes are necessary at the end of material removal.

a. Cutting speed of the grinding wheel should range from **25–30 meters/sec**. Flute and thread grinding wheels however, run at higher speeds ranging from 40–60 m/sec. Metal bond diamond or boron wheels too can be operated at 60 m/sec speed. Actual cutting speed depends upon the grinding wheel diameter [D] and the spindle RPM [N].

$$V_g [m/s] = \frac{\Pi D N}{1000} \quad \textbf{(Eqn. 8.7)}$$

b. Feed for table traverse $[f_t]$ depends upon the grinding wheel width. It ranges from $\frac{1}{4}$ to $\frac{1}{2}$ the wheel width in rough grinding. The finishing feed ranges from 1/8–1/6 of the wheel width.

Plunge grinding feed $[f_p]$ ranges from 0.002–0.01/Rev of the workpiece.

c. Workpiece speed $[V_w]$ in cylindrical grinding, and table traverse speed in horizontal spindle surface grinding $[f_1]$, ranges from 15–21 m/min for hardened work. For annealed workpiece or vertical spindle grinder, it can be increased to 30 m/min.

d. Metal removal rate $[Q \text{ cm}^3/\text{min}]$ depends upon the type of the operation. For a vertical spindle surface grinder, having a wheel diameter [D] bigger than the workpiece front to back width $[b_w]$,

$$Q_s [cm^3/min] = \frac{b_w d f_t}{1000} \quad [1 \text{ cm}^3 = 0.061 \text{ Inch}^3] \quad \textbf{(Eqn. 8.8)}$$

d = depth of cut [mm]; b_w = workpiece width [mm]

f_t = Table traverse speed [mm/min]

For **Cylindrical longitudinal grinding** of a workpiece, with a grinding wheel of diameter [D],

$$Q_{ct} [cm^3/min] = \frac{\Pi D d f_t}{1000} \quad \textbf{(Eqn. 8.9)}$$

For **Cylindrical plunge grinding** with wheel width [b] and plunge feed $[f_p]$ mm per rev,

$$Q_{cp} = \frac{\Pi D b f_p}{1000} \quad \textbf{(Eqn. 8.10)}$$

The above formulae are also applicable to centerless grinding with the following changes:

D = Workpiece diameter [mm]; f_t = Thru feed speed [mm]

e. Power required [PkW] depends upon the material removal rate [Q] and specific grinding power $[P_g]$ for removing 1 cm³/min. It depends upon the workpiece material [Table 8.25]. $[1 \text{ cm}^3 = 0.061 \text{ Inch}^3]$

Table 8.25: Specific grinding power [P_g kw] for removal of 1 cm³/min of various materials, at different feeds [Multiply by 22 for H.P. for Inch³/min]

Workpiece material	Specific Grinding Power [P_g kw] at feed of f_t/f_p [mm/traverse or mm/rev]			
workpiece material	0.0125 [.0005"]	0.025 [.001"]	0.05 [.002"]	0.075 [.003"]
M.S./Medium Carbon steel	1.4	0.88	0.7	0.6
Alloy steels	1.3	0.85	0.68	0.58
Tool steel/C.I.	1.5	0.82	0.65	0.56
Stainless steel	1.4	0.84	0.65	0.58
Aluminum and its alloys	0.58	0.45	0.35	0.29
Titanium alloys	0.93	0.79	0.60	0.56

8.6.6 Tool geometry

8.6.6.1 Tool angles

Tool angles [**Fig. 8.5**] depend upon the tool/workpiece material combination and the type of cutting process. First, let us define the angles. Turning tool is appropriate for defining all the angles.

- **Rake angle** [**Fig. 8.5a**] is the angle between the tool cutting-face and the plane, perpendicular to the direction of cutting. In an ISO system [**Fig. 8.5b**], **normal rake** [α_n] is the angle that the side-cutting edge makes with the mounting base, i.e. the horizontal plane. **Side or velocity rake** [α_s/α_v] is the angle of the cutting-face with the horizontal plane, measured in the plane perpendicular to the side-cutting edge.
- **Obliquity angle** [**I**] is the angle between the horizontal plane and the cutting edge, measured in a plane passing through the cutting edge.

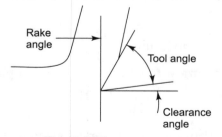

Fig. 8.5(a): Tool angles

An increase in the rake angle by 1°, reduces the required cutting force and power by 1% [up to a certain limit].

- **Approach angle** [θ_m] is the angle that the approach side of the tool makes with the spindle/workpiece axis.

Tool Engineering

- **Side cutting angle** $[90 - \theta_m]$ is the angle made by the side-cutting edge with the plane, square to the axis.
- **End cutting angle** $[\theta_s]$ is the angle between the workpiece/spindle axis and the trailing-cutting edge of the tool.
- **Clearance angle** [C] prevents friction between the machined surface and the tool, thereby reducing the required force and preventing damage to the machined surface.
- **Side clearance angle** $[C_s]$ **[Fig. 8.5b]** is the angle between the vertical plane through the side-cutting edge and the approach side-end face of the tool.

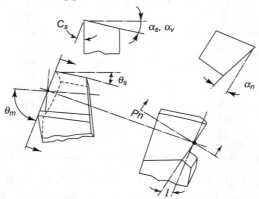

Fig. 8.5(b): I.S.O. normal rake system

- **End clearance angle** $[C_e]$ is the angle between the vertical plane through the end-cutting edge and the trailing-end face of the tool.

The above nomenclature, although designed for turning tools, is also applicable to milling, drilling, planing, and broaching tools.

Tool angles depend upon the workpiece/tool materials combination and the cutting process.

8.6.6.2 Turning tools' angles

Turning tools' angles have been standardized. The following tables give the standard angles for 'ISO 1' tool **[Fig. 8.5c]**.

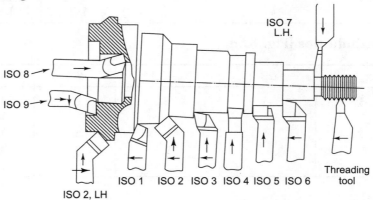

Fig. 8.5(c): I.S.O. standard turning tools

Table 8.26: Cutting angles for tools [ISO 1]

(a) High speed steel tools

Sr. No.	Normal or back rake $\alpha°_n$	Side rake $\alpha°_s$	Side cutting angle $[90 - \theta_m]°$	End cutting angle $\theta°_s$	For workpiece materials
1	10	12	15	15	Free-cutting steel
2	8	10	15	15	M.S.
3	0	10	15	15	Medium carbon steel, tool steels, stainless steel
4	0	5	15	15	titanium alloys
5	5	10	15	15	C.I. 100–200 HB
6	5	10	5	5	Copper alloys
7	5	8	15	15	C.I. 200–300 HB
8	5	3	15	15	C.I. 300–400 HB
9	20	15	5	5	Aluminum alloys, Magnesium alloys

Clearance angles $[C_s/C_p]$ for all ferrous materials and titanium alloys is 5°. However, the clearance angles should be increased to 8° for copper alloys, and 12° for aluminum and magnesium alloys.

(b) Brazed carbide tools [Fig. 8.5c for ISO 1]

Sr. No.	$\alpha°_n$	$\alpha°_s$	$[90-\theta_m]°$	$\theta°_s$	Workpiece material
1	0	6	15	15	All steels up to 400 HB, stainless steels, C.I. upto 300 HB
2	−5	−5	15	15	Tool steels up to RC 50, C.I. 300–400 HB
3	3	15	15	15	Aluminum alloys
4	0	8	15	15	Copper alloys
5	0	6	5	5	Titanium alloys

(c) Clamped carbide tips [Fig. 8.5d]

Sr. No.	$\alpha°_n$	$\alpha°_s$	$[90-\theta_m]°$	$\theta°_s$	Workpiece Material
1	−5	−5	15	15	All steels, tool steels. Stainless steels, cast iron
2	−5	−5	5	5	Titanium alloys
3	0	5	5	5	Aluminium, Copper, and Magnesium Alloys

Nose radius = 1.2 (when not specified)

Various shapes of turning tools shown in **[Figs. 8.5a, b, c,** and **d]**, have been standardized for various applications. One tool invariably needed for workpieces made from round bars, is the **Parting [cut-off]** tool.

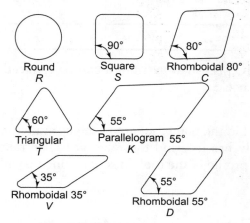

Fig. 8.5(d): Standard tip shapes

Parting tool [Fig. 8.5e]

The length of the parting tool should exceed the radius of the workpiece, to be cut-off. At the same time, the length should be proportional to the height [h], i.e. the tool shank height.

$$\text{Parting tool length } [L_p] = 0.75 - 0.9 \, h \qquad \textbf{(Eqn. 8.11)}$$

$$h = \text{height of the tool}$$

$$\text{Parting tool width } [b_p] = \frac{h}{6} \text{ max} \qquad \textbf{(Eqn. 8.12)}$$

As shown in the figure, the cutting edge is given the end clearance or side-edge angle $[\theta_s]$, to eliminate burr on the end of the face of the following workpiece.

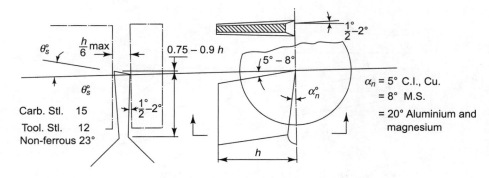

Fig. 8.5(e): Parting tool design

Table 8.26d: Parting tool side [end] edge angle [q_s]

Material	M.S. and Med. carbon steel	Tool steels	Non-ferrous
θ_s	15°	12°	23°

Clearance on the sides [C_s] is usually 2°. But it can be reduced to 1/2°, to avoid thinning and weakening of the tool neck.

Form tools

Usually, the form is ground in the plane, square to the clearance angle 'C' **[Fig. 8.5f]**, so that for re-sharpening, one only has to grind the rake face. The rake angle decreases the tool profile depth. The ground profile of the tool does not coincide with the workpiece profile (as can be seen in the figure).

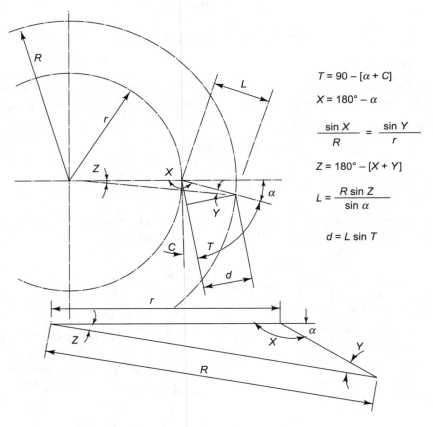

$T = 90 - [\alpha + C]$

$X = 180° - \alpha$

$$\frac{\sin X}{R} = \frac{\sin Y}{r}$$

$Z = 180° - [X + Y]$

$$L = \frac{R \sin Z}{\sin \alpha}$$

$d = L \sin T$

Fig. 8.5(f): Straight form tool with positive top rake angle

Tool Engineering

Automatic lathes use economic, long-life, **dovetailed tools,** with standardized tool holders [**Fig. 8.5g**]. Auto tools use higher rake angles to reduce power consumption and increase productivity; sacrificing tool life to some extent. Even the speeds and feeds in autos are increased by 25–50%, at the cost of the tool life.

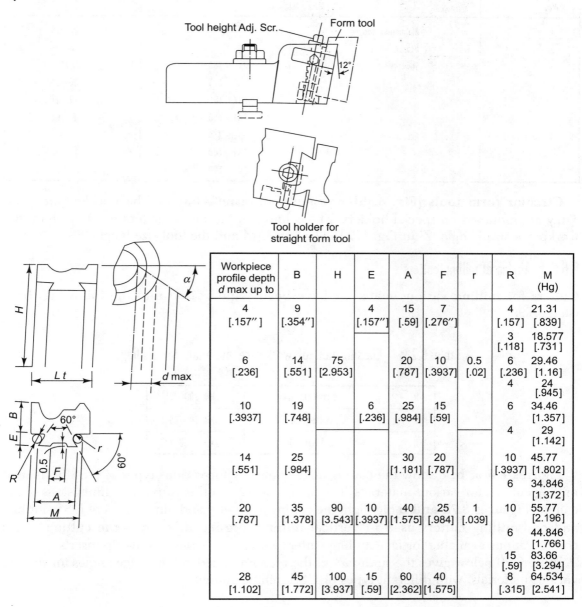

Workpiece profile depth d max up to	B	H	E	A	F	r	R	M (Hg)
4 [.157"]	9 [.354"]		4 [.157"]	15 [.59]	7 [.276"]		4 [.157]	21.31 [.839]
							3 [.118]	18.577 [.731]
6 [.236]	14 [.551]	75 [2.953]		20 [.787]	10 [.3937]	0.5 [.02]	6 [.236]	29.46 [1.16]
							4	24 [.945]
10 [.3937]	19 [.748]		6 [.236]	25 [.984]	15 [.59]		6	34.46 [1.357]
							4	29 [1.142]
14 [.551]	25 [.984]			30 [1.181]	20 [.787]		10 [.3937]	45.77 [1.802]
							6	34.846 [1.372]
20 [.787]	35 [1.378]	90 [3.543]	10 [.3937]	40 [1.575]	25 [.984]	1 [.039]	10	55.77 [2.196]
							6	44.846 [1.766]
							15 [.59]	83.66 [3.294]
28 [1.102]	45 [1.772]	100 [3.937]	15 [.59]	60 [2.362]	40 [1.575]		8 [.315]	64.534 [2.541]

Fig. 8.5(g): Standard straight form tools with dovetail

Table 8.36h: Rake angles for form tools

Sr. No.	Workpiece material	Hardness [BHN]	Rake angle [$\alpha°$]
1	Aluminium, copper		20–25
2	Bronze, brass		0–5
3	Steel	Up to 150	25
		151–235	20–25
		236–280	12–20
		281–350	8–12
4	Cast iron	Up to 150	15
		151–200	12
		201–250	8

Circular form tools [**Fig. 8.5h**] are simpler to manufacture and have a long life span. They are mounted on special holders. The cylindrical profile of the tool depends upon the workpiece size [Angle 'Z' in **Fig. 8.5i**], rake angle [α] and the tool size [angle 'S'].

8.6.6.3 Angles of milling cutters

Cutters for milling various materials have been standardized into the following three categories:

Table 8.28: Classification of workpiece materials for milling

N -	Normal Materials -	HB 200–325
S -	Soft Materials -	HB 30–200
H -	Hard Materials -	Above HB 325

Figures 8.6 a, b, c show the cutting edge geometry for various types of HSS horizontal milling cutters, for various materials. The figures also give other important dimensions of the cutters. **Figsures 8.7a and b** highlight the cutting geometry and dimensions of HSS slotting, and end-milling cutters used in vertical milling. **Figure 8.7c** gives the cutting edges configurations available for face milling cutters with replaceable carbide tip inserts.

Table 8.29 below gives the summary of the recommended cutting edge angles for milling various materials, with different types of HSS milling cutters.

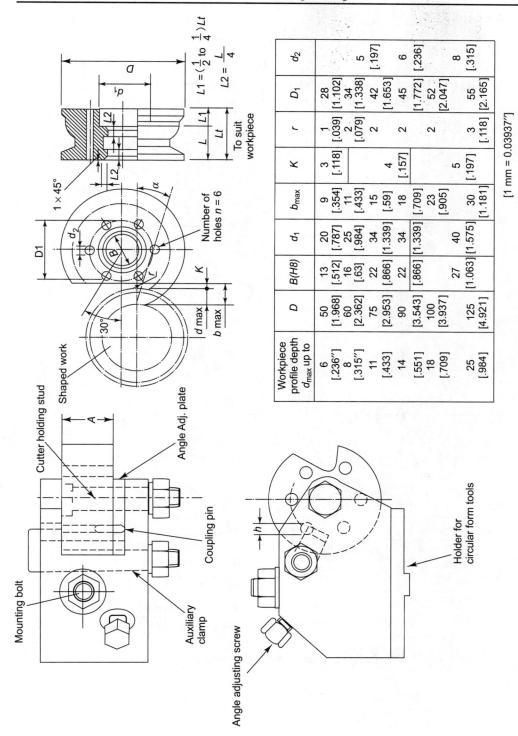

Fig. 8.5(h): Standard circular form tools

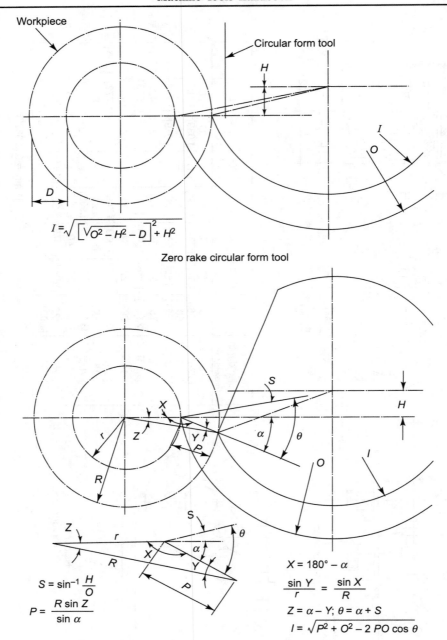

Fig. 8.5(i): Circular form tools geometry

Tool Engineering

Shell End Mills

D	40	50	63	80	100	125	160	Tool Type	α	β	γ
d	16	22	27	27	32	40	50				
Z N	6	8	8	10	10	12	14	N	5-8	10-12	25
Z H	8	10	10	12	14	16	18	H	4-5	6-8	20
Z S	4	4	4	6	6	8	10	S	6-10	20-25	30
F / f	1 / 0.12	1 / 0.12	1.2 / 0.12	1.4 / 0.12	1.4 / 0.12	1.5 / 0.18	1.6 / 0.18				
V	0.039	0.039	0.046	0.046	0.054	0.063	0.063				
V1	0.039	0.039	0.046	0.046	0.054	0.063	0.063				

*f = Cylindrical land

50° Shell end single angle cutter (θ ± 20′ – 25′)

D	40	50	63	80	100	125	160	Tool Type	α	β	γ
d	10	13	16	22	27	32	40				
Z N	–	–	–	–	–	–	–	N	–	–	–
Z H	14	16	18	20	22	24	26	H	3-5	0	–
Z S	–	–	–	–	–	–	–	S	–	–	–
F / f	1 / 0.12	1 / 0.12	1 / 0.12	1 / 0.12	1 / 0.12	1 / 0.18	1.4 / 0.18				
V	0.039	0.039	0.046	0.046	0.054	0.063	0.063				
V1	0.039	0.039	0.046	0.046	0.054	0.063	0.063				

Equal angle cutters (θ + 1°)

D	56	63	80	100	–	–	–	Tool Type	α	β	γ
d	16	22	22	27	–	–	–				
Z N	14	16	18	20	–	–	–	N	5-7	0	–
Z H	20	22	24	26	–	–	–	H	3-5	0	–
Z S	–	–	–	–	–	–	–	S	–	–	–
F / f	1 / 0.12	1 / 0.12	1.2 / 0.12	1.6 / 0.12	–	–	–				
V	0.046	0.046	0.046	0.054	–	–	–				
V1	0.046	0.046	0.046	0.054	–	–	–				

Angle spur milling cutters (θ ± 30)

D	16	20	25	16	20	25	–	Tool Type	α	β	γ
4 – θ	45°	45°	45°	60° & 75°	60° & 75°	60° & 75°	–				
Z N	12	14	14	10	12	12	–	N	5-8	0	0
Z H	–	–	–	–	–	–	–	H	–	–	–
F / f	0.4 / 0.08	0.5 / 0.08	0.6 / 0.12	0.4 / 0.08	0.5 / 0.08	0.6 / 0.12	–	S	–	–	–
V	0.027	0.033	0.033	0.027	0.033	0.033	–				
V1	0.027	0.033	0.033	0.027	0.033	0.033	–				

Note: Z = No. of teeth, N = Tool type for normal material, H = Tool type for hard and tough material, S = Tool type for soft material, V = True running, V1 = Wobble.

[1″ = 25.4 mm] [1 mm = 0.03937″]

Fig. 8.6(a): Number of teeth, run out, wobble and tooth geometry for milling cutters

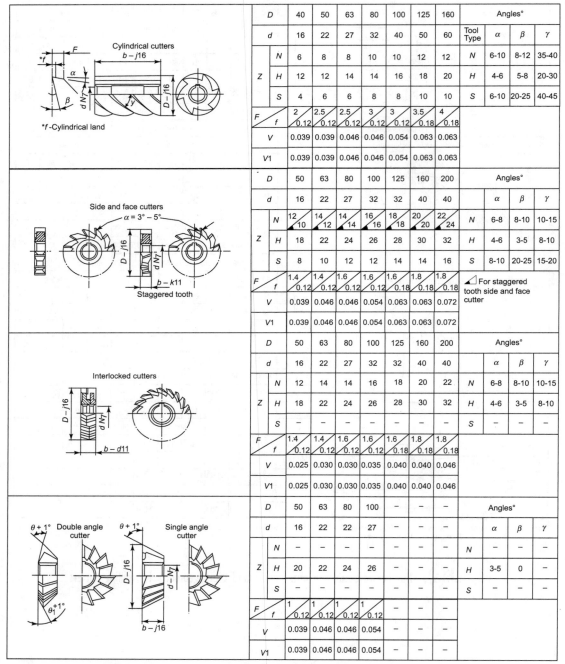

Note: Z = No. of teeth, N = Tool type for normal material, H = Tool type for hard and tough material, S = Tool type for soft material, V = True running, V1 = Wobble

Fig. 8.6(b): Suggested number of teeth, true running, wobble and dims for elements of milling cutters

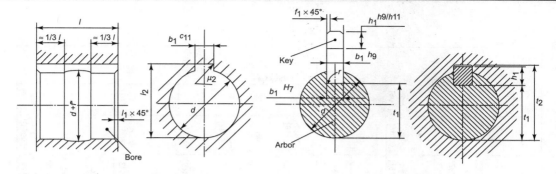

Fig. 8.6(c): Arbor bores and keyways for milling cutters [IS 1831] Dimensions on p684

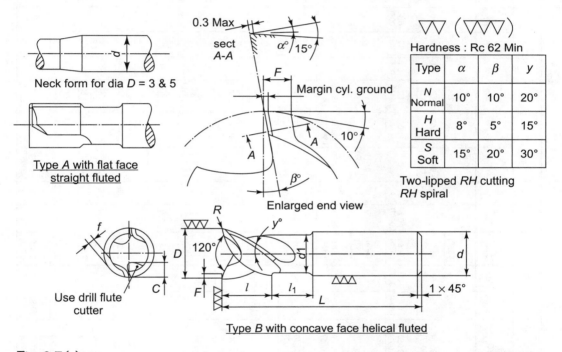

Fig. 8.7(a): Parallel shank slot milling cutters Type A and Type B – Tool type H [Reference IS: 1891–1961]

Dia	Width	Bore							Arbor						Key					
		d (H7)		b1 (C11)	r2		t2		l1	d (h6)		b1 (H7)	t1	r1		b1 (h9)		h1	h1 (h11)		f1
d	b1	Size	Tol. on	Tol. on	Size	Tol.	Size	Tol.	≈	Size	Tol. on	Tol. on	Size	Size	Tol.	Size	Tol. on	Size	Tol. on	Size	Tol.
8 [.315"]	2 [.0787"]	+0.015 [.00059]		+0.120 [.0047]	0.4 [.016"]		8.9 [.35"]		0.6 [.024]	6.7 [.264]	+0	+0.009					-0.025 [.00098]	2 [.0787"]	+0	0.2 [.008]	+0.1 [.004] -0
10 [.3937]	3 [.1181]	-0 [.00071]		+0.060 [.0024]	0.6 [.024]	+0	11.5 [.453"]	-0		8.2 [.323]	-0.009 [.00035]							3 [.1181]			
13 [.5118]	3 [.1181]	+0.018 [.00071]		+0.060 [.0024]	0.6 [.024]	-0.1 [.004"]	14.6 [.575]	+0.2 [.008"]		11.2 [.441]	+0	0		0.2 [.008]	-0 [.004]			4 [.1575]			
16 [.63]	4 [.1575]	-0		+0.145 [.0057]	1 [.039]		17.7 [.697]	-0	1 [.039]	13.2 [.52]	-0.011 [.00043]							5 [.1968]	-0.030 [.00118]		
(19) [.748]	5 [.1968]						21.1 [.831]			15.6 [.614]		+0.012 [.00047]		0.4 [.016]				6 [.2362]			
22 [.8661]	6 [.2362]	+0.021 [.00083]		+0.070 [.0027]		+0	24.1 [.949]	+0.3 [.012"]		17.6 [.693]	-0.013 [.0005]				+0		-0.030 [.00118]	7 [.2756]		0.4 [.016]	+0.2 [.008] -0
27 [1.063]	7 [.2756]	-0		+0.170 [.0067]	1.2 [.047]	-0.3 [.012]	29.8 [1.173]		1.6 [.063]	22.0 [.866]	+0	+0.015 [.0006]						8 [.315]	-0.036 [.00142]		
32 [1.26]	8 [.315]	+0.025 [.00098]		+0.080 [.0031]			34.8 [1.37]	-0		27 [1.063]	-0.016 [.0006]						-0.036 [.00142]				
40 [1.4758]	10 [.3937]	-0		+0.205 [.0081]	1.6 [.063]	-0.5 [.02]	43.5 [1.713]		2.5 [.098]	34.5 [1.358]	+0	+0.018 [.0007]		0.5 [.2]	-0.2 [.008]			9 [.3543]	-0.090 [.0035]	0.5 [.02]	
50 [1.9685]	12 [.4724]	+0.030 [.00118]		+0.095 [.0037]			53.5 [2.106]			44.5 [1.752]	-0.019 [.00075]						-0.043 [.00169]	10 [.3937]			
60 [2.3622]	14 [.5512]	-0		+0.205 [.0081]	2 [.079]		64.2 [2.527]			54.0 [2.126]	+0							11 [.4331]	+0		
70 [2.756]	16 [.63]	+0.030 [.00118]		+0.095 [.0037]		+0	75.0 [2.953]	+0		63.5 [2.5]	-0.022 [.00087]	+0.021 [.0008]		0.6 [.024]	-0					0.6 [.024]	
80 [3.1496]	18 [.7087]	+0.035 [.00138]		+0.240 [.0094]			85.5 [3.366]		3 [.118]	73.0 [2.874]							-0.052 [.00205]	14 [.55127]	-0.110 [.0043]		
100 [3.937]	24 [.9449]			+0.110 [.0043]	2.5 [.098]		107.0 [4.213]			91.0 [3.583]											

Refer Fig. 8.6(c)

*d + l shall be provided if width of Cutter 'L' is more than 16 mm
Non-preferred diameter 'd' in brackets.

[1" = 25.4 mm]   [1 mm = 0.03937"]

Tool Engineering

Dia-D		Dia-d		Dia-d1		L	l	l1	f	c	F	Rad-R		Lead
Size	Tol. e8	Size	Tol. h8	Size	Tol. h11							Size	Tol.	
3 [.1181"]	−0.014 −0.028 [.00110]	[0.00055"]						4 [.16"]	0.2 [.008"]	0.9 [.035"]	0.4 [.016]		0	41 [1.61"]
4 [.1575"]		4 [.1575]	0 −0.018 [.00071]	—	—	40 [1.85"]	6 [.24]	—	0.3 [.012]	1.2 [.047]	0.5 [.02]	0.2 [.008]	−0.1 [.004"]	54 [2.12]
5 [.1968]	−0.020 [.00079]							4	0.4 [.016]	1.5 [.06]	0.6 [.024]			68 [2.69]
6 [.2362]	−0.038 [.0015]	6 [.2362]				45 [1.77]	8 [0.31]	—	0.5 [.02]	1.8 [.071]	0.7 [.028]			82 [3.23]
(7) [.2756]				6.6 [.026]					0.6 [.024]	2.0 [.079]	0.8 [.031]			95 [3.74]
8 [.315]	−0.025 [.00098]	8 [.315]	0 −0.022 [.00087]		0 −0.090 [.0035]	50 [1.97]	10 [.39]	6 [.24]	0.8 [.031]	2.4 [.094]	1.0 [.039]	0.4 [.016]		109 [4.29]
(9) [.3543]	−0.047 [.00185]			7.6 [.299]					0.9 [.035]	2.6 [.102]			0	122 [4.8]
10 [.3937]										3.0 [.118]				136 [5.35]
(11) [.4331]		10 [.3937]		9.4 [.37]		56 [2.20]	12 [.47]	8 [.31]	1.0 [.039]		1.2 [.047]		−0.2 [.008]	150 [5.9]
12 [.4724]	−0.025 [.00098]		0		0				1.2 [.047]	3.2 [.126]	1.5 [.06]			163 [6.42]
										3.6 [.141]		0.5 [.02]		
14 [.5512]	−0.059 [.002323]	12 [.4724]	−0.027 [.00106]	11.6 [.457]	−0.110 [.0043]	63 [2.48]	16 [.59]	9 [.39]	1.4 [.055]	4.0 [.16]	1.8 [.071]			191 [7.52]
(15) [.59]										4.4 [.173]				204 [8.03]

Refer Fig. 8.7(a)

Note: (1) Non-preferred sizes are in brackets.
(2) RH cutters with RH spiral shall be made, unless otherwise specified
(3) 1" = 25.4mm   1mm = 0.03937"

Hardness: Rc 62 - 65 (VPN 800 - 900)

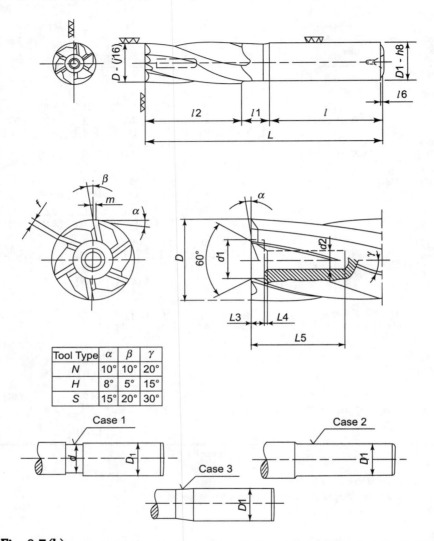

Fig. 8.7(b): Parallel shank end mills Tool type N, H, and S [Reference: Is: 1831–1961]

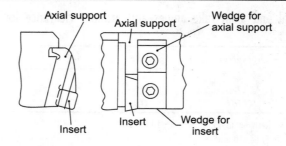

Face mill cartridge axial support and wedge clamp

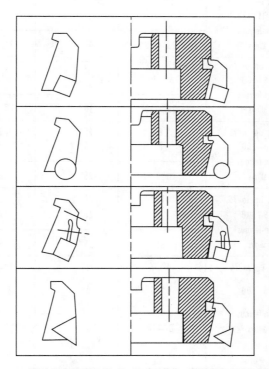

Fig. 8.7(c): Face mill inserts and cartridges

Table 8.29: Radial and axial rake angles and primary clearance for HSS milling cutters

Sr. No.	Cutter type	Workpiece strength kgf/mm² tensile	Rake Angle Radial α_r	Rake Angle Axial α_a	Primary clearance C	Remarks
1	Slab mills, slotting	Steels up to 75 [1,06,650 psi] [HB 211]	10°–16°	35°–38°	7°–12°	Higher value for down milling
		76–100 [1,08,070–1,42,200] [HB 212–282]	5°–12°	30°–35°	4°–8°	
		Non-ferrous alloys	25°–30°	45°	8°–14°	
2	Side and face	Steels up to 75	12°–18°	15°	7°–12°	
		76–100	6°–14°	10°–12°	5°–8°	
		Non-ferrous alloys	25°–30°	30°	8°–14°	
3	End mills, slot drills	Steels up to 75	8°	15°	7°	
		76–100	6°	15°	4°	
		Non-ferrous alloys	20°	25°	8°	
4	Face mills, shell end mills	Steels up to 75	10°	20°	7°	
		76–100	5°	20°	4°	
		Non-ferrous alloys	25°	35°	8°	
5	Form relieved cutters	Steel up to 75	4°–6°	0°	7°–12°	h = Cam fall $h = \frac{\pi D}{z} \tan C$ C = Clearance angle D = Cutter Φ Z = No. of teeth
		76–100	0°–2°	0°	5°–8°	
		Non-ferrous alloys	8°–10°	0°	8°–14°	

[1 kg/mm² = 1422 psi]

8.6.6.4 Drill geometry [Fig. 8.8]

The most important angles of a drill are: Point angle [B], Lip clearance angle [C], Chisel edge angle [E], and Rake [Helix] angle [α_a]. The following tables give the various angles for H.S.S. twist drills, for cutting various workpiece materials.

Tool Engineering

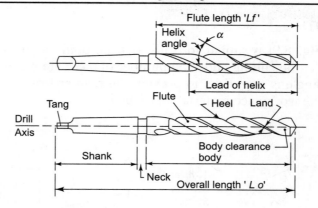

Fig. 8.8(a): Twist drill with tapered shank

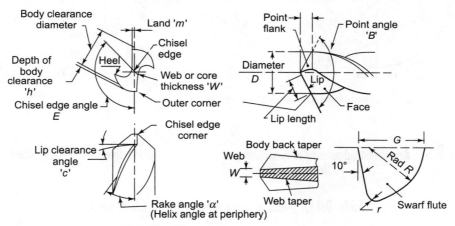

Fig. 8.8(b): Tool geometry of twist drill

Table 8.30 [Fig. 8.8] H.S.S. twist drill angles for various workpiece materials

Sr. No.	Material	Hardness HB	Point angle $B°*$	Lip clearance $C°$	Chisel edge Angle $E°$	Rake [Helix] Angle $\alpha_a°$
1	Steels, C.I., brass, copper, nickel, zinc	< 211	118°	12°	125–135°	30°
		212 – 282	118°	12°	125–135°	25°
		>283		10°	125–135°	20°
2	Aluminum alloys	30–150	90°	15°	125–135°	35–45°
3	Bronzes plastics	< 100	118°	15°	125–135°	8°–12°
		> 100		15°	125–135°	15°–20°

* The recommended point angles give good results for thru holes. For forming conical seats for Countersunk Head screws, the point angle will be 90° for all the materials. It can be changed, to suit the cone angle required, on the workpiece.

8.6.6.5 Reamer geometry

Reamer geometry [**Fig. 8.9**] is similar to drills, but the point angle is replaced by a small chamfer [lead angle] at the cutting end. The number of teeth are more for harder materials. **Axial rake** [Helix] angle is **negative,** to prevent the reamer from screwing in. The radial rake depends upon the workpiece material [hardness].

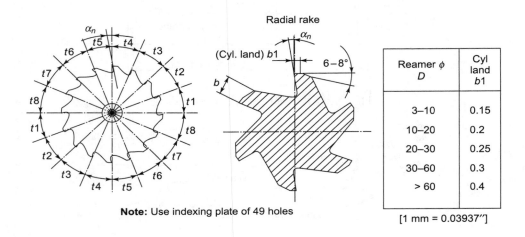

Fig. 8.9: Recommended spacing of flutes and geometry for reamers [Reference: Werkzeug Hand book]

Table 8.31: Radial rake angles for H.S.S. reamers

Workpiece material	Axial rake α	Radial rake α_n	Lead chamfer Length × Degree
Aluminum, brass HB 30–150	(–35)–(–45)	25°–30°	1.6 × 45 $\left[\frac{1''}{16} \times 45°\right]$
Steel — low carbon HB 85–211	(–12)–(–20)	15°–20°	1.6 × 45
— medium carbon HB 212–280	(–7)–(–8)	8°–12°	1.6 × 30
— high carbon HB 280–400		0°–5°	1.6 × 5
Cast iron — HB up to 200	(–12)–(–20)	6°–10°	1.6 × 45
— HB above 220	(–7)–(–8)	0°–5°	1.6 × 30

Tool Engineering

No. of Flutes	Angle t1	Indexing plate Rotations	Indexing plate No. of Holes	Angle t2	Indexing plate Rotations	Indexing plate No. of Holes	Angle t3	Indexing plate Rotations	Indexing plate No. of Holes	Angle t4	Indexing plate Rotations	Indexing plate No. of Holes	Angle t5	Indexing plate Rotations	Indexing plate No. of Holes	Angle t6	Indexing plate Rotations	Indexing plate No. of Holes	Angle t7	Indexing plate Rotations	Indexing plate No. of Holes	Angle t8
6	58°2'	6	22	59°53'	6	22	62°5'	6	32	—	—	44	—	—	—	—	—	—	—	—	—	—
8	42°	4	32	44°	4	32	46°	5	44	48°	5	16	—	—	—	—	—	—	—	—	—	—
10	33°	3	34	34°30'	3	34	36°	4	41	37°30'	4	8	39°	4	15	—	—	—	—	—	—
12	27°30'	3	3	28°30'	3	3	29°30'	3	8	30°30'	3	19	31°30'	3	24	32°30'	3	30	—	—	—
14	23°30'	2	30	24°15'	2	30	25°	2	34	25°45'	2	43	26°30'	2	46	27°	—	—	28°	3	5	—
16	20°30'	2	14	21°	2	14	21°30'	2	17	22°15'	2	20	22°45'	2	23	23°15'	2	26	24°	2	32	24°45'*

*for t8 Ind. Plate rot. 2, holes 35. Refer Fig. 8.9

Dia of reamer (mm)	From	3 [.118"]	6 [.236"]	9 [.354"]	14 [.55"]	20 [.787"]	28 [1.1"]	35 [1.38"]	41 [1.61"]	57 [2.24"]	69 [2.72"]	85 [3.35"]
	To	5 [.2"]	8 [.31"]	13 [.51"]	19 [.75"]	27 [1.06"]	34 [1.34"]	40 [1.57"]	56 [2.2"]	68 [2.68"]	84 [3.31"]	100 [3.94"]
Width b (mm)	From	0.25 [.01]	0.5 [.02]	0.7 [.03]	0.9 [.035]	1.1 [.043]	1.4 [.055]	1.7 [.067]	2 [.078]	2.2 [.087]	2.4 [.094]	2.6 [.102]
	To	0.4 [.016]	0.6 [.024]	0.8 [.031]								

Table 8.32: Cutting edge angles for carbide reamers

	Workpiece hardness HB	Steels and cast irons < HB 225	HB 226–325	HB 326-425	> HB 426	Ductile materials Al, Cu, Mg, malleable C. I.
1	Radial rake [α_n]	7°–10°	5°–7°	5°–7°	0°–5°	7°
2	Axial rake Helix [α_a] (Negative)	5°–8°	5°–8°	0°	0°	5°–8°
3	Cylindrical Land [f]	0.125–0.25 [.005–.01″]	0.05–0.125 [.002–.005″]	0.05–0.125 [.002–.005″]	0.05–0.125 [.002–.005″]	0.375–0.5 [.015–.020″]

8.6.6.6 Threading taps

Threading taps too should have different hook [rake] and lead chamfer angles, for different materials.

Table 8.33: Rake, and relief angles for taps [Figs. 8.10a, b]

Workpiece material	Radial rake α	Relief angle R
Steel 6t up to 60 kgf/mm² [HB 176] [85,300 psi]	15°	8°
60 to 90 kgf/mm² [HB 176 – 270] [85,300–1,28,000]	10°	8°
above 90 kgf/mm² [>HB270] [1,28,000]	5°	4°–8°
Cast iron	0°–5°	6°
Bronze	0°	6°
Aluminum	20°–30°	12°

Lead chamfer at the cutting end should be 3–4 thread pitches long. The chamfer too is relieved radially.

Material:	Steels	Stainless Stl. and copper alloys	Aluminum, titanium and magnesium alloys
Chamfer radial relief:	8°	10°	12°

Chamfer angle [Ψ] should be such that it reaches the thread minor diameter in the chamfer length.

$$\text{Lead angle}[\Psi] = \tan^{-1}\left(\frac{\text{Thread depth}}{\text{Chamfer length}}\right)$$

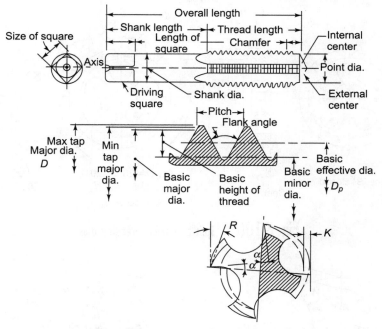

Fig. 8.10(a): Threading tap geometry

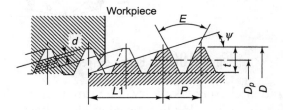

Fig. 8.10(b): Lead chamfer on taps

8.6.6.7 Button dies

Button dies can be conveniently used for threading diameters up to 68 mm. **Figure 8.10(c)** shows standard button dies. The radial split allows for adjusting thread diameter for roughing cut. **Figure 8.10(d)** shows the cutting edge geometry. Table 8.23b gives the rake angles for various ultimate strengths of workpiece materials.

Table 8.23b: Rake angle [α] for threading dies [Fig. 8.10 (d)]

Ultimate tensile strength (kg/mm^2)	< 60 [85,300 psi]	60–90 [85,300–1,28,000]	> 90 [1,28,000]
Rake angle α°	20°–25°	15°–20°	10°–15°

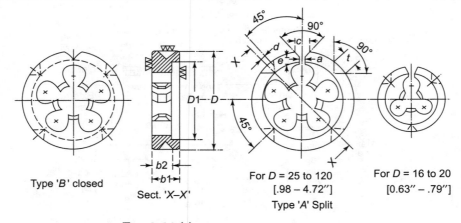

Fig. 8.10(c): Split and solid threading dies

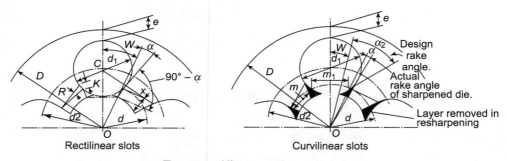

Fig. 8.10(d): Button die geometry

Figure 8.10e shows radial chasers geometry. The throat angle determines the minimum undercut necessary on for threading. Higher throat angle reduces the tool life.

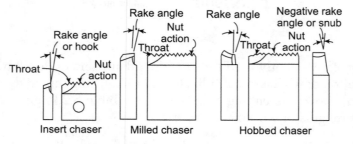

Fig. 8.10(e): Radial chasers

Throat angle = 10°–12° for lead screw machines
= 15°–35° for hand or spring-started chasers

Table. 8.23c gives chip thickness for various thread lengths, covered by the throat angle.

Tool Engineering

Dimensions for circular thread cutting dies for metric threads [Fig. 8.10c] (All dimension in millimetres)

Nominal size	D	D1	For metric coarse pitch		For metric fine pitch P										a	c	d	t
					0.2	0.25	0.35	0.5	0.75	1.00	1.25	1.5	2.00	3.00	4.00				
			b_1	b_2	b_1 b_2	b_1 b_2	b_1 b_2	b_1 b_2	b_1 b_2	b_1 b_2	b_1	b_1 b_2	b_1 b_2	b_1 b_2	b_1				
M 1			5	2	5 2														
(M 1.1)			5	2	5 2														
M 1.2			5	2	5 2														
(M 1.4)			5	2.5	5 2														
M 1.6	16	11	5	2.5	5 2														
(M 1.8)			5	2.5	5 2														
M 2			5	3		5 2													
(M 2.2)			5	3		5 2													
M 2.5			5	3			5 3												
M 3			5	—			5 3									1	4.5	0.5	3.8
(M 3.5)			5	—			5 3												
M 4	20	15	5	4				5 3.5											
(M 4.5)			5	—				5 —											
M 5			7	—				5 —									5	0.6	4.5
M 6	25	20	7	—					5 3.5										
(M 7)			9	—					9 6										
M 8	30	24	11	8					9 6	11 8								0.8	5.2
M 10	38	30	14	10					11 7	12 —	10 —	10 8				1.5	5.5	1	6
M 12			14	10						12 —	10 —	— —					6	1.2	6.5
(M 14)	45	36	18	12						14 10	10 —	10 —	14 10				7		7.5
M 16			18	—						14 10	—	14 10	14 10			2	8	1.5	8.5
(M 18)	55	45	22	16						16 10		14 10	16 10	10 —					
M 20			22	16						16 10		16 10	16 10	14 —					
(M 22)			25	18						18 12		16 10	18 12	14 —					
M 24			25	18						18 12		18 12	18 12	16 —					
(M 27)	65	54	25	—								18 12	18 12	18 —	(25) —		9.5	1.8	10
M 30			25	—								18 12	20 14	18 —	(25) —	2.5			
(M 33)	75	63	30	25								20 14	20 14	20 14	30 25		11		11.5
M 36			30	—											30 25				
(M 39)	90	75	36	28								22 16	22 16	22 —	30 25	3	12	2	13
M 42			36	—								22 16	22 16	22 —	36 28				
(M 45)			36	—											36 28				
M 48			36	—										36 22	36 28				
(M 52)	105	90	36	—										36 25	36 28				
M 56			36	—										36 25	36 28				
(M 60)	120	100	36	—										36 25	36 28	4	14	2.5	15
M 64			36	—										36 25	36 28				
(M 68)			36	—											36 28				

Non-preferred sizes in brackets   [1mm = 0.03937"; 1" = 25.4"]   IS:1859-1961

Designations

Designation of a circular split die of Type A for cutting metric coarse pitch threads M 10, made of carbon tool steel shall be: Thread cutting die AM 10 IS: 1859–CS

Designation of a circular closed die of Type B for cutting metric fine threads M 10 × 1, made of high speed steel shall be: Thread Cutting Die BM 10x1 IS: 1859–HS

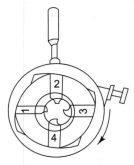

Fig. 8.10(f): Radial

	Chip thickness	No. of threads in throat
30°	0.31	1.2
20°	0.21	1.9
15°	0.16	2.6
12°	0.13	3.2
10°	0.11	3.9

Fig. 8.10(g): Effect of throat angle

Table 8.23c: Chip thickness for various lengths of throat angle in No. of threads

No. of threads in throat	1.2	1.9	2.6	3.2	3.9
Chip thickness [mm]	0.31 [.0122″]	0.21 [.0083″]	0.16 [.0063″]	0.13 [.0051″]	0.11 [.0043″]

8.6.6.8 Broaching

Broaching [**Fig. 8.11a,e**] means opening out, or enlarging an opening to a desired shape. Cutting material, to shape from a round hole to a square, can be done quickly, using the broaching process. Broaching is also used to machine surfaces, particularly a number of surfaces, with precise geometrical and dimensional relationships. **Figure 8.11a** shows some workpieces, which can be economically broached, to precise dimensions.

Workpiece hardness range for broaching = HB 194 – 237 [RC 12 – 22]

Workpieces with lesser hardness get torn; while workpieces of up to HB 327 [RC 35] hardness, can be broached. Broach life is shorter for harder work materials.

Broach, the tool used on broaching machines is a long multi-tooth cutter [**Fig. 8.11b**]. The cutting teeth are divided in 3 parts: roughing, semi-finishing, and finishing. Although roughing teeth remove the maximum material, the cut for each tooth rarely exceeds 0.2 mm.

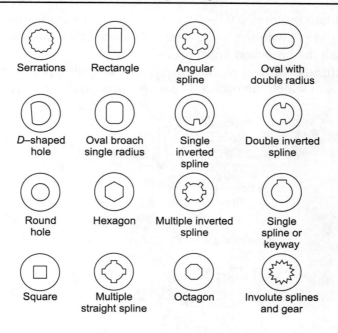

Fig. 8.11(a): Workpieces for internal broaching

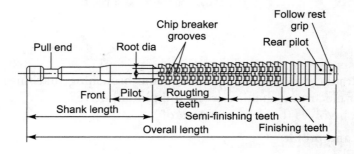

Fig. 8.11(b): Shank and pilot

The entering end of the broach, i.e. the shank, is made suitable for the broaching machine puller [**Fig. 8.11c**]. The spring-actuated, retainer halves of the puller enter the grooved part of the shank. This couples the broach to the machine ram, which is pulled, to pass the broach through/onto the workpiece.

Each successive tooth removes 0.025–0.2 material from the workpiece. There is a location pilot which is close slide fit with the workpiece hole. Even at the tail end of the broach, there is a pilot close fit with the broached finish profile.

Broaching machines, vertical or horizontal, can be pull-type, in which the broach is pulled, or push-type, in which the broach is pushed through the workpiece.

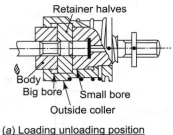

(a) Loading unloading position

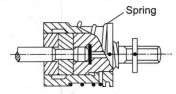

(b) Working position: Retainer halves engaged in broach shank grooves

Fig. 8.11(c): Spring-actuated puller

Pull-type broaches

Although the hydraulic drive is widely used, some broaching machines have mechanical [screw and nut], or electro-mechanical [rack and pinion] drives.

Broach design involves the determination of:

1. Tooth angles and profile
2. Tooth pitch [P]
3. Cut per tooth for roughing, semi-finishing, and finishing teeth
1. **Tooth profile [Fig. 8.11d]**: The rake angle should suit the workpiece material and operation, i.e. roughing/finishing/semi-finishing. Table 8.33a gives the rake [face] angles for various workpiece materials, and Table 8.33b gives the back-off [clearance] angles for various types of broaches. The choice of the tooth profile will depend upon the tooth pitch and the type of broaching.

Tool Engineering

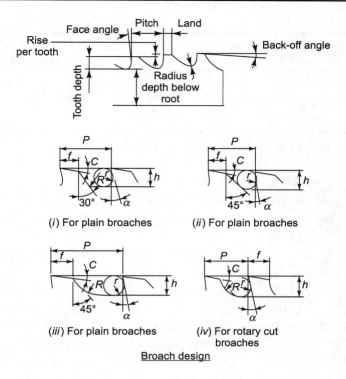

Fig. 8.11(d): Dimensions for broach tooth profiles

Table 8.33a: Rake angles $|\alpha|$ for broaches [Fig. 8.11]

Sr. No.	Workpiece material	Hardness [BHN]	Face (rake) roughing and semi-finishing	Angle finishing
1	Steel	< 200	15°–20°	5°
		200–230	18°	–5°
		>230	8°–12°	–5°
2	Aluminum and alloys		20°	20°
3	Grey cast iron		6°–8°	5°
4	Malleable iron		10°	5°
5	Brasses and bronzes		–5° + 5°	–10°

Ref. [Fig 8.11(d)]   [1mm = 0.03937"] [1mm² = .00155 Inch²]

Broach pitch p	Curvilinear shape of gullet Figs. 8.11d [iii] and [iv]					Rectilinear shape of gullet Figs. 8.11d [I] and [ii]			
	h	f	r	R	Gullet Area A_g, mm² Fig. 8.11d (iv)	h	f	r	Gullet Area A_g, mm² Fig. 8.11d [ii]
	mm					mm			
4 [.16"]	1.6 [.063]	1.5 [.059]	0.8 [.03]	2.5 [.1]	1.91 [3 × 10⁻³ Inch²]	—	—	—	—
5 [.2"]	2.0 [.079]	1.5 [.059]	1.0 [.04]	3.5 [.14]	3.14 [4.9 × 10⁻³]	—	—	—	—
6 [.24"]	2.5 [.098]	2 [.079]	1.25 [.05]	4 [.16]	4.91 [7.6 × 10⁻³]	2.0 [.079]	2.5 [.098]	1.0 [.04]	3.0 [4.6 × 10⁻³]
7 [.27"]		2.5 [.098]			7.06 [11 × 10⁻³]	2.3 [.09]	3.0 [.118]	1.25 [.05]	5.8 [9 × 10⁻³]
8 [.31"]	3 [.118]		1.5 [.06]	5 [.2]	7.06 [11 × 10⁻³]	2.7 [.106]	3.5 [.138]	1.5 [.06]	7.0 [10.8 × 10⁻³]
10 [.39"]	4 [.16]	3 [.118]	2.0 [.08]	7 [.27]	12.56 [19.5 × 10⁻³]	3.6 [.142]	4.0 [.157]	2.0 [.08]	12.5 [19.4 × 10⁻³]
12 [.47]	5 [.2]		2.5 [.1]	8 [.31]	19.62 [30 × 10⁻³]	4.5 [.177]	4.5 [.177]	2.5 [.10"]	19.3 [30 × 10⁻³]
14 [.55]	6 [.24]	4 [.16]	3.0 [.12]	10 [.39]	28.25 [43.8 × 10⁻³]	5.4 [.212]	5.0 [.197]	3.0 [.12]	27.9 [43 × 10⁻³]
16 [.63]	7 [.27]				38.46 [59.6 × 10⁻³]	6.3 [.248]	5.5 [.216]	3.5 [.14]	38.0 [58.9 × 10⁻³]
18 [.71]	8 [.31]		3.5 [.14]	12 [.47]	50.0 [77.5 × 10⁻³]	7.2 [.283]	6.0 [.236]	4.0 [.16]	49.6 [76.9 × 10⁻³]
20 [.79]	9 [.35]	6 [.24]	4.5 [.18]	14 [.55]	63.58 [98.5 × 10⁻³]	8.1 [.319]	6.5 [.256]	4.5 [.18]	62.7 [97.2 × 10⁻³]
22 [.87]	10 [.39]		5.0 [.2]	16 [.63]	78.5 [121.7 × 10⁻³]	9.0 [.354]	7.0 [.275]	5.0 [.2]	78.0 [120.9 × 10⁻³]

Notes: 1) Flat bottom gullet with elongated pitch **[Figure iii and iv]** may be used for each gullet depth h, provided the other elements are same.

2) The curvilinear shape of gullet (sketches i and ii) provides good conditions for chip-curling, and is utilized firstly with progressive cutting broaches and other broaches of intricate cutting scheme.

Land and Back-off (Primary clearance) [C]

A small portion (about 0.25 to 0.3 mm) following the cutting edge is ground with a back-off angle, ranging from 1/2° to 4°. The angle depends upon the type of broach as well as the operation: roughing, semi-finishing, or finishing.

Table 8.33: Back-off angles for broaches

Broach type	Back-off angle		
	Roughing	Semi-finishing	Finishing
1. Keyway, splines, round	3°	2°	½° – 1°
2. Surface	3° – 4°	2°	1° – 2°

Tool Engineering

Tooth pitch [P] depends upon the workpiece length [L] and the type of the broach.

Operation	Keyway squares broaching	Round broaching	Surface broaching
Pitch [P]	$1.25 - 1.5\sqrt{L}$	$1.45 - 2\sqrt{L}$	$\sqrt[3]{\dfrac{L.R.\ Ag}{C}}$

R = Rise/Tooth [Table 8.18 on p 667]

Ag = Gullet area [mm^2]; C = Chip cross sect [mm^2]

$\dfrac{Ag}{C} = 3 - 5$, for roughing

$\phantom{\dfrac{Ag}{C}} = 8$ for finishing.

The computation of broaching force has already been covered under para 8.140 [p 670]. **Figure 8.11e** gives the dimensions of the standard round shanks for pull-type internal broaches. **Figure 8.25** in the next section, shows a keyway broaching fixture.

d_1 (e8)	d_2 (c11)	d_4 (0.5–1)	c	l_1	l_2	l_3	l_4	r_1	r_2	α degree
					mm					
12 [.472"]	8 [.31"]	12 [.472"]						0.2 [.008"]		10
14 [.551"]	9.5 [.37]	14 [.551"]		120 [4.72"]	20 [.787]	20	12 [.472]		0.6 [.024"]	
16 [.63]	11 [.433]	16 [.63]	0.5 [.020"]							
18 [.709]	13 [.512]	18 [.709]								20
20 [.787]	15 [.59]	20 [.787]						0.3 [.012"]		
22 [.866]	17 [.669]	22 [.866]		140 [5.51"]						
25 [.984]	19 [.748]	25 [.984]			25 [.984]	25	16 [.63]		1.0 [.039"]	
28 [1.102]	22 [.866]	28 [1.102]	1 [.039"]							
32 [1.26]	25 [.984]	32 [1.26]		160 [6.23"]				0.4 [.016"]	1.6 [.63"]	
36 [1.417]	28 [1.102]	36 [1.417]			32 [1.26]	32	20 [.787]			30
40 [1.575]	32 [1.26]	40 [1.575]								
45 [1.772]	34 [1.338]	45 [1.772]		180 [709"]				0.5 [.020"]	2.5 [.098"]	
50 [1.968]	38 [1.496]	50 [1.968]	1.5 [.059"]							
56 [2.205]	42 [1.653]	56 [2.205]								
63 [2.48]	48 [1.89]	63 [2.48]		210 [8.27"]	40 [1.575]	40	25 [.984]	0.6 [.024"]	4.0 [.16"]	
70 [2.756]	53 [2.087]	70 [2.756]								
80 [3.15]	60 [2.362]	80 [3.15]								
90 [3.543]	70 [2.756]	90 [3.543]	2	240 [9.45"]	50 [1.968]	50	32 [1.26]	0.8 [.031"]	6.0 [.236"]	
100 [3.937]	75 [2.953]	100 [3.937]								

Refer Fig. 8.11 (e).

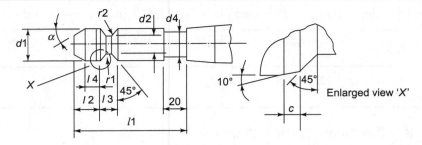

Fig. 8.11(e): Std. ends for round pull-type broaches

8.7 TOOLING

Tooling enables the wielding of tools, as required. They are the paraphernalia, or accessories necessary for usage of the tools, for a specific purpose [machining particular workpiece]. Tooling can be broadly divided into 2 categories:

1. Tool holders:
   (a) Standard: Chucks, collets, arbors, tool posts
   (b) Special: Non-standard collets, multiple toolholders
2. Workpiece holders
   (a) Standard: Chucks, collets, vices
   (b) Special: Special jaws, collets, mandrels
   (c) Jigs and fixtures

Holders can also be classified as those for cylindrical, and rectangular objects.

Standard tool/workholders for popular sizes, are commercially available. If a workpiece calls for some special size (of say collet), particularly for second operations, these can be produced expeditiously by grinding off the nearest, standard undersize holder [collet]. Similarly, jaws of standard chucks can be modified to suit particular workpieces.

Sophisticated workpieces however, call for specially-tailored jigs and fixtures, to suit the peculiarities of the workpiece.

8.7.1 Toolholders

Toolholders can be classified according to the shape of the tool: cylindrical and rectangular. **Cylindrical tools** [drills, reamers, taps, milling cutters] are held in chucks, collets, or on arbors.

8.7.2 Chucks

Straight shank drills, reamers, and taps up to 20, can be held in drill chucks. However, usually tools up to 12 diameter are held in chucks, while tools with bigger diameters have **morse taper shanks**, which give better concentricity. Tapered to approximately $1\frac{1}{2}°$, the self-holding taper can hold the drill against gravitational fall. The tapered shank is provided with **flats**, which get engaged in the slot provided in the drilling m/c spindle, to transmit the **drilling/reaming torque**.

Similarly, tap-holders engage with the two opposite sides of the squared end of the tap, to transmit the requisite torque.

8.7.3 Collets

These are rarely used on drilling machines but are commonly used for holding straight-shank, vertical milling cutters such as end-mills, and slot drills. Collets are more widely used for rolled bar stock, used as a raw material for workpieces. These are push-out collets **[Fig. 8.12a]** which tend to push the bar towards the center /turret/capstan.

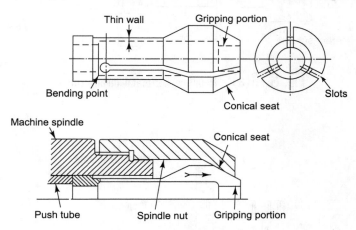

Fig. 8.12(a): Push-out collet

[Included with permission from Tata McGraw-Hill Pub. Co. Ltd.]

For the second operation, usually a pull-in type collet **[Fig. 8.12b]** is required. It tends to pull the collet away from the center/turret. The bore of the collet should suit the workpiece diameter, machined in the previous operation, and used as a location for the subsequent operation.

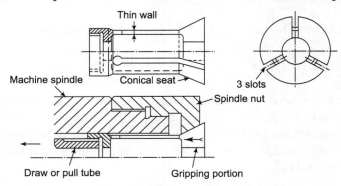

Fig. 8.12(b): Pull-in collet

8.7.4 Arbors

Bigger milling cutters are provided with the location bore and drive keyway. They are mounted on the keyed arbor. Even bigger end mills are manufactured as shells, and require stub arbor [**Fig. 8.13b**] for mounting. Usually, big face mills are also mounted on arbors.

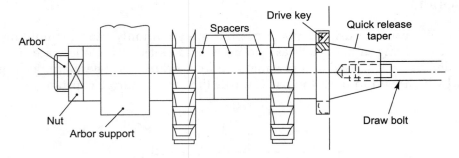

Fig. 8.13(a): Mounting horizontal milling cutters

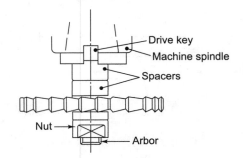

Fig. 8.13(b): Use of stub arbor

8.7.5 Tool posts

Tool posts used on lathes and turning machines, are an integral part of the machine.

8.7.6 Special toolholders

Special toolholders are necessary when the existing toolholders cannot accommodate as many tools as necessary, at the spacing required, for the particular workpiece [**Fig. 8.14a**]. Some special toolholders have become so popular that they have become a standard part of capstan/turret/automatic lathes, e.g. toolholders for simultaneous drilling and turning, one or two diameters [**Fig. 8.14b**].

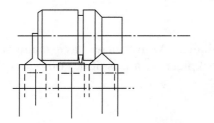

Fig. 8.14(a): Special toolholder for particular workpiece

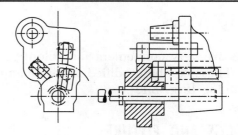

Fig. 8.14(b): Combination toolholder for 3 turning and 1 boring tool

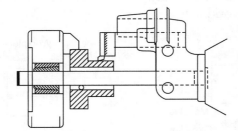

Fig. 8.14(c): Knee toolholder for 1 turning and 1 boring tool

8.7.7 Workpiece holders

Workpiece holders can be standard, modified standard or tailor-made, or fixtures, to suit the features of a particular workpiece. When equipped with tool-guiding elements, such as bushes for drills, they are called jigs. Templates guiding gas-cutting torches, or a tailor's marking chalk are jigs.

8.7.8 Chucks

Chucks for holding workpieces, are much bigger than drill chucks. Also, a wide variety of these is available. There are two/four jaw chucks, with an independent jaw operation for non-cylindrical workpieces. Three jaw chucks are usually self-centering. Some chucks have a provision for changing jaws. Special jaws, made to suit workpieces facilitate the gripping of eccentric or non-cylindrical objects.

8.7.9 Collets

Collets [Fig. 8.12a] for workpieces, are similar to collets used for holding vertical milling cutters. But the operation is speeded up by using a draw-tube, often operated pneumatically [Fig. 8.12].

8.7.10 Vices

Vices are very convenient for holding rectangular objects. Many vices have changeable jaws. These can be modified to locate and grip the workpiece in a position, convenient for machining [**Fig. 8.15**].

8.8 JIGS AND FIXTURES

Odd-shaped workpieces, or the need for rapid production at an economic cost, and high productivity, may call for the use of a jig or fixture, to suit a particular workpiece, or a similarly-shaped group of workpieces.

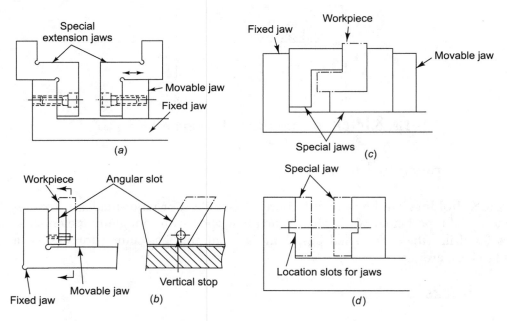

Fig. 8.15: Special vice jaws

8.8.1 Drawing conventions for jigs and fixtures

In jig/fixture drawings, the workpiece is drawn in a chain-dotted line, to distinguish it from fixture delineation. Drawn in red, green, or blue, the workpiece is considered transparent. Studs, and locators passing through the workpiece, are drawn using unbroken lines, instead of dotted lines [**Fig. 8.16**].

Tool Engineering

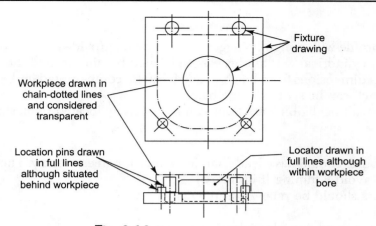

Fig. 8.16: Presentation of workpiece

Jigs/fixtures comprise the following elements:
1. Locators
2. Clamps
3. Tool [cutter]-guiding [bushes], or [cutter]-setting elements

8.8.2 Locators

Position the workpiece correctly, with respect to the cutting tool.

8.8.2.1 Location principles

1. Relate location to dimensional requirements of workpiece
2. Use the most accurately-machined surface for location
3. Prevent motions along and around the X, Y, and Z axes
4. Provide for easy and quick loading and unloading of workpiece, with minimum motion.
5. Avoid redundant locators
6. Prevent wrong loading of workpiece, by foolproofing.

8.8.2.2 Location methods

A Plane

1. A flat surface can be located on three equal height pins with the spherical surface at the resting point
2. A rough surface should be located on three adjustable height pins with a spherical resting point
3. Additional adjustable supports might be necessary to contain vibrations or distortion during the machining operation. The force for adjusting the supports should not dislocate or lift the workpiece from its location
4. A machined surface can be located by flat pads
5. Provide ample clearance for burr or dirt, so that the workpiece is well-located.
6. Six pads can prevent the linear movement of a cube along, and rotary motion around the three (X, Y, and Z) axes.

A Profile

1. Aligning a profile with a slightly bigger sighting plate, provides an approximate location
2. A profile, or cylindrical workpieces can be located by pins around the periphery
3. Variations in dimensions from batch to batch can be compensated for by using eccentric locators, which can be set to suit the batch
4. Workpieces with negligible variations can be precisely-located by nesting plates.

Cylinder

1. Spigots for bores should have lead chamfer/radius for an easy entry. Their length should be short, to avoid jamming **[Fig. 8.17b]**
2. Long locators should be relieved at the center **[Fig. 8.17a]**
3. Locators also used for clamping, must be retained by a nut or a grub screw **[Fig. 8.17b]**

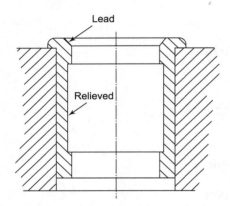

Fig. 8.17(a): Female locators

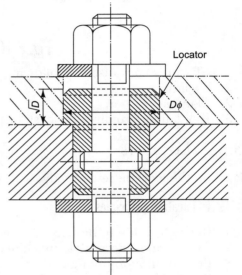

Fig. 8.17(b): Locators subjected to axial pull

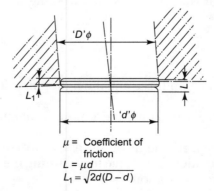

μ = Coefficient of friction
$L = \mu d$
$L_1 = \sqrt{2d(D-d)}$

Fig. 8.17(c): Jamming-prevention lead

4. If two location pins are used, the less important one should be diamond-shaped. The full pin should be longer, to facilitate easy loading

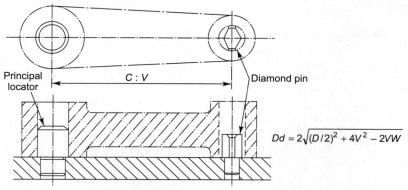

Fig. 8.17(d): Diamond pin application

$$Dd = 2\sqrt{(D/2)^2 + 4V^2 - 2VW}$$

5. Locate rough holes and bosses, by conical posts. These often have an integral clamping arrangement
6. Fixed 'V' blocks approximately locate a cylinder exterior [Fig. 8.18a]

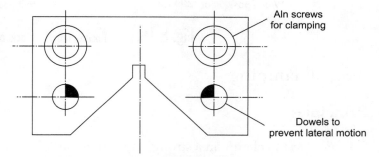

Fig. 8.18(a): Fixed 'V' block

7. For precise location, use an adjustable, guided 'V' block. It can be adjusted by screw [Fig. 8.18b] or cam, and withdrawn quickly by providing a swinging eyebolt [No Fig.].

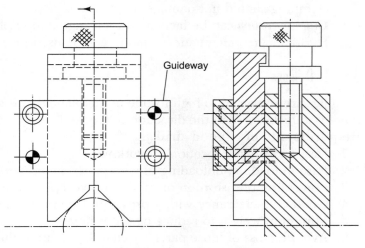

Fig. 8.18(b): Screw-adjusted 'V' locator

8. Position 'V' correctly, to prevent workpiece variations from affecting the location accuracy, e.g. for central hole/keyway, 'V' axis should align with the drill/cutter axis [**Fig. 8.18c**].

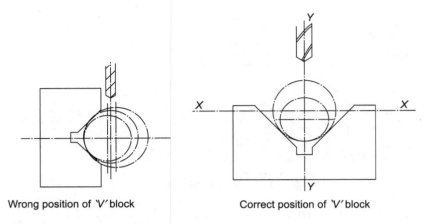

Wrong position of 'V' block   Correct position of 'V' block

Fig. 8.18(c): Positioning 'V' block correctly

8.8.3 Clamping

Principles

1. *Position* the clamp on a strong, supported part of the workpiece, clear of the tool path or the workpiece loading/unloading way.
2. *Strength* of the clamp should be adequate to enable it withstand operational forces. But clamping should not damage the workpiece.
3. *Quick operation* can be facilitated by using hand knobs, levers, etc. instead of spanners. Hydraulic, or pneumatic actuation can be used for simultaneous multiple-clamping.

8.8.3.1 Drill jigs

Drill jigs are subjected to high torque in the plane that is square to the axis of the drill. Also, there is a thrust force in the direction of the feed.

Requirements of a good drill jig
1. Fast, and precise location of workpiece
2. Ease in loading, unloading of workpiece. Avoid wrong loading by foolproofing
3. Prevention of distortion or motion of workpiece during drilling
4. Good chip-clearance with a provision for swarf-removal
5. Minimum weight to reduce operator fatigue
6. Avoiding loss of loose parts, by chaining them to the jig body.

Drill bushes

1. Press fit bushes [Fig. 8.19a]: For drilling in batch production.

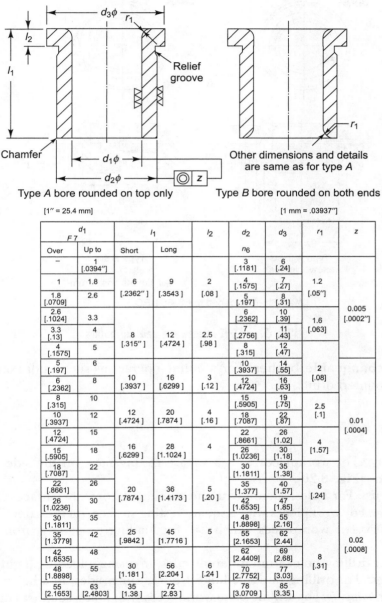

Type A bore rounded on top only   Type B bore rounded on both ends

[1″ = 25.4 mm]   [1 mm = .03937″]

d_1 F7		l_1		l_2	d_2 n6	d_3	r_1	z
Over	Up to	Short	Long					
–	1 [.0394″]				3 [.1181]	6 [.24]		
1	1.8	6 [.2362″]	9 [.3543]	2 [.08]	4 [.1575]	7 [.27]	1.2 [.05″]	
1.8 [.0709]	2.6				5 [.197]	8 [.31]		0.005 [.0002″]
2.6 [.1024]	3.3				6 [.2362]	10 [.39]	1.6 [.063]	
3.3 [.13]	4	8 [.315″]	12 [.4724]	2.5 [.98]	7 [.2756]	11 [.43]		
4 [.1575]	5				8 [.315]	12 [.47]		
5 [.197]	6				10 [.3937]	14 [.55]	2 [.08]	
6 [.2362]	8	10 [.3937]	16 [.6299]	3 [.12]	12 [.4724]	16 [.63]		
8 [.315]	10				15 [.5905]	19 [.75]	2.5 [.1]	
10 [.3937]	12	12 [.4724]	20 [.7874]	4 [.16]	18 [.7087]	22 [.87]		0.01 [.0004]
12 [.4724]	15				22 [.8661]	26 [1.02]	4 [1.57]	
15 [.5905]	18	16 [.6299]	28 [1.1024]	4	26 [1.0236]	30 [1.18]		
18 [.7087]	22				30 [1.1811]	35 [1.38]		
22 [.8661]	26	20 [.7874]	36 [1.4173]	5 [.20]	35 [1.377]	40 [1.57]	6 [.24]	
26 [1.0236]	30				42 [1.6535]	47 [1.85]		
30 [1.1811]	35				48 [1.8898]	55 [2.16]		
35 [1.3779]	42	25 [.9842]	45 [1.7716]	5	55 [2.1653]	62 [2.44]		0.02 [.0008]
42 [1.6535]	48				62 [2.4409]	69 [2.68]	8 [.31]	
48 [1.8898]	55	30 [1.181]	56 [2.204]	6 [.24]	70 [2.7752]	77 [3.03]		
55 [2.1653]	63 [2.4803]	35 [1.38]	72 [2.83]	6	78 [3.0709]	85 [3.35]		

Mat: Steel hardened to 62–65 HRC

Fig. 8.19(a): Headed jig bushes

2. **Slip renewable bushes [Fig. 8.19b]**: For drilling in continuous production. These are retained by the shoulder screws.

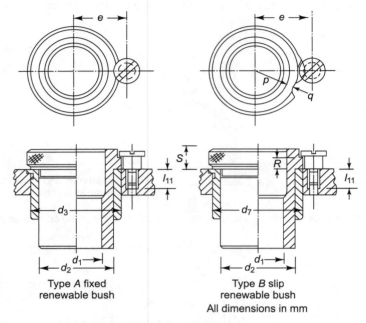

Type *A* fixed renewable bush

Type *B* slip renewable bush

All dimensions in mm

Mat: Steel hardened to 62–65 HRC

Fig. 8.19(b): Renewable and slip bushes

3. **Slip interchangeable bushes:** For multiple operations such as drilling and reaming/tapping/counterboring.

Drill jig types

1. *Plate-type jigs:* For workpieces with a square resting face, on the side opposite to the drilled holes **[Fig. 8.20]**.
2. *Angle plate jigs:* For holes, square with the machined bore/flat surfaces **[Fig. 8.21]**.
3. *Turn over jigs:* For workpieces with a square resting surface, on the same side as the holes.
4. *Leaf/latch jigs:* For workpieces which call for loading/unloading, from the side to be drilled.
5. *Box jigs:* For drilling holes from different sides, in a single loading [light workpiece].
6. *Trunnion jigs:* For drilling multiple holes from various sides, in a heavy workpiece.
7. *Sandwich or pump jig:* For drilling plate workpieces with holes, square to the plate surfaces.

Tool Engineering

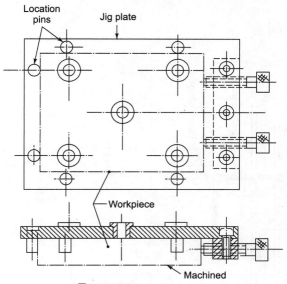

Fig. 8.20: Plate drill jig

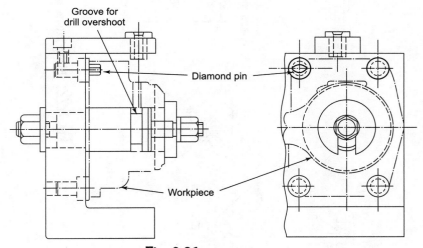

Fig. 8.21: Angle plate jig

8.8.3.2 Milling fixtures

Milling subjects the workpiece to severe vibrations and heavy thrust. Vertical/face milling also generates high torque, whereas horizontal milling generates a lifting force in up-milling, and a downward force and severe vibrations, in down-milling.

Essentials of milling fixtures [Fig. 8.22]

1. *Strength:* Milling fixtures should be robust enough to withstand heavy thrust and severe vibrations. Cast iron has vibrations-damping properties.

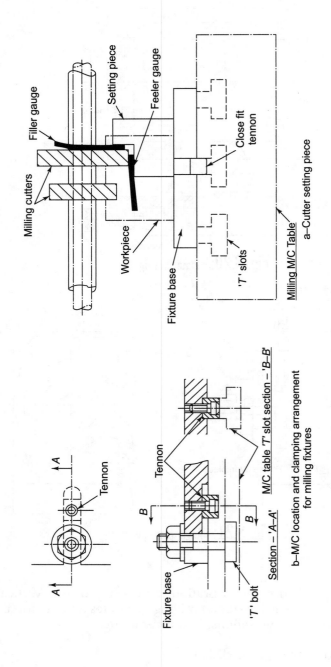

Fig. 8.22: Milling fixtures essentials

Tool Engineering 715

2. *Thrust:* Heavy thrust, generated in milling, must be directed to a strong, solid structure. It should not be directed towards clamp or a movable part. In a vice, the thrust must be directed towards the fixed jaw instead of the movable one.
3. *Cutter setting:* A setting piece should be provided to facilitate accurate cutter setting. A gap of 0.40 to 0.5 mm must be kept between the cutter and setting piece, to facilitate setting with a feeler gauge **[Fig. 8.22a]**.
4. *Location tennons:* Milling fixtures must be fixed with location tennons (tongues) in slots, milled on the resting face of the base **[Fig. 8.22b]**. Tennons are made to a close sliding fit in machine 'T' slots. They align the fixture with the longitudinal stroke of the milling machine.
5. *Rigid clamping:* Clamp the fixture well, to the machine table, to avoid slip due to cutter thrust, by providing a minimum of two 'U' slots in the fixture base **[Fig. 8.22b–plan]**. In case of a heavy cut and thrust, the number of clamping slots can be increased to four or six.
6. *Motion economy:* Workpiece clamping time should be reduced, by multiple clamping. In continuous production, pneumatic or hydraulic power clamping should be used. Clamping/unclamping time can be overlapped with milling time, by using two or more fixtures.
7. *Swarf disposal* milling chips are discontinuous. If swarf is likely to fall in a closed space, chip-removal gates should be provided.

Types of milling fixtures

1. Facing fixtures: Straddle milling fixtures, gang milling fixtures, indexing fixtures, rotary continuous milling fixtures, reciprocal [pendulum] milling fixtures.
2. Slotting fixtures

8.8.3.3 Turning fixtures

Used for facing, boring and turning, these fixtures clamp the workpiece, after aligning the diameter to be machined, with the axis of the lathe spindle. In facing, the surface to be machined should be square with the lathe spindle axis.

1. *Standard chucks:* Self-centering 3-jaw chuck, independent 4-jaw chuck, combination chuck with individual jaw adjustment. Standard jaws can be replaced by special jaws or soft jaws **[Fig. 8.23]**.
2. *Spring collets:* **[Fig. 8.12]** Push-out collets [bar work], pull-in collets [2nd operation], dead length collets, split-liner collets.
3. *Mandrels:* Tapered mandrel [0.4–0.6 mm/meter], axial clamping mandrels, expanding mandrels **[Fig. 8.24a]**, threaded mandrels **[Fig. 8.24b]**.
4. *Face plate fixtures* with balance weights.

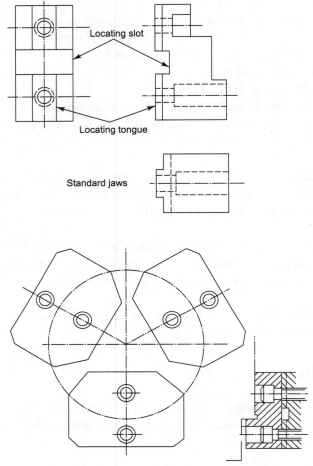

Fig. 8.23: Special jaws to increase clamping area

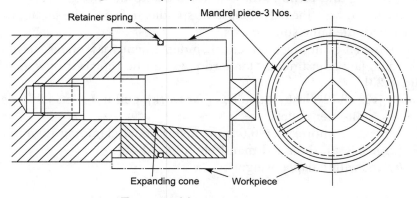

Fig. 8.24(a): Expanding mandrel

Tool Engineering

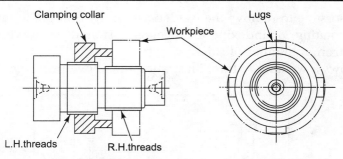

Fig. 8.24(b): Mandrel for threaded workpiece

8.8.3.4 Broaching fixtures

A fast and accurate method of metal-cutting, broaching, reproduces the precision that is built into the broach in the workpiece. A broaching fixture serves the following functions
1. It positions the workpiece precisely, with respect to the machine.
2. Broaching fixture is also used to guide the broach suitably, to meet dimensional requirements of the workpiece. **Figure 8.25** shows a keyway broaching fixture.

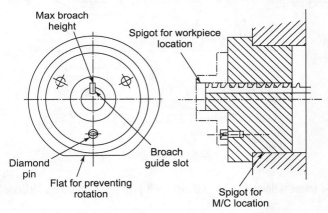

Fig. 8.25: Keyway broaching fixture

8.8.3.5 Welding fixtures

Hot joining fixtures call for the following considerations

1. Expansion of workpiece and locator due to heat, calls for more clearance between the locator and workpiece to facilitate easy unloading.

2. Handles, and knobs, subjected to heat, should be made of insulating materials like wood, to avoid the operator from burning her/his hands.

3. Parts within the spatter fall area should not be threaded. It is advisable to use a clamping system, which is not vulnerable to contact with the welding spatter, e.g. toggle clamps.

4. Suitable spatter grooves should be provided, to avoid the workpiece from getting welded to fixture.

5. It should be possible to remove the welded assembly from the fixture.

6. Easy tilting, rotating, or indexing arrangements should be provided, for workpieces requiring welding from a number of sides.

Figure 8.26 shows a welding fixture, illustrating the above principles.

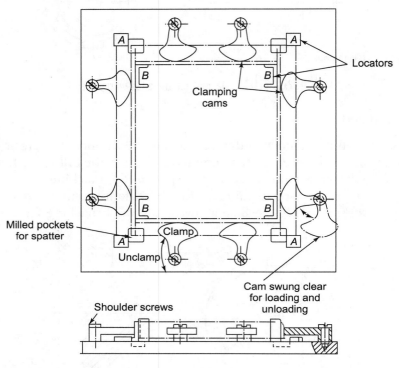

Fig. 8.26: Angle frame welding fixture [Included with permission from Tata McGraw-Hill publ. co. Ltd.]

Process planning

In small enterprises, tool engineers are also entrusted with process planning: determining the sequence of operations for economic precision manufacture. The following principles will help in the process of sequencing:

1. Establish reference datum surfaces at an early stage. For cubical parts, the first operation is generally machining resting face, which can be used as a datum for the following operations. In cylindrical workpieces, the first operation is generally turning. The turned portion can be used as a datum for the following operations.

2. Choose the cheapest machine/process which can accomplish the tolerances specified in the drawing. Workpieces with manufacturing tolerances above 0.05 mm should be

finished on lathes rather than on a cylindrical grinder. Unimportant flat faces can be machined on a shaping machine. Use milling for better finish and flatness. For perfect parallelism and good surface finish, use surface grinding.
3. The size of the machine should match the size of the workpiece.
4. The finishing operations should be carried out at later stages to prevent damage during transit.
5. The sequence of operations should be logical and practicable. For instance, lighting a stove will precede cooking and cleaning must follow eating.

Bibliography

1) Oberg, E., F. D. Jones and H. Horton, *Machinery's Handbook*, 23rd Ed., Industrial Press Inc., New York.
2) Central Machine Tool Institute, Bangalore, *Machine Tool Design Handbook*, Tata McGraw-Hill Publ. Co. Ltd., New Delhi.
3) Acherkan N., *Machine Tool Design,* Mir Publishers, Moscow.
4) Mehta N. K., *Machine Tool Design and Numerical Control,* 2nd Edition, Tata McGraw-Hill Publ. Co. Ltd., New Delhi.
5) Juneja B. L. and G. S. Sekhon, *Fundamentals of Metal Cutting and Machine Tools*, Wiley Eastern Ltd., New Delhi.
6) Berezovsky Yu., D. Chernilevsky and M. Petrov, *Machine Design,* Mir Publishers, Moscow.
7) Chapman W. A. J., *Workshop Technology,* Edward Arnold (Publishers) Ltd., London.
8) PSG College of Technology, Coimbatore, *Design Data,* 3rd Edition.
9) Joshi P. H., *Jigs and Fixtures*, 2nd Edition, Tata McGraw-Hill Publ. Co. Ltd., New Delhi.
10) Joshi P. H., *Cutting Tools,* Wheeler Publishing Co., New Delhi.

INDEX

A

Additon circuit for
    discrete function comparator   *585*
Alloy steel castings   *655–656*
Alloy steel elements   *651–652*
Analog comparator   *587*
Analog displacement measurement   *582*
Antifriction bearings   *401, 414*
Arc of contact of belts   *107*
Automobile differential gear   *232–234*
Automatically programmed tools APT   *607*
Autos-bar-single spindle   *23*
    multi spindle   *26*
    chucking   *27*
    single spindle bar   *23*
        accessories   *25*
    single spindle, cross slide cams   *25*

B

Ball clutches   *318–320*
Bar autos—single spindle   *23*
    multi spindle   *26*
Basic machine tools   *7*
Bearings: Antifriction [roller contact]   *401*
    : Ball—Angular contact   *401, 433–436*
        clearance & contact Angle   *411*
        correction for load angle for combined
            load   *415–416*
        deep grove   *411, 427–431*
        elastic deformation [deflection]   *408*
        fits   *421–422*
        sizes, standard   *408*
        spherical—self aligning   *401, 431–491*
        thrust   *397, 454–458*
    : Mounting   *416*
    : Needle Roller   *401, 445–449*
        : Bushes   *450–451*
        : cages   *449*
        : cages, flat   *449–450*
    : combined radial & thrust   *452*
    : double row   *438–439*
    : Roller, clearance & contact angle   *412*
    : single row   *401, 437–438*
    : double row   *439–440*
    : ring gauges   *440*
        : spherical roller   *443*
        : tapered roller   *401, 441–442*
Bearings Antifriction: preloading, radial   *420*
    : preloading, thrust   *415*
Bearings—clearance adjustment   *391*
    clearance ratio   *385*
Bearings comparison of sliding & rolling
    contact   *385*
Bearings, hydrodynamic   *386, 388*
    hydrodynamic load capacity   *391*
    hydrodynamic pressure   *389*
Bearing: sliding friction—lubrication   *386, 388*
    materials   *386*
    sommerfield number   *391–393*
    surface finish   *389*
    temperature rise   *394–395*
    thrust bearing   *397*
Bearings yield   *345*
Bi-directional clutches   *310–311*
Belts flat   *123*
Belts, flat—Initial and effective stress   *123*
Belts, Vee—Std sizes and min pulley diameter
    and recommended torque   *108, 111*
    inside and pitch length   *108*
    Std inside lengths with variation   *109*
Belt transmission [drives]   *107*
    Arc of contact   *107*
    Center distance limits   *108*
    Long center distance-pulley dias   *123*
    Efficiency   *123*
    Power [kW] capacity   *118*
    Service factor   *118*
    Slip   *107*
    Tensioning   *116*
    Velocity coefficient   *108*

Bevel Gears [Also see Gear, bevel]   110
Binary disc coded   582
Binary scale coded   582
    uncoded   581
Boring Jig: Machine [see Jig Boring Machine]   47
Boring machines   28
    horizontal   28
    vertical   30
Broach   696–697
    shank for round broaches   701–702
Broach design   698–702
Broaching fixture   62, 717
Broaching force, unit   670
Broaching machines   62, 698
    horizontal   62
    pull head   61
    mass production   63
    surface   63
    turret   62
    vertical   63

C

Cam and follower   284
Cam, circular arc   285
Cam, cross-slide of auto   25
Cam material—hardness, stress   287
Cam switches   545
Canned cycles in CNC program   630
Canned cycles in CNC turning machines   626
Carbon chromium steels   646
Cast irons, grey   657
Castings—alloy steel, carbon steel   658–659
    meehanite   658–659
Center—ball bearing   20
Center distance for belts   108
    Adjustment for assembly and tensioning   116
Center lathe   8
    face plate work   17
    headstock   8
    locating diameter   9
    miscellaneous operations—milling, slotting   17
    Tailstock   8
Centerless grinding   73
    axial feed   74
    end feed   75

grinding wheel   74
    stop   75
    regulation wheel   74
    through feed   74
Chain drives   125
    center distance   127
    elongation   133
    length   128
    service factor   133
    Speed limits   135
    transmission ratio limits   135
Chain sprockets design   130
    tolerance   132
    lubrication   133
    materials   133
    pitch Dia   128
Chains, Standard   127
Chatter   513
Chatter regenerative   516
Chuck-drill [Quick change]   35
Chucks   15
    3-jaw-self centering   15
    4- independent jaw   16
Column drilling machine   34
Clamping principles   710
Clutches   320–340
Clutches, ball   318–320
Clutches, bi-directional   311–312
    cone   329
    electromagnetic   339–340
    ferro-magnetic   340
    friction   320–331
    jaw [positive]   310–317
    roller   331–338
    safety [slip]   312–331
    unidirectional   311, 313, 331–338
Comparator for checking moved distance in NC machines   585
Comparison of speed, transmission ratio, and efficiency of machine   102
Computer aided numerical control [CNC] programming   607–642
CNC/Automatically Programmed Tools [APT]   607
Computer aided numerical control [CNC] auxiliary statements   615
Computer aided numerical control [CNC] canned cycles program   630–638

Index

CNC canned cycles turning machines *626*
Computer aided numerical control [CNC] continuous path program *609, 618–619*
Computer aided numerical control [CNC] contouring & mirroring program *640–642*
Computer aided numerical control [CNC] machines: direction of turn *610*
Computer aided numerical control [CNC] D-loop-drilling & milling *636*
Computer aided numerical control [CNC] machines drilling program *639*
Computer aided numerical control [CNC] machining centers *624*
Computer aided numerical control [CNC] milling & drilling programs *632*
Computer aided numerical control [CNC] mirror imaging *640*
Computer aided numerical control [CNC] motion statements *608*
CNC Operators for arithmetical & geometrical functions *616*
Computer aided numerical control [CNC] m/cs: Pattern *608–621*
   point to point program *617*
   post processor statements *614*
Computer aided numerical control [CNC] m/cs: setting zero *626–627*
CNC m/cs: specifying line *612*
Computer aided numerical control [CNC] m/cs specifying planes *613*
Computer aided numerical control [CNC] m/cs specifying a point *611*
Computer aided numerical control [CNC] m/cs: sub routines *636*
Computer aided numerical control [CNC] m/cs: taper turning *632*
CNC m/cs; tool diameter compensation - RH/LH *632–633*
Computer aided numerical control [CNC] m/cs: length offset *628, 633–634*
Computer aided numerical control [CNC] m/cs: nose radius & position *628*
Computer aided numerical control [CNC] m/cs: turning centers *624–632*
Computer aided numerical control [CNC] m/cs: turning centers program *631–633*

Computer aided numerical control [CNC] m/cs: turning program *622–623*
Contactors, electrical *546*
Continuous function comparator *586*
Control elements—electrical *544*
Control panels—electrical: symbols *535*
Control systems *520*
Control systems, convenience *520, 528*
Copying lathes *29*
Couplings, flexible *305–306*
   Oldham *307*
   muff *303–304*
   rigid *303–305*
   universal *307–310*
Crank and rocker mechanism *283*
Crank and slider mechanism *282*
Cutting coefficient *512*
Cutting power *668–670*
Cutting, rotary *1*
Cutting, rectilinear [linear] *1, 56*
Cutting speed *668–669*
Cutting speeds—Common materials *93*
   geometric progression *95*
   range ratio [R] *96*
   range ratio, no of speeds, & Speed ratio *97*
   ray diagrams *94, 105*
   standard spindle speeds [RPM] *99*
   speed ratio [ô] *95–96*
   speed ratio, recommended *100*
   stepless variation *94*
   stepped variation *94*
   transmission ratio [I] *106*
Cycle starting point [A] in CNC turning machines *626*
Cylindrical grinding *73*

D

Damping effect of cutting process *513*
Degree of freedom in m/c tool dynamics *509*
Deflection, resultant *510*
Diamond pin applications *709*
Diameter programming: CNC machines *626*
Die head, radial for chasers *694*
Digital analog converter for continuous function comparator *586–587*

724 Index

Digital measurement devices, NC machines 579
Direction discriminator for NC m/cs 580
Discrete function comparator: NC machines 585
Distributive numerical control [DNC] 623
Dividing head for milling machines 44–45
Drawing conventions for jigs & fixtures 706–707
Drill bushes 710–711
Drill chuck [Quick Change] 35
Drill head-multispindle 40
Drill jigs 710–713
Drilling machines 34
    column [single spindle] 34
    gang 34
    handheld, powered 38
    radial 36
    special purpose 38
    turret 34
    upright 34
Drive transmission & manipulation 303–342
Drives-chain [also see chain drives] 125
    efficiency: resultant 103
    infinitely variable stepless 246
Drives infinitely variable—electrical 254
    epicyclic speed variator 251
    hydraulic 256
    Face Plate Variators 247
    cone Variators 247
    variable diameter pulleys 253
Drives for machine tools: comparison 102
Drives mechanical 102
    frictional 102
    positive 102
    speed limits 102
Drives for NC systems 577
Drives for NC systems, continuous path 578
Drives for NC systems, point-to-point 577
Drives, rectilinear 282
    ball screw 291
    cam & follower 284
    circular arc cam 285
    crank & rocker mechanism 283
    crank & slider mechanism 282
    cylindrical cam 286–287
    face cam 284
    generating intermittent periodic motion 292

    geneva mechanism 296
    nut & screw combination 291
    plate cam 284
    rack & pinion combination 291
    ratchet & pawl mechanism 292
    dynamic characteristic 508, 514
    dynamic characteristics polar curves 514
Dynamics of machine tools 507
Dynamic stability of system 507

E

Economic considerations for machine tools 3
Efficiency: belt, chain, electrical, gear, hydraulic drives 101–103
Efficiency: resultant of many drives 103
Elastic system of machine tools 507
Electrical automation 540
Electrical fuses & conductors capacities 544, 548
Electrical no-voltage protection for motors 547
Electrical overload protection 543
Electrical panels 548
Electrical safety 543
Electrical signal lamps 548
Electrical starters & circuit breakers 548
Electrical symbols 549–551
Electrical wires 548
Electrical motors: Considerations 101
    multispeed 240
    power rating required for multidrive system 103
    specifications 102
    speeds [rpm] theoretical & actual 101
    two speed motors 101
    standard power ratings 101
Electromagnetic clutches 339–340
Elements of jigs & fixtures 707–708
Elements of machine tools 343
    base 504
    bearings [also see Bearings] 385–456
    columns 497–500
    guideways 459–460
        antifriction 483, 486–489
        clearance adjustment 469–470
        combination 460, 463

forces on   *472–489*
gibs   *463–466*
guide strips   *462–463*
hard wearpieces   *466*
hydraulic support   *479*
hydrostatic lubrication   *479–481*
protection   *489*
housing   *490*
shafts   *343*
spindles [also see spindles]   *343–384*
structures   *490*
structures, aperture: effect of   *495*
beams   *490*
design   *495*
distortion permissible   *495*
dynamic stiffness   *492*
economy of material   *491*
material choice   *490*
ribs stiffener   *495*
sections   *493–994*
size factor   *492*
static stiffness   *492*
stiffness enhancing   *495–496*
tables   *504*
Equivalent elastic system, single loop   *507*
Ergonomics - [work laws]   *528*
Error elimination in reading, by direction discriminator   *580*
Error elimination for eliminating false pulses   *281*
Evaluator circuit for wrong feedback   *581*

F

Face milling machine   *52*
Feed cutting   *666–667*
Feed milling, per tooth   *667*
Ferro-magnetic clutches   *340*
Fixture, broaching   *717*
Flanged couplings   *304–305*
Flat belts   *113*
Flexible coupling   *305–306*
Flow control valve-hydraulic   *246–267*
Fluid power automation   *554*
Forced vibrations   *517*
Forced vibrations amplitude   *518*
Friction clutches   *320–331*

cone   *329–331*
disc   *320–328*

G

Gear box—feed   *240*
cluster gears   *242*
cone gears   *242*
Meander mechanism   *244*
Norton tumbler gear   *243*
Gear box—speed-design   *169*
layout   *173*
transmission ratios   *168*
Gear chordal thickness   *116*
Gear class choice   *144*
Gear correction   *138*
Gear design for bending load   *145–146*
Gear design for wearing resistance   *147–148*
Gear drives—types   *135*
Gear form factor   *145, 152*
Gear measurement   *140, 142–143*
Gear module [m]   *136–137*
Gear permissible errors   *145–150*
Gear pressure angle   *135*
Gear rack   *134–135*
Gear rack equivalent pitch diameter   *142*
Gear tooth proportions parameters   *138*
Gear width factor   *145, 152*
Gear width in terms of module and center distance   *152*
Gearing [trains]   *134*
center distance tolerances   *145, 151*
change gears [lathes]   *167*
clusters   *167, 242*
cone gears   *242*
contact ratio [E]   *155*
design for given center distance   *158*
dynamic load factor   *167*
elasticity modulus for mating gear materials   *148*
internal   *180*
materials   *151*
mating Gears—important dimensions   *137*
mounting considerations   *165*
worm reduction   *135, 184*

    design *186*
    important dimensions *187-188*
    interference *189*
    lead angle *187*
    load capacity *190*
    manufacture *185*
    minimum no. of teeth *189*
    pressure angle *187*
    speed factor *190*
    tooth correction *189*
    standard *193*
    strength *191*
    strength factor *192*
    stress factor for mating materials *191*
    tooth correction *189*
    transverse lead angle *189-190*
    zone factor *192*
Gears—bevel *135, 137, 202*
    accuracy *202*
    backslash *204*
    barrelling *198*
    bearing load *234*
    crown gear *197-198*
    design graph *213*
    material factor *214*
    spiral *124*
    straight tooth *198-199, 216*
    tolerances *202-203, 205*
    tooth profile *206-207*
    tooth thickness *205-206*
    zerol *201*
Gears - epicyclic [planetary] *218*
    applications *218-219*
differential - automobile} *222-223*
differential - machine tool *223-224*
Gears
    epicyclic differential simple - bevel gear *220*
    epicyclic differential - spur gear *221*
    epicyclic simple *218*
    epicyclic train, compound *219*
    manufacture *225*
        hobbing *231*
        hobs, standard *232*
        milling *225*
        shaping *228*
        shaping cutter *230*
Gears - spur - helical *135, 156*

    helical crossed *182*
    helical transverse lead, module, pressure angle *156*
    straight teeth *135*
Grey cast irons *657*
Grey code for coded binary disc *582*
Grinding *67, 671-672*
Grinding, angular *72*
Grinding
    centerless [also see Centerless Grinding] *73*
    coolant *72*
    cut *68-70*
    cutters, clearance *79-80*
    facing wheel *73*
    internal *73*
    machines, pedestal *66*
    machines, special purpose *73*
    metal removal rate *671*
    specific power $[P_g]$ *672*
    speed - wheel [grinding] *67*
    speed - workpiece *70*
    surface *76*
    surface, feed *77*
    surface, holding non-magnetic materials *78*
Grinding surface, machines *75-76*
Grinding surface, magnetic tables *77*
Grinding wheel *80*
    abrasive *82*
    bond *82*
    grit *82*
    popular choices *82*
    segmental *81*
    shapes *81*
    structure *81*

H

Harmonious vibrations theory *512*
High carbon high chromium *644*
High carbon steel *653-654*
High frequency operation of motors *548*
High speed steel *644*
High tensile steel *652-654*
Hobbing *231*
Hook's joint *307-310*
Horizontal boring machine *28*
Horizontal broaching machine *63*

Index

Hydraulic actuators *272*
   cylinders *272*
   cushioning *274*
Functional classification *275*
Mounting styles *276*
Motors *272*
Hydraulic automation - Hydrocoping *561–564*
   interlocking actuators *561*
   sequencing [also see sequencing] *559–560*
   synchronizing [also see synchronization] *560*
Hydraulic automation *559–566*
   circuits *555–566*
   cross feed *564–566*
   force & pressure control *555*
   multi-pressure circuits *555*
   pressure, maintaining minimum *556*
   speed control *556*
   rapid approach by bypass *556–557*
   by composite cylinder *558*
   by regenerative action *557–559, 564*
   by two pumps *558*
Hydraulic circuits - symbols *266*
Hydraulic piping: classification, selection *277*
Hydraulic system *256*
   design *279*
   elements *256*
   fluid power generation *258*
   flow control *264, 267*
   oil conditioning *257*
   oil requisites *257*
   oil velocities, recommended *279*
   pressure choice *257*
   pumps types *259*
      Comparison *260*
      Gear pump *260*
      Piston - axial *263*
         radial *262*
      Vane—dual vane *261*
         eccentric rotor *261*
         variable discharge *261*
         solenoid *265*
         pilot *265*
   direction control *264, 267–270*
      2way-3port *268*
      4way-4port *268–269*
   flow control *264, 267*
   flow control, location *267*
   poppet *265*
   pressure control *264*
      pressure reducing valve *271*
      relief valve *271*
   rotating spool *265*
   sequence valve *271*
   sliding pool *264*

I

Inductosyn for measurements in CNC machines *584*
Interlocks for safety—hydraulic *561*
   mechanical *521*
International equivalent steels *656*

J

Jaw clutches *310–317*
Jigs & fixtures *706–718*
   broaching fixtures *717*
   clamping principles *710*
   diamond pin application *709*
   drill bushes *711–712*
   drill jigs *710–713*
   elements *707*
   location principles & methods *707*
   milling fixtures *713–715*
   process planning principles *718–719*
   turning fixtures *715–717*
   welding fixtures *717–718*
Jig boring machine *47, 51*
   exclusive accessories *50*
   slotting head *50*
   straight cut control system *51*
   universal work table *50*

K

Kudinov's expression *512*

L

Lathe, automatic *23*
   Single spindle *23*
   Multispindle *26*
Lathe, capstan *19*

Index

Lathe, centre 8
Lathe, chucking 27
Lathe, copying 27
Lathe relieving 29
Limit switches 545
Location diameter of lathes 9

M

Machinability 664–665
Machine tools—Appearance 3
   basic 7
   classification 1
   criteria for buying new 5
   definition 1
   drives comparison 102
   economic considerations 3
   investment cost 4
   rotary 1
   precision requisites for 2
   production cost 3
   productivity maximization 2
   requisites 1
   safety 3
   space requirement 4
Machining 665
Machining
   feed 666–667
   feed factor 669
   feed, milling, per tooth 667
   power 668–670
   speed 665–666
Machines—broaching 698
Mass production 18
   operations 13–14
   relieving 29
   tailstock 8
Mass production surface broaching machines 66
Mass production lathes 18
Material choice 663–665
Materials: Aluminium 660
   bearing alloys 661–663
   brasses & bronzes applications 663
   brasses & bronzes composition 661
   cast irons 651
   common 651
   copper 660
   non-ferrous 659–661
   steels [also see steels] 651–656
   tool [also see tool materials] 643–647
Measurement modes,
   analog displacement 582
   absolute 581, 584, 588
   incremental 688
   in NC machines 689
Material choice / comparison 385
Metal removal rates: turning, drilling, milling,
   broaching 667–668
Micro switches 545–546
Milling cutter angles 668, 681, 686, 688
Milling feed/tooth 667
Milling machines accessories – Dividing
   head 44–45
   motorized overarm 45
   slotting 46
   spiral attachment 46
Milling machines 39
 Face 50–52
   horizontal 38
   universal 45
   vertical 42
Milling–head–vertical 41
Milling, Up 41
Milling, Down 40
Mnemonics [memory aiding] 534–535
Morse tapers 375
Motion reversal mechanisms 340–342
Motion reversal mechanisms bevel gear 342
Motion reversal mechanisms spur gear 341–342
Motor protection- no voltage 547
Motor stepper 577
Motor switching: star-delta 548
Muff coupling 303–304

N

Numerical Control [NC] Programming 587–606
Numerical Control Programming:
   coding systems 567–569
   : decoding 572–573
   : displacement [point-to-point] 591
   : displacement measurement 578

Index

: drives *577–578, 596–599*
: feed magic *3* code *593*
: feed rate number FRN *600*
: formats: fixed block, tab sequential, word address *590*
: function codes *593*
: Interpolation *573*
: lathes *599*
: machine zero *590–596*
: miscellaneous functions codes *595*
: preparatory functions codes *594*
: readers *571–572*
: set point *588*
: speed magic *3* code *593*
: stepper motor *577–578*
: storage; buffer *572–573*
: tool nose radius effect *600*
: tool path length *600*
: turning cone *599*
: turning radii *603*
: workpiece zero *588*
: zero disposition *588–590*
Nyquist criterion *508*

O

Oil, hydraulic—conditioning *257*
    requisites *257*
    velocities recommended *279*
Oldham couplings *306–307*
Open loop system *513*
Operating system—convenience *528*
Orthogonal cutting mode - coupled freedom of 2 degrees *610*
Overload protection - electrical *543*

P

Phase sensitive transducer *583*
Piping for hydraulic systems *277*
Piston pumps hydraulic *262*
Pitch error - bevel gears *202, 205*
Planetary gears *218–219*
Pneumatic and hydraulic systems comparison *555*
Polar curves: cutting *514*

Polar curves: dynamic characteristics of cutting *508*
Polar curves: multi degree freedom *515*
: Open loop system *513*
Polar curves: positive and negative deflections *510*
Polar curves:
    resultant of positive and negative polar curves *510*
    : single degree of freedom *516*
    : transfer function *519*
    : two degrees of freedom *517*
Poppet valves, hydraulic *265*
Positive drives, mechanical *102*
Power rating, multi-drive m/cs *103*
Power ratings, standard for electrical motors *101*
Pressure limiting *554*
Pressure switches—electrical *546*
Precision requisites for machine tools *2*
Pressure choice—hydraulic *257*
Pressure control valves hydraulic *264*
Pressure reducing valve *271*
Pressure relief valve *271*
Production cost *3*
Pulleys for flat belts *114*
Pulley for 'V' belts, groove angle, size margin *108, 111*
Pulleys for very long center distance *112*
Pulleys small for flat belts *116*
Pull head for broaching machine *63*
Pumps, hydraulic [also see Hydraulic Pumps] *259–262*
Push buttons *546*

Q

Quick change drill chuck *35*
Quick release tapers for milling machines *5/39-40*

R

Ray Diagrams *94, 105*
Rapid approach & return thru hydraulics *556*
Rectilinear cutting machines *53*
    broaching *59*

planing   *58-60*
shaping   *56*
slotting   *57*
Regenerative action in hydraulic circuit   *557*
Regenerative chatter   *516*
Relays - electrical   *544-547*
Relieving lathes   *29*
Resonance frequency   *509*
Resultant deflection   *510*
Rigid couplings   *303-305*
Rigidity ellipse for lathes   *510*
Roller clutches—unidirectional   *311, 313, 331-338*

S

Safety clutches   *312-331*
Safety in machine tools   *3*
Safety of control systems for machine tools   *520*
Safety clutches   *526*
Seals, oil   *417-420*
Segmental grinding wheel   *81*
Self holding tapers—drilling machines   *378*
    milling machines   *379-380*
Sequence valve - hydraulic   *272*
Sequencing hydraulic actuators   *559-560*
Sequencing hydraulic actuators by electrical limit switches   *560*
Sequencing hydraulic actuators by pressure switches   *560*
Sequencing hydraulic actuators by sequence valve   *559*
Service factors for chains   *133*
Shank for round broaches   *61*
Shank with self holding tapers   *378-380*
Shaping machines   *54*
    gears   *228*
    hydraulic   *56*
    mechanism   *55*
    speed   *56*
Shaping vertical face   *53*
Shear pins/keys, safety   *25*
Single spindle auto   *23*
Sliding spool hydraulic valves   *264*
Slip of belt   *107*
Slotting machines   *57-60*
Slotting tools   *57*

Space requirement for machine tools   *57*
Specific power [K] for various materials   *102*
Speeds, cutting   *665-666*
Spindles for machine tools   *343-384*
    angular deflection [twist]   *357*
    deflection by B.M. Area moment   *353*
    drilling machines   *375*
    face plates   *370-374*
    functions   *343*
    grinding machines   *382-384*
    materials   *343*
    Noses—drilling machines   *375*
        grinding machines   *382-384*
        lathes   *363*
        milling machines   *381*
        precision   *424*
        purposes   *362*
        rigidity   *344*
        requisites   *343*
        runout   *355*
Spindle speeds - critical   *358-361*
    stiffness   *344-355*
    stress   *344*
    supports [also see bearing]   *401-452*
Speed factor for wear strength for worm reduction gears   *190-191*
Speed ratio [φ]   *95, 96, 100*
Speeds for electrical motors   *101*
Spiral bevel gears   *216*
Spur gear differential gear   *221*
Stability analysis   *513*
Stability of a system   *507*
Standard inside lengths of V belts   *108*
Standard power ratings for electrical motors   *101*
Standard worm reduction gears   *193*
Static characteristics   *512-513*
Steels   *651-656*
    alloy   *655-656*
    alloying elements   *651-652*
    carbon   *652-654*
    carbon-chromium   *644*
    castings   *658*
    high carbon high chromium (HCHC)   *644*
    high speed   *644*
    high tensile   *652-654*
    international equivalents   *656*
    medium carbon   *652-654*

Index

mild *652–654*
non-shrinking OHNS *644*
Stepless speed variation *94, 244–253*
Stop, axial for cross slide *20*
Stroke reversal in cylindrical grinders *69*
Structure of grinding wheel *80*
Subtraction circuit for discrete function comparator *585*
Surface broaching *63*
Surface grinding *75*
Switching devices - electrical *544*
Symbols for control panels *535*
Symbols for electrical circuits *548–551*
Synchro for measurements in CNC machines *583*
Synchronization of hydraulic actuators *560*
Synchronization of hydraulic actuators by coupled motors *560*
Synchronization of hydraulic actuators by flow control valves *560*
Synchronization of hydraulic actuators by rigid coupling *560*
Synchronization of hydraulic actuators two variable discharge pumps *560–561*

T

Tailstock of lathe *8*
Taper grinding *69–71*
Tapers - quick release *381–382*
    self-holding *379*
Teeth minimum number for gears *189*
Tensioning belts by pulleys center distance adjustment *116*
Through feed in centerless grinding *74*
Tolerances for IT grades *424*
Tool angles *672–673*
Tool angles,
    forming tools - circular *678–680*
    forming tools - straight *676–678*
    parting tools *675–677*
    turning tools *673–676*
Tool geometry: Broaches *698–701*
    Drills *688–689*
    Forming tools, circular *678–680*
    Forming tools, straight *677–678*
    Milling cutters *678, 681, 686, 688*

Reamers *690–692*
Threading chasers *694–696*
Threading dies *693–694*
Threading taps *692–693*
Toolholders -
    arbors *704*
    boring bar *21*
    box turning *21–22*
    chucks *702*
    clamping by padbolt *19*
    collets *703*
    combination *19*
    drill/reamer—floating *21*
    knee *19*
    special *20, 704*
Tooling *702*
Tool materials *643–647*
    applications *647*
    carbides *645*
    high carbon steels *644*
    high speed steels *644*
Tooth proportion for gears *137*
Tooth rest for grinding cutters *79*
Transmission by belts *107*
Transmission ratio *106*
Transducer phase—sensitive *583*
Transducer Selsyn *583*
Transverse lead, module, pitch, pressure angle for helical gears *156*
Turret drilling machine *34*
Turning between centers *14*
Turning machine, auto *23–25*
    special purpose *29*
Turret lathes *20*
    vertical spindle *30*
Types of worm reduction gears *184–185*
Truing workpiece in chuck—cylindrical grinding *589*

U

Unidirectional clutches *311, 313, 331–338*
Universal couplings *307–310*
Universal milling machines *48*
Universal vice *80*
Universal worktable for jig boring machines *51*
Upright drilling machines *33*

V

Valves operation—pilot *554*
    solenoid [electrical] *554*
Vane pumps *261*
Variable [stepless] speed drives *246*
Variator, speed *247–249*
Vee belts *108, 111–112*
Velocities recommended for hydraulic systems *279*
Vertical milling machines *42*
Vibrations *513*
Vibrations, forced *517*

W

Width of gear in terms of center distance and module *152*
Workpiece holders *705–706*
    chucks *705*
    jigs & fixtures [also see Jigs and Fixtures] *706–709*
    standard *705–706*
    vices *706*
Workpiece materials: Aluminium *660*
    bearing alloys *661–663*
    brasses & bronzes *661*
    cast irons *657*
    copper *660*
    non-ferrous *659*
    steels [also see steels] *651–656*
Worm and wheel reduction gears *135, 184*
Worm and wheel reduction gears - standard *193*
Worm types *184–185*

Z

Zone factor for worm reduction gears *198*